Java™

A Beginner's Guide

Ninth Edition

About the Author

Best-selling author **Herbert Schildt** has written extensively about programming for over three decades and is a leading authority on the Java language. Called "one of the world's foremost authors of books about programming" by *International Developer* magazine, his books have sold millions of copies worldwide and have been translated into all major foreign languages. He is the author of numerous books on Java, including *Java: The Complete Reference; Herb Schildt's Java Programming Cookbook; Introducing JavaFX 8 Programming;* and *Swing: A Beginner's Guide*. He has also written extensively about C, C++, and C#. Featured as one of the rock star programmers in Ed Burns' book *Secrets of the Rock Star Programmers: Riding the IT Crest*, Schildt is interested in all facets of computing, but his primary focus is computer languages. Schildt holds both BA and MCS degrees from the University of Illinois. His website is **www.HerbSchildt.com**.

About the Technical Editor

Dr. Danny Coward has worked on all editions of the Java platform. He led the definition of Java Servlets into the first version of the Java EE platform and beyond, web services into the Java ME platform, and the strategy and planning for Java SE 7. He founded JavaFX technology and, most recently, designed the largest addition to the Java EE 7 standard, the Java WebSocket API. From coding in Java, to designing APIs with industry experts, to serving for several years as an executive to the Java Community Process, he has a uniquely broad perspective into multiple aspects of Java technology. In addition, he is the author of two books on Java programming: *Java WebSocket Programming* and *Java EE 7: The Big Picture*. Most recently, he has been applying his knowledge of Java to helping scale massive Java-based services for one of the world's most successful software companies. Dr. Coward holds a bachelor's, master's, and doctorate in mathematics from the University of Oxford.

Java™
A Beginner's Guide

Ninth Edition

Herbert Schildt

New York   Chicago   San Francisco
Athens   London   Madrid   Mexico City
Milan   New Delhi   Singapore   Sydney   Toronto

Java™: A Beginner's Guide, Ninth Edition

1 2 3 4 5 6 7 8 9   LCR   25 24 23 22 21

Library of Congress Control Number: 2021949339

ISBN   978-1-260-46355-2
MHID   1-260-46355-9

Sponsoring Editor Lisa McClain		**Indexer** Sherry Schildt	
Editorial Supervisor Patty Mon		**Production Supervisor** Thomas Somers	
Acquisitions Coordinator Emily Walters		**Composition** KnowledgeWorks Global Ltd.	
Technical Editor Danny Coward		**Illustration** KnowledgeWorks Global Ltd.	
Copy Editor Lisa McCoy		**Art Director, Cover** Jeff Weeks	

Contents

Introduction

The purpose of this book is to teach you the fundamentals of Java programming. It uses a step-by-step approach complete with numerous examples, self tests, and projects. It assumes no previous programming experience. The book starts with the basics, such as how to compile and run a Java program. It then discusses the keywords, features, and constructs that form the core of the Java language. You'll also find coverage of some of Java's most advanced features, including multithreaded programming, generics, lambda expressions, records, and modules. An introduction to the fundamentals of Swing concludes the book. By the time you finish, you will have a firm grasp of the essentials of Java programming.

It is important to state at the outset that this book is just a starting point. Java is more than just the elements that define the language. Java also includes extensive libraries and tools that aid in the development of programs. To be a top-notch Java programmer implies mastery of these areas, too. After completing this book, you will have the knowledge to pursue any and all other aspects of Java.

The Evolution of Java

Only a few languages have fundamentally reshaped the very essence of programming. In this elite group, one stands out because its impact was both rapid and widespread. This language is, of course, Java. It is not an overstatement to say that the original release of Java 1.0 in 1995 by Sun Microsystems, Inc., caused a revolution in programming. This revolution radically transformed the Web into a highly interactive environment. In the process, Java set a new standard in computer language design.

Over the years, Java has continued to grow, evolve, and otherwise redefine itself. Unlike many other languages, which are slow to incorporate new features, Java has often been at the forefront of computer language development. One reason for this is the culture of innovation and change that came to surround Java. As a result, Java has gone through several upgrades—some relatively small, others more significant.

The first major update to Java was version 1.1. The features added by Java 1.1 were more substantial than the increase in the minor revision number would have you think. For example, Java 1.1 added many new library elements, redefined the way events are handled, and reconfigured many features of the original 1.0 library.

The next major release of Java was Java 2, where the 2 indicates "second generation." The creation of Java 2 was a watershed event, marking the beginning of Java's "modern age." The first release of Java 2 carried the version number 1.2. It may seem odd that the first release of Java 2 used the 1.2 version number. The reason is that it originally referred to the internal version number of the Java libraries but then was generalized to refer to the entire release itself. With Java 2, Sun repackaged the Java product as J2SE (Java 2 Platform Standard Edition), and the version numbers began to be applied to that product.

The next upgrade of Java was J2SE 1.3. This version of Java was the first major upgrade to the original Java 2 release. For the most part, it added to existing functionality and "tightened up" the development environment. The release of J2SE 1.4 further enhanced Java. This release contained several important new features, including chained exceptions, channel-based I/O, and the **assert** keyword.

The release of J2SE 5 created nothing short of a second Java revolution. Unlike most of the previous Java upgrades, which offered important but incremental improvements, J2SE 5 fundamentally expanded the scope, power, and range of the language. To give you an idea of the magnitude of the changes caused by J2SE 5, here is a list of its major new features:

- Generics

- Autoboxing/unboxing

- Enumerations

- The enhanced "for-each" style **for** loop

- Variable-length arguments (varargs)

- Static import

- Annotations

This is not a list of minor tweaks or incremental upgrades. Each item in the list represents a significant addition to the Java language. Some, such as generics, the enhanced **for** loop, and varargs, introduced new syntax elements. Others, such as autoboxing and auto-unboxing, altered the semantics of the language. Annotations added an entirely new dimension to programming.

The importance of these new features is reflected in the use of the version number "5." The next version number for Java would normally have been 1.5. However, the new features were so significant that a shift from 1.4 to 1.5 just didn't seem to express the magnitude of the change. Instead, Sun elected to increase the version number to 5 as a way of emphasizing that a major event was taking place. Thus, it was named J2SE 5, and the Java Development Kit (JDK) was called JDK 5. In order to maintain consistency, however, Sun decided to use 1.5 as its internal version number, which is also referred to as the developer version number. The "5" in J2SE 5 is called the product version number.

The next release of Java was called Java SE 6, and Sun once again decided to change the name of the Java platform. First, notice that the "2" has been dropped. Thus, the platform now had the name Java SE, and the official product name was Java Platform, Standard Edition 6, with the development kit being called JDK 6. As with J2SE 5, the 6 in Java SE 6 is the product version number. The internal, developer version number is 1.6.

Java SE 6 built on the base of J2SE 5, adding incremental improvements. Java SE 6 added no major features to the Java language proper, but it did enhance the API libraries, added several new packages, and offered improvements to the run time. It also went through several updates during its long (in Java terms) life cycle, with several upgrades added along the way. In general, Java SE 6 served to further solidify the advances made by J2SE 5.

The next release of Java was called Java SE 7, with the development kit being called JDK 7. It has an internal version number of 1.7. Java SE 7 was the first major release of Java after Sun Microsystems was acquired by Oracle. Java SE 7 added several new features, including significant additions to the language and the API libraries. Some of the most important features added by Java SE 7 were those developed as part of Project Coin. The purpose of Project Coin was to identify a number of small changes to the Java language that would be incorporated into JDK 7, including

- A **String** can control a **switch** statement.

- Binary integer literals.

- Underscores in numeric literals.

- An expanded **try** statement, called **try**-*with-resources*, that supports automatic resource management.

- Type inference (via the *diamond* operator) when constructing a generic instance.

- Enhanced exception handling in which two or more exceptions can be caught by a single **catch** (multicatch) and better type checking for exceptions that are rethrown.

As you can see, even though the Project Coin features were considered to be small changes to the language, their benefits were much larger than the qualifier "small" would suggest. In particular, the **try**-with-resources statement profoundly affects the way that a substantial amount of code is written.

The next release of Java was Java SE 8, with the development kit being called JDK 8. It has an internal version number of 1.8. JDK 8 represented a very significant upgrade to the Java language because of the inclusion of a far-reaching new language feature: the *lambda expression*. The impact of lambda expressions was, and continues to be, quite profound, changing both the way that programming solutions are conceptualized and how Java code is written. In the process, lambda expressions can simplify and reduce the amount of source code needed to create certain constructs. The addition of lambda expressions also caused a new operator (the –>) and a new syntax element to be added to the language. In addition to lambda expressions, JDK 8 added many other important new features. For example, beginning with JDK 8, it is now possible to define a default implementation for a method specified by an interface. In the final analysis, Java SE 8 was a major release that profoundly expanded the capabilities of the language and changed the way that Java code is written.

The next release of Java was Java SE 9. The developer's kit was called JDK 9. With the release of JDK 9, the internal version number is also 9. JDK 9 represented a major Java release, incorporating significant enhancements to both the Java language and its libraries. The primary new feature was *modules,* which enable you to specify the relationships and dependencies of the code that comprises an application. Modules also add another dimension to Java's access control features. The inclusion of modules caused a new syntax element, several new keywords, and various tool enhancements to be added to Java. Modules had a profound effect on the API library because, beginning with JDK 9, the library packages are now organized into modules.

In addition to modules, JDK 9 included several other new features. One of particular interest is JShell, which is a tool that supports interactive program experimentation and learning. (An introduction to JShell is found in Appendix D.) Another interesting upgrade is support for private interface methods. Their inclusion further enhanced JDK 8's support for default methods in interfaces. JDK 9 added a search feature to the **javadoc** tool and a new tag called **@index** to support it. As with previous releases, JDK 9 contains a number of updates and enhancements to Java's API libraries.

As a general rule, in any Java release, it is the new features that receive the most attention. However, there is one high-profile aspect of Java that was deprecated by JDK 9: applets. Beginning with JDK 9, applets are no longer recommended for new projects. As will be explained in greater detail in Chapter 1, because of waning browser support for applets (and other factors), JDK 9 deprecated the entire applet API.

The next release of Java was Java SE 10 (JDK 10). However, prior to its release, a major change occurred in the Java release schedule. In the past, major releases were often separated by two or more years. However, beginning with JDK 10, the time between releases was significantly shortened. Releases are now expected to occur on a strict time-based schedule, with the anticipated time between major releases (now called *feature releases*) to be just six months. As a result, JDK 10 was released in March 2018, which is six months after the release of JDK 9. This more rapid *release cadence* enables new features and improvements to be quickly available to Java programmers. Instead of waiting two or more years, when a new feature is ready, it becomes part of the next scheduled release.

Another facet of the changes to the Java release schedule is the *long-term support* (LTS) release. It is now anticipated that an LTS release will take place every three years. An LTS release will be supported (and thus remain viable) for a period of time longer than six months. The first LTS release was JDK 11. The second LTS release was JDK 17, for which this book has been updated. Because of the stability that an LTS release offers, it is likely that its feature set will define a baseline of functionality for a number of years. Consult Oracle for the latest information concerning long-term support and the LTS release schedule.

The primary new language feature added by JDK 10 was support for *local variable type inference.* With local variable type inference, it is now possible to let the type of a local variable be inferred from the type of its initializer, rather than being explicitly specified.

To support this new capability, the context-sensitive keyword **var** was added to Java. Type inference can streamline code by eliminating the need to redundantly specify a variable's type when it can be inferred from its initializer. It can also simplify declarations in cases in which the type is difficult to discern or cannot be explicitly specified. Local variable type inference has become a common part of the contemporary programming environment. Its inclusion in Java helps keep Java up-to-date with evolving trends in language design. Along with a number of other changes, JDK 10 also redefined the Java version string, changing the meaning of the version numbers so they better align with the new time-based release schedule.

The next version of Java was Java SE 11 (JDK 11). It was released in September 2018, which is six months after JDK 10. It was an LTS release. The primary new language feature in JDK 11 was its support for the use of **var** in a lambda expression. Also, another execution mode was added to the Java launcher that enables it to directly execute simple single-file programs. JDK 11 also removed some features. Perhaps of greatest interest, because of its historical significance, is the removal of support for applets. Recall that applets were first deprecated by JDK 9. With the release of JDK 11, applet support has been removed. Support for another deployment-related technology called Java Web Start was also removed from JDK 11. There is one other high-profile removal in JDK 11: JavaFX. This GUI framework is no longer part of the JDK, becoming a separate open-source project instead. Because these features have been removed from the JDK, they are not discussed in this book.

Between the JDK 11 LTS and the next LTS release (JDK 17) were five feature releases: JDK 12 through JDK 16. JDK 12 and JDK 13 did not add any new language features. JDK 14 added support for the **switch** expression, which is a **switch** that produces a value. Other enhancements to **switch** were also included. Text blocks, which are essentially string literals that can span more than one line, were added by JDK 15. JDK 16 enhanced **instanceof** with pattern matching and added a new type of class called a *record* along with the new context-sensitive keyword **record**. A record provides a convenient means of aggregating data. JDK 16 also supplied a new application packaging tool called **jpackage**.

At the time of this writing, Java SE 17 (JDK 17) is the latest version of Java. As mentioned, it is the second LTS Java release. Thus, it is of particular importance. Its major new feature is the ability to seal classes and interfaces. Sealing gives you control over the inheritance of a class and the inheritance and implementation of an interface. Towards this end, it adds a new context-sensitive keyword **sealed**. It also adds the context-sensitive keyword **non-sealed**, which is the first hyphenated Java keyword. JDK 17 marks the applet API as deprecated for removal. As explained, support of applets was removed several years ago. However, the applet API was simply deprecated, which allowed vestigial code that relied on this API to still compile. With the release of JDK 17, the applet API is now subject to removal by a future release.

One other point about the evolution of Java: Beginning in 2006, the process of open-sourcing Java began. Today, open-source implementations of the JDK are available. Open sourcing further contributes to the dynamic nature of Java development. In the final analysis, Java's legacy of innovation is secure. Java remains the vibrant, nimble language that the programming world has come to expect.

The material in this book has been updated through JDK 17. As the preceding discussion has highlighted, however, the history of Java programming is marked by dynamic change. As you advance in your study of Java, you will want to watch for new features of each subsequent Java release. Simply put: The evolution of Java continues!

How This Book Is Organized

This book presents an evenly paced tutorial in which each section builds upon the previous one. It contains 17 chapters, each discussing an aspect of Java. This book is unique because it includes several special elements that reinforce what you are learning.

Key Skills & Concepts

Each chapter begins with a set of critical skills that you will be learning.

Self Test

Each chapter concludes with a Self Test that lets you test your knowledge. The answers are in Appendix A.

Ask the Expert

Sprinkled throughout the book are special "Ask the Expert" boxes. These contain additional information or interesting commentary about a topic. They use a question/answer format.

Try This Elements

Each chapter contains one or more Try This elements, which are projects that show you how to apply what you are learning. In many cases, these are real-world examples that you can use as starting points for your own programs.

No Previous Programming Experience Required

This book assumes no previous programming experience. Thus, if you have never programmed before, you can use this book. If you do have some previous programming experience, you will be able to advance a bit more quickly. Keep in mind, however, that Java differs in several key ways from other popular computer languages. It is important not to jump to conclusions. Thus, even for the experienced programmer, a careful reading is advised.

Required Software

To compile and run all of the programs in this book, you will need the latest Java Development Kit (JDK), which, at the time of this writing, is JDK 17. This is the JDK for Java SE 17. Instructions for obtaining the Java JDK are given in Chapter 1.

If you are using an earlier version of Java, you will still be able to use this book, but you won't be able to compile and run the programs that use Java's newer features.

Don't Forget: Code on the Web

Remember, the source code for all of the examples and projects in this book is available free of charge on the Web at **www.mhprofessional.com**.

Special Thanks

Special thanks to Danny Coward, the technical editor for this edition of the book. Danny has worked on several of my books, and his advice, insights, and suggestions have always been of great value and much appreciated.

For Further Study

Java: A Beginner's Guide is your gateway to the Herb Schildt series of Java programming books. Here are some others that you will find of interest:

Java: The Complete Reference

Herb Schildt's Java Programming Cookbook

The Art of Java

Swing: A Beginner's Guide

Introducing JavaFX 8 Programming

Chapter 1

Java Fundamentals

Key Skills & Concepts

- Know the history and philosophy of Java

- Understand Java's contribution to the Internet

- Understand the importance of bytecode

- Know the Java buzzwords

- Understand the foundational principles of object-oriented programming

- Create, compile, and run a simple Java program

- Use variables

- Use the **if** and **for** control statements

- Create blocks of code

- Understand how statements are positioned, indented, and terminated

- Know the Java keywords

- Understand the rules for Java identifiers

In computing, few technologies have had the impact of Java. Its creation in the early days of the Web helped shape the modern form of the Internet, including both the client and server sides. Its innovative features advanced the art and science of programming, setting a new standard in computer language design. The forward-thinking culture that grew up around Java ensured it would remain vibrant and alive, adapting to the often rapid and varied changes in the computing landscape. Simply put: not only is Java one of the world's most important computer languages, it is a force that revolutionized programming and, in the process, changed the world.

Although Java is a language often associated with Internet programming, it is by no means limited in that regard. Java is a powerful, full-featured, general-purpose programming language. Thus, if you are new to programming, Java is an excellent language to learn. Moreover, to be a professional programmer today implies the ability to program in Java—it is that important. In the course of this book, you will learn the basic skills that will help you master it.

The purpose of this chapter is to introduce you to Java, beginning with its history, its design philosophy, and several of its most important features. By far, the hardest thing about learning a programming language is the fact that no element exists in isolation. Instead, the components of the language work in conjunction with each other. This interrelatedness is especially pronounced in Java. In fact, it is difficult to discuss one aspect of Java without involving others. To help overcome this problem, this chapter provides a brief overview of several Java features, including the general form of a Java program, some basic control structures, and simple operators. It does not go into too many details, but, rather, concentrates on general concepts common to any Java program.

The History and Philosophy of Java

Before one can fully appreciate the unique aspects of Java, it is necessary to understand the forces that drove its creation, the programming philosophy that it embodies, and key concepts of its design. As you advance through this book, you will see that many aspects of Java are either a direct or indirect result of the historical forces that shaped the language. Thus, it is fitting that we begin our examination of Java by exploring how Java relates to the larger programming universe.

The Origins of Java

Java was conceived by James Gosling, Patrick Naughton, Chris Warth, Ed Frank, and Mike Sheridan at Sun Microsystems in 1991. This language was initially called "Oak" but was renamed "Java" in 1995. Somewhat surprisingly, the original impetus for Java was not the Internet! Instead, the primary motivation was the need for a platform-independent language that could be used to create software to be embedded in various consumer electronic devices, such as toasters, microwave ovens, and remote controls. As you can probably guess, many different types of CPUs are used as controllers. The trouble was that (at the time) most computer languages were designed to be compiled into machine code that was targeted for a specific type of CPU. For example, consider the C++ language.

Although it was possible to compile a C++ program for just about any type of CPU, to do so required a full C++ compiler targeted for that CPU. The problem, however, is that compilers are expensive and time consuming to create. In an attempt to find a better solution, Gosling and the others worked on a portable, cross-platform language that could produce code that would run on a variety of CPUs under differing environments. This effort ultimately led to the creation of Java.

About the time that the details of Java were being worked out, a second, and ultimately more important, factor emerged that would play a crucial role in the future of Java. This second force was, of course, the World Wide Web. Had the Web not taken shape at about the same time that Java was being implemented, Java might have remained a useful but obscure language for programming consumer electronics. However, with the emergence of the Web, Java was propelled to the forefront of computer language design, because the Web, too, demanded portable programs.

Most programmers learn early in their careers that portable programs are as elusive as they are desirable. While the quest for a way to create efficient, portable (platform-independent) programs is nearly as old as the discipline of programming itself, it had taken a back seat to other, more pressing problems. However, with the advent of the Internet and the Web, the old problem of portability returned with a vengeance. After all, the Internet consisted of a diverse, distributed universe populated with many types of computers, operating systems, and CPUs.

What was once an irritating but low-priority problem had become a high-profile necessity. By 1993 it became obvious to members of the Java design team that the problems of portability frequently encountered when creating code for embedded controllers are also found when attempting to create code for the Internet. This realization caused the focus of Java to switch from consumer electronics to Internet programming. So, although it was the desire for an architecture-neutral programming language that provided the initial spark, it was the Internet that ultimately led to Java's large-scale success.

Java's Lineage: C and C++

The history of computer languages is not one of isolated events. Rather, it is a continuum in which each new language is influenced in one way or another by what has come before. In this regard, Java is no exception. Before moving on, it is useful to understand where Java fits into the family tree of computer languages.

The two languages that form Java's closest ancestors are C and C++. As you may know, C and C++ are among the most important computer languages ever invented and are still in widespread use today. From C, Java inherits its syntax. Java's object model is adapted from C++. Java's relationship to C and C++ is important for a number of reasons. First, at the time of Java's creation, many programmers were familiar with the C/C++ syntax. Because Java uses a similar syntax, it was relatively easy for a C/C++ programmer to learn Java. This made it possible for Java to be readily utilized by the pool of existing programmers, thus facilitating Java's acceptance by the programming community.

Second, Java's designers did not "reinvent the wheel." Instead, they further refined an already highly successful programming paradigm. The modern age of programming began with C. It moved to C++ and then to Java. By inheriting and building on that rich heritage, Java provides a powerful, logically consistent programming environment that takes the best of the past and adds new features related to the online environment and advances in the art of programming. Perhaps most important, because of their similarities, C, C++, and Java define a common, conceptual framework for the professional programmer. Programmers do not face major rifts when switching from one language to another.

Java has another attribute in common with C and C++: it was designed, tested, and refined by real working programmers. It is a language grounded in the needs and experiences of the people who devised it. There is no better way to produce a top-flight professional programming language.

One last point: although C++ and Java are related, especially in their support for object-oriented programming, Java is *not* simply the "Internet version of C++." Java has significant practical and philosophical differences from C++. Furthermore, Java is *not* an enhanced version of C++. For example, it is neither upwardly nor downwardly compatible with C++. Moreover, Java was not designed to replace C++. Java was designed to solve a certain set of problems. C++ was designed to solve a different set of problems. They will coexist for many years to come.

How Java Impacted the Internet

The Internet helped catapult Java to the forefront of programming, and Java, in turn, had a profound effect on the Internet. First, the creation of Java simplified Internet programming in general, acting as a catalyst that drew legions of programmers to the Web. Second, Java innovated a new type of networked program called the *applet* that changed the way the online world thought about content. Finally, and perhaps most importantly, Java addressed some of the thorniest issues associated with the Internet: portability and security.

Java Simplified Web-Based Programming

From the start, Java simplified web-based programming in a number of ways. Arguably the most important is found in its ability to create portable, cross-platform programs. Of nearly equal importance is Java's support for networking. Its library of ready-to-use functionality enabled programmers to easily write programs that accessed or made use of the Internet. It also provided mechanisms that enabled programs to be readily delivered over the Internet. Although the details are beyond the scope of this book, it is sufficient to know that Java's support for networking was a key factor in its rapid rise.

Java Applets

At the time of Java's creation, one of its most exciting features was the applet. An *applet* is a special kind of Java program that is designed to be transmitted over the Internet and automatically executed inside a Java-compatible web browser. If the user clicks a link that contains an applet, the applet will download and run in the browser automatically. Applets were intended to be small programs, typically used to display data provided by the server, handle user input, or provide simple functions, such as a loan calculator. The key feature of applets is that they execute locally, rather than on the server. In essence, the applet allowed some functionality to be moved from the server to the client.

The creation of the applet was important because, at the time, it expanded the universe of objects that could move about freely in cyberspace. In general, there are two very broad categories of objects that are transmitted between the server and the client, passive information and dynamic active programs. For example, when you read your e-mail, you are viewing passive data. Even when you download a program, the program's code is still only passive data until you execute it. By contrast, the applet is a dynamic, self-executing program. Such a program is an active agent on the client computer, yet it is delivered by the server.

In the early days of Java, applets were a crucial part of Java programming. They illustrated the power and benefits of Java, added an exciting dimension to web pages, and enabled programmers to explore the full extent of what was possible with Java. Although it is likely that there are still applets in use today, over time they became less important, and for reasons that will be explained shortly, JDK 9 began their phase-out process. Finally, applet support was removed by JDK 11.

Ask the Expert

Q: What is C# and how does it relate to Java?

A: A few years after the creation of Java, Microsoft developed the C# language. This is important because C# is closely related to Java. In fact, many of C#'s features directly parallel Java. Both Java and C# share the same general C++-style syntax, support distributed programming, and utilize a similar object model. There are, of course, differences between Java and C#, but the overall "look and feel" of these languages is very similar. This means that if you already know C#, then learning Java will be especially easy. Conversely, if C# is in your future, then your knowledge of Java will come in handy.

Security

As desirable as dynamic, networked programs are, they also present serious problems in the areas of security and portability. Obviously, a program that downloads and executes on the client computer must be prevented from doing harm. It must also be able to run in a variety of different environments and under different operating systems. As you will see, Java addressed these problems in an effective and elegant way. Let's look a bit more closely at each, beginning with security.

As you are likely aware, every time that you download a program, you are taking a risk because the code you are downloading might contain a virus, Trojan horse, or other harmful code. At the core of the problem is the fact that malicious code can cause damage because it has gained unauthorized access to system resources. For example, a virus program might gather private information, such as credit card numbers, bank account balances, and passwords, by searching the contents of your computer's local file system. In order for Java to enable programs to be safely downloaded and executed on the client computer, it was necessary to prevent them from launching such an attack.

Java achieved this protection by enabling you to confine an application to the Java execution environment and prevent it from accessing other parts of the computer. (You will see how this is accomplished shortly.) The ability to download an application with a high level of confidence that no harm will be done contributed significantly to Java's early success.

Portability

Portability is a major aspect of the Internet because there are many different types of computers and operating systems connected to it. If a Java program were to be run on virtually any computer connected to the Internet, there needed to be some way to enable that program to execute on different types of systems. In other words, a mechanism that allows the same application to be downloaded and executed by a wide variety of CPUs, operating systems, and browsers is required. It is not practical to have different versions of the same application for different computers. The *same* application code must work in *all* computers. Therefore, some means of generating portable code was needed. As you will soon see, the same mechanism that helps ensure security also helps create portability.

Java's Magic: The Bytecode

The key that allowed Java to address both the security and the portability problems just described is that the output of a Java compiler is not executable code. Rather, it is bytecode. *Bytecode* is a highly optimized set of instructions designed to be executed by what is called the *Java Virtual Machine (JVM)*, which is part of the Java Runtime Environment (JRE). In essence, the original JVM was designed as an *interpreter for bytecode.* This may come as a bit of a surprise because many modern languages are designed to be compiled into CPU-specific, executable code due to performance concerns. However, the fact that a Java program is executed by the JVM helps solve the major problems associated with web-based programs. Here is why.

Translating a Java program into bytecode makes it much easier to run a program in a wide variety of environments because only the JRE (which includes the JVM) needs to be implemented for each platform. Once a JRE exists for a given system, any Java program can run on it. Remember, although the details of the JRE will differ from platform to platform, all JREs understand the same Java bytecode. If a Java program were compiled to native code, then different versions of the same program would have to exist for each type of CPU connected to the Internet. This is, of course, not a feasible solution. Thus, the execution of bytecode by the JVM is the easiest way to create truly portable programs.

The fact that a Java program is executed by the JVM also helps to make it secure. Because the JVM is in control, it manages program execution. Thus, it was possible for the JVM to create a restricted execution environment, called the *sandbox*, that contains the program, preventing unrestricted access to the machine. Safety is also enhanced by certain restrictions that exist in the Java language.

When a program is interpreted, it generally runs slower than the same program would run if compiled to executable code. However, with Java, the differential between the two is not so great. Because bytecode has been highly optimized, the use of bytecode enables the JVM to execute programs much faster than you might expect.

Although Java was designed as an interpreted language, there is nothing about Java that prevents on-the-fly compilation of bytecode into native code in order to boost performance. For this reason, the HotSpot JVM was introduced not long after Java's initial release. HotSpot includes a just-in-time (JIT) compiler for bytecode. When a JIT compiler is part of the JVM, selected portions of bytecode are compiled into executable code in real time on a piece-by-piece demand basis. That is, a JIT compiler compiles code as it is needed during execution. Furthermore, not all sequences of bytecode are compiled—only those that will benefit from compilation. The remaining code is simply interpreted. However, the just-in-time approach still yields a significant performance boost. Even when dynamic compilation is applied to bytecode, the portability and safety features still apply because the JVM is still in charge of the execution environment.

One other point: There has been experimentation with an *ahead-of-time* compiler for Java. Such a compiler can be used to compile bytecode into native code *prior* to execution by the JVM, rather than on-the-fly. Some previous versions of the JDK supplied an experimental ahead-of-time compiler; however, JDK 17 has removed it. Ahead-of-time compilation is a specialized feature and it does not replace Java's traditional approach just described. Because of the highly sophisticated nature of ahead-of-time compilation, it is not something that you will use when learning Java, and it is not discussed further in this book.

Ask the Expert

Q: I have heard about a special type of Java program called a servlet. What is it?

A: A Java *servlet* is a small program that executes on a server. Servlets dynamically extend the functionality of a web server. It is helpful to understand that as useful as client-side applications can be, they are just one half of the client/server equation. Not long after the initial release of Java, it became obvious that Java would also be useful on the server side. One result was the servlet. Thus, with the advent of the servlet, Java spanned both sides of the client/server connection. Although the topic of servlets, and server-side programming in general, is beyond the scope of this beginner's guide, they are something that you will likely find of interest as you advance in Java programming.

Moving Beyond Applets

At the time of this writing, it has been more than two decades since Java's original release. Over those years, many changes have taken place. At the time of Java's creation, the Internet was a new and exciting innovation; web browsers were undergoing rapid development and refinement; the modern form of the smartphone had not yet been invented; and the near ubiquitous use of computers was still a few years off. As you would expect, Java has also changed and so, too, has the way that Java is used. Perhaps nothing illustrates the ongoing evolution of Java better than the applet.

As explained previously, in the early years of Java, applets were a crucial part of Java programming. They not only added excitement to a web page, they were a highly visible part of Java, which added to its charisma. However, applets rely on a Java browser plug-in. Thus, for an applet to work, it must be supported by the browser. Over the past few years support for the Java browser plug-in has been waning. Simply put, without browser support, applets are not viable. Because of this, beginning with JDK 9, the phase-out of applets was begun, with support for applets being deprecated. In the language of Java, *deprecated* means that a feature is still available but flagged as obsolete. Thus, a deprecated feature should not be used for new code. The phase-out became complete with the release of JDK 11 because run-time support for applets was removed. Beginning with JDK 17, the entire Applet API was *deprecated for removal*, which means that it will be removed from the JDK at some point in the future.

As a point of interest, a few years after Java's creation an alternative to applets was added. Called Java Web Start, it enabled an application to be dynamically downloaded from a web page. It was a deployment mechanism that was especially useful for larger Java applications that were not appropriate for applets. The difference between an applet and a Web Start application is that a Web Start application runs on its own, not inside the browser. Thus, it looks much like a "normal" application. It does, however, require that a stand-alone JRE that supports Web Start is available on the host system. Beginning with JDK 11, support for Java Web Start has been removed.

Given that neither applets nor Java Web Start are viable options for modern versions of Java, you might wonder what mechanism should be used to deploy a Java application. At the time of this writing, one part of the answer is to use the **jlink** tool added by JDK 9. It can create a complete run-time image that includes all necessary support for your program, including the JRE. Another part of the answer is the **jpackage** tool. Added by JDK 16, it can be used to create a ready-to-install application. As you might guess, deployment is a rather advanced topic that is outside the scope of this book. Fortunately, you won't need to worry about deployment to use this book because all of the sample programs run directly on your computer. They are not deployed over the Internet.

A Faster Release Schedule

Not long ago, another major change occurred in Java, but it does not involve changes to the language or the run-time environment. Rather, it relates to the way that Java releases are scheduled. In the past, major Java releases were typically separated by two or more years. However, subsequent to the release of JDK 9, the time between major Java releases has been decreased. Today, it is anticipated that a major release will occur on a strict time-based schedule, with the expected time between major releases being just six months.

Each major release, now called a *feature release,* will include those features ready at the time of the release. This increased *release cadence* enables new features and enhancements to be available to Java programmers in a timely fashion. Furthermore, it allows Java to respond quickly to the demands of an ever-changing programming environment. Simply put, the faster release schedule promises to be a very positive development for Java programmers.

At three-year intervals it is anticipated that a *long-term support* (LTS) release will take place. An LTS release will be supported (and thus remain viable) for a period of time longer than six months. The first LTS release was JDK 11. The second LTS release was JDK 17, for which this book has been updated. Because of the stability that an LTS release offers, it is likely that its feature-set will define a baseline of functionality for a number of years. Consult Oracle for the latest information concerning long-term support and the LTS release schedule.

Currently, feature releases are scheduled for March and September of each year. As a result, JDK 10 was released in March 2018, which was six months after the release of JDK 9. The next release (JDK 11) was in September 2018. It was an LTS release. This was followed by JDK 12 In March 2019, JDK 13 in September 2019, and so on. At the time of this writing, the latest release is JDK 17, which is an LTS release. Again, it is anticipated that every six months a new feature release will take place. Of course, you will want to consult the latest release schedule information.

At the time of this writing, there are a number of new Java features on the horizon. Because of the faster release schedule, it is very likely that several of them will be added to Java over the next few years. You will want to review the information and release notes provided by each six-month release in detail. It is truly an exciting time to be a Java programmer!

The Java Buzzwords

No history of Java is complete without a look at the Java buzzwords. Although the fundamental forces that necessitated the invention of Java are portability and security, other factors played an important role in molding the final form of the language. The key considerations were summed up by the Java design team in the following list of buzzwords.

Simple	Java has a concise, cohesive set of features that makes it easy to learn and use.
Secure	Java provides a secure means of creating Internet applications.
Portable	Java programs can execute in any environment for which there is a Java run-time system.
Object-oriented	Java embodies the modern object-oriented programming philosophy.
Robust	Java encourages error-free programming by being strictly typed and performing run-time checks.
Multithreaded	Java provides integrated support for multithreaded programming.
Architecture-neutral	Java is not tied to a specific machine or operating system architecture.
Interpreted	Java supports cross-platform code through the use of Java bytecode.
High performance	The Java bytecode is highly optimized for speed of execution.
Distributed	Java was designed with the distributed environment of the Internet in mind.
Dynamic	Java programs carry with them substantial amounts of run-time type information that is used to verify and resolve access to objects at run time.

Object-Oriented Programming

At the center of Java is object-oriented programming (OOP). The object-oriented methodology is inseparable from Java, and all Java programs are, to at least some extent, object-oriented. Because of OOP's importance to Java, it is useful to understand in a general way OOP's basic principles before you write even a simple Java program. Later in this book, you will see how to put these concepts into practice.

OOP is a powerful way to approach the job of programming. Programming methodologies have changed dramatically since the invention of the computer, primarily to accommodate the increasing complexity of programs. For example, when computers were first invented, programming was done by toggling in the binary machine instructions using the computer's front panel. As long as programs were just a few hundred instructions long, this approach worked. As programs grew, assembly language was invented so that a programmer could deal with larger, increasingly complex programs, using symbolic representations of the machine instructions. As programs continued to grow, high-level languages were introduced that gave the programmer more tools with which to handle complexity. The first widespread language was, of course, FORTRAN. Although FORTRAN was a very impressive first step, it was hardly a language that encouraged clear, easy-to-understand programs.

The 1960s gave birth to structured programming. This is the method encouraged by languages such as C and Pascal. The use of structured languages made it possible to write moderately complex programs fairly easily. Structured languages are characterized by their support for stand-alone subroutines, local variables, rich control constructs, and their lack of reliance upon the GOTO. Although structured languages are a powerful tool, even they reach their limit when a project becomes too large.

Consider this: At each milestone in the development of programming, techniques and tools were created to allow the programmer to deal with increasingly greater complexity. Each step of the way, the new approach took the best elements of the previous methods and moved forward. Prior to the invention of OOP, many projects were nearing (or exceeding) the point where the structured approach no longer works. Object-oriented methods were created to help programmers break through these barriers.

Object-oriented programming took the best ideas of structured programming and combined them with several new concepts. The result was a different way of organizing a program. In the most general sense, a program can be organized in one of two ways: around its code (what is happening) or around its data (what is being affected). Using only structured programming techniques, programs are typically organized around code. This approach can be thought of as "code acting on data."

Object-oriented programs work the other way around. They are organized around data, with the key principle being "data controlling access to code." In an object-oriented language, you define the data and the routines that are permitted to act on that data. Thus, a data type defines precisely what sort of operations can be applied to that data.

To support the principles of object-oriented programming, all OOP languages, including Java, have three traits in common: encapsulation, polymorphism, and inheritance. Let's examine each.

Encapsulation

Encapsulation is a programming mechanism that binds together code and the data it manipulates, and that keeps both safe from outside interference and misuse. In an object-oriented language, code and data can be bound together in such a way that a self-contained *black box* is created. Within the box are all necessary data and code. When code and data are linked together in this fashion, an object is created. In other words, an object is the device that supports encapsulation.

Within an object, code, data, or both may be *private* to that object or *public*. Private code or data is known to and accessible by only another part of the object. That is, private code or data cannot be accessed by a piece of the program that exists outside the object. When code or data is public, other parts of your program can access it even though it is defined within an object. Typically, the public parts of an object are used to provide a controlled interface to the private elements of the object.

Java's basic unit of encapsulation is the *class*. Although the class will be examined in great detail later in this book, the following brief discussion will be helpful now. A class defines the form of an object. It specifies both the data and the code that will operate on that data. Java uses a class specification to construct *objects*. Objects are instances of a class. Thus, a class is essentially a set of plans that specify how to build an object.

The code and data that constitute a class are called *members* of the class. Specifically, the data defined by the class are referred to as *member variables* or *instance variables*. The code that operates on that data is referred to as *member methods* or just *methods. Method* is Java's term for a subroutine. If you are familiar with C/C++, it may help to know that what a Java programmer calls a *method*, a C/C++ programmer calls a *function*.

Polymorphism

Polymorphism (from Greek, meaning "many forms") is the quality that allows one interface to access a general class of actions. The specific action is determined by the exact nature of the situation. A simple example of polymorphism is found in the steering wheel of an automobile. The steering wheel (i.e., the interface) is the same no matter what type of actual steering mechanism is used. That is, the steering wheel works the same whether your car has manual steering, power steering, or rack-and-pinion steering. Therefore, once you know how to operate the steering wheel, you can drive any type of car.

The same principle can also apply to programming. For example, consider a stack (which is a first-in, last-out list). You might have a program that requires three different types of stacks. One stack is used for integer values, one for floating-point values, and one for characters. In this case, the algorithm that implements each stack is the same, even though the data being stored differs. In a non-object-oriented language, you would be required to create three different sets of stack routines, with each set using different names. However, because of polymorphism, in Java you can create one general set of stack routines that works for all three specific situations. This way, once you know how to use one stack, you can use them all.

More generally, the concept of polymorphism is often expressed by the phrase "one interface, multiple methods." This means that it is possible to design a generic interface to a group of related activities. Polymorphism helps reduce complexity by allowing the same interface to be used to specify a *general class of action.* It is the compiler's job to select the *specific action* (i.e., method) as it applies to each situation. You, the programmer, don't need to do this selection manually. You need only remember and utilize the general interface.

Inheritance

Inheritance is the process by which one object can acquire the properties of another object. This is important because it supports the concept of hierarchical classification. If you think about it, most knowledge is made manageable by hierarchical (i.e., top-down) classifications. For example, a Red Delicious apple is part of the classification *apple,* which in turn is part of the *fruit* class, which is under the larger class *food.* That is, the *food* class possesses certain qualities (edible, nutritious, etc.) which also, logically, apply to its subclass, *fruit.* In addition to these qualities, the *fruit* class has specific characteristics (juicy, sweet, etc.) that distinguish it from other food. The *apple* class defines those qualities specific to an apple (grows on trees, not tropical, etc.). A Red Delicious apple would, in turn, inherit all the qualities of all preceding classes, and would define only those qualities that make it unique.

Without the use of hierarchies, each object would have to explicitly define all of its characteristics. Using inheritance, an object need only define those qualities that make it unique within its class. It can inherit its general attributes from its parent. Thus, it is the inheritance mechanism that makes it possible for one object to be a specific instance of a more general case.

The Java Development Kit

Now that the theoretical underpinning of Java has been explained, it is time to start writing Java programs. Before you can compile and run those programs, you must have a Java Development Kit (JDK). At the time of this writing, the current release of the JDK is JDK 17. This is the version for Java SE 17. (*SE* stands for *Standard Edition.*) It is also the version described in this book. Because JDK 17 contains features that are not supported by earlier versions of Java, it is recommended that you use JDK 17 (or later) to compile and run the programs in this book. (Remember, because of Java's faster release schedule, JDK feature releases are expected at six-month intervals. Thus, don't be surprised by a JDK with a higher release number.) However, depending on the environment in which you are working, an earlier JDK may already be installed. If this is the case, then newer Java features will not be available.

If you need to install the JDK on your computer, be aware that for modern versions of Java, both Oracle JDKs and open source OpenJDKs are available for download. In general, you should first find the JDK you want to use. For example, at the time of this writing, the Oracle JDK can be downloaded from **www.oracle.com/java/technologies/downloads/**. Also at the time of this writing, an open source version is available at **jdk.java.net**. Next, download the JDK of your choice and follow its instructions to install it on your computer. After you have installed the JDK, you will be able to compile and run programs.

The JDK supplies two primary programs. The first is **javac**, which is the Java compiler. The second is **java**, which is the standard Java interpreter and is also referred to as the *application launcher.* One other point: The JDK runs in the command-prompt environment and uses command-line tools. It is not a windowed application. It is also not an integrated development environment (IDE).

Ask the Expert

Q: You state that object-oriented programming is an effective way to manage large programs. However, it seems that it might add substantial overhead to relatively small ones. Since you say that all Java programs are, to some extent, object-oriented, does this impose a penalty for smaller programs?

A: No. As you will see, for small programs, Java's object-oriented features are nearly transparent. Although it is true that Java follows a strict object model, you have wide latitude as to the degree to which you employ it. For smaller programs, their "object-orientedness" is barely perceptible. As your programs grow, you will integrate more object-oriented features effortlessly.

NOTE

In addition to the basic command-line tools supplied with the JDK, there are several high-quality IDEs available for Java, such as NetBeans and Eclipse. An IDE can be very helpful when developing and deploying commercial applications. As a general rule, you can also use an IDE to compile and run the programs in this book if you so choose. However, the instructions presented in this book for compiling and running a Java program describe only the JDK command-line tools. The reasons for this are easy to understand. First, the JDK is readily available to all readers. Second, the instructions for using the JDK will be the same for all readers. Furthermore, for the simple programs presented in this book, using the JDK command-line tools is usually the easiest approach. If you are using an IDE, you will need to follow its instructions. Because of differences between IDEs, no general set of instructions can be given.

A First Simple Program

Let's start by compiling and running the short sample program shown here:

```
/*
   This is a simple Java program.

   Call this file Example.java.
*/
class Example {
  // A Java program begins with a call to main().
  public static void main(String[] args) {
    System.out.println("Java drives the Web.");
  }
}
```

You will follow these three steps:

1. Enter the program.

2. Compile the program.

3. Run the program.

Entering the Program

The programs shown in this book are available from **www.mhprofessional.com**. However, if you want to enter the programs by hand, you are free to do so. In this case, you must enter the program into your computer using a text editor, not a word processor. Word processors typically store format information along with text. This format information will confuse the Java compiler. If you are using a Windows platform, you can use Notepad or any other programming editor that you like.

For most computer languages, the name of the file that holds the source code to a program is arbitrary. However, this is not the case with Java. The first thing that you must learn about Java is that *the name you give to a source file is very important.* For this example, the name of the source file should be **Example.java**. Let's see why.

In Java, a source file is officially called a *compilation unit.* It is a text file that contains (among other things) one or more class definitions. (For now, we will be using source files that contain only one class.) The Java compiler requires that a source file use the **.java** filename extension. As you can see by looking at the program, the name of the class defined by the program is also **Example**. This is not a coincidence. In Java, all code must reside inside a class. By convention, the name of the main class should match the name of the file that holds the program. You should also make sure that the capitalization of the filename matches the class name. The reason for this is that Java is case sensitive. At this point, the convention that filenames correspond to class names may seem arbitrary. However, this convention makes it easier to maintain and organize your programs. Furthermore, as you will see later in this book, in some cases, it is required.

Compiling the Program

To compile the **Example** program, execute the compiler, **javac**, specifying the name of the source file on the command line, as shown here:

```
javac Example.java
```

The **javac** compiler creates a file called **Example.class** that contains the bytecode version of the program. Remember, bytecode is not executable code. Bytecode must be executed by a Java Virtual Machine. Thus, the output of **javac** is not code that can be directly executed.

To actually run the program, you must use the Java interpreter, **java**. To do so, pass the class name **Example** as a command-line argument, as shown here:

```
java Example
```

When the program is run, the following output is displayed:

```
Java drives the Web.
```

When Java source code is compiled, each individual class is put into its own output file named after the class and using the **.class** extension. This is why it is a good idea to give your Java source files the same name as the class they contain—the name of the source file will match the name of the .class file. When you execute the Java interpreter as just shown, you are actually specifying the name of the class that you want the interpreter to execute. It will automatically search for a file by that name that has the **.class** extension. If it finds the file, it will execute the code contained in the specified class.

Before moving on, it is important to mention that beginning with JDK 11, Java provides a way to run some types of simple programs directly from a source file, without explicitly invoking **javac**. This technique, which can be useful in some situations, is described in Appendix C. For the purposes of this book, it is assumed that you are using the normal compilation process just described.

NOTE

If, when you try to compile the program, the computer cannot find **javac** (and assuming that you have installed the JDK correctly), you may need to specify the path to the command-line tools. In Windows, for example, this means that you will need to add the path to the command-line tools to the paths defined for the **PATH** environmental variable. For example, if JDK 17 was installed under the Program Files directory, then the path to the command-line tools will be similar to **C:\Program Files\Java\jdk-17\bin**. (Of course, you will need to find the path to Java on your computer, which may differ from the one just shown. Also the specific version of the JDK may differ.) You will need to consult the documentation for your operating system on how to set the path, because this procedure differs between OSes.

The First Sample Program Line by Line

Although **Example.java** is quite short, it includes several key features that are common to all Java programs. Let's closely examine each part of the program.

The program begins with the following lines:

```
/*
   This is a simple Java program.

   Call this file Example.java.
*/
```

This is a *comment*. Like most other programming languages, Java lets you enter a remark into a program's source file. The contents of a comment are ignored by the compiler. Instead, a comment describes or explains the operation of the program to anyone who is reading its source code. In this case, the comment describes the program and reminds you that the source file should be called **Example.java**. Of course, in real applications, comments generally explain how some part of the program works or what a specific feature does.

Java supports three styles of comments. The one shown at the top of the program is called a *multiline comment*. This type of comment must begin with /* and end with */. Anything between these two comment symbols is ignored by the compiler. As the name suggests, a multiline comment may be several lines long.

The next line of code in the program is shown here:

```
class Example {
```

This line uses the keyword **class** to declare that a new class is being defined. As mentioned, the class is Java's basic unit of encapsulation. **Example** is the name of the class. The class definition begins with the opening curly brace ({) and ends with the closing curly brace (}). The elements between the two braces are members of the class. For the moment, don't worry too much about the details of a class except to note that in Java, all program activity occurs within one. This is one reason why all Java programs are (at least a little bit) object-oriented.

The next line in the program is the *single-line comment,* shown here:

```
// A Java program begins with a call to main().
```

This is the second type of comment supported by Java. A single-line comment begins with a **//** and ends at the end of the line. As a general rule, programmers use multiline comments for longer remarks and single-line comments for brief, line-by-line descriptions.

The next line of code is shown here:

```
public static void main (String[] args) {
```

This line begins the **main( )** method. As mentioned earlier, in Java, a subroutine is called a *method.* As the comment preceding it suggests, this is the line at which the program will begin executing. In general, Java applications begin execution by calling **main( )**. The exact meaning of each part of this line cannot be given now, since it involves a detailed understanding of several other of Java's features. However, since many of the examples in this book will use this line of code, let's take a brief look at each part now.

The **public** keyword is an *access modifier.* An access modifier determines how other parts of the program can access the members of the class. When a class member is preceded by **public**, then that member can be accessed by code outside the class in which it is declared. (The opposite of **public** is **private**, which prevents a member from being used by code defined outside of its class.) In this case, **main( )** must be declared as **public**, since it must be called by code outside of its class when the program is started. The keyword **static** allows **main( )** to be called before an object of the class has been created. This is necessary because **main( )** is called by the JVM before any objects are made. The keyword **void** simply tells the compiler that **main( )** does not return a value. As you will see, methods may also return values. If all this seems a bit confusing, don't worry. All of these concepts will be discussed in detail in subsequent chapters.

As stated, **main( )** is the method called when a Java application begins. Any information that you need to pass to a method is received by variables specified within the set of parentheses that follow the name of the method. These variables are called *parameters.* If no parameters are required for a given method, you still need to include the empty parentheses. In **main( )** there is only one parameter, **String[ ] args**, which declares a parameter named **args**. This is an array of objects of type **String**. (*Arrays* are collections of similar objects.) Objects of type **String** store sequences of characters. In this case, **args** receives any command-line arguments present when the program is executed. This program does not make use of this information, but other programs shown later in this book will.

The last character on the line is the {. This signals the start of **main( )**'s body. All of the code included in a method will occur between the method's opening curly brace and its closing curly brace.

The next line of code is shown here. Notice that it occurs inside **main( )**.

```
System.out.println("Java drives the Web.");
```

This line outputs the string "Java drives the Web." followed by a new line on the screen. Output is actually accomplished by the built-in **println( )** method. In this case, **println( )** displays the string that is passed to it. As you will see, **println( )** can be used to display other types of information, too. The line begins with **System.out**. While too complicated to explain in detail at this time, briefly, **System** is a predefined class that provides access to the system, and **out** is the output stream that is connected to the console. Thus, **System.out** is an object that encapsulates console output. The fact that Java uses an object to define console output is further evidence of its object-oriented nature.

As you have probably guessed, console output (and input) is not used frequently in real-world Java applications. Since most modern computing environments are windowed and graphical in nature, console I/O is used mostly for simple utility programs, for demonstration programs, and for server-side code. Later in this book, you will learn other ways to generate output using Java, but for now, we will continue to use the console I/O methods.

Notice that the **println( )** statement ends with a semicolon. Many statements in Java end with a semicolon. As you will see, the semicolon is an important part of the Java syntax.

The first } in the program ends **main( )**, and the last } ends the **Example** class definition.

One last point: Java is case sensitive. Forgetting this can cause you serious problems. For example, if you accidentally type **Main** instead of **main**, or **PrintLn** instead of **println**, the preceding program will be incorrect. Furthermore, although the Java compiler *will* compile classes that do not contain a **main( )** method, it has no way to execute them. So, if you had mistyped **main**, the compiler would still compile your program. However, the Java interpreter would report an error because it would be unable to find the **main( )** method.

Handling Syntax Errors

If you have not yet done so, enter, compile, and run the preceding program. As you may know from your previous programming experience, it is quite easy to accidentally type something incorrectly when entering code into your computer. Fortunately, if you enter something incorrectly into your program, the compiler will report a *syntax error* message when it tries to compile it. The Java compiler attempts to make sense out of your source code no matter what you have written. For this reason, the error that is reported may not always reflect the actual cause of the problem. In the preceding program, for example, an accidental omission of the opening curly brace after the **main( )** method causes the compiler to report the following two errors:

```
Example.java:8: error: ';' expected
  public static void main(String[] args)
                                        ^
Example.java:11: error: class, interface, enum, or record expected
}
^
```

Clearly, the first error message is completely wrong because what is missing is not a semicolon, but a curly brace.

The point of this discussion is that when your program contains a syntax error, you shouldn't necessarily take the compiler's messages at face value. The messages may be misleading. You may need to "second-guess" an error message in order to find the real problem. Also, look at the last few lines of code in your program that precede the line being flagged. Sometimes an error will not be reported until several lines after the point at which the error actually occurred.

A Second Simple Program

Perhaps no other construct is as important to a programming language as the assignment of a value to a variable. A *variable* is a named memory location that can be assigned a value. Further, the value of a variable can be changed during the execution of a program. That is, the content of a variable is changeable, not fixed. The following program creates two variables called **myVar1** and **myVar2**:

```java
/*
   This demonstrates a variable.

   Call this file Example2.java.
*/
class Example2 {
  public static void main(String[] args) {
    int myVar1; // this declares a variable ◄─────── Declare variables.
    int myVar2; // this declares another variable

    myVar1 = 1024; // this assigns 1024 to myVar1 ◄── Assign a variable a value.

    System.out.println("myVar1 contains " + myVar1);

    myVar2 = myVar1 / 2;

    System.out.print("myVar2 contains myVar1 / 2: ");
    System.out.println(myVar2);
  }
}
```

When you run this program, you will see the following output:

```
myVar1 contains 1024
myVar2 contains myVar1 / 2: 512
```

This program introduces several new concepts. First, the statement

```
int myVar1; // this declares a variable
```

declares a variable called **myVar1** of type integer. In Java, all variables must be declared before they are used. Further, the type of values that the variable can hold must also be specified. This is called the *type* of the variable. In this case, **myVar1** can hold integer values. These are whole number values. In Java, to declare a variable to be of type integer, precede its name with the keyword **int**. Thus, the preceding statement declares a variable called **myVar1** of type **int**.

The next line declares a second variable called **myVar2**:

```
int myVar2; // this declares another variable
```

Notice that this line uses the same format as the first line except that the name of the variable is different.

In general, to declare a variable you will use a statement like this:

type var-name;

Here, *type* specifies the type of variable being declared, and *var-name* is the name of the variable. In addition to **int**, Java supports several other data types.

The following line of code assigns **myVar1** the value 1024:

```
myVar1 = 1024; // this assigns 1024 to var1
```

In Java, the assignment operator is the single equal sign. It copies the value on its right side into the variable on its left.

The next line of code outputs the value of **myVar1** preceded by the string "myVar1 contains ":

```
System.out.println("myVar1 contains " + myVar1);
```

In this statement, the plus sign causes the value of **myVar1** to be displayed after the string that precedes it. This approach can be generalized. Using the **+** operator, you can chain together as many items as you want within a single **println( )** statement.

The next line of code assigns **myVar2** the value of **myVar1** divided by 2:

```
myVar2 = myVar1 / 2;
```

This line divides the value in **myVar1** by 2 and then stores that result in **myVar2**. Thus, after the line executes, **myVar2** will contain the value 512. The value of **myVar1** will be unchanged. Like most other computer languages, Java supports a full range of arithmetic operators, including those shown here:

+	Addition
–	Subtraction
*	Multiplication
/	Division

Here are the next two lines in the program:

```
System.out.print("myVar2 contains myVar1 / 2: ");
System.out.println(myVar2);
```

Two new things are occurring here. First, the built-in method **print( )** is used to display the string "myVar2 contains myVar1 / 2: ". This string is *not* followed by a new line. This means that when the next output is generated, it will start on the same line. The **print( )** method is just like **println( )**, except that it does not output a new line after each call. Second, in the call to **println( )**, notice that **myVar2** is used by itself. Both **print( )** and **println( )** can be used to output values of any of Java's built-in types.

One more point about declaring variables before we move on: It is possible to declare two or more variables using the same declaration statement. Just separate their names by commas. For example, **myVar1** and **myVar2** could have been declared like this:

```
int myVar1, myVar2; // both declared using one statement
```

Another Data Type

In the preceding program, a variable of type **int** was used. However, a variable of type **int** can hold only whole numbers. Thus, it cannot be used when a fractional component is required. For example, an **int** variable can hold the value 18, but not the value 18.3. Fortunately, **int** is only one of several data types defined by Java. To allow numbers with fractional components, Java defines two floating-point types: **float** and **double**, which represent single- and double-precision values, respectively. Of the two, **double** is the most commonly used.

To declare a variable of type **double**, use a statement similar to that shown here:

```
double x;
```

Here, **x** is the name of the variable, which is of type **double**. Because **x** has a floating-point type, it can hold values such as 122.23, 0.034, or –19.0.

To better understand the difference between **int** and **double**, try the following program:

```
/*
    This program illustrates the differences
    between int and double.

    Call this file Example3.java.
*/
class Example3 {
  public static void main(String[] args) {
    int v; // this declares an int variable
    double x; // this declares a floating-point variable

    v = 10; // assign v the value 10

    x = 10.0; // assign x the value 10.0

    System.out.println("Original value of v: " + v);
    System.out.println("Original value of x: " + x);
    System.out.println(); // print a blank line   ◄─────────── Output a blank line.

    // now, divide both by 4
    v = v / 4;
    x = x / 4;

    System.out.println("v after division: " + v);
    System.out.println("x after division: " + x);
  }
}
```

Ask the Expert

Q: Why does Java have different data types for integers and floating-point values? That is, why aren't all numeric values just the same type?

A: Java supplies different data types so that you can write efficient programs. For example, integer arithmetic is faster than floating-point calculations. Thus, if you don't need fractional values, then you don't need to incur the overhead associated with types **float** or **double**. Second, the amount of memory required for one type of data might be less than that required for another. By supplying different types, Java enables you to make best use of system resources. Finally, some algorithms require (or at least benefit from) the use of a specific type of data. In general, Java supplies a number of built-in types to give you the greatest flexibility.

The output from this program is shown here:

```
Original value of v: 10
Original value of x: 10.0

v after division: 2  ◄────────── Fractional component lost
x after division: 2.5 ◄────────── Fractional component preserved
```

As you can see, when **v** is divided by 4, a whole-number division is performed, and the outcome is 2—the fractional component is lost. However, when the **double** variable **x** is divided by 4, the fractional component is preserved, and the proper answer is displayed.

There is one other new thing to notice in the program. To print a blank line, simply call **println( )** without any arguments.

Try This 1-1   Converting Gallons to Liters

GalToLit.java   Although the preceding sample programs illustrate several important features of the Java language, they are not very useful. Even though you do not know much about Java at this point, you can still put what you have learned to work to create a practical program. In this project, we will create a program that converts gallons to liters. The program will work by declaring two **double** variables. One will hold the number of the gallons, and the second will hold the number of liters after the conversion. There are 3.7854

(continued)

liters in a gallon. Thus, to convert gallons to liters, the gallon value is multiplied by 3.7854. The program displays both the number of gallons and the equivalent number of liters.

1. Create a new file called **GalToLit.java**.

2. Enter the following program into the file:

```
/*
   Try This 1-1

   This program converts gallons to liters.

   Call this program GalToLit.java.
*/
class GalToLit {
  public static void main(String[] args) {
    double gallons; // holds the number of gallons
    double liters; // holds conversion to liters

    gallons = 10; // start with 10 gallons

    liters = gallons * 3.7854; // convert to liters

     System.out.println(gallons + " gallons is " + liters + " liters.");
  }
}
```

3. Compile the program using the following command line:

```
javac GalToLit.java
```

4. Run the program using this command:

```
java GalToLit
```

You will see this output:

```
10.0 gallons is 37.854 liters.
```

5. As it stands, this program converts 10 gallons to liters. However, by changing the value assigned to **gallons**, you can have the program convert a different number of gallons into its equivalent number of liters.

Two Control Statements

Inside a method, execution proceeds from one statement to the next, top to bottom. However, it is possible to alter this flow through the use of the various program control statements supported by Java. Although we will look closely at control statements later, two are briefly introduced here because we will be using them to write sample programs.

The if Statement

You can selectively execute part of a program through the use of Java's conditional statement: the **if**. The Java **if** statement works much like the IF statement in any other language. It determines the flow of program execution based on whether some condition is true or false. Its simplest form is shown here:

if(*condition*) statement;

Here, *condition* is a Boolean expression. (A Boolean expression is one that evaluates to either true or false.) If *condition* is true, then the statement is executed. If *condition* is false, then the statement is bypassed. Here is an example:

```
if(10 < 11) System.out.println("10 is less than 11");
```

In this case, since 10 is less than 11, the conditional expression is true, and **println( )** will execute. However, consider the following:

```
if(10 < 9) System.out.println("this won't be displayed");
```

In this case, 10 is not less than 9. Thus, the call to **println( )** will not take place.

Java defines a full complement of relational operators that may be used in a conditional expression. They are shown here:

Operator	Meaning
<	Less than
<=	Less than or equal
>	Greater than
>=	Greater than or equal
==	Equal to
!=	Not equal

Notice that the test for equality is the double equal sign.

Here is a program that illustrates the **if** statement:

```
/*
   Demonstrate the if.

   Call this file IfDemo.java.
*/
class IfDemo {
  public static void main(String[] args) {
    int a, b, c;

    a = 2;
    b = 3;

    if(a < b) System.out.println("a is less than b");
```

```
        // this won't display anything
        if(a == b) System.out.println("you won't see this");

        System.out.println();

        c = a - b; // c contains -1

        System.out.println("c contains -1");
        if(c >= 0) System.out.println("c is non-negative");
        if(c < 0) System.out.println("c is negative");

        System.out.println();

        c = b - a; // c now contains 1

        System.out.println("c contains 1");
        if(c >= 0) System.out.println("c is non-negative");
        if(c < 0) System.out.println("c is negative");

    }
}
```

The output generated by this program is shown here:

```
a is less than b

c contains -1
c is negative

c contains 1
c is non-negative
```

Notice one other thing in this program. The line

```
int a, b, c;
```

declares three variables, **a**, **b**, and **c**, by use of a comma-separated list. As mentioned earlier, when you need two or more variables of the same type, they can be declared in one statement. Just separate the variable names by commas.

The for Loop

You can repeatedly execute a sequence of code by creating a *loop*. Loops are used whenever you need to perform a repetitive task because they are much simpler and easier than trying to write the same statement sequence over and over again. Java supplies a powerful assortment of loop constructs. The one we will look at here is the **for** loop. The simplest form of the **for** loop is shown here:

for(*initialization; condition; iteration*) *statement;*

In its most common form, the *initialization* portion of the loop sets a loop control variable to an initial value. The *condition* is a Boolean expression that tests the loop control variable.

If the outcome of that test is true, *statement* executes and the **for** loop continues to iterate. If it is false, the loop terminates. The *iteration* expression determines how the loop control variable is changed each time the loop iterates. Here is a short program that illustrates the **for** loop:

```
/*
  Demonstrate the for loop.

  Call this file ForDemo.java.
*/
class ForDemo {
  public static void main(String[] args) {
    int count;

    for(count = 0; count < 5; count = count+1)          This loop iterates five times.
      System.out.println("This is count: " + count);

    System.out.println("Done!");
  }
}
```

The output generated by the program is shown here:

```
This is count: 0
This is count: 1
This is count: 2
This is count: 3
This is count: 4
Done!
```

In this example, **count** is the loop control variable. It is set to zero in the initialization portion of the **for**. At the start of each iteration (including the first one), the conditional test **count < 5** is performed. If the outcome of this test is true, the **println( )** statement is executed, and then the iteration portion of the loop is executed, which increases **count** by 1. This process continues until the conditional test is false, at which point execution picks up at the bottom of the loop. As a point of interest, in professionally written Java programs, you will almost never see the iteration portion of the loop written as shown in the preceding program. That is, you will seldom see statements like this:

```
count = count + 1;
```

The reason is that Java includes a special increment operator that performs this operation more efficiently. The increment operator is **++** (that is, two plus signs back to back). The increment operator increases its operand by one. By use of the increment operator, the preceding statement can be written like this:

```
count++;
```

Thus, the **for** in the preceding program will usually be written like this:

```
for(count = 0; count < 5; count++)
```

You might want to try this. As you will see, the loop still runs exactly the same as it did before.

Java also provides a decrement operator, which is specified as – –. This operator decreases its operand by one.

Create Blocks of Code

Another key element of Java is the *code block*. A code block is a grouping of two or more statements. This is done by enclosing the statements between opening and closing curly braces. Once a block of code has been created, it becomes a logical unit that can be used any place that a single statement can. For example, a block can be a target for Java's **if** and **for** statements. Consider this **if** statement:

```
if(w < h) {            Start of block
  v = w * h;
  w = 0;
}            End of block
```

Here, if **w** is less than **h**, both statements inside the block will be executed. Thus, the two statements inside the block form a logical unit, and one statement cannot execute without the other also executing. The key point here is that whenever you need to logically link two or more statements, you do so by creating a block. Code blocks allow many algorithms to be implemented with greater clarity and efficiency.

Here is a program that uses a block of code to prevent a division by zero:

```
/*
  Demonstrate a block of code.

  Call this file BlockDemo.java.
*/
class BlockDemo {
  public static void main(String[] args) {
    double i, j, d;

    i = 5;
    j = 10;

    // the target of this if is a block
    if(i != 0) {
      System.out.println("i does not equal zero");
      d = j / i;
      System.out.println("j / i is " + d);
    }
  }
}
```

The target of the **if** is this entire block.

The output generated by this program is shown here:

```
i does not equal zero
j / i is 2.0
```

Ask the Expert

Q: Does the use of a code block introduce any run-time inefficiencies? In other words, does Java actually execute the { and }?

A: No. Code blocks do not add any overhead whatsoever. In fact, because of their ability to simplify the coding of certain algorithms, their use generally increases speed and efficiency. Also, the { and } exist only in your program's source code. Java does not, per se, execute the { or }.

In this case, the target of the **if** statement is a block of code and not just a single statement. If the condition controlling the **if** is true (as it is in this case), the three statements inside the block will be executed. Try setting **i** to zero and observe the result. You will see that the entire block is skipped.

As you will see later in this book, blocks of code have additional properties and uses. However, the main reason for their existence is to create logically inseparable units of code.

Semicolons and Positioning

In Java, the semicolon is a *separator*. It is often used to terminate a statement. In essence, the semicolon indicates the end of one logical entity.

As you know, a block is a set of logically connected statements that are surrounded by opening and closing braces. A block is *not* terminated with a semicolon. Instead, the end of the block is indicated by the closing brace.

Java does not recognize the end of the line as a terminator. For this reason, it does not matter where on a line you put a statement. For example,

```
x = y;
y = y + 1;
System.out.println(x + " " + y);
```

is the same as the following, to Java:

```
x = y; y = y + 1; System.out.println(x + " " + y);
```

Furthermore, the individual elements of a statement can also be put on separate lines. For example, the following is perfectly acceptable:

```
System.out.println("This is a long line of output" +
                x + y + z +
                "more output");
```

Breaking long lines in this fashion is often used to make programs more readable. It can also help prevent excessively long lines from wrapping.

Indentation Practices

You may have noticed in the previous examples that certain statements were indented. Java is a free-form language, meaning that it does not matter where you place statements relative to each other on a line. However, over the years, a common and accepted indentation style has developed that allows for very readable programs. This book follows that style, and it is recommended that you do so as well. Using this style, you indent one level after each opening brace, and move back out one level after each closing brace. Certain statements encourage some additional indenting; these will be covered later.

Try This 1-2   Improving the Gallons-to-Liters Converter

GalToLitTable.java

You can use the **for** loop, the **if** statement, and code blocks to create an improved version of the gallons-to-liters converter that you developed in the first project. This new version will print a table of conversions, beginning with 1 gallon and ending at 100 gallons. After every 10 gallons, a blank line will be output. This is accomplished through the use of a variable called **counter** that counts the number of lines that have been output. Pay special attention to its use.

1. Create a new file called **GalToLitTable.java**.

2. Enter the following program into the file:

```
/*
    Try This 1-2

    This program displays a conversion
    table of gallons to liters.

    Call this program "GalToLitTable.java".
*/
class GalToLitTable {
  public static void main(String[] args) {
    double gallons, liters;
    int counter;

    counter = 0;                            Line counter is initially set to zero.
    for(gallons = 1; gallons <= 100; gallons++) {
      liters = gallons * 3.7854; // convert to liters
      System.out.println(gallons + " gallons is " +
                         liters + " liters.");

      counter++;                            Increment the line counter
      // every 10th line, print a blank line   with each loop iteration.
      if(counter == 10) {                   If counter is 10,
        System.out.println();               output a blank line.
        counter = 0; // reset the line counter
      }
    }
  }
}
```

3. Compile the program using the following command line:

```
javac GalToLitTable.java
```

4. Run the program using this command:

```
java GalToLitTable
```

Here is a portion of the output that you will see:

```
1.0 gallons is 3.7854 liters.
2.0 gallons is 7.5708 liters.
3.0 gallons is 11.356200000000001 liters.
4.0 gallons is 15.1416 liters.
5.0 gallons is 18.927 liters.
6.0 gallons is 22.712400000000002 liters.
7.0 gallons is 26.4978 liters.
8.0 gallons is 30.2832 liters.
9.0 gallons is 34.0686 liters.
10.0 gallons is 37.854 liters.

11.0 gallons is 41.6394 liters.
12.0 gallons is 45.424800000000005 liters.
13.0 gallons is 49.2102 liters.
14.0 gallons is 52.9956 liters.
15.0 gallons is 56.781 liters.
16.0 gallons is 60.5664 liters.
17.0 gallons is 64.3518 liters.
18.0 gallons is 68.1372 liters.
19.0 gallons is 71.9226 liters.
20.0 gallons is 75.708 liters.

21.0 gallons is 79.49340000000001 liters.
22.0 gallons is 83.2788 liters.
23.0 gallons is 87.0642 liters.
24.0 gallons is 90.84960000000001 liters.
25.0 gallons is 94.635 liters.
26.0 gallons is 98.4204 liters.
27.0 gallons is 102.2058 liters.
28.0 gallons is 105.9912 liters.
29.0 gallons is 109.7766 liters.
30.0 gallons is 113.562 liters.
```

The Java Keywords

Sixty-seven keywords are currently defined in the Java language (see Table 1-1). These keywords, combined with the syntax of the operators and separators, form the definition of the Java language. In general, keywords cannot be used as names for a variable, class, or method. However, 16 of the keywords are context-sensitive, which means that they are only keywords

abstract	assert	boolean	break	byte	case
catch	char	class	const	continue	default
do	double	else	enum	exports	extends
final	finally	float	for	goto	if
implements	import	instanceof	int	interface	long
module	native	new	non-sealed	open	opens
package	permits	private	protected	provides	public
record	requires	return	sealed	short	static
strictfp	super	switch	synchronized	this	throw
throws	to	transient	transitive try	uses	var
void	volatile	while	with	yield	_

Table 1-1   The Java Keywords

when used with the feature to which they relate. They support features added to Java over the past few years. Ten relate to modules: **exports**, **module**, **open**, **opens**, **provides**, **requires**, **to**, **transitive**, **uses**, and **with**. Records are declared by **record**; sealed classes and interfaces use **sealed**, **non-sealed**, and **permits**; **yield** is used by the enhanced **switch**; and **var** supports local variable type inference. Because they are context-sensitive, existing programs were unaffected by their addition. Also, beginning with JDK 9, an underscore by itself is considered a keyword in order to prevent its use as the name of something in your program. Beginning with JDK 17, **strictfp** no longer has any effect and is unnecessary. It is, however, still a Java keyword.

The keywords **const** and **goto** are reserved but not used. In the early days of Java, several other keywords were reserved for possible future use. However, the current specification for Java defines only the keywords shown in Table 1-1.

In addition to the keywords, Java reserves three other names that have been part of Java since the start: **true**, **false**, and **null**. These are values defined by Java. You may not use these words for the names of variables, classes, and so on.

Identifiers in Java

In Java an identifier is, essentially, a name given to a method, a variable, or any other user-defined item. Identifiers can be from one to several characters long. Variable names may start with any letter of the alphabet, an underscore, or a dollar sign. (The $ is not intended for general use.) Next may be either a letter, a digit, a dollar sign, or an underscore. The underscore can be used to enhance the readability of a variable name, as in **line_count**. Uppercase and lowercase are different; that is, to Java, **myvar** and **MyVar** are separate names. Here are some examples of legal identifiers:

Test	x	y2	MaxLoad
up	_top	my_var	sample23

Remember, you can't start an identifier with a digit. Thus, **12x** is invalid, for example.

In general, you cannot use the Java keywords as identifier names. Also, you should not use the name of any standard method, such as **println**, as an identifier. Beyond these two restrictions, good programming practice dictates that you use identifier names that reflect the meaning or usage of the items being named.

The Java Class Libraries

The sample programs shown in this chapter make use of two of Java's built-in methods: **println( )** and **print( )**. These methods are accessed through **System.out**. **System** is a class predefined by Java that is automatically included in your programs. In the larger view, the Java environment relies on several built-in class libraries that contain many built-in methods that provide support for such things as I/O, string handling, networking, and graphics. The standard classes also provide support for a graphical user interface (GUI). Thus, Java as a totality is a combination of the Java language itself, plus its standard classes. As you will see, the class libraries provide much of the functionality that comes with Java. Indeed, part of becoming a Java programmer is learning to use the standard Java classes. Throughout this book, various elements of the standard library classes and methods are described. However, the Java library is something that you will also want to explore more on your own.

Chapter 1 Self Test

1. What is bytecode and why is it important to Java's use for Internet programming?

2. What are the three main principles of object-oriented programming?

3. Where do Java programs begin execution?

4. What is a variable?

5. Which of the following variable names is invalid?

   A. count

   B. $count

   C. count27

   D. 67count

6. How do you create a single-line comment? How do you create a multiline comment?

7. Show the general form of the **if** statement. Show the general form of the **for** loop.

8. How do you create a block of code?

9. The moon's gravity is about 17 percent that of earth's. Write a program that computes your effective weight on the moon.

10. Adapt Try This 1-2 so that it prints a conversion table of inches to meters. Display 12 feet of conversions, inch by inch. Output a blank line every 12 inches. (One meter equals approximately 39.37 inches.)

11. If you make a typing mistake when entering your program, what sort of error will result?

12. Does it matter where on a line you put a statement?

Chapter 2

Introducing Data Types
and Operators

Key Skills & Concepts

- Know Java's primitive types

- Use literals

- Initialize variables

- Know the scope rules of variables within a method

- Use the arithmetic operators

- Use the relational and logical operators

- Understand the assignment operators

- Use shorthand assignments

- Understand type conversion in assignments

- Cast incompatible types

- Understand type conversion in expressions

At the foundation of any programming language are its data types and operators, and Java is no exception. These elements define the limits of a language and determine the kind of tasks to which it can be applied. Fortunately, Java supports a rich assortment of both data types and operators, making it suitable for any type of programming.

Data types and operators are a large subject. We will begin here with an examination of Java's foundational data types and its most commonly used operators. We will also take a closer look at variables and examine the expression.

Why Data Types Are Important

Data types are especially important in Java because it is a strongly typed language. This means that all operations are type-checked by the compiler for type compatibility. Illegal operations will not be compiled. Thus, strong type checking helps prevent errors and enhances reliability. To enable strong type checking, all variables, expressions, and values have a type. There is no concept of a "type-less" variable, for example. Furthermore, the type of a value determines what operations are allowed on it. An operation allowed on one type might not be allowed on another.

Java's Primitive Types

Java contains two general categories of built-in data types: object-oriented and non-object-oriented. Java's object-oriented types are defined by classes, and a discussion of classes is

Type	Meaning
boolean	Represents true/false values
byte	8-bit integer
char	Character
double	Double-precision floating point
float	Single-precision floating point
int	Integer
long	Long integer
short	Short integer

Table 2-1   Java's Built-in Primitive Data Types

deferred until later. However, at the core of Java are eight primitive (also called elemental or simple) types of data, which are shown in Table 2-1. The term *primitive* is used here to indicate that these types are not objects in an object-oriented sense, but rather, normal binary values. These primitive types are not objects because of efficiency concerns. All of Java's other data types are constructed from these primitive types.

Java strictly specifies a range and behavior for each primitive type, which all implementations of the Java Virtual Machine must support. Because of Java's portability requirement, Java is uncompromising on this account. For example, an **int** is the same in all execution environments. This allows programs to be fully portable. There is no need to rewrite code to fit a specific platform. Although strictly specifying the range of the primitive types may cause a small loss of performance in some environments, it is necessary in order to achieve portability.

Integers

Java defines four integer types: **byte**, **short**, **int**, and **long**, which are shown here:

Type	Width in Bits	Range
byte	8	−128 to 127
short	16	−32,768 to 32,767
int	32	−2,147,483,648 to 2,147,483,647
long	64	−9,223,372,036,854,775,808 to 9,223,372,036,854,775,807

As the table shows, all of the integer types are signed positive and negative values. Java does not support unsigned (positive-only) integers. Many other computer languages support both signed and unsigned integers. However, Java's designers felt that unsigned integers were unnecessary.

NOTE

Technically, the Java run-time system can use any size it wants to store a primitive type. However, in all cases, types must act as specified.

The most commonly used integer type is **int**. Variables of type **int** are often employed to control loops, to index arrays, and to perform general-purpose integer math.

When you need an integer that has a range greater than **int**, use **long**. For example, here is a program that computes the number of cubic inches contained in a cube that is one mile by one mile, by one mile:

```java
/*
   Compute the number of cubic inches
   in 1 cubic mile.
*/
class Inches {
  public static void main(String[] args) {
    long ci;
    long im;

    im = 5280 * 12;

    ci = im * im * im;

    System.out.println("There are " + ci +
                       " cubic inches in cubic mile.");

  }
}
```

Here is the output from the program:

```
There are 254358061056000 cubic inches in cubic mile.
```

Clearly, the result could not have been held in an **int** variable.

The smallest integer type is **byte**. Variables of type **byte** are especially useful when working with raw binary data that may not be directly compatible with Java's other built-in types. The **short** type creates a short integer. Variables of type **short** are appropriate when you don't need the larger range offered by **int**.

Ask the Expert

Q: You say that there are four integer types: int, short, long, **and** byte. **However, I have heard that** char **can also be categorized as an integer type in Java. Can you explain?**

A: The formal specification for Java defines a type category called integral types, which includes **byte**, **short**, **int**, **long**, and **char**. They are called integral types because they all hold whole-number, binary values. However, the purpose of the first four is to represent numeric integer quantities. The purpose of **char** is to represent characters. Therefore, the principal uses of **char** and the principal uses of the other integral types are fundamentally different. Because of the differences, the **char** type is treated separately in this book.

Floating-Point Types

As explained in Chapter 1, the floating-point types can represent numbers that have fractional components. There are two kinds of floating-point types, **float** and **double**, which represent single- and double-precision numbers, respectively. Type **float** is 32 bits wide and type **double** is 64 bits wide.

Of the two, **double** is the most commonly used, and many of the math functions in Java's class library use **double** values. For example, the **sqrt( )** method (which is defined by the standard **Math** class) returns a **double** value that is the square root of its **double** argument. Here, **sqrt( )** is used to compute the length of the hypotenuse, given the lengths of the two opposing sides:

```
/*
   Use the Pythagorean theorem to
   find the length of the hypotenuse
   given the lengths of the two opposing
   sides.
*/
class Hypot {
  public static void main(String[] args) {
    double x, y, z;

    x = 3;
    y = 4;                  ——— Notice how sqrt( ) is called. It is preceded by
                                the name of the class of which it is a member.
    z = Math.sqrt(x*x + y*y);

    System.out.println("Hypotenuse is " +z);
  }
}
```

The output from the program is shown here:

```
Hypotenuse is 5.0
```

One other point about the preceding example: As mentioned, **sqrt( )** is a member of the standard **Math** class. Notice how **sqrt( )** is called; it is preceded by the name **Math**. This is similar to the way **System.out** precedes **println( )**. Although not all standard methods are called by specifying their class name first, several are.

Characters

In Java, characters are not 8-bit quantities like they are in many other computer languages. Instead, Java uses Unicode. Unicode defines a character set that can represent all of the characters found in all human languages. In Java, **char** is an unsigned 16-bit type having a range of 0 to 65,535. The standard 8-bit ASCII character set is a subset of Unicode and ranges from 0 to 127. Thus, the ASCII characters are still valid Java characters.

A character variable can be assigned a value by enclosing the character in single quotes. For example, this assigns the variable **ch** the letter X:

```
char ch;
ch = 'X';
```

You can output a **char** value using a **println( )** statement. For example, this line outputs the value in **ch**:

```
System.out.println("This is ch: " + ch);
```

Since **char** is an unsigned 16-bit type, it is possible to perform various arithmetic manipulations on a **char** variable. For example, consider the following program:

```
// Character variables can be handled like integers.
class CharArithDemo {
  public static void main(String[] args) {
    char ch;

    ch = 'X';
    System.out.println("ch contains " + ch);

    ch++; // increment ch  ◄──────── A char can be incremented.
    System.out.println("ch is now " + ch);

    ch = 90; // give ch the value Z ◄──────── A char can be assigned an integer value.
    System.out.println("ch is now " + ch);
  }
}
```

The output generated by this program is shown here:

```
ch contains X
ch is now Y
ch is now Z
```

In the program, **ch** is first given the value X. Next, **ch** is incremented. This results in **ch** containing Y, the next character in the ASCII (and Unicode) sequence. Next, **ch** is assigned the value 90, which is the ASCII (and Unicode) value that corresponds to the letter Z. Since the ASCII character set occupies the first 127 values in the Unicode character set, all the "old tricks" that you may have used with characters in other languages will work in Java, too.

Ask the Expert

Q: Why does Java use Unicode?

A: Java was designed for worldwide use. Thus, it needs to use a character set that can represent all the world's languages. Unicode is the standard character set designed expressly for this purpose. Of course, the use of Unicode is inefficient for languages such as English, German, Spanish, or French, whose characters can be contained within 8 bits. But such is the price that must be paid for global portability.

The Boolean Type

The **boolean** type represents true/false values. Java defines the values true and false using the reserved words **true** and **false**. Thus, a variable or expression of type **boolean** will be one of these two values.

Here is a program that demonstrates the **boolean** type:

```
// Demonstrate boolean values.
class BoolDemo {
  public static void main(String[] args) {
    boolean b;

    b = false;
    System.out.println("b is " + b);
    b = true;
    System.out.println("b is " + b);

    // a boolean value can control the if statement
    if(b) System.out.println("This is executed.");

    b = false;
    if(b) System.out.println("This is not executed.");

    // outcome of a relational operator is a boolean value
    System.out.println("10 > 9 is " + (10 > 9));
  }
}
```

The output generated by this program is shown here:

```
b is false
b is true
This is executed.
10 > 9 is true
```

There are three interesting things to notice about this program. First, as you can see, when a **boolean** value is output by **println( )**, "true" or "false" is displayed. Second, the value of a **boolean** variable is sufficient, by itself, to control the **if** statement. There is no need to write an **if** statement like this:

```
if(b == true) ...
```

Third, the outcome of a relational operator, such as **<**, is a **boolean** value. This is why the expression **10 > 9** displays the value "true." Further, the extra set of parentheses around **10 > 9** is necessary because the **+** operator has a higher precedence than the **>**.

Try This 2-1   How Far Away Is the Lightning?

`Sound.java`   In this project, you will create a program that computes how far away, in feet, a listener is from a lightning strike. Sound travels approximately 1,100 feet per second through air. Thus, knowing the interval between the time you see a lightning bolt and the time the sound reaches you enables you to compute the distance to the lightning. For this project, assume that the time interval is 7.2 seconds.

1. Create a new file called **Sound.java**.

2. To compute the distance, you will need to use floating-point values. Why? Because the time interval, 7.2, has a fractional component. Although it would be permissible to use a value of type **float**, we will use **double** in the example.

3. To compute the distance, you will multiply 7.2 by 1,100. You will then assign this value to a variable.

4. Finally, you will display the result.

   Here is the entire **Sound.java** program listing:

```
/*
   Try This 2-1
   Compute the distance to a lightning
   strike whose sound takes 7.2 seconds
   to reach you.
*/
class Sound {
  public static void main(String[] args) {
    double dist;

    dist = 7.2 * 1100;

    System.out.println("The lightning is " + dist +
                       " feet away.");

  }
}
```

5. Compile and run the program. The following result is displayed:

```
The lightning is 7920.0 feet away.
```

6. Extra challenge: You can compute the distance to a large object, such as a rock wall, by timing the echo. For example, if you clap your hands and time how long it takes for you to hear the echo, then you know the total round-trip time. Dividing this value by two yields the time it takes the sound to go one way. You can then use this value to compute the distance to the object. Modify the preceding program so that it computes the distance, assuming that the time interval is that of an echo.

Literals

In Java, *literals* refer to fixed values that are represented in their human-readable form. For example, the number 100 is a literal. Literals are also commonly called *constants*. For the most part, literals, and their usage, are so intuitive that they have been used in one form or another by all the preceding sample programs. Now the time has come to explain them formally.

Java literals can be of any of the primitive data types. The way each literal is represented depends upon its type. As explained earlier, character constants are enclosed in single quotes. For example, 'a' and ' %' are both character constants.

Integer literals are specified as numbers without fractional components. For example, 10 and −100 are integer literals. Floating-point literals require the use of the decimal point followed by the number's fractional component. For example, 11.123 is a floating-point literal. Java also allows you to use scientific notation for floating-point numbers.

By default, integer literals are of type **int**. If you want to specify a **long** literal, append an l or an L. For example, 12 is an **int**, but 12L is a **long**.

By default, floating-point literals are of type **double**. To specify a **float** literal, append an F or f to the constant. For example, 10.19F is of type **float**.

Although integer literals create an **int** value by default, they can still be assigned to variables of type **char**, **byte**, or **short** as long as the value being assigned can be represented by the target type. An integer literal can always be assigned to a **long** variable.

You can embed one or more underscores into an integer or floating-point literal. Doing so can make it easier to read values consisting of many digits. When the literal is compiled, the underscores are simply discarded. Here is an example:

```
123_45_1234
```

This specifies the value 123,451,234. The use of underscores is particularly useful when encoding things like part numbers, customer IDs, and status codes that are commonly thought of as consisting of subgroups of digits.

Hexadecimal, Octal, and Binary Literals

As you may know, in programming it is sometimes easier to use a number system based on 8 or 16 instead of 10. The number system based on 8 is called *octal*, and it uses the digits 0 through 7. In octal the number 10 is the same as 8 in decimal. The base 16 number system is called *hexadecimal* and uses the digits 0 through 9 plus the letters A through F, which stand for 10, 11, 12, 13, 14, and 15. For example, the hexadecimal number 10 is 16 in decimal. Because of the frequency with which these two number systems are used, Java allows you to specify integer literals in hexadecimal or octal instead of decimal. A hexadecimal literal must begin with **0x** or **0X** (a zero followed by an x or X). An octal literal begins with a zero. Here are some examples:

```
hex = 0xFF; // 255 in decimal
oct = 011; // 9 in decimal
```

As a point of interest, Java also allows hexadecimal floating-point literals, but they are seldom used.

It is possible to specify an integer literal by use of binary. To do so, precede the binary number with a **0b** or **0B**. For example, this specifies the value 12 in binary: **0b1100**.

Character Escape Sequences

Enclosing character constants in single quotes works for most printing characters, but a few characters, such as the carriage return, pose a special problem when a text editor is used. In addition, certain other characters, such as the single and double quotes, have special meaning in Java, so you cannot use them directly. For these reasons, Java provides special *escape sequences*, sometimes referred to as backslash character constants, shown in Table 2-2. These sequences are used in place of the characters that they represent.

Escape Sequence	Description
\'	Single quote
\"	Double quote
\\	Backslash
\r	Carriage return
\n	New line
\f	Form feed
\t	Horizontal tab
\b	Backspace
\ddd	Octal constant (where *ddd* is an octal constant)
\uxxxx	Hexadecimal constant (where *xxxx* is a hexadecimal constant)
\s	Space (added by JDK 15)
\endofline	Continue line (applies to only text block; added by JDK 15)

Table 2-2 Character Escape Sequences

For example, this assigns **ch** the tab character:

```
ch = '\t';
```

The next example assigns a single quote to **ch**:

```
ch = '\'';
```

String Literals

Java supports another type of literal: the string. A *string* is a set of characters enclosed by double quotes. For example,

```
"this is a test"
```

is a string. You have seen examples of strings in many of the **println( )** statements in the preceding sample programs.

In addition to normal characters, a string literal can also contain one or more of the escape sequences just described. For example, consider the following program. It uses the **\n** and **\t** escape sequences.

```
// Demonstrate escape sequences in strings.
class StrDemo {
  public static void main(String[] args) {
    System.out.println("First line\nSecond line");
    System.out.println("A\tB\tC");
    System.out.println("D\tE\tF");
  }
}
```

— Use **\n** to generate a new line.

— Use tabs to align output.

The output is shown here:

```
First line
Second line
A       B       C
D       E       F
```

Ask the Expert

Q: Is a string consisting of a single character the same as a character literal? For example, is "k" the same as 'k'?

A: No. You must not confuse strings with characters. A character literal represents a single letter of type **char**. A string containing only one letter is still a string. Although strings consist of characters, they are not the same type.

Notice how the **\n** escape sequence is used to generate a new line. You don't need to use multiple **println( )** statements to get multiline output. Just embed **\n** within a longer string at the points where you want the new lines to occur. One other point: As you will see in Chapter 5, there is a feature called a *text block* that was recently added to Java. A text block offers more control and flexibility when you need multiple lines of text.

A Closer Look at Variables

Variables were introduced in Chapter 1. Here, we will take a closer look at them. As you learned earlier, variables are declared using this form of statement,

type var-name;

where *type* is the data type of the variable, and *var-name* is its name. You can declare a variable of any valid type, including the simple types just described, and every variable will have a type. Thus, the capabilities of a variable are determined by its type. For example, a variable of type **boolean** cannot be used to store floating-point values. Furthermore, the type of a variable cannot change during its lifetime. An **int** variable cannot turn into a **char** variable, for example.

All variables in Java must be declared prior to their use. This is necessary because the compiler must know what type of data a variable contains before it can properly compile any statement that uses the variable. It also enables Java to perform strict type checking.

Initializing a Variable

In general, you must give a variable a value prior to using it. One way to give a variable a value is through an assignment statement, as you have already seen. Another way is by giving it an initial value when it is declared. To do this, follow the variable's name with an equal sign and the value being assigned. The general form of initialization is shown here:

type var = value;

Here, *value* is the value that is given to *var* when *var* is created. The value must be compatible with the specified type. Here are some examples:

```
int count = 10; // give count an initial value of 10
char ch = 'X'; // initialize ch with the letter X
float f = 1.2F; // f is initialized with 1.2
```

When declaring two or more variables of the same type using a comma-separated list, you can give one or more of those variables an initial value. For example:

```
int a, b = 8, c = 19, d; // b and c have initializations
```

In this case, only **b** and **c** are initialized.

Dynamic Initialization

Although the preceding examples have used only constants as initializers, Java allows variables to be initialized dynamically, using any expression valid at the time the variable is declared. For example, here is a short program that computes the volume of a cylinder given the radius of its base and its height:

```
// Demonstrate dynamic initialization.
class DynInit {
    public static void main(String[] args) {
        double radius = 4, height = 5;

        // dynamically initialize volume
        double volume = 3.1416 * radius * radius * height;

        System.out.println("Volume is " + volume);
    }
}
```

volume is dynamically initialized at run time.

Here, three local variables—**radius**, **height**, and **volume**—are declared. The first two, **radius** and **height**, are initialized by constants. However, **volume** is initialized dynamically to the volume of the cylinder. The key point here is that the initialization expression can use any element valid at the time of the initialization, including calls to methods, other variables, or literals.

The Scope and Lifetime of Variables

So far, all of the variables that we have been using were declared at the start of the **main( )** method. However, Java allows variables to be declared within any block. As explained in Chapter 1, a block is begun with an opening curly brace and ended by a closing curly brace. A block defines a *scope*. Thus, each time you start a new block, you are creating a new scope. A scope determines what objects are visible to other parts of your program. It also determines the lifetime of those objects.

In general, every declaration in Java has a scope. As a result, Java defines a powerful, finely grained concept of scope. Two of the most common scopes in Java are those defined by a class and those defined by a method. A discussion of class scope (and variables declared within it) is deferred until later in this book, when classes are described. For now, we will examine only the scopes defined by or within a method.

The scope defined by a method begins with its opening curly brace. However, if that method has parameters, they too are included within the method's scope. A method's scope ends with its closing curly brace. This block of code is called the *method body*.

As a general rule, variables declared inside a scope are not visible (that is, accessible) to code that is defined outside that scope. Thus, when you declare a variable within a scope, you are localizing that variable and protecting it from unauthorized access and/or modification. Indeed, the scope rules provide the foundation for encapsulation. A variable declared within a block is called a *local variable*.

Scopes can be nested. For example, each time you create a block of code, you are creating a new, nested scope. When this occurs, the outer scope encloses the inner scope. This means that objects declared in the outer scope will be visible to code within the inner scope. However, the reverse is not true. Objects declared within the inner scope will not be visible outside it.

To understand the effect of nested scopes, consider the following program:

```
// Demonstrate block scope.
class ScopeDemo {
  public static void main(String[] args) {
    int x; // known to all code within main

    x = 10;
    if(x == 10) { // start new scope

      int y = 20; // known only to this block

      // x and y both known here.

      System.out.println("x and y: " + x + " " + y);
      x = y * 2;
    }
    // y = 100; // Error! y not known here          Here, y is outside of its scope.

    // x is still known here.
    System.out.println("x is " + x);
  }
}
```

As the comments indicate, the variable **x** is declared at the start of **main( )**'s scope and is accessible to all subsequent code within **main( )**. Within the **if** block, **y** is declared. Since a block defines a scope, **y** is visible only to other code within its block. This is why outside of its block, the line **y = 100;** is commented out. If you remove the leading comment symbol, a compile-time error will occur, because **y** is not visible outside of its block. Within the **if** block, **x** can be used because code within a block (that is, a nested scope) has access to variables declared by an enclosing scope.

Within a block, variables can be declared at any point, but are valid only after they are declared. Thus, if you define a variable at the start of a method, it is available to all of the code within that method. Conversely, if you declare a variable at the end of a block, it is effectively useless, because no code will have access to it.

Here is another important point to remember: variables are created when their scope is entered, and destroyed when their scope is left. This means that a variable will not hold its value once it has gone out of scope. Therefore, variables declared within a method will not hold their values between calls to that method. Also, a variable declared within a block will lose its value when the block is left. Thus, the lifetime of a variable is confined to its scope.

If a variable declaration includes an initializer, that variable will be reinitialized each time the block in which it is declared is entered. For example, consider this program:

```
// Demonstrate lifetime of a variable.
class VarInitDemo {
  public static void main(String[] args) {
    int x;
```

```
    for(x = 0; x < 3; x++) {
      int y = -1; // y is initialized each time block is entered
      System.out.println("y is: " + y); // this always prints -1
      y = 100;
      System.out.println("y is now: " + y);
    }
  }
}
```

The output generated by this program is shown here:

```
y is: -1
y is now: 100
y is: -1
y is now: 100
y is: -1
y is now: 100
```

As you can see, **y** is reinitialized to –1 each time the inner **for** loop is entered. Even though it is subsequently assigned the value 100, this value is lost.

There is one quirk to Java's scope rules that may surprise you: although blocks can be nested, no variable declared within an inner scope can have the same name as a variable declared by an enclosing scope. For example, the following program, which tries to declare two separate variables with the same name, will not compile.

```
/*
   This program attempts to declare a variable
   in an inner scope with the same name as one
   defined in an outer scope.

   *** This program will not compile. ***
*/
class NestVar {
  public static void main(String[] args) {
    int count;  ◄─────────────────────────────┐
                                               │
    for(count = 0; count < 10; count = count+1) {
      System.out.println("This is count: " + count);
                                               │
      int count; // illegal!!! ◄───────────────┘  Can't declare count again because
      for(count = 0; count < 2; count++)          it's already declared.
        System.out.println("This program is in error!");
    }
  }
}
```

Operators

Java provides a rich operator environment. An *operator* is a symbol that tells the compiler to perform a specific mathematical or logical manipulation. Java has four general classes of operators: arithmetic, bitwise, relational, and logical. Java also defines some additional operators that handle certain special situations. This chapter will examine the arithmetic, relational, and logical operators. We will also examine the assignment operator. The bitwise and other special operators are examined later.

Arithmetic Operators

Java defines the following arithmetic operators:

Operator	Meaning
+	Addition (also unary plus)
−	Subtraction (also unary minus)
*	Multiplication
/	Division
%	Modulus
++	Increment
− −	Decrement

The operators +, −, *, and / all work the same way in Java as they do in any other computer language (or algebra, for that matter). These can be applied to any built-in numeric data type. They can also be used on objects of type **char**.

Although the actions of arithmetic operators are well known to all readers, a few special situations warrant some explanation. First, remember that when / is applied to an integer, any remainder will be truncated; for example, 10/3 will equal 3 in integer division. You can obtain the remainder of this division by using the modulus operator %. It yields the remainder of an integer division. For example, 10 % 3 is 1. In Java, the % can be applied to both integer and floating-point types. Thus, 10.0 % 3.0 is also 1. The following program demonstrates the modulus operator.

```
// Demonstrate the % operator.
class ModDemo {
  public static void main(String[] args) {
    int iresult, irem;
    double dresult, drem;

    iresult = 10 / 3;
    irem = 10 % 3;
```

```
    dresult = 10.0 / 3.0;
    drem = 10.0 % 3.0;

    System.out.println("Result and remainder of 10 / 3: " +
                       iresult + " " + irem);
    System.out.println("Result and remainder of 10.0 / 3.0: " +
                       dresult + " " + drem);

  }
}
```

The output from the program is shown here:

```
Result and remainder of 10 / 3: 3 1
Result and remainder of 10.0 / 3.0: 3.3333333333333335 1.0
```

As you can see, the % yields a remainder of 1 for both integer and floating-point operations.

Increment and Decrement

Introduced in Chapter 1, the **++** and the **− −** are Java's increment and decrement operators. As you will see, they have some special properties that make them quite interesting. Let's begin by reviewing precisely what the increment and decrement operators do.

The increment operator adds 1 to its operand, and the decrement operator subtracts 1. Therefore,

```
x = x + 1;
```

is the same as

```
x++;
```

and

```
x = x - 1;
```

is the same as

```
x--;
```

Both the increment and decrement operators can either precede (prefix) or follow (postfix) the operand. For example,

```
x = x + 1;
```

can be written as

```
++x; // prefix form
```

or as

```
x++; // postfix form
```

In the foregoing example, there is no difference whether the increment is applied as a prefix or a postfix. However, when an increment or decrement is used as part of a larger expression, there is an important difference. When an increment or decrement operator precedes its operand, Java will perform the corresponding operation prior to obtaining the operand's value for use by the rest of the expression. If the operator follows its operand, Java will obtain the operand's value before incrementing or decrementing it. Consider the following:

```
x = 10;
y = ++x;
```

In this case, **y** will be set to 11. However, if the code is written as

```
x = 10;
y = x++;
```

then **y** will be set to 10. In both cases, **x** is still set to 11; the difference is when it happens. There are significant advantages in being able to control when the increment or decrement operation takes place.

Relational and Logical Operators

In the terms *relational operator* and *logical operator, relational* refers to the relationships that values can have with one another, and *logical* refers to the ways in which true and false values can be connected together. Since the relational operators produce true or false results, they often work with the logical operators. For this reason they will be discussed together here.

The relational operators are shown here:

Operator	Meaning
==	Equal to
!=	Not equal to
>	Greater than
<	Less than
>=	Greater than or equal to
<=	Less than or equal to

The logical operators are shown next:

Operator	Meaning
&	AND
|	OR
^	XOR (exclusive OR)
||	Short-circuit OR
&&	Short-circuit AND
!	NOT

The outcome of the relational and logical operators is a **boolean** value.

In Java, all objects can be compared for equality or inequality using = = and !=. However, the comparison operators, **<**, **>**, **<=**, or **>=**, can be applied only to those types that support an ordering relationship. Therefore, all of the relational operators can be applied to all numeric types and to type **char**. However, values of type **boolean** can only be compared for equality or inequality, since the **true** and **false** values are not ordered. For example, **true > false** has no meaning in Java.

For the logical operators, the operands must be of type **boolean**, and the result of a logical operation is of type **boolean**. The logical operators, **&**, **|**, **^**, and **!**, support the basic logical operations AND, OR, XOR, and NOT, according to the following truth table:

p	q	p & q	p | q	p ^ q	!p
False	False	False	False	False	True
True	False	False	True	True	False
False	True	False	True	True	True
True	True	True	True	False	False

As the table shows, the outcome of an exclusive OR operation is true when exactly one and only one operand is true.

Here is a program that demonstrates several of the relational and logical operators:

```java
// Demonstrate the relational and logical operators.
class RelLogOps {
  public static void main(String[] args) {
    int i, j;
    boolean b1, b2;

    i = 10;
    j = 11;
    if(i < j) System.out.println("i < j");
    if(i <= j) System.out.println("i <= j");
    if(i != j) System.out.println("i != j");
    if(i == j) System.out.println("this won't execute");
    if(i >= j) System.out.println("this won't execute");
    if(i > j) System.out.println("this won't execute");

    b1 = true;
    b2 = false;
    if(b1 & b2) System.out.println("this won't execute");
    if(!(b1 & b2)) System.out.println("!(b1 & b2) is true");
    if(b1 | b2) System.out.println("b1 | b2 is true");
    if(b1 ^ b2) System.out.println("b1 ^ b2 is true");
  }
}
```

The output from the program is shown here:

```
i < j
i <= j
i != j
!(b1 & b2) is true
b1 | b2 is true
b1 ^ b2 is true
```

Short-Circuit Logical Operators

Java supplies special *short-circuit* versions of its AND and OR logical operators that can be used to produce more efficient code. To understand why, consider the following. In an AND operation, if the first operand is false, the outcome is false no matter what value the second operand has. In an OR operation, if the first operand is true, the outcome of the operation is true no matter what the value of the second operand. Thus, in these two cases there is no need to evaluate the second operand. By not evaluating the second operand, time is saved and more efficient code is produced.

The short-circuit AND operator is **&&**, and the short-circuit OR operator is **||**. Their normal counterparts are **&** and **|**. The only difference between the normal and short-circuit versions is that the normal operands will always evaluate each operand, but short-circuit versions will evaluate the second operand only when necessary.

Here is a program that demonstrates the short-circuit AND operator. The program determines whether the value in **d** is a factor of **n**. It does this by performing a modulus operation. If the remainder of **n / d** is zero, then **d** is a factor. However, since the modulus operation involves a division, the short-circuit form of the AND is used to prevent a divide-by-zero error.

```
// Demonstrate the short-circuit operators.
class SCops {
  public static void main(String[] args) {
    int n, d, q;

    n = 10;
    d = 2;
    if(d != 0 && (n % d) == 0)
      System.out.println(d + " is a factor of " + n);

    d = 0; // now, set d to zero

    // Since d is zero, the second operand is not evaluated.
    if(d != 0 && (n % d) == 0)               The short-circuit
      System.out.println(d + " is a factor of " + n);    operator prevents
                                              a division by zero.

    /* Now, try same thing without short-circuit operator.
```

```
    This will cause a divide-by-zero error.
  */
  if(d != 0 & (n % d) == 0)                          Now both
    System.out.println(d + " is a factor of " + n);  expressions
  }                                                   are evaluated,
}                                                     allowing a division
                                                      by zero to occur.
```

To prevent a divide-by-zero, the **if** statement first checks to see if **d** is equal to zero. If it is, the short-circuit AND stops at that point and does not perform the modulus division. Thus, in the first test, **d** is 2 and the modulus operation is performed. The second test fails because **d** is set to zero, and the modulus operation is skipped, avoiding a divide-by-zero error. Finally, the normal AND operator is tried. This causes both operands to be evaluated, which leads to a run-time error when the division by zero occurs.

One last point: The formal specification for Java refers to the short-circuit operators as the *conditional-or* and the *conditional-and* operators, but the term "short-circuit" is commonly used.

The Assignment Operator

You have been using the assignment operator since Chapter 1. Now it is time to take a formal look at it. The *assignment operator* is the single equal sign, =. This operator works in Java much as it does in any other computer language. It has this general form:

var = expression;

Here, the type of *var* must be compatible with the type of *expression*.

The assignment operator does have one interesting attribute that you may not be familiar with: it allows you to create a chain of assignments. For example, consider this fragment:

```
int x, y, z;

x = y = z = 100; // set x, y, and z to 100
```

This fragment sets the variables **x**, **y**, and **z** to 100 using a single statement. This works because the = is an operator that yields the value of the right-hand expression. Thus, the value of **z = 100** is 100, which is then assigned to **y**, which in turn is assigned to **x**. Using a "chain of assignment" is an easy way to set a group of variables to a common value.

Shorthand Assignments

Java provides special *shorthand* assignment operators that simplify the coding of certain assignment statements. Let's begin with an example. The assignment statement shown here

```
x = x + 10;
```

can be written, using Java shorthand, as

```
x += 10;
```

Ask the Expert

Q: Since the short-circuit operators are, in some cases, more efficient than their normal counterparts, why does Java still offer the normal AND and OR operators?

A: In some cases you will want both operands of an AND or OR operation to be evaluated because of the side effects produced. Consider the following:

```
// Side effects can be important.
class SideEffects {
  public static void main(String[] args) {
    int i;

    i = 0;

    /* Here, i is still incremented even though
       the if statement fails. */
    if(false & (++i < 100))
       System.out.println("this won't be displayed");
    System.out.println("if statement executed: " + i); // displays 1

    /* In this case, i is not incremented because
       the short-circuit operator skips the increment. */
    if(false && (++i < 100))
      System.out.println("this won't be displayed");
    System.out.println("if statement executed: " + i); // still 1 !!
  }
}
```

As the comments indicate, in the first **if** statement, **i** is incremented whether the **if** succeeds or not. However, when the short-circuit operator is used, the variable **i** is not incremented when the first operand is false. The lesson here is that if your code expects the right-hand operand of an AND or OR operation to be evaluated, you must use Java's non-short-circuit forms of these operations.

The operator pair **+=** tells the compiler to assign to **x** the value of **x** plus 10. Here is another example. The statement

```
x = x - 100;
```

is the same as

```
x -= 100;
```

Both statements assign to **x** the value of **x** minus 100.

This shorthand will work for all the binary operators in Java (that is, those that require two operands). The general form of the shorthand is

var op = expression;

Thus, the arithmetic and logical shorthand assignment operators are the following:

+=	-=	*=	/=
%=	&=	|=	^=

Because these operators combine an operation with an assignment, they are formally referred to as *compound assignment* operators.

The compound assignment operators provide two benefits. First, they are more compact than their "longhand" equivalents. Second, in some cases, they are more efficient. For these reasons, you will often see the compound assignment operators used in professionally written Java programs.

Type Conversion in Assignments

In programming, it is common to assign one type of variable to another. For example, you might want to assign an **int** value to a **float** variable, as shown here:

```
int i;
float f;

i = 10;
f = i; // assign an int to a float
```

When compatible types are mixed in an assignment, the value of the right side is automatically converted to the type of the left side. Thus, in the preceding fragment, the value in **i** is converted into a **float** and then assigned to **f**. However, because of Java's strict type checking, not all types are compatible, and thus, not all type conversions are implicitly allowed. For example, **boolean** and **int** are not compatible.

When one type of data is assigned to another type of variable, an *automatic type conversion* will take place if

- The two types are compatible.

- The destination type is larger than the source type.

When these two conditions are met, a *widening conversion* takes place. For example, the **int** type is always large enough to hold all valid **byte** values, and both **int** and **byte** are integer types, so an automatic conversion from **byte** to **int** can be applied.

For widening conversions, the numeric types, including integer and floating-point types, are compatible with each other. For example, the following program is perfectly valid since **long** to **double** is a widening conversion that is automatically performed.

```
// Demonstrate automatic conversion from long to double.
class LtoD {
  public static void main(String[] args) {
    long L;
    double D;

    L = 100123285L;
    D = L; ◄──────── Automatic conversion from long to double

    System.out.println("L and D: " + L + " " + D);

  }
}
```

Although there is an automatic conversion from **long** to **double**, there is no automatic conversion from **double** to **long**, since this is not a widening conversion. Thus, the following version of the preceding program is invalid.

```
// *** This program will not compile. ***
class LtoD {
  public static void main(String[] args) {
    long L;
    double D;

    D = 100123285.0;
    L = D; // Illegal!!! ◄──────── No automatic conversion from double to long

    System.out.println("L and D: " + L + " " + D);

  }
}
```

There are no automatic conversions from the numeric types to **char** or **boolean**. Also, **char** and **boolean** are not compatible with each other. However, an integer literal can be assigned to **char**.

Casting Incompatible Types

Although the automatic type conversions are helpful, they will not fulfill all programming needs because they apply only to widening conversions between compatible types. For all other cases you must employ a cast. A *cast* is an instruction to the compiler to convert one type into another. Thus, it requests an explicit type conversion. A cast has this general form:

(target-type) expression

Here, *target-type* specifies the desired type to convert the specified expression to. For example, if you want to convert the type of the expression **x/y** to **int**, you can write

```
double x, y;
// ...
(int) (x / y)
```

Here, even though **x** and **y** are of type **double**, the cast converts the outcome of the expression to **int**. The parentheses surrounding **x / y** are necessary. Otherwise, the cast to **int** would apply only to the **x** and not to the outcome of the division. The cast is necessary here because there is no automatic conversion from **double** to **int**.

When a cast involves a *narrowing conversion,* information might be lost. For example, when casting a **long** into a **short**, information will be lost if the **long**'s value is greater than the range of a **short** because its high-order bits are removed. When a floating-point value is cast to an integer type, the fractional component will also be lost due to truncation. For example, if the value 1.23 is assigned to an integer, the resulting value will simply be 1. The 0.23 is lost.

The following program demonstrates some type conversions that require casts:

```
// Demonstrate casting.
class CastDemo {
  public static void main(String[] args) {
    double x, y;
    byte b;
    int i;
    char ch;

    x = 10.0;
    y = 3.0;
                                                    Truncation will occur in this conversion.
    i = (int) (x / y); // cast double to int
    System.out.println("Integer outcome of x / y: " + i);

    i = 100;
    b = (byte) i;          No loss of info here. A byte can hold the value 100.
    System.out.println("Value of b: " + b);

    i = 257;
    b = (byte) i;          Information loss this time. A byte cannot hold the value 257.
    System.out.println("Value of b: " + b);

    b = 88; // ASCII code for X
    ch = (char) b;         Cast between incompatible types
    System.out.println("ch: " + ch);
  }
}
```

The output from the program is shown here:

```
Integer outcome of x / y: 3
Value of b: 100
Value of b: 1
ch: X
```

In the program, the cast of (**x** / **y**) to **int** results in the truncation of the fractional component, and information is lost. Next, no loss of information occurs when **b** is assigned the value 100 because a **byte** can hold the value 100. However, when the attempt is made to assign **b** the value 257, information loss occurs because 257 exceeds a **byte**'s maximum value. Finally, no information is lost, but a cast is needed when assigning a **byte** value to a **char**.

Operator Precedence

Table 2-3 shows the order of precedence for all Java operators, from highest to lowest. This table includes several operators that will be discussed later in this book. Although technically separators, the [], (), and . can also act like operators. In that capacity, they would have the highest precedence.

Highest						
++ (postfix)	– – (postfix)					
++ (prefix)	– – (prefix)	~	!	+ (unary)	– (unary)	(*type-cast*)
*	/	%				
+	–					
>>	>>>	<<				
>	>=	<	<=	instanceof		
==	!=					
&						
^						
|						
&&						
||						
?:						
->						
=	op=					
Lowest						

Table 2-3 The Precedence of the Java Operators

Try This 2-2   Display a Truth Table for the Logical Operators

LogicalOpTable.java

In this project, you will create a program that displays the truth table for Java's logical operators. You must make the columns in the table line up. This project makes use of several features covered in this chapter, including one of Java's escape sequences and the logical operators. It also illustrates the differences in the precedence between the arithmetic + operator and the logical operators.

1. Create a new file called **LogicalOpTable.java**.

2. To ensure that the columns line up, you will use the \t escape sequence to embed tabs into each output string. For example, this **println( )** statement displays the header for the table:

```
System.out.println("P\tQ\LAND\tOR\tXOR\tNOT");
```

3. Each subsequent line in the table will use tabs to position the outcome of each operation under its proper heading.

4. Here is the entire **LogicalOpTable.java** program listing. Enter it at this time.

```
// Try This 2-2: a truth table for the logical operators.
class LogicalOpTable {
  public static void main(String[] args) {

    boolean p, q;

    System.out.println("P\tQ\tAND\tOR\tXOR\tNOT");

    p = true, q = true;
    System.out.print(p + "\t" + q +"\t");
    System.out.print((p&q) + "\t" + (p|q) + "\t");
    System.out.println((p^q) + "\t" + (!p));

    p = true; q = false;
    System.out.print(p + "\t" + q +"\t");
    System.out.print((p&q) + "\t" + (p|q) + "\t");
    System.out.println((p^q) + "\t" + (!p));

    p = false; q = true;
    System.out.print(p + "\t" + q +"\t");
    System.out.print((p&q) + "\t" + (p|q) + "\t");
    System.out.println((p^q) + "\t" + (!p));
```

(continued)

```
        p = false; q = false;
        System.out.print(p + "\t" + q +"\t");
        System.out.print((p&q) + "\t" + (p|q) + "\t");
        System.out.println((p^q) + "\t" + (!p));
    }
}
```

Notice the parentheses surrounding the logical operations inside the **println( )** statements. They are necessary because of the precedence of Java's operators. The **+** operator is higher than the logical operators.

5. Compile and run the program. The following table is displayed.

```
P         Q         AND       OR        XOR       NOT
true      true      true      true      false     false
true      false     false     true      true      false
false     true      false     true      true      true
false     false     false     false     false     true
```

6. On your own, try modifying the program so that it uses and displays 1's and 0's, rather than true and false. This may involve a bit more effort than you might at first think!

Expressions

Operators, variables, and literals are constituents of *expressions*. You probably already know the general form of an expression from your other programming experience, or from algebra. However, a few aspects of expressions will be discussed now.

Type Conversion in Expressions

Within an expression, it is possible to mix two or more different types of data as long as they are compatible with each other. For example, you can mix **short** and **long** within an expression because they are both numeric types. When different types of data are mixed within an expression, they are all converted to the same type. This is accomplished through the use of Java's *type promotion rules*.

First, all **char**, **byte**, and **short** values are promoted to **int**. Then, if one operand is a **long**, the whole expression is promoted to **long**. If one operand is a **float** operand, the entire expression is promoted to **float**. If any of the operands is **double**, the result is **double**.

It is important to understand that type promotions apply only to the values operated upon when an expression is evaluated. For example, if the value of a **byte** variable is promoted to **int** inside an expression, outside the expression, the variable is still a **byte**. Type promotion only affects the evaluation of an expression.

Type promotion can, however, lead to somewhat unexpected results. For example, when an arithmetic operation involves two **byte** values, the following sequence occurs: First, the **byte** operands are promoted to **int**. Then the operation takes place, yielding an **int** result.

Thus, the outcome of an operation involving two **byte** values will be an **int**. This is not what you might intuitively expect. Consider the following program:

```
// A promotion surprise!
class PromDemo {
  public static void main(String[] args) {
    byte b;
    int i;

    b = 10;
    i = b * b; // OK, no cast needed

    b = 10;
    b = (byte) (b * b); // cast needed!!

    System.out.println("i and b: " + i + " " + b);
  }
}
```

No cast needed because result is already elevated to **int**.

Cast is needed here to assign an **int** *to a* **byte**!

Somewhat counterintuitively, no cast is needed when assigning **b*b** to **i**, because **b** is promoted to **int** when the expression is evaluated. However, when you try to assign **b * b** to **b**, you do need a cast—back to **byte**! Keep this in mind if you get unexpected type-incompatibility error messages on expressions that would otherwise seem perfectly OK.

This same sort of situation also occurs when performing operations on **char**s. For example, in the following fragment, the cast back to **char** is needed because of the promotion of **ch1** and **ch2** to **int** within the expression:

```
char ch1 = 'a', ch2 = 'b';

ch1 = (char) (ch1 + ch2);
```

Without the cast, the result of adding **ch1** to **ch2** would be **int**, which can't be assigned to a **char**.

Casts are not only useful when converting between types in an assignment. For example, consider the following program. It uses a cast to **double** to obtain a fractional component from an otherwise integer division.

```
// Using a cast.
class UseCast {
  public static void main(String[] args) {
    int i;

    for(i = 0; i < 5; i++) {
      System.out.println(i + " / 3: " + i / 3);
      System.out.println(i + " / 3 with fractions: "
                         + (double) i / 3);
      System.out.println();
    }
  }
}
```

The output from the program is shown here:

```
0 / 3: 0
0 / 3 with fractions: 0.0

1 / 3: 0
1 / 3 with fractions: 0.3333333333333333

2 / 3: 0
2 / 3 with fractions: 0.6666666666666666

3 / 3: 1
3 / 3 with fractions: 1.0

4 / 3: 1
4 / 3 with fractions: 1.3333333333333333
```

Spacing and Parentheses

An expression in Java may have tabs and spaces in it to make it more readable. For example, the following two expressions are the same, but the second is easier to read:

```
x=10/y*(127/x);
```

```
x = 10 / y * (127/x);
```

Parentheses increase the precedence of the operations contained within them, just like in algebra. Use of redundant or additional parentheses will not cause errors or slow down the execution of the expression. You are encouraged to use parentheses to make clear the exact order of evaluation, both for yourself and for others who may have to figure out your program later. For example, which of the following two expressions is easier to read?

```
x = y/3-34*temp+127;
```

```
x = (y/3) - (34*temp) + 127;
```

Chapter 2 Self Test

1. Why does Java strictly specify the range and behavior of its primitive types?

2. What is Java's character type, and how does it differ from the character type used by some other programming languages?

3. A **boolean** value can have any value you like because any non-zero value is true. True or False?

4. Given this output,

```
One
Two
Three
```

using a single string, show the **println( )** statement that produced it.

5. What is wrong with this fragment?

```
for(i = 0; i < 10; i++) {
   int sum;

   sum = sum + i;
}
System.out.println("Sum is: " + sum);
```

6. Explain the difference between the prefix and postfix forms of the increment operator.

7. Show how a short-circuit AND can be used to prevent a divide-by-zero error.

8. In an expression, what type are **byte** and **short** promoted to?

9. In general, when is a cast needed?

10. Write a program that finds all of the prime numbers between 2 and 100.

11. Does the use of redundant parentheses affect program performance?

12. Does a block define a scope?

Chapter 3

Program Control Statements

Key Skills & Concepts

- Input characters from the keyboard

- Know the complete form of the **if** statement

- Use the **switch** statement

- Know the complete form of the **for** loop

- Use the **while** loop

- Use the **do-while** loop

- Use **break** to exit a loop

- Use **break** as a form of goto

- Apply **continue**

- Nest loops

In this chapter, you will learn about the statements that control a program's flow of execution. There are three categories of program control statements: *selection* statements, which include the **if** and the **switch**; *iteration* statements, which include the **for**, **while**, and **do-while** loops; and *jump* statements, which include **break**, **continue**, and **return**. Except for **return**, which is discussed later in this book, the remaining control statements, including the **if** and **for** statements to which you have already had a brief introduction, are examined in detail here. The chapter begins by explaining how to perform some simple keyboard input.

Input Characters from the Keyboard

Before examining Java's control statements, we will make a short digression that will allow you to begin writing interactive programs. Up to this point, the sample programs in this book have displayed information *to* the user, but they have not received information *from* the user. Thus, you have been using console output, but not console (keyboard) input. The main reason for this is that Java's input capabilities rely on or make use of features not discussed until later in this book. Also, most real-world Java applications will be graphical and window based, not console based. For these reasons, not much use of console input is found in this book. However, there is one type of console input that is relatively easy to use: reading a character from the keyboard. Since several of the examples in this chapter will make use of this feature, it is discussed here.

To read a character from the keyboard, we will use **System.in.read( )**. **System.in** is the complement to **System.out**. It is the input object attached to the keyboard. The **read( )** method waits until the user presses a key and then returns the result. The character is returned as an integer, so it must be cast into a **char** to assign it to a **char** variable. By default, console input is *line buffered*. Here, the term *buffer* refers to a small portion of memory that is used to hold the characters before they are read by your program. In this case, the buffer holds a complete line of text. As a result, you must press ENTER before any character that you type will be sent to your program. Here is a program that reads a character from the keyboard:

```
// Read a character from the keyboard.
class KbIn {
  public static void main(String[] args)
    throws java.io.IOException {

    char ch;

    System.out.print("Press a key followed by ENTER: ");

    ch = (char) System.in.read(); // get a char  ◄──────── Read a character
                                                            from the keyboard.
    System.out.println("Your key is: " + ch);
  }
}
```

Here is a sample run:

```
Press a key followed by ENTER: t
Your key is: t
```

In the program, notice that **main( )** begins like this:

```
public static void main(String[] args)
  throws java.io.IOException {
```

Because **System.in.read( )** is being used, the program must specify the **throws java.io.IOException** clause. This line is necessary to handle input errors. It is part of Java's exception handling mechanism, which is discussed in Chapter 9. For now, don't worry about its precise meaning.

The fact that **System.in** is line buffered is a source of annoyance at times. When you press ENTER, a carriage return, line feed sequence is entered into the input stream. Furthermore, these characters are left pending in the input buffer until you read them. Thus, for some applications, you may need to remove them (by reading them) before the next input operation. You will see an example of this later in this chapter.

The if Statement

Chapter 1 introduced the **if** statement. It is examined in detail here. The complete form of the **if** statement is

if(*condition*) *statement;*
else *statement;*

where the targets of the **if** and **else** are single statements. The **else** clause is optional. The targets of both the **if** and **else** can be blocks of statements. The general form of the **if**, using blocks of statements, is

```
if(condition)
{
  statement sequence
}
else
{
  statement sequence
}
```

If the conditional expression is true, the target of the **if** will be executed; otherwise, if it exists, the target of the **else** will be executed. At no time will both of them be executed. The conditional expression controlling the **if** must produce a **boolean** result.

To demonstrate the **if** (and several other control statements), we will create and develop a simple computerized guessing game that would be suitable for young children. In the first version of the game, the program asks the player for a letter between A and Z. If the player presses the correct letter on the keyboard, the program responds by printing the message **** Right ****. The program is shown here:

```
// Guess the letter game.
class Guess {
  public static void main(String[] args)
    throws java.io.IOException {

    char ch, answer = 'K';

    System.out.println("I'm thinking of a letter between A and Z.");
    System.out.print("Can you guess it: ");

    ch = (char) System.in.read(); // read a char from the keyboard

    if(ch == answer) System.out.println("** Right **");
  }
}
```

This program prompts the player and then reads a character from the keyboard. Using an **if** statement, it then checks that character against the answer, which is K in this case. If K was entered, the message is displayed. When you try this program, remember that the K must be entered in uppercase.

Taking the guessing game further, the next version uses the **else** to print a message when the wrong letter is picked.

```
// Guess the letter game, 2nd version.
class Guess2 {
  public static void main(String[] args)
    throws java.io.IOException {
```

```
      char ch, answer = 'K';

      System.out.println("I'm thinking of a letter between A and Z.");
      System.out.print("Can you guess it: ");

      ch = (char) System.in.read(); // get a char

      if(ch == answer) System.out.println("** Right **");
      else System.out.println("...Sorry, you're wrong.");
   }
}
```

Nested ifs

A *nested if* is an **if** statement that is the target of another **if** or **else**. Nested **if**s are very common in programming. The main thing to remember about nested **if**s in Java is that an **else** statement always refers to the nearest **if** statement that is within the same block as the **else** and not already associated with an **else**. Here is an example:

```
if(i == 10) {
  if(j < 20) a - b;
  if(k > 100) c = d;
  else a = c; // this else refers to if(k > 100)
}
else a = d; // this else refers to if(i == 10)
```

As the comments indicate, the final **else** is not associated with **if(j < 20)**, because it is not in the same block (even though it is the nearest **If** without an **else**). Rather, the final **else** is associated with **if(i == 10)**. The inner **else** refers to **if(k > 100)**, because it is the closest **if** within the same block.

You can use a nested **if** to add a further improvement to the guessing game. This addition provides the player with feedback about a wrong guess.

```
// Guess the letter game, 3rd version.
class Guess3 {
  public static void main(String[] args)
    throws java.io.IOException {

    char ch, answer = 'K';

    System.out.println("I'm thinking of a letter between A and Z.");
    System.out.print("Can you guess it: ");

    ch = (char) System.in.read(); // get a char

    if(ch == answer) System.out.println("** Right **");
    else {
      System.out.print("...Sorry, you're ");
```

This is a nested **if**.

```
        // a nested if
      if(ch < answer) System.out.println("too low");
      else System.out.println("too high");
    }
  }
}
```

A sample run is shown here:

```
I'm thinking of a letter between A and Z.
Can you guess it: Z
...Sorry, you're too high
```

The if-else-if Ladder

A common programming construct that is based upon the nested **if** is the **if-else-if** *ladder*. It looks like this:

if(*condition*)
   statement;
else if(*condition*)
   statement;
else if(*condition*)
   statement;
.
.
.
else
   statement;

The conditional expressions are evaluated from the top downward. As soon as a true condition is found, the statement associated with it is executed, and the rest of the ladder is bypassed. If none of the conditions are true, the final **else** statement will be executed. The final **else** often acts as a default condition; that is, if all other conditional tests fail, the last **else** statement is performed. If there is no final **else** and all other conditions are false, no action will take place.

The following program demonstrates the **if-else-if** ladder:

```
// Demonstrate an if-else-if ladder.
class Ladder {
  public static void main(String[] args) {
    int x;

    for(x=0; x<6; x++) {
      if(x==1)
        System.out.println("x is one");
      else if(x==2)
        System.out.println("x is two");
      else if(x==3)
```

```
      System.out.println("x is three");
    else if(x==4)
      System.out.println("x is four");
    else
      System.out.println("x is not between 1 and 4");  ◄——— This is the
    }                                                        default statement.
  }
}
```

The program produces the following output:

```
x is not between 1 and 4
x is one
x is two
x is three
x is four
x is not between 1 and 4
```

As you can see, the default **else** is executed only if none of the preceding **if** statements succeeds.

The Traditional switch Statement

The second of Java's selection statements is the **switch**. The **switch** provides for a multiway branch. Thus, it enables a program to select among several alternatives. Although a series of nested **if** statements can perform multiway tests, for many situations the **switch** is a more efficient approach.

Before we continue, an important point needs to be made. Beginning with JDK 14, the **switch** has been significantly enhanced and expanded with several new features that go far beyond its original capabilities. Because of the substantial nature of the recent **switch** enhancements, they are described in Chapter 16, in the context of other recent additions to Java. Here, the **switch** is introduced in its traditional form. This is the form of **switch** that has been part of Java from the start and is in widespread use. It is also the form that will work in all Java development environments. The traditional **switch** works like this: the value of an expression is successively tested against a list of constants. When a match is found, the statement sequence associated with that match is executed.

The general form of the traditional **switch** statement is

```
switch(expression) {
  case constant1:
    statement sequence
    break;
  case constant2:
    statement sequence
    break;
  case constant3:
    statement sequence
    break;
```

.
.
.

```
  default:
    statement sequence
}
```

For versions of Java prior to JDK 7, the *expression* controlling the **switch** must resolve to type **byte**, **short**, **int**, **char**, or an enumeration. (Enumerations are described in Chapter 12.) However, today, *expression* can also be of type **String**. This means that modern versions of Java can use a string to control a **switch**. (This technique is demonstrated in Chapter 5, when **String** is described.) Frequently, the expression controlling a **switch** is simply a variable rather than a larger expression.

Each value specified in the **case** statements must be a unique constant expression (such as a literal value). Duplicate **case** values are not allowed. The type of each value must be compatible with the type of *expression*.

The **default** statement sequence is executed if no **case** constant matches the expression. The **default** is optional; if it is not present, no action takes place if all matches fail. When a match is found, the statements associated with that **case** are executed until the **break** is encountered or, in the case of **default** or the last **case**, until the end of the **switch** is reached.

The following program demonstrates the **switch**:

```java
// Demonstrate the switch.
class SwitchDemo {
  public static void main(String[] args) {
    int i;

    for(i=0; i<10; i++)
      switch(i) {
        case 0:
          System.out.println("i is zero");
          break;
        case 1:
          System.out.println("i is one");
          break;
        case 2:
          System.out.println("i is two");
          break;
        case 3:
          System.out.println("i is three");
          break;
        case 4:
          System.out.println("i is four");
          break;
        default:
          System.out.println("i is five or more");
      }
  }
}
```

The output produced by this program is shown here:

```
i is zero
i is one
i is two
i is three
i is four
i is five or more
i is five or more
i is five or more
i is five or more
i is five or more
```

As you can see, each time through the loop, the statements associated with the **case** constant that matches **i** are executed. All others are bypassed. When **i** is five or greater, no **case** statements match, so the **default** statement is executed.

Technically, the **break** statement is optional, although most applications of the **switch** will use it. When encountered within the statement sequence of a **case**, the **break** statement causes program flow to exit from the entire **switch** statement and resume at the next statement outside the **switch**. However, if a **break** statement does not end the statement sequence associated with a **case**, then all the statements *at and following* the matching **case** will be executed until a **break** (or the end of the **switch**) is encountered. Thus, a **case** without a **break** will "fall through" to the next **case**.

For example, study the following program carefully. Before looking at the output, can you figure out what it will display on the screen?

```
// Demonstrate the switch without break statements.
class NoBreak {
  public static void main(String[] args) {
    int i;

    for(i=0; i<=5; i++) {
      switch(i) {
        case 0:
          System.out.println("i is less than one");
        case 1:
          System.out.println("i is less than two");
        case 2:
          System.out.println("i is less than three");
        case 3:
          System.out.println("i is less than four");
        case 4:
          System.out.println("i is less than five");
      }
      System.out.println();
    }
  }
}
```

The **case** statements fall through here.

This program displays the following output:

```
i is less than one
i is less than two
i is less than three
i is less than four
i is less than five

i is less than two
i is less than three
i is less than four
i is less than five

i is less than three
i is less than four
i is less than five

i is less than four
i is less than five

i is less than five
```

As this program illustrates, execution will continue into the next **case** if no **break** statement is present.

You can have empty **case**s, as shown in this example:

```
switch(i) {
  case 1:
  case 2:
  case 3: System.out.println("i is 1, 2 or 3");
    break;
  case 4: System.out.println("i is 4");
    break;
}
```

In this fragment, if **i** has the value 1, 2, or 3, the first **println( )** statement executes. If it is 4, the second **println( )** statement executes. The "stacking" of **case**s, as shown in this example, is common when several **case**s share common code.

REMEMBER

Recently, the capabilities and features of **switch** have been substantially expanded beyond those offered by the traditional **switch** just described. Refer to Chapter 16 for details on the enhanced **switch**.

Nested switch Statements

It is possible to have a **switch** as part of the statement sequence of an outer **switch**. This is called a nested **switch**. Even if the **case** constants of the inner and outer **switch** contain common values, no conflicts will arise. For example, the following code fragment is perfectly acceptable:

```
switch(ch1) {
  case 'A': System.out.println("This A is part of outer switch.");
    switch(ch2) {
      case 'A':
        System.out.println("This A is part of inner switch");
        break;
      case 'B': // ...
    } // end of inner switch
    break;
  case 'B': // ...
```

Try This 3-1 Start Building a Java Help System

Help.java

This project builds a simple help system that displays the syntax for the Java control statements. The program displays a menu containing the control statements and then waits for you to choose one. After one is chosen, the syntax of the statement is displayed. In this first version of the program, help is available for only the **if** and traditional **switch** statements. The other control statements are added in subsequent projects.

1. Create a file called **Help.java**.

2. The program begins by displaying the following menu:

```
Help on:
  1. if
  2. switch
Choose one:
```

To accomplish this, you will use the statement sequence shown here:

```
System.out.println("Help on:");
System.out.println("  1. if");
System.out.println("  2. switch");
System.out.print("Choose one: ");
```

3. Next, the program obtains the user's selection by calling **System.in.read( )**, as shown here:

```
choice = (char) System.in.read();
```

(continued)

4. Once the selection has been obtained, the program uses the **switch** statement shown here to display the syntax for the selected statement.

```java
switch(choice) {
  case '1':
    System.out.println("The if:\n");
    System.out.println("if(condition) statement;");
    System.out.println("else statement;");
    break;
  case '2':
    System.out.println("The traditional switch:\n");
    System.out.println("switch(expression) {");
    System.out.println("  case constant:");
    System.out.println("    statement sequence");
    System.out.println("    break;");
    System.out.println("  // ...");
    System.out.println("}");
    break;
  default:
    System.out.print("Selection not found.");
}
```

Notice how the **default** clause catches invalid choices. For example, if the user enters 3, no **case** constants will match, causing the **default** sequence to execute.

5. Here is the entire **Help.java** program listing:

```java
/*
    Try This 3-1

    A simple help system.
*/
class Help {
  public static void main(String[] args)
    throws java.io.IOException {
    char choice;

    System.out.println("Help on:");
    System.out.println("  1. if");
    System.out.println("  2. switch");
    System.out.print("Choose one: ");
    choice = (char) System.in.read();

    System.out.println("\n");

    switch(choice) {
      case '1':
        System.out.println("The if:\n");
        System.out.println("if(condition) statement;");
```

```
        System.out.println("else statement;");
        break;
      case '2':
        System.out.println("The traditional switch:\n");
        System.out.println("switch(expression) {");
        System.out.println("  case constant:");
        System.out.println("    statement sequence");
        System.out.println("    break;");
        System.out.println("  // ...");
        System.out.println("}");
        break;
      default:
        System.out.print("Selection not found.");
    }
  }
}
```

6. Here is a sample run.

```
Help on:
  1. if
  2. switch
Choose one: 1

The if.

if(condition) statement;
else statement;
```

The for Loop

You have been using a simple form of the **for** loop since Chapter 1. You might be surprised at just how powerful and flexible the **for** loop is. Let's begin by reviewing the basics, starting with the most traditional forms of the **for**.

The general form of the **for** loop for repeating a single statement is

for(*initialization; condition; iteration*) *statement;*

For repeating a block, the general form is

for(*initialization; condition; iteration*)
{
  statement sequence
}

The *initialization* is usually an assignment statement that sets the initial value of the *loop control variable,* which acts as the counter that controls the loop. The *condition* is a Boolean expression that determines whether or not the loop will repeat. The *iteration* expression defines the amount

Ask the Expert

Q: Under what conditions should I use an if-else-if **ladder rather than a** switch **when coding a multiway branch?**

A: In general, use an **if-else-if** ladder when the conditions controlling the selection process do not rely upon a single value. For example, consider the following **if-else-if** sequence:

```
if(x < 10) // ...
else if(y != 0) // ...
else if(!done) // ...
```

This sequence cannot be recoded into a **switch** because all three conditions involve different variables—and differing types. What variable would control the **switch**? Also, you will need to use an **if-else-if** ladder when testing floating-point values or other objects that are not of types valid for use in the expression controlling the **switch**.

by which the loop control variable will change each time the loop is repeated. Notice that these three major sections of the loop must be separated by semicolons. The **for** loop will continue to execute as long as the condition tests true. Once the condition becomes false, the loop will exit, and program execution will resume on the statement following the **for**.

The following program uses a **for** loop to print the square roots of the numbers between 1 and 99. It also displays the rounding error present for each square root.

```java
// Show square roots of 1 to 99 and the rounding error.
class SqrRoot {
  public static void main(String[] args) {
    double num, sroot, rerr;

    for(num = 1.0; num < 100.0; num++) {
      sroot = Math.sqrt(num);
      System.out.println("Square root of " + num +
                         " is " + sroot);

      // compute rounding error
      rerr = num - (sroot * sroot);
      System.out.println("Rounding error is " + rerr);
      System.out.println();
    }
  }
}
```

Notice that the rounding error is computed by squaring the square root of each number. This result is then subtracted from the original number, thus yielding the rounding error.

The **for** loop can proceed in a positive or negative fashion, and it can change the loop control variable by any amount. For example, the following program prints the numbers 100 to –95, in decrements of 5:

```
// A negatively running for loop.
class DecrFor {
  public static void main(String[] args) {
    int x;

    for(x = 100; x > -100; x -= 5)          Loop control variable is
      System.out.println(x);                decremented by 5 each time.
  }
}
```

An important point about **for** loops is that the conditional expression is always tested at the top of the loop. This means that the code inside the loop may not be executed at all if the condition is false to begin with. Here is an example:

```
for(count=10; count < 5; count++)
  x += count; // this statement will not execute
```

This loop will never execute because its control variable, **count**, is greater than 5 when the loop is first entered. This makes the conditional expression, **count < 5**, false from the outset; thus, not even one iteration of the loop will occur.

Some Variations on the for Loop

The **for** is one of the most versatile statements in the Java language because it allows a wide range of variations. For example, multiple loop control variables can be used. Consider the following program:

```
// Use commas in a for statement.
class Comma {
  public static void main(String[] args) {
    int i, j;

    for(i=0, j=10; i < j; i++, j--)          Notice the two loop
      System.out.println("i and j: " + i + " " + j);   control variables.
  }
}
```

The output from the program is shown here:

```
i and j: 0 10
i and j: 1 9
i and j: 2 8
i and j: 3 7
i and j: 4 6
```

Here, commas separate the two initialization statements and the two iteration expressions. When the loop begins, both **i** and **j** are initialized. Each time the loop repeats, **i** is incremented and **j** is decremented. Multiple loop control variables are often convenient and can simplify certain algorithms. You can have any number of initialization and iteration statements, but in practice, more than two or three make the **for** loop unwieldy.

The condition controlling the loop can be any valid Boolean expression. It does not need to involve the loop control variable. In the next example, the loop continues to execute until the user types the letter S at the keyboard:

```
// Loop until an S is typed.
class ForTest {
  public static void main(String[] args)
    throws java.io.IOException {

    int i;

    System.out.println("Press S to stop.");

    for(i = 0; (char) System.in.read() != 'S'; i++)
      System.out.println("Pass #" + i);
  }
}
```

Missing Pieces

Some interesting **for** loop variations are created by leaving pieces of the loop definition empty. In Java, it is possible for any or all of the initialization, condition, or iteration portions of the **for** loop to be blank. For example, consider the following program:

```
// Parts of the for can be empty.
class Empty {
  public static void main(String[] args) {
    int i;

    for(i = 0; i < 10; ) {  ◄───────────── The iteration expression is missing.
      System.out.println("Pass #" + i);
      i++; // increment loop control var
    }
  }
}
```

Here, the iteration expression of the **for** is empty. Instead, the loop control variable **i** is incremented inside the body of the loop. This means that each time the loop repeats, **i** is tested to see whether it equals 10, but no further action takes place. Of course, since **i** is still incremented within the body of the loop, the loop runs normally, displaying the following output:

```
Pass #0
Pass #1
Pass #2
Pass #3
Pass #4
Pass #5
Pass #6
```

```
Pass #7
Pass #8
Pass #9
```

In the next example, the initialization portion is also moved out of the **for**:

```
// Move more out of the for loop.
class Empty2 {
  public static void main(String[] args) {
    int i;

    i = 0; // move initialization out of loop        ──── The initialization expression
    for(; i < 10; ) {                                      is moved out of the loop.
      System.out.println("Pass #" + i);
      i++; // increment loop control var
    }
  }
}
```

In this version, **i** is initialized before the loop begins, rather than as part of the **for**. Normally, you will want to initialize the loop control variable inside the **for**. Placing the initialization outside of the loop is generally done only when the initial value is derived through a complex process that does not lend itself to containment inside the **for** statement.

The Infinite Loop

You can create an *infinite loop* (a loop that never terminates) using the **for** by leaving the conditional expression empty. For example, the following fragment shows the way many Java programmers create an infinite loop:

```
for(;;) // intentionally infinite loop
{
  //...
}
```

This loop will run forever. Although there are some programming tasks, such as operating system command processors, that require an infinite loop, most "infinite loops" are really just loops with special termination requirements. Near the end of this chapter, you will see how to halt a loop of this type. (Hint: It's done using the **break** statement.)

Loops with No Body

In Java, the body associated with a **for** loop (or any other loop) can be empty. This is because a *null statement* is syntactically valid. Body-less loops are often useful. For example, the following program uses one to sum the numbers 1 through 5:

```
// The body of a loop can be empty.
class Empty3 {
  public static void main(String[] args) {
    int i;
    int sum = 0;
```

```
   // sum the numbers through 5
   for(i = 1; i <= 5; sum += i++) ;    ◄————— No body in this loop!

   System.out.println("Sum is " + sum);
  }
}
```

The output from the program is shown here:

```
Sum is 15
```

Notice that the summation process is handled entirely within the **for** statement, and no body is needed. Pay special attention to the iteration expression:

```
sum += i++
```

Don't be intimidated by statements like this. They are common in professionally written Java programs and are easy to understand if you break them down into their parts. In other words, this statement says, "Add to **sum** the value of **sum** plus **i**, then increment **i**." Thus, it is the same as this sequence of statements:

```
sum = sum + i;
i++;
```

Declaring Loop Control Variables Inside the for Loop

Often the variable that controls a **for** loop is needed only for the purposes of the loop and is not used elsewhere. When this is the case, it is possible to declare the variable inside the initialization portion of the **for**. For example, the following program computes both the summation and the factorial of the numbers 1 through 5. It declares its loop control variable **i** inside the **for**.

```
// Declare loop control variable inside the for.
class ForVar {
  public static void main(String[] args) {
    int sum = 0;
    int fact = 1;

    // compute the factorial of the numbers through 5
    for(int i = 1; i <= 5; i++) {    ◄——————— The variable i is declared
      sum += i; // i is known throughout the loop       inside the for statement.
      fact *= i;
    }

    // but, i is not known here

    System.out.println("Sum is " + sum);
    System.out.println("Factorial is " + fact);
  }
}
```

When you declare a variable inside a **for** loop, there is one important point to remember: the scope of that variable ends when the **for** statement does. (That is, the scope of the variable is limited to the **for** loop.) Outside the **for** loop, the variable will cease to exist. Thus, in the preceding example, **i** is not accessible outside the **for** loop. If you need to use the loop control variable elsewhere in your program, you will not be able to declare it inside the **for** loop.

Before moving on, you might want to experiment with your own variations on the **for** loop. As you will find, it is a fascinating loop.

The Enhanced for Loop

There is another form of the **for** loop, called the *enhanced for*. The enhanced **for** provides a streamlined way to cycle through the contents of a collection of objects, such as an array. The enhanced **for** loop is discussed in Chapter 5, after arrays have been introduced.

The while Loop

Another of Java's loops is the **while**. The general form of the **while** loop is

while(*condition*) *statement*;

where *statement* may be a single statement or a block of statements, and *condition* defines the condition that controls the loop. The condition may be any valid Boolean expression. The loop repeats while the condition is true. When the condition becomes false, program control passes to the line immediately following the loop.

Here is a simple example in which a **while** is used to print the alphabet:

```
// Demonstrate the while loop.
class WhileDemo {
  public static void main(String[] args) {
    char ch;

    // print the alphabet using a while loop
    ch = 'a';
    while(ch <= 'z') {
      System.out.print(ch);
      ch++;
    }
  }
}
```

Here, **ch** is initialized to the letter a. Each time through the loop, **ch** is output and then incremented. This process continues until **ch** is greater than z.

As with the **for** loop, the **while** checks the conditional expression at the top of the loop, which means that the loop code may not execute at all. This eliminates the need for performing a separate test before the loop. The following program illustrates this characteristic of the **while** loop. It computes the integer powers of 2, from 0 to 9.

```
// Compute integer powers of 2.
class Power {
  public static void main(String[] args) {
    int e;
    int result;

    for(int i=0; i < 10; i++) {
      result = 1;
      e = i;
      while(e > 0) {
        result *= 2;
        e--;
      }

      System.out.println("2 to the " + i +
                          " power is " + result);
    }
  }
}
```

The output from the program is shown here:

```
2 to the 0 power is 1
2 to the 1 power is 2
2 to the 2 power is 4
2 to the 3 power is 8
2 to the 4 power is 16
2 to the 5 power is 32
2 to the 6 power is 64
2 to the 7 power is 128
2 to the 8 power is 256
2 to the 9 power is 512
```

Notice that the **while** loop executes only when **e** is greater than 0. Thus, when **e** is zero, as it is in the first iteration of the **for** loop, the **while** loop is skipped.

Ask the Expert

Q: Given the flexibility inherent in all of Java's loops, what criteria should I use when selecting a loop? That is, how do I choose the right loop for a specific job?

A: Use a **for** loop when performing a known number of iterations based on the value of a loop control variable. Use the **do-while** when you need a loop that will always perform at least one iteration. The **while** is best used when the loop will repeat until some condition becomes false.

The do-while Loop

The last of Java's loops is the **do-while**. Unlike the **for** and the **while** loops, in which the condition is tested at the top of the loop, the **do-while** loop checks its condition at the bottom of the loop. This means that a **do-while** loop will always execute at least once. The general form of the **do-while** loop is

```
do {
    statements;
} while(condition);
```

Although the braces are not necessary when only one statement is present, they are often used to improve readability of the **do-while** construct, thus preventing confusion with the **while**. The **do-while** loop executes as long as the conditional expression is true.

The following program loops until the user enters the letter q:

```
// Demonstrate the do-while loop.
class DWDemo {
  public static void main(String[] args)
    throws java.io.IOException {

    char ch;

    do {
      System.out.print("Press a key followed by ENTER: ");
      ch = (char) System.in.read(); // get a char
    } while(ch != 'q');
  }
}
```

Using a **do-while** loop, we can further improve the guessing game program from earlier in this chapter. This time, the program loops until you guess the letter.

```
// Guess the letter game, 4th version.
class Guess4 {
  public static void main(String[] args)
    throws java.io.IOException {

    char ch, ignore, answer = 'K';

    do {
      System.out.println("I'm thinking of a letter between A and Z.");
      System.out.print("Can you guess it: ");

      // read a character
      ch = (char) System.in.read();

      // discard any other characters in the input buffer
      do {
        ignore = (char) System.in.read();
```

```
    } while(ignore != '\n');

    if(ch == answer) System.out.println("** Right **");
    else {
      System.out.print("...Sorry, you're ");
      if(ch < answer) System.out.println("too low");
      else System.out.println("too high");
      System.out.println("Try again!\n");
    }
  } while(answer != ch);
 }
}
```

Here is a sample run:

```
I'm thinking of a letter between A and Z.
Can you guess it: A
...Sorry, you're too low
Try again!

I'm thinking of a letter between A and Z.
Can you guess it: Z
...Sorry, you're too high
Try again!

I'm thinking of a letter between A and Z.
Can you guess it: K
** Right **
```

Notice one other thing of interest in this program. There are two **do-while** loops in the program. The first loops until the user guesses the letter. Its operation and meaning should be clear. The second **do-while** loop, shown again here, warrants some explanation:

```
// discard any other characters in the input buffer
do {
  ignore = (char) System.in.read();
} while(ignore != '\n');
```

As explained earlier, console input is line buffered—you have to press ENTER before characters are sent. Pressing ENTER causes a carriage return and a line feed (newline) sequence to be generated. These characters are left pending in the input buffer. Also, if you typed more than one key before pressing ENTER, they too would still be in the input buffer. This loop discards those characters by continuing to read input until the end of the line is reached. If they were not discarded, then those characters would also be sent to the program as guesses, which is not what is wanted. (To see the effect of this, you might try removing the inner **do-while** loop.) In Chapter 10, after you have learned more about Java, some other, higher-level ways of handling console input are described. However, the use of **read( )** here gives you insight into how the foundation of Java's I/O system operates. It also shows another example of Java's loops in action.

Try This 3-2 Improve the Java Help System

Help2.java

This project expands on the Java help system that was created in Try This 3-1. This version adds the syntax for the **for**, **while**, and **do-while** loops. It also checks the user's menu selection, looping until a valid response is entered.

1. Copy **Help.java** to a new file called **Help2.java**.

2. Change the first part of **main( )** so that it uses a loop to display the choices, as shown here:

```java
public static void main(String[] args)
   throws java.io.IOException {
   char choice, ignore;

   do {
     System.out.println("Help on:");
     System.out.println("  1. if");
     System.out.println("  2. switch");
     System.out.println("  3. for");
     System.out.println("  4. while");
     System.out.println("  5. do-while\n");
     System.out.print("Choose one: ");

     choice = (char) System.in.read();

     do {
       ignore = (char) System.in.read();
     } while(ignore != '\n');
   } while( choice < '1' | choice > '5');
```

Notice that a nested **do-while** loop is used to discard any unwanted characters remaining in the input buffer. After making this change, the program will loop, displaying the menu until the user enters a response that is between 1 and 5.

3. Expand the **switch** statement to include the **for**, **while**, and **do-while** loops, as shown here:

```java
switch(choice) {
  case '1':
    System.out.println("The if:\n");
    System.out.println("if(condition) statement;");
    System.out.println("else statement;");
    break;
  case '2':
    System.out.println("The traditional switch:\n");
    System.out.println("switch(expression) {");
    System.out.println("  case constant:");
    System.out.println("    statement sequence");
    System.out.println("    break;");
```

(continued)

```
      System.out.println("   // ...");
      System.out.println("}");
      break;
    case '3':
      System.out.println("The for:\n");
      System.out.print("for(init; condition; iteration)");
      System.out.println(" statement;");
      break;
    case '4':
      System.out.println("The while:\n");
      System.out.println("while(condition) statement;");
      break;
    case '5':
      System.out.println("The do-while:\n");
      System.out.println("do {");
      System.out.println("  statement;");
      System.out.println("} while (condition);");
      break;
  }
```

Notice that no **default** statement is present in this version of the **switch**. Since the menu loop ensures that a valid response will be entered, it is no longer necessary to include a **default** statement to handle an invalid choice.

4. Here is the entire **Help2.java** program listing:

```
/*
    Try This 3-2

    An improved Help system that uses a
    do-while to process a menu selection.
*/
class Help2 {
  public static void main(String[] args)
    throws java.io.IOException {
    char choice, ignore;

    do {
      System.out.println("Help on:");
      System.out.println("  1. if");
      System.out.println("  2. switch");
      System.out.println("  3. for");
      System.out.println("  4. while");
      System.out.println("  5. do-while\n");
      System.out.print("Choose one: ");
```

```
      choice = (char) System.in.read();

      do {
        ignore = (char) System.in.read();
      } while(ignore != '\n');
    } while( choice < '1' | choice > '5');

    System.out.println("\n");

    switch(choice) {
      case '1':
        System.out.println("The if:\n");
        System.out.println("if(condition) statement;");
        System.out.println("else statement;");
        break;
      case '2':
        System.out.println("The traditional switch:\n");
        System.out.println("switch(expression) {");
        System.out.println("  case constant:");
        System.out.println("    statement sequence");
        System.out.println("    break;");
        System.out.println("  // ...");
        System.out.println("}");
        break;
      case '3':
        System.out.println("The for:\n");
        System.out.print("for(init; condition; iteration)");
        System.out.println(" statement;");
        break;
      case '4':
        System.out.println("The while:\n");
        System.out.println("while(condition) statement;");
        break;
      case '5':
        System.out.println("The do-while:\n");
        System.out.println("do {");
        System.out.println("  statement;");
        System.out.println("} while (condition);");
        break;
    }
  }
}
```

Use break to Exit a Loop

It is possible to force an immediate exit from a loop, bypassing any remaining code in the body of the loop and the loop's conditional test, by using the **break** statement. When a **break** statement is encountered inside a loop, the loop is terminated and program control resumes at the next statement following the loop. Here is a simple example:

```
// Using break to exit a loop.
class BreakDemo {
  public static void main(String[] args) {
    int num;

    num = 100;

    // loop while i-squared is less than num
    for(int i=0; i < num; i++) {
      if(i*i >= num) break; // terminate loop if i*i >= 100
      System.out.print(i + " ");
    }
    System.out.println("Loop complete.");
  }
}
```

This program generates the following output:

```
0 1 2 3 4 5 6 7 8 9 Loop complete.
```

As you can see, although the **for** loop is designed to run from 0 to **num** (which in this case is 100), the **break** statement causes it to terminate early, when **i** squared is greater than or equal to **num**.

The **break** statement can be used with any of Java's loops, including intentionally infinite loops. For example, the following program simply reads input until the user types the letter q:

```
// Read input until a q is received.
class Break2 {
  public static void main(String[] args)
    throws java.io.IOException {

    char ch;

    for( ; ; ) {                              ← This "infinite" loop is
      ch = (char) System.in.read(); // get a char   terminated by the break.
      if(ch == 'q') break; ←
    }
    System.out.println("You pressed q!");
  }
}
```

When used inside a set of nested loops, the **break** statement will break out of only the innermost loop. For example:

```
// Using break with nested loops.
class Break3 {
  public static void main(String[] args) {

    for(int i=0; i<3; i++) {
      System.out.println("Outer loop count: " + i);
      System.out.print("    Inner loop count: ");

      int t = 0;
      while(t < 100) {
        if(t == 10) break; // terminate loop if t is 10
        System.out.print(t + " ");
        t++;
      }
      System.out.println();
    }
    System.out.println("Loops complete.");
  }
}
```

This program generates the following output:

```
Outer loop count: 0
    Inner loop count: 0 1 2 3 4 5 6 7 8 9
Outer loop count: 1
    Inner loop count: 0 1 2 3 4 5 6 7 8 9
Outer loop count: 2
    Inner loop count: 0 1 2 3 4 5 6 7 8 9
Loops complete.
```

As you can see, the **break** statement in the inner loop causes the termination of only that loop. The outer loop is unaffected.

Here are two other points to remember about **break**. First, more than one **break** statement may appear in a loop. However, be careful. Too many **break** statements have the tendency to destructure your code. Second, the **break** that terminates a **switch** statement affects only that **switch** statement and not any enclosing loops.

Use break as a Form of goto

In addition to its uses with the **switch** statement and loops, the **break** statement can be employed by itself to provide a "civilized" form of the goto statement. Java does not have a goto statement, because it provides an unstructured way to alter the flow of program execution. Programs that make extensive use of the goto are usually hard to understand and hard to maintain. There are, however, a few places where the goto is a useful and legitimate device.

For example, the goto can be helpful when exiting from a deeply nested set of loops. To handle such situations, Java defines an expanded form of the **break** statement. By using this form of **break**, you can, for example, break out of one or more blocks of code. These blocks need not be part of a loop or a **switch**. They can be any block. Further, you can specify precisely where execution will resume, because this form of **break** works with a label. As you will see, **break** gives you the benefits of a goto without its problems.

The general form of the labeled **break** statement is shown here:

break *label*;

Typically, *label* is the name of a label that identifies a block of code. When this form of **break** executes, control is transferred out of the named block of code. The labeled block of code must enclose the **break** statement, but it does not need to be the immediately enclosing block. This means that you can use a labeled **break** statement to exit from a set of nested blocks. But you cannot use **break** to transfer control to a block of code that does not enclose the **break** statement.

To name a block, put a label at the start of it. The block being labeled can be a stand-alone block, or a statement that has a block as its target. A *label* is any valid Java identifier followed by a colon. Once you have labeled a block, you can then use this label as the target of a **break** statement. Doing so causes execution to resume at the *end* of the labeled block. For example, the following program shows three nested blocks:

```
// Using break with a label.
class Break4 {
  public static void main(String[] args) {
    int i;

    for(i=1; i<4; i++) {
one:    {
two:      {
three:      {
            System.out.println("\ni is " + i);
            if(i==1) break one;  ←———— Break to a label.
            if(i==2) break two;
            if(i==3) break three;

            // this is never reached
            System.out.println("won't print");
          }
          System.out.println("After block three.");
        }
        System.out.println("After block two.");
      }
      System.out.println("After block one.");
    }
    System.out.println("After for.");
  }
}
```

The output from the program is shown here:

```
i is 1
After block one.

i is 2
After block two.
After block one.

i is 3
After block three.
After block two.
After block one.
After for.
```

Let's look closely at the program to understand precisely why this output is produced. When **i** is 1, the first **if** statement succeeds, causing a **break** to the end of the block of code defined by label **one**. This causes **After block one.** to print. When **i** is 2, the second **if** succeeds, causing control to be transferred to the end of the block labeled by **two**. This causes the messages **After block two.** and **After block one.** to be printed, in that order. When **i** is 3, the third **if** succeeds, and control is transferred to the end of the block labeled by **three**. Now, all three messages are displayed.

Here is another example. This time, **break** is being used to jump outside of a series of nested **for** loops. When the **break** statement in the inner loop is executed, program control jumps to the end of the block defined by the outer **for** loop, which is labeled by **done**. This causes the remainder of all three loops to be bypassed.

```
// Another example of using break with a label.
class Break5 {
  public static void main(String[] args) {

done:
    for(int i=0; i<10; i++) {
      for(int j=0; j<10; j++) {
        for(int k=0; k<10; k++) {
          System.out.println(k + " ");
          if(k == 5) break done; // jump to done
        }
        System.out.println("After k loop"); // won't execute
      }
      System.out.println("After j loop"); // won't execute
    }
    System.out.println("After i loop");
  }
}
```

The output from the program is shown here:

```
0
1
2
3
4
5
After i loop
```

Precisely where you put a label is very important—especially when working with loops. For example, consider the following program:

```java
// Where you put a label is important.
class Break6 {
  public static void main(String[] args) {
    int x=0, y=0;

// here, put label before for statement.
stop1: for(x=0; x < 5; x++) {
        for(y = 0; y < 5; y++) {
          if(y == 2) break stop1;
          System.out.println("x and y: " + x + " " + y);
        }
      }

      System.out.println();

// now, put label immediately before {
      for(x=0; x < 5; x++)
stop2: {
        for(y = 0; y < 5; y++) {
          if(y == 2) break stop2;
          System.out.println("x and y: " + x + " " + y);
        }
      }
  }
}
```

The output from this program is shown here:

```
x and y: 0 0
x and y: 0 1

x and y: 0 0
x and y: 0 1
```

```
x and y: 1 0
x and y: 1 1
x and y: 2 0
x and y: 2 1
x and y: 3 0
x and y: 3 1
x and y: 4 0
x and y: 4 1
```

In the program, both sets of nested loops are the same except for one point. In the first set, the label precedes the outer **for** loop. In this case, when the **break** executes, it transfers control to the end of the entire **for** block, skipping the rest of the outer loop's iterations. In the second set, the label precedes the outer **for**'s opening curly brace. Thus, when **break stop2** executes, control is transferred to the end of the outer **for**'s block, causing the next iteration to occur.

Keep in mind that you cannot **break** to any label that is not defined for an enclosing block. For example, the following program is invalid and will not compile:

```java
// This program contains an error.
class BreakErr {
  public static void main(String[] args) {

    one: for(int i=0; i<3; i++) {
      System.out.print("Pass " + i + ": ");
    }

    for(int j=0; j<100; j++) {
      if(j == 10) break one; // WRONG
      System.out.print(j + " ");
    }
  }
}
```

Since the loop labeled **one** does not enclose the **break** statement, it is not possible to transfer control to that block.

Ask the Expert

Q: You say that the goto is unstructured and that the break with a label offers a better alternative. But really, doesn't breaking to a label, which might be many lines of code and levels of nesting removed from the break, also destructure code?

A: The short answer is yes! However, in those cases in which a jarring change in program flow is required, breaking to a label still retains some structure. A **goto** has none!

Use continue

It is possible to force an early iteration of a loop, bypassing the loop's normal control structure. This is accomplished using **continue**. The **continue** statement forces the next iteration of the loop to take place, skipping any code between itself and the conditional expression that controls the loop. Thus, **continue** is essentially the complement of **break**. For example, the following program uses **continue** to help print the even numbers between 0 and 100:

```
// Use continue.
class ContDemo {
  public static void main(String[] args) {
    int i;

    // print even numbers between 0 and 100
    for(i = 0; i<=100; i++) {
      if((i%2) != 0) continue; // iterate
      System.out.println(i);
    }
  }
}
```

Only even numbers are printed, because an odd one will cause the loop to iterate early, bypassing the call to **println( )**.

In **while** and **do-while** loops, a **continue** statement will cause control to go directly to the conditional expression and then continue the looping process. In the case of the **for**, the iteration expression of the loop is evaluated, then the conditional expression is executed, and then the loop continues.

As with the **break** statement, **continue** may specify a label to describe which enclosing loop to continue. Here is an example program that uses **continue** with a label:

```
// Use continue with a label.
class ContToLabel {
  public static void main(String[] args) {

outerloop:
    for(int i=1; i < 10; i++) {
      System.out.print("\nOuter loop pass " + i +
                        ", Inner loop: ");
      for(int j = 1; j < 10; j++) {
        if(j == 5) continue outerloop; // continue outer loop
        System.out.print(j);
      }
    }
  }
}
```

The output from the program is shown here:

```
Outer loop pass 1, Inner loop: 1234
Outer loop pass 2, Inner loop: 1234
Outer loop pass 3, Inner loop: 1234
Outer loop pass 4, Inner loop: 1234
Outer loop pass 5, Inner loop: 1234
Outer loop pass 6, Inner loop: 1234
Outer loop pass 7, Inner loop: 1234
Outer loop pass 8, Inner loop: 1234
Outer loop pass 9, Inner loop: 1234
```

As the output shows, when the **continue** executes, control passes to the outer loop, skipping the remainder of the inner loop.

Good uses of **continue** are rare. One reason is that Java provides a rich set of loop statements that fit most applications. However, for those special circumstances in which early iteration is needed, the **continue** statement provides a structured way to accomplish it.

Try This 3-3   Finish the Java Help System

Help3.java

This project puts the finishing touches on the Java help system that was created in the previous projects. This version adds the syntax for **break** and **continue**. It also allows the user to request the syntax for more than one statement. It does this by adding an outer loop that runs until the user enters **q** as a menu selection.

1. Copy **Help2.java** to a new file called **Help3.java**.

2. Surround all of the program code with an infinite **for** loop. Break out of this loop, using **break**, when a letter **q** is entered. Since this loop surrounds all of the program code, breaking out of this loop causes the program to terminate.

3. Change the menu loop as shown here:

```
do {
  System.out.println("Help on:");
  System.out.println("  1. if");
  System.out.println("  2. switch");
  System.out.println("  3. for");
  System.out.println("  4. while");
  System.out.println("  5. do-while");
  System.out.println("  6. break");
  System.out.println("  7. continue\n");
  System.out.print("Choose one (q to quit): ");

  choice = (char) System.in.read();
```

(continued)

```
do {
  ignore = (char) System.in.read();
} while(ignore != '\n');
} while( choice < '1' | choice > '7' & choice != 'q');
```

Notice that this loop now includes the **break** and **continue** statements. It also accepts the letter **q** as a valid choice.

4. Expand the **switch** statement to include the **break** and **continue** statements, as shown here:

```
case '6':
  System.out.println("The break:\n");
  System.out.println("break; or break label;");
  break;
case '7':
  System.out.println("The continue:\n");
  System.out.println("continue; or continue label;");
  break;
```

5. Here is the entire **Help3.java** program listing:

```
/*
   Try This 3-3

   The finished Java statement Help system
   that processes multiple requests.
*/
class Help3 {
  public static void main(String[] args)
    throws java.io.IOException {
    char choice, ignore;

    for(;;) {
      do {
        System.out.println("Help on:");
        System.out.println("  1. if");
        System.out.println("  2. switch");
        System.out.println("  3. for");
        System.out.println("  4. while");
        System.out.println("  5. do-while");
        System.out.println("  6. break");
        System.out.println("  7. continue\n");
        System.out.print("Choose one (q to quit): ");

        choice = (char) System.in.read();

        do {
          ignore = (char) System.in.read();
        } while(ignore != '\n');
```

```
  } while( choice < '1' | choice > '7' & choice != 'q');

  if(choice == 'q') break;

  System.out.println("\n");

  switch(choice) {
    case '1':
      System.out.println("The if:\n");
      System.out.println("if(condition) statement;");
      System.out.println("else statement;");
      break;
    case '2':
      System.out.println("The traditional switch:\n");
      System.out.println("switch(expression) {");
      System.out.println("  case constant:");
      System.out.println("    statement sequence");
      System.out.println("    break;");
      System.out.println("  // ...");
      System.out.println("}");
      break;
    case '3':
      System.out.println("The for:\n");
      System.out.print("for(init; condition; iteration)");
      System.out.println(" statement;");
      break;
    case '4':
      System.out.println("The while:\n");
      System.out.println("while(condition) statement;");
      break;
    case '5':
      System.out.println("The do-while:\n");
      System.out.println("do {");
      System.out.println("  statement;");
      System.out.println("} while (condition);");
      break;
    case '6':
      System.out.println("The break:\n");
      System.out.println("break; or break label;");
      break;
    case '7':
      System.out.println("The continue:\n");
      System.out.println("continue; or continue label;");
      break;
  }
  System.out.println();
```

(continued)

```
        }
      }
    }
```

6. Here is a sample run:

```
Help on:
  1. if
  2. switch
  3. for
  4. while
  5. do-while
  6. break
  7. continue

Choose one (q to quit): 1

The if:

if(condition) statement;
else statement;

Help on:
  1. if
  2. switch
  3. for
  4. while
  5. do-while
  6. break
  7. continue

Choose one (q to quit): 6

The break:

break; or break label;

Help on:
  1. if
  2. switch
  3. for
  4. while
  5. do-while
  6. break
  7. continue

Choose one (q to quit): q
```

Nested Loops

As you have seen in some of the preceding examples, one loop can be nested inside of another. Nested loops are used to solve a wide variety of programming problems and are an essential part of programming. So, before leaving the topic of Java's loop statements, let's look at one more nested loop example. The following program uses a nested **for** loop to find the factors of the numbers from 2 to 100:

```
/*
    Use nested loops to find factors of numbers
    between 2 and 100.
*/
class FindFac {
  public static void main(String[] args) {

    for(int i=2; i <= 100; i++) {
      System.out.print("Factors of " + i + ": ");
      for(int j = 2; j < i; j++)
        if((i%j) == 0) System.out.print(j + " ");
      System.out.println();
    }
  }
}
```

Here is a portion of the output produced by the program:

```
Factors of 2:
Factors of 3:
Factors of 4: 2
Factors of 5:
Factors of 6: 2 3
Factors of 7:
Factors of 8: 2 4
Factors of 9: 3
Factors of 10: 2 5
Factors of 11:
Factors of 12: 2 3 4 6
Factors of 13:
Factors of 14: 2 7
Factors of 15: 3 5
Factors of 16: 2 4 8
Factors of 17:
Factors of 18: 2 3 6 9
Factors of 19:
Factors of 20: 2 4 5 10
```

In the program, the outer loop runs **i** from 2 through 100. The inner loop successively tests all numbers from 2 up to **i**, printing those that evenly divide **i**. Extra challenge: The preceding program can be made more efficient. Can you see how? (Hint: The number of iterations in the inner loop can be reduced.)

Chapter 3 Self Test

1. Write a program that reads characters from the keyboard until a period is received. Have the program count the number of spaces. Report the total at the end of the program.

2. Show the general form of the **if-else-if** ladder.

3. Given

```
if(x < 10)
  if(y > 100) {
    if(!done) x = z;
    else y = z;
  }
  else System.out.println("error"); // what if?
```

to what **if** does the last **else** associate?

4. Show the **for** statement for a loop that counts from 1000 to 0 by –2.

5. Is the following fragment valid?

```
for(int i = 0; i < num; i++)
  sum += i;

count = i;
```

6. Explain what **break** does. Be sure to explain both of its forms.

7. In the following fragment, after the **break** statement executes, what is displayed?

```
for(i = 0; i < 10; i++) {
  while(running) {
    if(x<y) break;
    // ...
  }
  System.out.println("after while");
}
System.out.println("After for");
```

8. What does the following fragment print?

```
for(int i = 0; i<10; i++) {
  System.out.print(i + " ");
  if((i%2) == 0) continue;
  System.out.println();
}
```

9. The iteration expression in a **for** loop need not always alter the loop control variable by a fixed amount. Instead, the loop control variable can change in any arbitrary way. Using this concept, write a program that uses a **for** loop to generate and display the progression 1, 2, 4, 8, 16, 32, and so on.

10. The ASCII lowercase letters are separated from the uppercase letters by 32. Thus, to convert a lowercase letter to uppercase, subtract 32 from it. Use this information to write a program that reads characters from the keyboard. Have it convert all lowercase letters to uppercase, and all uppercase letters to lowercase, displaying the result. Make no changes to any other character. Have the program stop when the user enters a period. At the end, have the program display the number of case changes that have taken place.

11. What is an infinite loop?

12. When using **break** with a label, must the label be on a block that contains the **break**?

Chapter 4

Introducing Classes, Objects, and Methods

Key Skills & Concepts

- Know the fundamentals of the class
- Understand how objects are created
- Understand how reference variables are assigned
- Create methods, return values, and use parameters
- Use the **return** keyword
- Return a value from a method
- Add parameters to a method
- Utilize constructors
- Create parameterized constructors
- Understand **new**
- Understand garbage collection
- Use the **this** keyword

Before you can go much further in your study of Java, you need to learn about the class. The class is the essence of Java. It is the foundation upon which the entire Java language is built because the class defines the nature of an object. As such, the class forms the basis for object-oriented programming in Java. Within a class are defined data and code that acts upon that data. The code is contained in methods. Because classes, objects, and methods are fundamental to Java, they are introduced in this chapter. Having a basic understanding of these features will allow you to write more sophisticated programs and better understand certain key Java elements described in the following chapter.

Class Fundamentals

Since all Java program activity occurs within a class, we have been using classes since the start of this book. Of course, only extremely simple classes have been used, and we have not taken advantage of the majority of their features. As you will see, classes are substantially more powerful than the limited ones presented so far.

Let's begin by reviewing the basics. A class is a template that defines the form of an object. It specifies both the data and the code that will operate on that data. Java uses a class specification to construct *objects*. Objects are *instances* of a class. Thus, a class is essentially a set of plans that specify how to build an object. It is important to be clear on one issue: a class is a logical abstraction. It is not until an object of that class has been created that a physical representation of that class exists in memory.

One other point: Recall that the methods and variables that constitute a class are called *members* of the class. The data members are also referred to as *instance variables*.

The General Form of a Class

When you define a class, you declare its exact form and nature. You do this by specifying the instance variables that it contains and the methods that operate on them. Although very simple classes might contain only methods or only instance variables, most real-world classes contain both.

A class is created by using the keyword **class**. A simplified general form of a **class** definition is shown here:

```
class classname {
  // declare instance variables
  type var1;
  type var2;
  // ...
  type varN;

  // declare methods
  type method1(parameters) {
    // body of method
  }
  type method2(parameters) {
    // body of method
  }
  // ...
  type methodN(parameters) {
    // body of method
  }
}
```

Although there is no syntactic rule that enforces it, a well-designed class should define one and only one logical entity. For example, a class that stores names and telephone numbers will not normally also store information about the stock market, average rainfall, sunspot cycles, or other unrelated information. The point here is that a well-designed class groups logically connected information. Putting unrelated information into the same class will quickly destructure your code!

Up to this point, the classes that we have been using have had only one method: **main( )**. Soon you will see how to create others. However, notice that the general form of a class does not specify a **main( )** method. A **main( )** method is required only if that class is the starting point for your program. Also, some types of Java applications don't require a **main( )**.

Defining a Class

To illustrate classes, we will develop a class that encapsulates information about vehicles, such as cars, vans, and trucks. This class is called **Vehicle**, and it will store three items of information about a vehicle: the number of passengers that it can carry, its fuel capacity, and its average fuel consumption (in miles per gallon).

The first version of **Vehicle** is shown next. It defines three instance variables: **passengers**, **fuelcap**, and **mpg**. Notice that **Vehicle** does not contain any methods. Thus, it is currently a data-only class. (Subsequent sections will add methods to it.)

```
class Vehicle {
  int passengers; // number of passengers
  int fuelcap;    // fuel capacity in gallons
  int mpg;        // fuel consumption in miles per gallon
}
```

A **class** definition creates a new data type. In this case, the new data type is called **Vehicle**. You will use this name to declare objects of type **Vehicle**. Remember that a **class** declaration is only a type description; it does not create an actual object. Thus, the preceding code does not cause any objects of type **Vehicle** to come into existence.

To actually create a **Vehicle** object, you will use a statement like the following:

```
Vehicle minivan = new Vehicle(); // create a Vehicle object called minivan
```

After this statement executes, **minivan** refers to an instance of **Vehicle**. Thus, it will have "physical" reality. For the moment, don't worry about the details of this statement.

Each time you create an instance of a class, you are creating an object that contains its own copy of each instance variable defined by the class. Thus, every **Vehicle** object will contain its own copies of the instance variables **passengers**, **fuelcap**, and **mpg**. To access these variables, you will use the dot (.) operator. The *dot operator* links the name of an object with the name of a member. The general form of the dot operator is shown here:

object.member

Thus, the object is specified on the left, and the member is put on the right. For example, to assign the **fuelcap** variable of **minivan** the value 16, use the following statement:

```
minivan.fuelcap = 16;
```

In general, you can use the dot operator to access both instance variables and methods.

Here is a complete program that uses the **Vehicle** class:

```
// A program that uses the Vehicle class.

class Vehicle {
  int passengers; // number of passengers
  int fuelcap;    // fuel capacity in gallons
  int mpg;        // fuel consumption in miles per gallon
}

// This class declares an object of type Vehicle.
class VehicleDemo {
```

```
public static void main(String[] args) {
  Vehicle minivan = new Vehicle();
  int range;

  // assign values to fields in minivan
  minivan.passengers = 7;
  minivan.fuelcap = 16;  ←——————— Notice the use of the dot
  minivan.mpg = 21;                operator to access a member.

  // compute the range assuming a full tank of gas
  range = minivan.fuelcap * minivan.mpg;
  System.out.println("Minivan can carry " + minivan.passengers +
                    " with a range of " + range);
  }
}
```

To try this program, you can put both the **Vehicle** and **VehicleDemo** classes in the same source file. For example, you could call the file that contains this program **VehicleDemo.java**. This name makes sense because the **main( )** method is in the class called **VehicleDemo**, not the class called **Vehicle**. Either class can be the first one in the file. When you compile this program using **javac**, you will find that two **.class** files have been created, one for **Vehicle** and one for **VehicleDemo**. The Java compiler automatically puts each class into its own **.class** file. It is important to understand that it is not necessary for both the **Vehicle** and the **VehicleDemo** class to be in the same source file. You could put each class in its own file, called **Vehicle.java** and **VehicleDemo.java**, respectively. If you do this, you can still compile the program by compiling **VehicleDemo.java**.

To run this program, you must execute **VehicleDemo.class**. The following output is displayed:

```
Minivan can carry 7 with a range of 336
```

Before moving on, let's review a fundamental principle: each object has its own copies of the instance variables defined by its class. Thus, the contents of the variables in one object can differ from the contents of the variables in another. There is no connection between the two objects except for the fact that they are both objects of the same type. For example, if you have two **Vehicle** objects, each has its own copy of **passengers**, **fuelcap**, and **mpg**, and the contents of these can differ between the two objects. The following program demonstrates this fact. (Notice that the class with **main( )** is now called **TwoVehicles**.)

```
// This program creates two Vehicle objects.

class Vehicle {
  int passengers; // number of passengers
  int fuelcap;    // fuel capacity in gallons
  int mpg;        // fuel consumption in miles per gallon
}

// This class declares an object of type Vehicle.
class TwoVehicles {
  public static void main(String[] args) {
```

```
Vehicle minivan = new Vehicle();
Vehicle sportscar = new Vehicle();

int range1, range2;

// assign values to fields in minivan
minivan.passengers = 7;
minivan.fuelcap = 16;
minivan.mpg = 21;

// assign values to fields in sportscar
sportscar.passengers = 2;
sportscar.fuelcap = 14;
sportscar.mpg = 12;

// compute the ranges assuming a full tank of gas
range1 = minivan.fuelcap * minivan.mpg;
range2 = sportscar.fuelcap * sportscar.mpg;

System.out.println("Minivan can carry " + minivan.passengers +
                   " with a range of " + range1);

System.out.println("Sportscar can carry " + sportscar.passengers +
                   " with a range of " + range2);
  }
}
```

Remember, **minivan** and **sportscar** refer to separate objects.

The output produced by this program is shown here:

```
Minivan can carry 7 with a range of 336
Sportscar can carry 2 with a range of 168
```

As you can see, **minivan**'s data is completely separate from the data contained in **sportscar**. The following illustration depicts this situation.

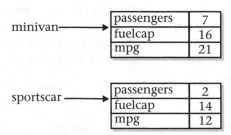

How Objects Are Created

In the preceding programs, the following line was used to declare an object of type **Vehicle**:

```
Vehicle minivan = new Vehicle();
```

This declaration performs two functions. First, it declares a variable called **minivan** of the class type **Vehicle**. This variable does not define an object. Instead, it is simply a variable that can *refer to* an object. Second, the declaration creates an instance of the object and assigns to **minivan** a reference to that object. This is done by using the **new** operator.

The **new** operator dynamically allocates (that is, allocates at run time) memory for an object and returns a reference to it. This reference is, essentially, the address in memory of the object allocated by **new**. This reference is then stored in a variable. Thus, in Java, all class objects must be dynamically allocated.

The two steps combined in the preceding statement can be rewritten like this to show each step individually:

```
Vehicle minivan; // declare reference to object
minivan = new Vehicle(); // allocate a Vehicle object
```

The first line declares **minivan** as a reference to an object of type **Vehicle**. Thus, **minivan** is a variable that can refer to an object, but it is not an object itself. At this point, **minivan** does not refer to an object. The next line creates a new **Vehicle** object and assigns a reference to it to **minivan**. Now, **minivan** is linked with an object.

Reference Variables and Assignment

In an assignment operation, object reference variables act differently than do variables of a primitive type, such as **int**. When you assign one primitive-type variable to another, the situation is straightforward. The variable on the left receives a *copy* of the *value* of the variable on the right. When you assign one object reference variable to another, the situation is a bit more complicated because you are changing the object that the reference variable refers to. The effect of this difference can cause some counterintuitive results. For example, consider the following fragment:

```
Vehicle car1 = new Vehicle();
Vehicle car2 = car1;
```

At first glance, it is easy to think that **car1** and **car2** refer to different objects, but this is not the case. Instead, **car1** and **car2** will both refer to the same object. The assignment of **car1** to **car2** simply makes **car2** refer to the same object as does **car1**. Thus, the object can be acted upon by either **car1** or **car2**. For example, after the assignment

```
car1.mpg = 26;
```

executes, both of these **println( )** statements

```
System.out.println(car1.mpg);
System.out.println(car2.mpg);
```

display the same value: 26.

Although **car1** and **car2** both refer to the same object, they are not linked in any other way. For example, a subsequent assignment to **car2** simply changes the object to which **car2** refers. For example:

```
Vehicle car1 = new Vehicle();
Vehicle car2 = car1;
Vehicle car3 = new Vehicle();

car2 = car3; // now car2 and car3 refer to the same object.
```

After this sequence executes, **car2** refers to the same object as **car3**. The object referred to by **car1** is unchanged.

Methods

As explained, instance variables and methods are constituents of classes. So far, the **Vehicle** class contains data, but no methods. Although data-only classes are perfectly valid, most classes will have methods. Methods are subroutines that manipulate the data defined by the class and, in many cases, provide access to that data. In most cases, other parts of your program will interact with a class through its methods.

A method contains one or more statements. In well-written Java code, each method performs only one task. Each method has a name, and it is this name that is used to call the method. In general, you can give a method whatever name you please. However, remember that **main( )** is reserved for the method that begins execution of your program. Also, don't use Java's keywords for method names.

When denoting methods in text, this book has used and will continue to use a convention that has become common when writing about Java. A method will have parentheses after its name. For example, if a method's name is **getVal**, it will be written **getVal( )** when its name is used in a sentence. This notation will help you distinguish variable names from method names in this book.

The general form of a method is shown here:

ret-type name(*parameter-list*) {
  // body of method
}

Here, *ret-type* specifies the type of data returned by the method. This can be any valid type, including class types that you create. If the method does not return a value, its return type must be **void**. The name of the method is specified by *name*. This can be any legal identifier other than those already used by other items within the current scope. The *parameter-list* is a sequence of type and identifier pairs separated by commas. Parameters are essentially variables that receive the value of the *arguments* passed to the method when it is called. If the method has no parameters, the parameter list will be empty.

Adding a Method to the Vehicle Class

As just explained, the methods of a class typically manipulate and provide access to the data of the class. With this in mind, recall that **main( )** in the preceding examples computed the range of a vehicle by multiplying its fuel consumption rate by its fuel capacity. While technically correct, this is not the best way to handle this computation. The calculation of a vehicle's range is something that is best handled by the **Vehicle** class itself. The reason for this conclusion is easy to understand: the range of a vehicle is dependent upon the capacity of the fuel tank and the rate of fuel consumption, and both of these quantities are encapsulated by **Vehicle**. By adding a method to **Vehicle** that computes the range, you are enhancing its object-oriented structure. To add a method to **Vehicle**, specify it within **Vehicle**'s declaration. For example, the following version of **Vehicle** contains a method called **range( )** that displays the range of the vehicle.

```
// Add range to Vehicle.

class Vehicle {
  int passengers; // number of passengers
  int fuelcap;     // fuel capacity in gallons
  int mpg;         // fuel consumption in miles per gallon

  // Display the range.
  void range() {  ◄——————— The range( ) method is contained within the Vehicle class.
    System.out.println("Range is " + fuelcap * mpg);
  }
}
                              ▲              ▲
                              └──────┬───────┘
                 Notice that fuelcap and mpg are used directly, without the dot operator.
class AddMeth {
  public static void main(String[] args) {
    Vehicle minivan = new Vehicle();
    Vehicle sportscar = new Vehicle();

    int range1, range2;

    // assign values to fields in minivan
    minivan.passengers = 7;
    minivan.fuelcap = 16;
    minivan.mpg = 21;

    // assign values to fields in sportscar
    sportscar.passengers = 2;
    sportscar.fuelcap = 14;
    sportscar.mpg = 12;


    System.out.print("Minivan can carry " + minivan.passengers +
                     ". ");

    minivan.range(); // display range of minivan

    System.out.print("Sportscar can carry " + sportscar.passengers +
                     ". ");

    sportscar.range(); // display range of sportscar.
  }
}
```

This program generates the following output:

```
Minivan can carry 7. Range is 336
Sportscar can carry 2. Range is 168
```

Let's look at the key elements of this program, beginning with the **range( )** method itself. The first line of **range( )** is

```
void range() {
```

This line declares a method called **range** that has no parameters. Its return type is **void**. Thus, **range( )** does not return a value to the caller. The line ends with the opening curly brace of the method body.

The body of **range( )** consists solely of this line:

```
System.out.println("Range is " + fuelcap * mpg);
```

This statement displays the range of the vehicle by multiplying **fuelcap** by **mpg**. Since each object of type **Vehicle** has its own copy of **fuelcap** and **mpg**, when **range( )** is called, the range computation uses the calling object's copies of those variables.

The **range( )** method ends when its closing curly brace is encountered. This causes program control to transfer back to the caller.

Next, look closely at this line of code from inside **main( )**:

```
minivan.range();
```

This statement invokes the **range( )** method on **minivan**. That is, it calls **range( )** relative to the **minivan** object, using the object's name followed by the dot operator. When a method is called, program control is transferred to the method. When the method terminates, control is transferred back to the caller, and execution resumes with the line of code following the call.

In this case, the call to **minivan.range( )** displays the range of the vehicle defined by **minivan**. In similar fashion, the call to **sportscar.range( )** displays the range of the vehicle defined by **sportscar**. Each time **range( )** is invoked, it displays the range for the specified object.

There is something very important to notice inside the **range( )** method: the instance variables **fuelcap** and **mpg** are referred to directly, without preceding them with an object name or the dot operator. When a method uses an instance variable that is defined by its class, it does so directly, without explicit reference to an object and without use of the dot operator. This is easy to understand if you think about it. A method is always invoked relative to some object of its class. Once this invocation has occurred, the object is known. Thus, within a method, there is no need to specify the object a second time. This means that **fuelcap** and **mpg** inside **range( )** implicitly refer to the copies of those variables found in the object that invokes **range( )**.

Returning from a Method

In general, there are two conditions that cause a method to return—first, as the **range( )** method in the preceding example shows, when the method's closing curly brace is encountered. The second is when a **return** statement is executed. There are two forms of **return**—one for use in

void methods (those that do not return a value) and one for returning values. The first form is examined here. The next section explains how to return values.

In a **void** method, you can cause the immediate termination of a method by using this form of **return**:

```
return ;
```

When this statement executes, program control returns to the caller, skipping any remaining code in the method. For example, consider this method:

```
void myMeth() {
  int i;

  for(i=0; i<10; i++) {
    if(i == 5) return; // stop at 5
    System.out.println();
  }
}
```

Here, the **for** loop will only run from 0 to 5, because once i equals 5, the method returns. It is permissible to have multiple **return** statements in a method, especially when there are two or more routes out of it. For example:

```
void myMeth() {
  // ...
  if(done) return;
  // ...
  if(error) return;
  // ...
}
```

Here, the method returns if it is done or if an error occurs. Be careful, however, because having too many exit points in a method can destructure your code; so avoid using them casually. A well-designed method has well-defined exit points.

To review: A **void** method can return in one of two ways—its closing curly brace is reached, or a **return** statement is executed.

Returning a Value

Although methods with a return type of **void** are not rare, most methods will return a value. In fact, the ability to return a value is one of the most useful features of a method. You have already seen one example of a return value: when we used the **sqrt( )** function to obtain a square root.

Return values are used for a variety of purposes in programming. In some cases, such as with **sqrt( )**, the return value contains the outcome of some calculation. In other cases, the return value may simply indicate success or failure. In still others, it may contain a status code. Whatever the purpose, using method return values is an integral part of Java programming.

Methods return a value to the calling routine using this form of **return**:

return *value*;

Here, *value* is the value returned. This form of **return** can be used only with methods that have a non-**void** return type. Furthermore, a non-**void** method *must* return a value by using this form of **return**.

You can use a return value to improve the implementation of **range( )**. Instead of displaying the range, a better approach is to have **range( )** compute the range and return this value. Among the advantages to this approach is that you can use the value for other calculations. The following example modifies **range( )** to return the range rather than displaying it.

```
// Use a return value.

class Vehicle {
  int passengers; // number of passengers
  int fuelcap;    // fuel capacity in gallons
  int mpg;        // fuel consumption in miles per gallon

  // Return the range.
  int range() {
    return mpg * fuelcap;   ←——————— Return the range for a given vehicle.
  }
}

class RetMeth {
  public static void main(String[] args) {
    Vehicle minivan = new Vehicle();
    Vehicle sportscar = new Vehicle();

    int range1, range2;

    // assign values to fields in minivan
    minivan.passengers = 7;
    minivan.fuelcap = 16;
    minivan.mpg = 21;

    // assign values to fields in sportscar
    sportscar.passengers = 2;
    sportscar.fuelcap = 14;
    sportscar.mpg = 12;

    // get the ranges
    range1 = minivan.range();
    range2 = sportscar.range();   ———— Assign the value returned to a variable.
```

```
System.out.println("Minivan can carry " + minivan.passengers +
                   " with range of " + range1 + " Miles");


System.out.println("Sportscar can carry " + sportscar.passengers +
                   " with range of " + range2 + " miles");

    }
}
```

The output is shown here:

```
Minivan can carry 7 with range of 336 Miles
Sportscar can carry 2 with range of 168 miles
```

In the program, notice that when **range( )** is called, it is put on the right side of an assignment statement. On the left is a variable that will receive the value returned by **range( )**. Thus, after

```
range1 = minivan.range();
```

executes, the range of the **minivan** object is stored in **range1**.

Notice that **range( )** now has a return type of **int**. This means that it will return an integer value to the caller. The return type of a method is important because the type of data returned by a method must be compatible with the return type specified by the method. Thus, if you want a method to return data of type **double**, its return type must be type **double**.

Although the preceding program is correct, it is not written as efficiently as it could be. Specifically, there is no need for the **range1** or **range2** variables. A call to **range( )** can be used in the **println( )** statement directly, as shown here:

```
System.out.println("Minivan can carry " + minivan.passengers +
                   " with range of " + minivan.range() + " Miles");
```

In this case, when **println( )** is executed, **minivan.range( )** is called automatically and its value will be passed to **println( )**. Furthermore, you can use a call to **range( )** whenever the range of a **Vehicle** object is needed. For example, this statement compares the ranges of two vehicles:

```
if(v1.range() > v2.range()) System.out.println("v1 has greater range");
```

Using Parameters

It is possible to pass one or more values to a method when the method is called. Recall that a value passed to a method is called an *argument*. Inside the method, the variable that receives the argument is called a *parameter*. Parameters are declared inside the parentheses that follow the method's name. The parameter declaration syntax is the same as that used for variables. A parameter is within the scope of its method, and aside from its special task of receiving an argument, it acts like any other local variable.

Here is a simple example that uses a parameter. Inside the **ChkNum** class, the method **isEven( )** returns **true** if the value that it is passed is even. It returns **false** otherwise. Therefore, **isEven( )** has a return type of **boolean**.

```java
// A simple example that uses a parameter.

class ChkNum {
  // return true if x is even
  boolean isEven(int x) {          Here, x is an integer parameter of isEven( ).
    if((x%2) == 0) return true;
    else return false;
  }
}

class ParmDemo {
  public static void main(String[] args) {
    ChkNum e = new ChkNum();
                                                        Pass arguments
    if(e.isEven(10)) System.out.println("10 is even.");  to isEven( ).

    if(e.isEven(9)) System.out.println("9 is even.");

    if(e.isEven(8)) System.out.println("8 is even.");

  }
}
```

Here is the output produced by the program:

```
10 is even.
8 is even.
```

In the program, **isEven( )** is called three times, and each time a different value is passed. Let's look at this process closely. First, notice how **isEven( )** is called. The argument is specified between the parentheses. When **isEven( )** is called the first time, it is passed the value 10. Thus, when **isEven( )** begins executing, the parameter **x** receives the value 10. In the second call, 9 is the argument, and **x**, then, has the value 9. In the third call, the argument is 8, which is the value that **x** receives. The point is that the value passed as an argument when **isEven( )** is called is the value received by its parameter, **x**.

A method can have more than one parameter. Simply declare each parameter, separating one from the next with a comma. For example, the **Factor** class defines a method called **isFactor( )** that determines whether the first parameter is a factor of the second.

```java
class Factor {
  boolean isFactor(int a, int b) {          This method has two parameters.
    if( (b % a) == 0) return true;
    else return false;
```

```
    }
}
class IsFact {
  public static void main(String[] args) {
    Factor x = new Factor();

    if(x.isFactor(2, 20)) System.out.println("2 is factor");
    if(x.isFactor(3, 20)) System.out.println("this won't be displayed");

  }
}
```

Pass two arguments
to **isFactor( )**.

Notice that when **isFactor( )** is called, the arguments are also separated by commas.

When using multiple parameters, each parameter specifies its own type, which can differ from the others. For example, this is perfectly valid:

```
int myMeth(int a, double b, float c) {
// ...
```

Adding a Parameterized Method to Vehicle

You can use a parameterized method to add a new feature to the **Vehicle** class: the ability to compute the amount of fuel needed for a given distance. This new method is called **fuelNeeded( )**. This method takes the number of miles that you want to drive and returns the number of gallons of gas required. The **fuelNeeded( )** method is defined like this:

```
double fuelNeeded(int miles) {
  return (double) miles / mpg;
}
```

Notice that this method returns a value of type **double**. This is useful since the amount of fuel needed for a given distance might not be a whole number. The entire **Vehicle** class that includes **fuelNeeded( )** is shown here:

```
/*
   Add a parameterized method that computes the
   fuel required for a given distance.
*/

class Vehicle {
  int passengers; // number of passengers
  int fuelcap;    // fuel capacity in gallons
  int mpg;        // fuel consumption in miles per gallon
```

```
  // Return the range.
  int range() {
    return mpg * fuelcap;
  }

  // Compute fuel needed for a given distance.
  double fuelNeeded(int miles) {
    return (double) miles / mpg;
  }
}

class CompFuel {
  public static void main(String[] args) {
    Vehicle minivan = new Vehicle();
    Vehicle sportscar = new Vehicle();
    double gallons;
    int dist = 252;

    // assign values to fields in minivan
    minivan.passengers = 7;
    minivan.fuelcap = 16;
    minivan.mpg = 21;

    // assign values to fields in sportscar
    sportscar.passengers = 2;
    sportscar.fuelcap = 14;
    sportscar.mpg = 12;

    gallons = minivan.fuelNeeded(dist);

    System.out.println("To go " + dist + " miles minivan needs " +
                        gallons + " gallons of fuel.");

    gallons = sportscar.fuelNeeded(dist);

    System.out.println("To go " + dist + " miles sportscar needs " +
                        gallons + " gallons of fuel.");

  }
}
```

The output from the program is shown here:

```
To go 252 miles minivan needs 12.0 gallons of fuel.
To go 252 miles sportscar needs 21.0 gallons of fuel.
```

Try This 4-1   Creating a Help Class

`HelpClassDemo.java`

If one were to try to summarize the essence of the class in one sentence, it might be this: a class encapsulates functionality. Of course, sometimes the trick is knowing where one "functionality" ends and another begins. As a general rule, you will want your classes to be the building blocks of your larger application. In order to do this, each class must represent a single functional unit that performs clearly delineated actions. Thus, you will want your classes to be as small as possible—but no smaller! That is, classes that contain extraneous functionality confuse and destructure code, but classes that contain too little functionality are fragmented. What is the balance? It is at this point that the science of programming becomes the *art* of programming. Fortunately, most programmers find that this balancing act becomes easier with experience.

To begin to gain that experience you will convert the help system from Try This 3-3 in the preceding chapter into a Help class. Let's examine why this is a good idea. First, the help system defines one logical unit. It simply displays the syntax for Java's control statements. Thus, its functionality is compact and well defined. Second, putting help in a class is an esthetically pleasing approach. Whenever you want to offer the help system to a user, simply instantiate a help-system object. Finally, because help is encapsulated, it can be upgraded or changed without causing unwanted side effects in the programs that use it.

1. Create a new file called **HelpClassDemo.java**. To save you some typing, you might want to copy the file from Try This 3-3, **Help3.java**, into **HelpClassDemo.java**.

2. To convert the help system into a class, you must first determine precisely what constitutes the help system. For example, in **Help3.java**, there is code to display a menu, input the user's choice, check for a valid response, and display information about the item selected. The program also loops until the letter q is pressed. If you think about it, it is clear that the menu, the check for a valid response, and the display of the information are integral to the help system. How user input is obtained, and whether repeated requests should be processed, are not. Thus, you will create a class that displays the help information, the help menu, and checks for a valid selection. Its methods will be called **helpOn( )**, **showMenu( )**, and **isValid( )**, respectively.

3. Create the **helpOn( )** method as shown here:

```java
void helpOn(int what) {
  switch(what) {
    case '1':
      System.out.println("The if:\n");
      System.out.println("if(condition) statement;");
      System.out.println("else statement;");
      break;
    case '2':
      System.out.println("The traditional switch:\n");
      System.out.println("switch(expression) {");
```

(continued)

```
          System.out.println("    case constant:");
          System.out.println("      statement sequence");
          System.out.println("      break;");
          System.out.println("    // ...");
          System.out.println("}");
          break;
        case '3':
          System.out.println("The for:\n");
          System.out.print("for(init; condition; iteration)");
          System.out.println(" statement;");
          break;
        case '4':
          System.out.println("The while:\n");
          System.out.println("while(condition) statement;");
          break;
        case '5':
          System.out.println("The do-while:\n");
          System.out.println("do {");
          System.out.println("  statement;");
          System.out.println("} while (condition);");
          break;
        case '6':
          System.out.println("The break:\n");
          System.out.println("break; or break label;");
          break;
        case '7':
          System.out.println("The continue:\n");
          System.out.println("continue; or continue label;");
          break;
      }
      System.out.println();
  }
```

4. Next, create the **showMenu( )** method:

```
void showMenu() {
  System.out.println("Help on:");
  System.out.println("  1. if");
  System.out.println("  2. switch");
  System.out.println("  3. for");
  System.out.println("  4. while");
  System.out.println("  5. do-while");
  System.out.println("  6. break");
  System.out.println("  7. continue\n");
  System.out.print("Choose one (q to quit): ");
}
```

5. Create the **isValid( )** method, shown here:

```
boolean isValid(int ch) {
  if(ch < '1' | ch > '7' & ch != 'q') return false;
  else return true;
}
```

6. Assemble the foregoing methods into the **Help** class, shown here:

```
class Help {
  void helpOn(int what) {
    switch(what) {
      case '1':
        System.out.println("The if:\n");
        System.out.println("if(condition) statement,");
        System.out.println("else statement;");
        break;
      case '2':
        System.out.println("The traditional switch:\n");
        System.out.println("switch(expression) {");
        System.out.println("  case constant:");
        System.out.println("    statement sequence");
        System.out.println("    break;");
        System.out.println("  // ...");
        System.out.println("}");
        break;
      case '3':
        System.out.println("The for:\n");
        System.out.print("for(init; condition; iteration)");
        System.out.println(" statement;");
        break;
      case '4':
        System.out.println("The while:\n");
        System.out.println("while(condition) statement;");
        break;
      case '5':
        System.out.println("The do-while:\n");
        System.out.println("do {");
        System.out.println("  statement;");
        System.out.println("} while (condition);");
        break;
      case '6':
        System.out.println("The break:\n");
        System.out.println("break; or break label;");
        break;
```

(continued)

```
          case '7':
            System.out.println("The continue:\n");
            System.out.println("continue; or continue label;");
            break;
        }
        System.out.println();
    }

    void showMenu() {
      System.out.println("Help on:");
      System.out.println("  1. if");
      System.out.println("  2. switch");
      System.out.println("  3. for");
      System.out.println("  4. while");
      System.out.println("  5. do-while");
      System.out.println("  6. break");
      System.out.println("  7. continue\n");
      System.out.print("Choose one (q to quit): ");
    }

    boolean isValid(int ch) {
      if(ch < '1' | ch > '7' & ch != 'q') return false;
      else return true;
    }

}
```

7. Finally, rewrite the **main( )** method from Try This 3-3 so that it uses the new **Help** class. Call this class **HelpClassDemo**. The entire listing for **HelpClassDemo.java** is shown here:

```
/*
    Try This 4-1

    Convert the help system from Try This 3-3 into
    a Help class.
*/

class Help {
  void helpOn(int what) {
    switch(what) {
      case '1':
        System.out.println("The if:\n");
        System.out.println("if(condition) statement;");
        System.out.println("else statement;");
        break;
```

```
      case '2':
        System.out.println("The traditional switch:\n");
        System.out.println("switch(expression) {");
        System.out.println("  case constant:");
        System.out.println("    statement sequence");
        System.out.println("    break;");
        System.out.println("  // ...");
        System.out.println("}");
        break;
      case '3':
        System.out.println("The for:\n");
        System.out.print("for(init; condition; iteration)");
        System.out.println(" statement;");
        break;
      case '4':
        System.out.println("The while:\n");
        System.out.println("while(condition) statement;");
        break;
      case '5':
        System.out.println("The do-while:\n");
        System.out.println("do {");
        System.out.println("  statement;");
        System.out.println("} while (condition);");
        break;
      case '6':
        System.out.println("The break:\n");
        System.out.println("break; or break label;");
        break;
      case '7':
        System.out.println("The continue:\n");
        System.out.println("continue; or continue label;");
        break;
    }
    System.out.println();
  }

  void showMenu() {
    System.out.println("Help on:");
    System.out.println("  1. if");
    System.out.println("  2. switch");
    System.out.println("  3. for");
    System.out.println("  4. while");
    System.out.println("  5. do-while");
    System.out.println("  6. break");
    System.out.println("  7. continue\n");
    System.out.print("Choose one (q to quit): ");
  }
```

(continued)

```java
      boolean isValid(int ch) {
        if(ch < '1' | ch > '7' & ch != 'q') return false;
        else return true;
      }

}

class HelpClassDemo {
  public static void main(String[] args)
    throws java.io.IOException {
    char choice, ignore;
    Help hlpobj = new Help();

    for(;;) {
      do {
        hlpobj.showMenu();

        choice = (char) System.in.read();

        do {
           ignore = (char) System.in.read();
         } while(ignore != '\n');

      } while( !hlpobj.isValid(choice) );

      if(choice == 'q') break;

      System.out.println("\n");

      hlpobj.helpOn(choice);
    }
  }
}
```

When you try the program, you will find that it is functionally the same as before. The advantage to this approach is that you now have a help system component that can be reused whenever it is needed.

Constructors

In the preceding examples, the instance variables of each **Vehicle** object had to be set manually using a sequence of statements, such as:

```java
minivan.passengers = 7;
minivan.fuelcap = 16;
minivan.mpg = 21;
```

An approach like this would never be used in professionally written Java code. Aside from being error prone (you might forget to set one of the fields), there is simply a better way to accomplish this task: the constructor.

A *constructor* initializes an object when it is created. It has the same name as its class and is syntactically similar to a method. However, constructors have no explicit return type. Typically, you will use a constructor to give initial values to the instance variables defined by the class, or to perform any other startup procedures required to create a fully formed object.

All classes have constructors, whether you define one or not, because Java automatically provides a default constructor. In this case, non-initialized member variables have their default values, which are zero, **null**, and **false**, for numeric types, reference types, and **boolean**s, respectively. Once you define your own constructor, the default constructor is no longer used.

Here is a simple example that uses a constructor:

```java
// A simple constructor.

class MyClass {
  int x;

  MyClass() {  ◄─────────── This is the constructor for MyClass.
    x = 10;
  }
}

class ConsDemo {
  public static void main(String[] args) {
    MyClass t1 = new MyClass();
    MyClass t2 = new MyClass();

    System.out.println(t1.x + " " + t2.x);
  }
}
```

In this example, the constructor for **MyClass** is

```java
MyClass() {
  x = 10;
}
```

This constructor assigns the instance variable **x** of **MyClass** the value 10. This constructor is called by **new** when an object is created. For example, in the line

```java
MyClass t1 = new MyClass();
```

the constructor **MyClass( )** is called on the **t1** object, giving **t1.x** the value 10. The same is true for **t2**. After construction, **t2.x** has the value 10. Thus, the output from the program is

```
10 10
```

Parameterized Constructors

In the preceding example, a parameter-less constructor was used. Although this is fine for some situations, most often you will need a constructor that accepts one or more parameters. Parameters are added to a constructor in the same way that they are added to a method: just declare them inside the parentheses after the constructor's name. For example, here, **MyClass** is given a parameterized constructor:

```java
// A parameterized constructor.

class MyClass {
  int x;

  MyClass(int i) {          ◄──────────── This constructor has a parameter.
    x = i;
  }
}

class ParmConsDemo {
  public static void main(String[] args) {
    MyClass t1 = new MyClass(10);
    MyClass t2 = new MyClass(88);

    System.out.println(t1.x + " " + t2.x);
  }
}
```

The output from this program is shown here:

```
10 88
```

In this version of the program, the **MyClass( )** constructor defines one parameter called **i**, which is used to initialize the instance variable, **x**. Thus, when the line

```java
MyClass t1 = new MyClass(10);
```

executes, the value 10 is passed to **i**, which is then assigned to **x**.

Adding a Constructor to the Vehicle Class

We can improve the **Vehicle** class by adding a constructor that automatically initializes the **passengers**, **fuelcap**, and **mpg** fields when an object is constructed. Pay special attention to how **Vehicle** objects are created.

```java
// Add a constructor.

class Vehicle {
  int passengers; // number of passengers
  int fuelcap;    // fuel capacity in gallons
  int mpg;        // fuel consumption in miles per gallon
```

```
   // This is a constructor for Vehicle.
   Vehicle(int p, int f, int m) {  ◄──────────── Constructor for Vehicle
     passengers = p;
     fuelcap = f;
     mpg = m;
   }

   // Return the range.
   int range() {
     return mpg * fuelcap;
   }

   // Compute fuel needed for a given distance.
   double fuelNeeded(int miles) {
     return (double) miles / mpg;
   }
}

class VehConsDemo {
  public static void main(String[] args) {

    // construct complete vehicles
    Vehicle minivan = new Vehicle(7, 16, 21);
    Vehicle sportscar = new Vehicle(2, 14, 12);
    double gallons;
    int dist = 252;

    gallons = minivan.fuelNeeded(dist);

    System.out.println("To go " + dist + " miles minivan needs " +
                       gallons + " gallons of fuel.");

    gallons = sportscar.fuelNeeded(dist);

    System.out.println("To go " + dist + " miles sportscar needs " +
                       gallons + " gallons of fuel.");

  }
}
```

Both **minivan** and **sportscar** are initialized by the **Vehicle( )** constructor when they are created. Each object is initialized as specified in the parameters to its constructor. For example, in the following line,

```
Vehicle minivan = new Vehicle(7, 16, 21);
```

the values 7, 16, and 21 are passed to the **Vehicle( )** constructor when **new** creates the object. Thus, **minivan**'s copy of **passengers**, **fuelcap**, and **mpg** will contain the values 7, 16, and 21, respectively. The output from this program is the same as the previous version.

The new Operator Revisited

Now that you know more about classes and their constructors, let's take a closer look at the **new** operator. In the context of an assignment, the **new** operator has this general form:

class-var = new *class-name*(*arg-list*);

Here, *class-var* is a variable of the class type being created. The *class-name* is the name of the class that is being instantiated. The class name followed by a parenthesized argument list (which can be empty) specifies the constructor for the class. If a class does not define its own constructor, **new** will use the default constructor supplied by Java. Thus, **new** can be used to create an object of any class type. The **new** operator returns a reference to the newly created object, which (in this case) is assigned to *class-var*.

Since memory is finite, it is possible that **new** will not be able to allocate memory for an object because insufficient memory exists. If this happens, a run-time exception will occur. (You will learn about exceptions in Chapter 9.) For the sample programs in this book, you won't need to worry about running out of memory, but you will need to consider this possibility in real-world programs that you write.

Garbage Collection

As you have seen, objects are dynamically allocated from a pool of free memory by using the **new** operator. As explained, memory is not infinite, and the free memory can be exhausted. Thus, it is possible for **new** to fail because there is insufficient free memory to create the desired object. For this reason, a key component of any dynamic allocation scheme is the recovery of free memory from unused objects, making that memory available for subsequent reallocation. In some programming languages, the release of previously allocated memory is handled manually. However, Java uses a different, more trouble-free approach: *garbage collection.*

Java's garbage collection system reclaims objects automatically—occurring transparently, behind the scenes, without any programmer intervention. It works like this: When no references to an object exist, that object is assumed to be no longer needed, and the memory occupied by the object is released. This recycled memory can then be used for a subsequent allocation.

Ask the Expert

Q: Why don't I need to use new **for variables of the primitive types, such as** int **or** float?

A: Java's primitive types are not implemented as objects. Rather, because of efficiency concerns, they are implemented as "normal" variables. A variable of a primitive type actually contains the value that you have given it. As explained, object variables are references to the object. This layer of indirection (and other object features) adds overhead to an object that is avoided by a primitive type.

Garbage collection occurs only sporadically during the execution of your program. It will not occur simply because one or more objects exist that are no longer used. For efficiency, the garbage collector will usually run only when two conditions are met: there are objects to recycle, and there is a reason to recycle them. Remember, garbage collection takes time, so the Java run-time system does it only when it is appropriate. Thus, you can't know precisely when garbage collection will take place.

The this Keyword

Before concluding this chapter, it is necessary to introduce **this**. When a method is called, it is automatically passed an implicit argument that is a reference to the invoking object (that is, the object on which the method is called). This reference is called **this**. To understand **this**, first consider a program that creates a class called **Pwr** that computes the result of a number raised to some integer power:

```
class Pwr {
  double b;
  int e;
  double val;

  Pwr(double base, int exp) {
    b = base;
    e = exp;

    val = 1;
    if(exp==0) return;
    for( ; exp>0; exp--) val = val * base;
  }

  double getVal() {
    return val;
  }
}

class DemoPwr {
  public static void main(String[] args) {
    Pwr x = new Pwr(4.0, 2);
    Pwr y = new Pwr(2.5, 1);
    Pwr z = new Pwr(5.7, 0);

    System.out.println(x.b + " raised to the " + x.e +
                       " power is " + x.getVal());
    System.out.println(y.b + " raised to the " + y.e +
                       " power is " + y.getVal());
    System.out.println(z.b + " raised to the " + z.e +
                       " power is " + z.getVal());
  }
}
```

As you know, within a method, the other members of a class can be accessed directly, without any object or class qualification. Thus, inside **getVal( )**, the statement

```
return val;
```

means that the copy of **val** associated with the invoking object will be returned. However, the same statement can also be written like this:

```
return this.val;
```

Here, **this** refers to the object on which **getVal( )** was called. Thus, **this.val** refers to that object's copy of **val**. For example, if **getVal( )** had been invoked on **x**, then **this** in the preceding statement would have been referring to **x**. Writing the statement without using **this** is really just shorthand.

Here is the entire **Pwr** class written using the **this** reference:

```
class Pwr {
  double b;
  int e;
  double val;

  Pwr(double base, int exp) {
    this.b = base;
    this.e = exp;

    this.val = 1;
    if(exp==0) return;
    for( ; exp>0; exp--) this.val = this.val * base;
  }

  double getVal() {
    return this.val;
  }
}
```

Actually, no Java programmer would write **Pwr** as just shown because nothing is gained, and the standard form is easier. However, **this** has some important uses. For example, the Java syntax permits the name of a parameter or a local variable to be the same as the name of an instance variable. When this happens, the local name *hides* the instance variable. You can gain access to the hidden instance variable by referring to it through **this**. For example, the following is a syntactically valid way to write the **Pwr( )** constructor.

```
Pwr(double b, int e) {
  this.b = b;
  this.e = e;
```

This refers to the **b** instance variable, not the parameter.

```
    val = 1;
    if(e==0) return;
    for( ; e>0; e--) val = val * b;
}
```

In this version, the names of the parameters are the same as the names of the instance variables, thus hiding them. However, **this** is used to "uncover" the instance variables.

Chapter 4 Self Test

1. What is the difference between a class and an object?

2. How is a class defined?

3. What does each object have its own copy of?

4. Using two separate statements, show how to declare an object called **counter** of a class called **MyCounter**.

5. Show how a method called **myMeth( )** is declared if it has a return type of **double** and has two **int** parameters called **a** and **b**.

6. How must a method return if it returns a value?

7. What name does a constructor have?

8. What does **new** do?

9. What is garbage collection, and how does it work?

10. What is **this**?

11. Can a constructor have one or more parameters?

12. If a method returns no value, what must its return type be?

Chapter 5

More Data Types
and Operators

Key Skills & Concepts

- Understand and create arrays

- Create multidimensional arrays

- Create irregular arrays

- Know the alternative array declaration syntax

- Assign array references

- Use the **length** array member

- Use the for-each style **for** loop

- Work with strings

- Apply command-line arguments

- Use type inference with local variables

- Use the bitwise operators

- Apply the **?** operator

This chapter returns to the subject of Java's data types and operators. It discusses arrays, the **String** type, local variable type inference, the bitwise operators, and the **?** ternary operator. It also covers Java's for-each style **for** loop. Along the way, command-line arguments are described.

Arrays

An *array* is a collection of variables of the same type, referred to by a common name. In Java, arrays can have one or more dimensions, although the one-dimensional array is the most common. Arrays are used for a variety of purposes because they offer a convenient means of grouping together related variables. For example, you might use an array to hold a record of the daily high temperature for a month, a list of stock price averages, or a list of your collection of programming books.

The principal advantage of an array is that it organizes data in such a way that it can be easily manipulated. For example, if you have an array containing the incomes for a selected group of households, it is easy to compute the average income by cycling through the array. Also, arrays organize data in such a way that it can be easily sorted.

Although arrays in Java can be used just like arrays in other programming languages, they have one special attribute: they are implemented as objects. This fact is one reason that a discussion of arrays was deferred until objects had been introduced. By implementing arrays as objects, several important advantages are gained, not the least of which is that unused arrays can be garbage collected.

One-Dimensional Arrays

A one-dimensional array is a list of related variables. Such lists are common in programming. For example, you might use a one-dimensional array to store the account numbers of the active users on a network. Another array might be used to store the current batting averages for a baseball team.

To declare a one-dimensional array, you can use this general form:

type[ ] *array-name* = new *type*[*size*];

Here, *type* declares the element type of the array. (The element type is also commonly referred to as the base type.) The element type determines the data type of each element contained in the array. The number of elements that the array will hold is determined by *size*. Since arrays are implemented as objects, the creation of an array is a two-step process. First, you declare an array reference variable. Second, you allocate memory for the array, assigning a reference to that memory to the array variable. Thus, arrays in Java are dynamically allocated using the **new** operator.

Here is an example. The following creates an **int** array of 10 elements and links it to an array reference variable named **sample**:

```
int[] sample = new int[10];
```

This declaration works just like an object declaration. The **sample** variable holds a reference to the memory allocated by **new**. This memory is large enough to hold 10 elements of type **int**. As with objects, it is possible to break the preceding declaration in two. For example:

```
int[] sample;
sample = new int[10];
```

In this case, when **sample** is first created, it refers to no physical object. It is only after the second statement executes that **sample** is linked with an array.

An individual element within an array is accessed by use of an index. An *index* describes the position of an element within an array. In Java, all arrays have zero as the index of their first element. Because **sample** has 10 elements, it has index values of 0 through 9. To index an array, specify the number of the element you want, surrounded by square brackets. Thus, the first element in **sample** is **sample[0]**, and the last element is **sample[9]**. For example, the following program loads **sample** with the numbers 0 through 9:

```
// Demonstrate a one-dimensional array.
class ArrayDemo {
  public static void main(String[] args) {
    int[] sample = new int[10];
    int i;

    for(i = 0; i < 10; i = i+1)          ◄───────────────┐
      sample[i] = i;
                                      Arrays are indexed from zero.
    for(i = 0; i < 10; i = i+1)   ◄───────────────────────┘
```

```
        System.out.println("This is sample[" + i + "]: " +
                              sample[i]);
    }
}
```

The output from the program is shown here:

```
This is sample[0]: 0
This is sample[1]: 1
This is sample[2]: 2
This is sample[3]: 3
This is sample[4]: 4
This is sample[5]: 5
This is sample[6]: 6
This is sample[7]: 7
This is sample[8]: 8
This is sample[9]: 9
```

Conceptually, the **sample** array looks like this:

0	1	2	3	4	5	6	7	8	9
Sample [0]	Sample [1]	Sample [2]	Sample [3]	Sample [4]	Sample [5]	Sample [6]	Sample [7]	Sample [8]	Sample [9]

Arrays are common in programming because they let you deal easily with large numbers of related variables. For example, the following program finds the minimum and maximum values stored in the **nums** array by cycling through the array using a **for** loop:

```
// Find the minimum and maximum values in an array.
class MinMax {
  public static void main(String[] args) {
    int[] nums = new int[10];
    int min, max;

    nums[0] = 99;
    nums[1] = -10;
    nums[2] = 100123;
    nums[3] = 18;
    nums[4] = -978;
    nums[5] = 5623;
    nums[6] = 463;
    nums[7] = -9;
    nums[8] = 287;
    nums[9] = 49;
```

```
   min = max = nums[0];
   for(int i=1; i < 10; i++) {
     if(nums[i] < min) min = nums[i];
     if(nums[i] > max) max = nums[i];
   }
   System.out.println("min and max: " + min + " " + max);
 }
}
```

The output from the program is shown here:

```
min and max: -978 100123
```

In the preceding program, the **nums** array was given values by hand, using 10 separate assignment statements. Although perfectly correct, there is an easier way to accomplish this. Arrays can be initialized when they are created. The general form for initializing a one-dimensional array is shown here:

type[ ] *array-name* = { *val1, val2, val3, ... , valN* };

Here, the initial values are specified by *val1* through *valN*. They are assigned in sequence, left to right, in index order. Java automatically allocates an array large enough to hold the initializers that you specify. There is no need to explicitly use the **new** operator. For example, here is a better way to write the **MinMax** program:

```
// Use array initializers.
class MinMax2 {
  public static void main(String[] args) {
    int[] nums = { 99, -10, 100123, 18, -978,
                   5623, 463, -9, 287, 49 };          ◄────── Array initializers
    int min, max;

    min = max = nums[0];
    for(int i=1; i < 10; i++) {
      if(nums[i] < min) min = nums[i];
      if(nums[i] > max) max = nums[i];
    }
    System.out.println("Min and max: " + min + " " + max);
  }
}
```

Array boundaries are strictly enforced in Java; it is a run-time error to overrun or underrun the end of an array. If you want to confirm this for yourself, try the following program that purposely overruns an array:

```
// Demonstrate an array overrun.
class ArrayErr {
  public static void main(String[] args) {
    int[] sample = new int[10];
    int i;
```

```
      // generate an array overrun
      for(i = 0; i < 100; i = i+1)
        sample[i] = i;
  }
}
```

As soon as **i** reaches 10, an **ArrayIndexOutOfBoundsException** is generated and the program is terminated.

Sorting an Array

Bubble.java

Because a one-dimensional array organizes data into an indexable linear list, it is the perfect data structure for sorting. In this project you will learn a simple way to sort an array. As you may know, there are a number of different sorting algorithms. There are the quick sort, the shaker sort, and the shell sort, to name just three. However, the best known, simplest, and easiest to understand is called the Bubble sort. Although the Bubble sort is not very efficient—in fact, its performance is unacceptable for sorting large arrays—it may be used effectively for sorting small arrays.

1. Create a file called **Bubble.java**.

2. The Bubble sort gets its name from the way it performs the sorting operation. It uses the repeated comparison and, if necessary, exchange of adjacent elements in the array. In this process, small values move toward one end and large ones toward the other end. The process is conceptually similar to bubbles finding their own level in a tank of water. The Bubble sort operates by making several passes through the array, exchanging out-of-place elements when necessary. The number of passes required to ensure that the array is sorted is equal to one less than the number of elements in the array.

    Here is the code that forms the core of the Bubble sort. The array being sorted is called **nums**.

    ```
    // This is the Bubble sort.
    for(a=1; a < size; a++)
      for(b=size-1; b >= a; b--) {
        if(nums[b-1] > nums[b]) { // if out of order
          // exchange elements
          t = nums[b-1];
          nums[b-1] = nums[b];
          nums[b] = t;
      }
    }
    ```

    Notice that sort relies on two **for** loops. The inner loop checks adjacent elements in the array, looking for out-of-order elements. When an out-of-order element pair is found, the two elements are exchanged. With each pass, the smallest of the remaining elements moves into its proper location. The outer loop causes this process to repeat until the entire array has been sorted.

3. Here is the entire **Bubble** program:

```
/*
    Try This 5-1

    Demonstrate the Bubble sort.
*/

class Bubble {
  public static void main(String[] args) {
    int[] nums = { 99, -10, 100123, 18, -978,
                   5623, 463, -9, 287, 49 };
    int a, b, t;
    int size;

    size = 10; // number of elements to sort

    // display original array
    System.out.print("Original array is:");
    for(int i=0; i < size; i++)
      System.out.print(" " + nums[i]);
    System.out.println();

    // This is the Bubble sort.
    for(a=1; a < size; a++)
      for(b=size-1; b >= a; b--) {
        if(nums[b-1] > nums[b]) { // if out of order
          // exchange elements
          t = nums[b-1];
          nums[b-1] = nums[b];
          nums[b] = t;
        }
      }

    // display sorted array
    System.out.print("Sorted array is:");
    for(int i=0; i < size; i++)
      System.out.print(" " + nums[i]);
    System.out.println();
  }
}
```

The output from the program is shown here:

```
Original array is: 99 -10 100123 18 -978 5623 463 -9 287 49
Sorted array is: -978 -10 -9 18 49 99 287 463 5623 100123
```

4. Although the Bubble sort is good for small arrays, it is not efficient when used on larger ones. A much better general-purpose sorting algorithm is the Quicksort. The Quicksort, however, relies on features of Java that you have not yet learned about.

Multidimensional Arrays

Although the one-dimensional array is often the most commonly used array in programming, multidimensional arrays (arrays of two or more dimensions) are certainly not rare. In Java, a multidimensional array is an array of arrays.

Two-Dimensional Arrays

The simplest form of the multidimensional array is the two-dimensional array. A two-dimensional array is, in essence, a list of one-dimensional arrays. To declare a two-dimensional integer array **table** of size 10, 20 you would write

```
int[][] table = new int[10][20];
```

Pay careful attention to the declaration. Unlike some other computer languages, which use commas to separate the array dimensions, Java places each dimension in its own set of brackets. Similarly, to access point 3, 5 of array **table**, you would use **table[3][5]**.

In the next example, a two-dimensional array is loaded with the numbers 1 through 12.

```java
// Demonstrate a two-dimensional array.
class TwoD {
  public static void main(String[] args) {
    int t, i;
    int[][] table = new int[3][4];

    for(t=0; t < 3; ++t) {
      for(i=0; i < 4; ++i) {
        table[t][i] = (t*4)+i+1;
        System.out.print(table[t][i] + " ");
      }
      System.out.println();
    }
  }
}
```

In this example, **table[0][0]** will have the value 1, **table[0][1]** the value 2, **table[0][2]** the value 3, and so on. The value of **table[2][3]** will be 12. Conceptually, the array will look like that shown in Figure 5-1.

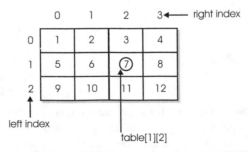

Figure 5-1 Conceptual view of the **table** array by the **TwoD** program

Irregular Arrays

When you allocate memory for a multidimensional array, you need to specify only the memory for the first (leftmost) dimension. You can allocate the remaining dimensions separately. For example, the following code allocates memory for the first dimension of **table** when it is declared. It allocates the second dimension manually.

```
int[][] table = new int[3][];
table[0] = new int[4];
table[1] - new int[4];
table[2] = new int[4];
```

Although there is no advantage to individually allocating the second dimension arrays in this situation, there may be in others. For example, when you allocate dimensions separately, you do not need to allocate the same number of elements for each index. Since multidimensional arrays are implemented as arrays of arrays, the length of each array is under your control. For example, assume you are writing a program that stores the number of passengers that ride an airport shuttle. If the shuttle runs 10 times a day during the week and twice a day on Saturday and Sunday, you could use the **riders** array shown in the following program to store the information. Notice that the length of the second dimension for the first five indices is 10 and the length of the second dimension for the last two indices is 2.

```
// Manually allocate differing size second dimensions.
class Ragged {
  public static void main(String[] args) {
    int[][] riders = new int[7][];
    riders[0] = new int[10];
    riders[1] = new int[10];
    riders[2] = new int[10];       Here, the second dimensions
    riders[3] = new int[10];       are 10 elements long.
    riders[4] = new int[10];
    riders[5] = new int[2];
    riders[6] = new int[2];        But here, they are
                                   2 elements long.

    int i, j;

    // fabricate some fake data
    for(i=0; i < 5; i++)
      for(j=0; j < 10; j++)
        riders[i][j] = i + j + 10;
    for(i=5; i < 7; i++)
      for(j=0; j < 2; j++)
        riders[i][j] = i + j + 10;

    System.out.println("Riders per trip during the week:");
    for(i=0; i < 5; i++) {
      for(j=0; j < 10; j++)
        System.out.print(riders[i][j] + " ");
      System.out.println();
    }
```

```
      System.out.println();

      System.out.println("Riders per trip on the weekend:");
      for(i=5; i < 7; i++) {
        for(j=0; j < 2; j++)
          System.out.print(riders[i][j] + " ");
        System.out.println();
      }
    }
}
```

The use of irregular (or ragged) multidimensional arrays does not, obviously, apply to all cases. However, irregular arrays can be quite effective in some situations. For example, if you need a very large two-dimensional array that is sparsely populated (that is, one in which not all of the elements will be used), an irregular array might be a perfect solution.

Arrays of Three or More Dimensions

Java allows arrays with more than two dimensions. Here is the general form of a multidimensional array declaration:

type[ ] [ ]...[ ] *name* = new *type*[*size1*][*size2*]...[*sizeN*];

For example, the following declaration creates a 4 × 10 × 3 three-dimensional integer array.

```
int[][][] multidim = new int[4][10][3];
```

Initializing Multidimensional Arrays

A multidimensional array can be initialized by enclosing each dimension's initializer list within its own set of curly braces. For example, the general form of array initialization for a two-dimensional array is shown here:

type-specifier[ ] [ ] *array_name* = {
  { *val, val, val, ..., val* },
  { *val, val, val, ..., val* },

  .
  .
  .

  { *val, val, val, ..., val* }
};

Here, *val* indicates an initialization value. Each inner block designates a row. Within each row, the first value will be stored in the first position of the subarray, the second value in the second position, and so on. Notice that commas separate the initializer blocks and that a semicolon follows the closing }.

For example, the following program initializes an array called **sqrs** with the numbers 1 through 10 and their squares:

```
// Initialize a two-dimensional array.
class Squares {
  public static void main(String[] args) {
    int[][] sqrs = {
      { 1, 1 },
      { 2, 4 },
      { 3, 9 },
      { 4, 16 },
      { 5, 25 },
      { 6, 36 },
      { 7, 49 },
      { 8, 64 },
      { 9, 81 },
      { 10, 100 }
    };
    int i, j;

    for(i=0; i < 10; i++) {
      for(j=0; j < 2; j++)
        System.out.print(sqrs[i][j] + " ");
      System.out.println();
    }
  }
}
```

Notice how each row has its own set of initializers.

Here is the output from the program:

```
1 1
2 4
3 9
4 16
5 25
6 36
7 49
8 64
9 81
10 100
```

Alternative Array Declaration Syntax

There is a second form that can be used to declare an array:

type var-name[ ];

Here, the square brackets follow the name of the array variable, not the type specifier. For example, the following two declarations are equivalent:

```
int counter[] = new int[3];
int[] counter = new int[3];
```

The following declarations are also equivalent:

```
char table[][] = new char[3][4];
char[][] table = new char[3][4];
```

This alternative declaration form offers convenience when converting code from C/C++ to Java. (In C/C++, arrays are declared in a fashion similar to Java's alternative form.) It also lets you declare both array and non-array variables in a single declaration statement. Today, the alternative form of array declaration is less commonly used, but it is still important that you are familiar with it because both forms of array declarations are legal in Java.

Assigning Array References

As with other objects, when you assign one array reference variable to another, you are simply changing what object that variable refers to. You are not causing a copy of the array to be made, nor are you causing the contents of one array to be copied to the other. For example, consider this program:

```
// Assigning array reference variables.
class AssignARef {
  public static void main(String[] args) {
    int i;

    int[] nums1 = new int[10];
    int[] nums2 = new int[10];

    for(i=0; i < 10; i++)
      nums1[i] = i;

    for(i=0; i < 10; i++)
      nums2[i] = -i;

    System.out.print("Here is nums1: ");
    for(i=0; i < 10; i++)
      System.out.print(nums1[i] + " ");
    System.out.println();

    System.out.print("Here is nums2: ");
    for(i=0; i < 10; i++)
      System.out.print(nums2[i] + " ");
    System.out.println();

    nums2 = nums1; // now nums2 refers to nums1   ◄──── Assign an array reference.
```

```
System.out.print("Here is nums2 after assignment: ");
for(i=0; i < 10; i++)
  System.out.print(nums2[i] + " ");
System.out.println();

// now operate on nums1 array through nums2
nums2[3] = 99;

System.out.print("Here is nums1 after change through nums2: ");
for(i=0; i < 10; i++)
  System.out.print(nums1[i] + " ");
System.out.println();
  }
}
```

The output from the program is shown here:

```
Here is nums1: 0 1 2 3 4 5 6 7 8 9
Here is nums2: 0 -1 -2 -3 -4 -5 -6 -7 -8 -9
Here is nums2 after assignment: 0 1 2 3 4 5 6 7 8 9
Here is nums1 after change through nums2: 0 1 2 99 4 5 6 7 8 9
```

As the output shows, after the assignment of **nums1** to **nums2**, both array reference variables refer to the same object.

Using the length Member

Recall that in Java, arrays are implemented as objects. One benefit of this approach is that each array has associated with it a **length** instance variable that contains the number of elements that the array can hold. (In other words, **length** contains the size of the array.) Here is a program that demonstrates this property:

```
// Use the length array member.
class LengthDemo {
  public static void main(String[] args) {
    int[] list = new int[10];
    int[] nums = { 1, 2, 3 };
    int[][] table = { // a variable-length table
      {1, 2, 3},
      {4, 5},
      {6, 7, 8, 9}
    };

    System.out.println("length of list is " + list.length);
    System.out.println("length of nums is " + nums.length);
    System.out.println("length of table is " + table.length);
    System.out.println("length of table[0] is " + table[0].length);
    System.out.println("length of table[1] is " + table[1].length);
    System.out.println("length of table[2] is " + table[2].length);
    System.out.println();
```

```
        // use length to initialize list
        for(int i=0; i < list.length; i++)
          list[i] = i * i;

        System.out.print("Here is list: ");
        // now use length to display list
        for(int i=0; i < list.length; i++)
          System.out.print(list[i] + " ");
        System.out.println();
    }
}
```

Use **length** to control a **for** loop.

This program displays the following output:

```
length of list is 10
length of nums is 3
length of table is 3
length of table[0] is 3
length of table[1] is 2
length of table[2] is 4

Here is list: 0 1 4 9 16 25 36 49 64 81
```

Pay special attention to the way **length** is used with the two-dimensional array **table**. As explained, a two-dimensional array is an array of arrays. Thus, when the expression

```
table.length
```

is used, it obtains the number of arrays stored in table, which is 3 in this case. To obtain the length of any individual array in table, you will use an expression such as this,

```
table[0].length
```

which, in this case, obtains the length of the first array.

One other thing to notice in **LengthDemo** is the way that **list.length** is used by the **for** loops to govern the number of iterations that take place. Since each array carries with it its own length, you can use this information rather than manually keeping track of an array's size. Keep in mind that the value of **length** has nothing to do with the number of elements that are actually in use. It contains the number of elements that the array is capable of holding.

The inclusion of the **length** member simplifies many algorithms by making certain types of array operations easier—and safer—to perform. For example, the following program uses **length** to copy one array to another while preventing an array overrun and its attendant run-time exception.

```
// Use length variable to help copy an array.
class ACopy {
  public static void main(String[] args) {
```

```
    int i;
    int[] nums1 = new int[10];
    int[] nums2 = new int[10];

    for(i=0; i < nums1.length; i++)
      nums1[i] = i;

    // copy nums1 to nums2
    if(nums2.length >= nums1.length)  ◄─────── Use length to compare array sizes.
      for(i = 0; i < nums1.length; i++)
        nums2[i] = nums1[i];

    for(i=0; i < nums2.length; i++)
      System.out.print(nums2[i] + " ");
  }
}
```

Here, **length** helps perform two important functions. First, it is used to confirm that the target array is large enough to hold the contents of the source array. Second, it provides the termination condition of the **for** loop that performs the copy. Of course, in this simple example, the sizes of the arrays are easily known, but this same approach can be applied to a wide range of more challenging situations.

Try This 5-2   A Queue Class

QDemo.java

As you may know, a data structure is a means of organizing data. The simplest data structure is the array, which is a linear list that supports random access to its elements. Arrays are often used as the underpinning for more sophisticated data structures, such as stacks and queues. A *stack* is a list in which elements can be accessed in first-in, last-out (FILO) order only. A *queue* is a list in which elements can be accessed in first-in, first-out (FIFO) order only. Thus, a stack is like a stack of plates on a table—the first down is the last to be used. A queue is like a line at a bank—the first in line is the first served.

What makes data structures such as stacks and queues interesting is that they combine storage for information with the methods that access that information. Thus, stacks and queues are *data engines* in which storage and retrieval are provided by the data structure itself, not manually by your program. Such a combination is, obviously, an excellent choice for a class, and in this project you will create a simple queue class.

In general, queues support two basic operations: put and get. Each put operation places a new element on the end of the queue. Each get operation retrieves the next element from the front of the queue. Queue operations are *consumptive*: once an element has been retrieved, it cannot be retrieved again. The queue can also become full, if there is no space available to store an item, and it can become empty, if all of the elements have been removed.

(continued)

One last point: There are two basic types of queues—circular and noncircular. A *circular queue* reuses locations in the underlying array when elements are removed. A *noncircular queue* does not reuse locations and eventually becomes exhausted. For the sake of simplicity, this example creates a noncircular queue, but with a little thought and effort, you can easily transform it into a circular queue.

1. Create a file called **QDemo.java**.

2. Although there are other ways to support a queue, the method we will use is based upon an array. That is, an array will provide the storage for the items put into the queue. This array will be accessed through two indices. The *put* index determines where the next element of data will be stored. The *get* index indicates at what location the next element of data will be obtained. Keep in mind that the get operation is consumptive, and it is not possible to retrieve the same element twice. Although the queue that we will be creating stores characters, the same logic can be used to store any type of object. Begin creating the **Queue** class with these lines:

```
class Queue {
  char[] q; // this array holds the queue
  int putloc, getloc; // the put and get indices
```

3. The constructor for the **Queue** class creates a queue of a given size. Here is the **Queue** constructor:

```
Queue(int size) {
  q = new char[size]; // allocate memory for queue
  putloc = getloc = 0;
}
```

Notice that the put and get indices are initially set to zero.

4. The **put( )** method, which stores elements, is shown next:

```
// put a character into the queue
void put(char ch) {
  if(putloc==q.length) {
    System.out.println(" - Queue is full.");
    return;
  }

  q[putloc++] = ch;
}
```

The method begins by checking for a queue-full condition. If **putloc** is equal to one past the last location in the **q** array, there is no more room in which to store elements. Otherwise, the new element is stored at that location and **putloc** is incremented. Thus, **putloc** is always the index where the next element will be stored.

5. To retrieve elements, use the **get( )** method, shown next:

```
// get a character from the queue
char get() {
  if(getloc == putloc) {
    System.out.println(" - Queue is empty.");
    return (char) 0;
  }

  return q[getloc++];
}
```

Notice first the check for queue-empty. If **getloc** and **putloc** both index the same element, the queue is assumed to be empty. This is why **getloc** and **putloc** were both initialized to zero by the **Queue** constructor. Then, the next element is returned. In the process, **getloc** is incremented. Thus, **getloc** always indicates the location of the next element to be retrieved.

6. Here is the entire **QDemo.java** program:

```
/*
   Try This 5-2

   A queue class for characters.
*/

class Queue {
  char[] q; // this array holds the queue
  int putloc, getloc; // the put and get indices

  Queue(int size) {
    q = new char[size]; // allocate memory for queue
    putloc = getloc = 0;
  }

  // put a character into the queue
  void put(char ch) {
    if(putloc==q.length) {
      System.out.println(" - Queue is full.");
      return;
    }

    q[putloc++] = ch;
  }

  // get a character from the queue
  char get() {
    if(getloc == putloc) {
      System.out.println(" - Queue is empty.");
      return (char) 0;
    }
```

(continued)

```java
      return q[getloc++];
  }
}

// Demonstrate the Queue class.
class QDemo {
  public static void main(String[] args) {
    Queue bigQ = new Queue(100);
    Queue smallQ = new Queue(4);
    char ch;
    int i;

    System.out.println("Using bigQ to store the alphabet.");
    // put some numbers into bigQ
    for(i=0; i < 26; i++)
      bigQ.put((char) ('A' + i));

    // retrieve and display elements from bigQ
    System.out.print("Contents of bigQ: ");
    for(i=0; i < 26; i++) {
      ch = bigQ.get();
      if(ch != (char) 0) System.out.print(ch);
    }

    System.out.println("\n");


    System.out.println("Using smallQ to generate errors.");
    // Now, use smallQ to generate some errors
    for(i=0; i < 5; i++) {
      System.out.print("Attempting to store " +
                        (char) ('Z' - i));

      smallQ.put((char) ('Z' - i));

      System.out.println();
    }
    System.out.println();

    // more errors on smallQ
    System.out.print("Contents of smallQ: ");
    for(i=0; i < 5; i++) {
      ch = smallQ.get();

      if(ch != (char) 0) System.out.print(ch);
    }
  }
}
```

7. The output produced by the program is shown here:

```
Using bigQ to store the alphabet.
Contents of bigQ: ABCDEFGHIJKLMNOPQRSTUVWXYZ

Using smallQ to generate errors.

Attempting to store Z
Attempting to store Y
Attempting to store X
Attempting to store W
Attempting to store V - Queue is full.


Contents of smallQ: ZYXW - Queue is empty.
```

8. On your own, try modifying **Queue** so that it stores other types of objects. For example, have it store **ints** or **doubles**.

The For-Each Style for Loop

When working with arrays, it is common to encounter situations in which each element in an array must be examined, from start to finish. For example, to compute the sum of the values held in an array, each element in the array must be examined. The same situation occurs when computing an average, searching for a value, copying an array, and so on. Because such "start to finish" operations are so common, Java defines a second form of the **for** loop that streamlines this operation.

The second form of the **for** implements a "for-each" style loop. A for-each loop cycles through a collection of objects, such as an array, in strictly sequential fashion, from start to finish. For reasons that will become clear, for-each style loops have become quite popular among both computer language designers and programmers. Interestingly, Java did not originally offer a for-each style loop. However, several years ago (beginning with JDK 5), the **for** loop was enhanced to provide this option. The for-each style of **for** is also referred to as the *enhanced for loop*. Both terms are used in this book.

The general form of the for-each style **for** is shown here.

for(*type itr-var* : *collection*) *statement-or-block*

Here, *type* specifies the type, and *itr-var* specifies the name of an *iteration variable* that will receive the elements from a collection, one at a time, from beginning to end. The collection being cycled through is specified by *collection*. There are various types of collections that can be used with the **for**, but the only type used in this book is the array. With each iteration of the loop, the next element in the collection is retrieved and stored in *itr-var*. The loop repeats until all elements in the collection have been obtained. Thus, when iterating over an array of size N, the enhanced **for** obtains the elements in the array in index order, from 0 to $N-1$.

Because the iteration variable receives values from the collection, *type* must be the same as (or compatible with) the elements stored in the collection. Thus, when iterating over arrays, *type* must be compatible with the element type of the array.

To understand the motivation behind a for-each style loop, consider the type of **for** loop that it is designed to replace. The following fragment uses a traditional **for** loop to compute the sum of the values in an array:

```
int[] nums = { 1, 2, 3, 4, 5, 6, 7, 8, 9, 10 };
int sum = 0;

for(int i=0; i < 10; i++) sum += nums[i];
```

To compute the sum, each element in **nums** is read, in order, from start to finish. Thus, the entire array is read in strictly sequential order. This is accomplished by manually indexing the **nums** array by **i**, the loop control variable. Furthermore, the starting and ending value for the loop control variable, and its increment, must be explicitly specified.

The for-each style **for** automates the preceding loop. Specifically, it eliminates the need to establish a loop counter, specify a starting and ending value, and manually index the array. Instead, it automatically cycles through the entire array, obtaining one element at a time, in sequence, from beginning to end. For example, here is the preceding fragment rewritten using a for-each version of the **for:**

```
int[] nums = { 1, 2, 3, 4, 5, 6, 7, 8, 9, 10 };
int sum = 0;

for(int x: nums) sum += x;
```

With each pass through the loop, **x** is automatically given a value equal to the next element in **nums**. Thus, on the first iteration, **x** contains 1, on the second iteration, **x** contains 2, and so on. Not only is the syntax streamlined, it also prevents boundary errors.

Ask the Expert

Q: Aside from arrays, what other types of collections can the for-each style for **loop** cycle through?

A: One of the most important uses of the for-each style **for** is to cycle through the contents of a collection defined by the Collections Framework. The Collections Framework is a set of classes that implement various data structures, such as lists, vectors, sets, and maps. A discussion of the Collections Framework is beyond the scope of this book, but detailed coverage of the Collections Framework can be found in *Java: The Complete Reference, Twelfth Edition* (McGraw Hill, 2022).

Here is an entire program that demonstrates the for-each version of the **for** just described:

```
// Use a for-each style for loop.
class ForEach {
  public static void main(String[] args) {
    int[] nums = { 1, 2, 3, 4, 5, 6, 7, 8, 9, 10 };
    int sum = 0;

    // Use for-each style for to display and sum the values.
    for(int x : nums) {  ◄──────── A for-each style for loop
      System.out.println("Value is: " + x);
      sum += x;
    }

    System.out.println("Summation: " + sum);
  }
}
```

The output from the program is shown here:

```
Value is: 1
Value is: 2
Value is: 3
Value is: 4
Value is: 5
Value is: 6
Value is: 7
Value is: 8
Value is: 9
Value is: 10
Summation: 55
```

As this output shows, the for-each style **for** automatically cycles through an array in sequence from the lowest index to the highest.

Although the for-each **for** loop iterates until all elements in an array have been examined, it is possible to terminate the loop early by using a **break** statement. For example, this loop sums only the first five elements of **nums**:

```
// Sum only the first 5 elements.
for(int x : nums) {
  System.out.println("Value is: " + x);
  sum += x;
  if(x == 5) break; // stop the loop when 5 is obtained
}
```

There is one important point to understand about the for-each style **for** loop. Its iteration variable is "read-only" as it relates to the underlying array. An assignment to the iteration variable has no effect on the underlying array. In other words, you can't change the contents of the array by assigning the iteration variable a new value. For example, consider this program:

```java
// The for-each loop is essentially read-only.
class NoChange {
  public static void main(String[] args) {
    int[] nums = { 1, 2, 3, 4, 5, 6, 7, 8, 9, 10 };

    for(int x : nums) {
      System.out.print(x + " ");
      x = x * 10; // no effect on nums  ◄────────── This does not change nums.
    }

    System.out.println();

    for(int x : nums)
      System.out.print(x + " ");

    System.out.println();
  }
}
```

The first **for** loop increases the value of the iteration variable by a factor of 10. However, this assignment has no effect on the underlying array **nums**, as the second **for** loop illustrates. The output, shown here, proves this point:

```
1 2 3 4 5 6 7 8 9 10
1 2 3 4 5 6 7 8 9 10
```

Iterating Over Multidimensional Arrays

The enhanced **for** also works on multidimensional arrays. Remember, however, that in Java, multidimensional arrays consist of *arrays of arrays*. (For example, a two-dimensional array is an array of one-dimensional arrays.) This is important when iterating over a multidimensional array because each iteration obtains the *next array*, not an individual element. Furthermore, the iteration variable in the **for** loop must be compatible with the type of array being obtained. For example, in the case of a two-dimensional array, the iteration variable must be a reference to a one-dimensional array. In general, when using the for-each **for** to iterate over an array of N dimensions, the objects obtained will be arrays of $N-1$ dimensions. To understand the implications of this, consider the following program. It uses nested **for** loops to obtain the elements of a two-dimensional array in row order, from first to last.

```java
// Use for-each style for on a two-dimensional array.
class ForEach2 {
  public static void main(String[] args) {
    int sum = 0;
    int[][] nums = new int[3][5];
```

```
   // give nums some values
   for(int i = 0; i < 3; i++)
      for(int j=0; j < 5; j++)
         nums[i][j] = (i+1)*(j+1);

   // Use for-each for loop to display and sum the values.
   for(int[] x : nums) {        ◄——————— Notice how x is declared.
      for(int y : x) {
         System.out.println("Value is: " + y);
         sum += y;
      }
   }
   System.out.println("Summation: " + sum);
   }
}
```

The output from this program is shown here:

```
Value is: 1
Value is: 2
Value is: 3
Value is: 4
Value is: 5
Value is: 2
Value is: 4
Value is: 6
Value is: 8
Value is: 10
Value is: 3
Value is: 6
Value is: 9
Value is: 12
Value is: 15
Summation: 90
```

In the program, pay special attention to this line:

```
for(int[] x : nums) {
```

Notice how **x** is declared. It is a reference to a one-dimensional array of integers. This is necessary because each iteration of the **for** obtains the next *array* in **nums**, beginning with the array specified by **nums[0]**. The inner **for** loop then cycles through each of these arrays, displaying the values of each element.

Applying the Enhanced for

Since the for-each style **for** can only cycle through an array sequentially, from start to finish, you might think that its use is limited. However, this is not true. A large number of algorithms require exactly this mechanism. One of the most common is searching.

For example, the following program uses a **for** loop to search an unsorted array for a value. It stops if the value is found.

```java
// Search an array using for-each style for.
class Search {
  public static void main(String[] args) {
    int[] nums = { 6, 8, 3, 7, 5, 6, 1, 4 };
    int val = 5;
    boolean found = false;

    // Use for-each style for to search nums for val.
    for(int x : nums) {
      if(x == val) {
        found = true;
        break;
      }
    }

    if(found)
      System.out.println("Value found!");
  }
}
```

The for-each style **for** is an excellent choice in this application because searching an unsorted array involves examining each element in sequence. (Of course, if the array were sorted, a binary search could be used, which would require a different style loop.) Other types of applications that benefit from for-each style loops include computing an average, finding the minimum or maximum of a set, looking for duplicates, and so on.

Now that the for-each style **for** has been introduced, it will be used where appropriate throughout the remainder of this book.

Strings

From a day-to-day programming standpoint, one of the most important of Java's data types is **String**. **String** defines and supports character strings. In some other programming languages, a string is an array of characters. This is not the case with Java. In Java, strings are objects.

Actually, you have been using the **String** class since Chapter 1, but you did not know it. When you create a string literal, you are actually creating a **String** object. For example, in the statement

```java
System.out.println("In Java, strings are objects.");
```

the string "In Java, strings are objects." is automatically made into a **String** object by Java. Thus, the use of the **String** class has been "below the surface" in the preceding programs.

In the following sections, you will learn to handle it explicitly. Be aware, however, that the **String** class is quite large, and we will only scratch its surface here. It is a class that you will want to explore on its own.

Constructing Strings

You can construct a **String** just like you construct any other type of object: by using **new** and calling the **String** constructor. For example:

```
String str = new String("Hello");
```

This creates a **String** object called **str** that contains the character string "Hello". You can also construct a **String** from another **String**. For example:

```
String str = new String("Hello");
String str2 = new String(str);
```

After this sequence executes, **str2** will also contain the character string "Hello".

Another easy way to create a **String** is shown here:

```
String str = "Java strings are powerful.";
```

In this case, **str** is initialized to the character sequence "Java strings are powerful."

Once you have created a **String** object, you can use it anywhere that a quoted string is allowed. For example, you can use a **String** object as an argument to **println( )**, as shown in this example:

```
// Introduce String.
class StringDemo {
  public static void main(String[] args) {
    // declare strings in various ways
    String str1 = new String("Java strings are objects.");
    String str2 = "They are constructed various ways.";
    String str3 = new String(str2);

    System.out.println(str1);
    System.out.println(str2);
    System.out.println(str3);
  }
}
```

The output from the program is shown here:

```
Java strings are objects.
They are constructed various ways.
They are constructed various ways.
```

Operating on Strings

The **String** class contains several methods that operate on strings. Here are the general forms for a few:

boolean equals(*str*)	Returns true if the invoking string contains the same character sequence as *str*.
int length( )	Obtains the length of a string.
char charAt(*index*)	Obtains the character at the index specified by *index*.
int compareTo(*str*)	Returns less than zero if the invoking string is less than *str*, greater than zero if the invoking string is greater than *str*, and zero if the strings are equal.
int indexOf(*str*)	Searches the invoking string for the substring specified by *str*. Returns the index of the first match or −1 on failure.
int lastIndexOf(*str*)	Searches the invoking string for the substring specified by *str*. Returns the index of the last match or −1 on failure.

Here is a program that demonstrates these methods:

```java
// Some String operations.
class StrOps {
  public static void main(String[] args) {
    String str1 =
      "When it comes to Web programming, Java is #1.";
    String str2 = new String(str1);
    String str3 = "Java strings are powerful.";
    int result, idx;
    char ch;

    System.out.println("Length of str1: " +
                       str1.length());

    // display str1, one char at a time.
    for(int i=0; i < str1.length(); i++)
      System.out.print(str1.charAt(i));
    System.out.println();

    if(str1.equals(str2))
      System.out.println("str1 equals str2");
    else
      System.out.println("str1 does not equal str2");

    if(str1.equals(str3))
      System.out.println("str1 equals str3");
    else
      System.out.println("str1 does not equal str3");
```

```
    result = str1.compareTo(str3);
    if(result == 0)
      System.out.println("str1 and str3 are equal");
    else if(result < 0)
      System.out.println("str1 is less than str3");
    else
      System.out.println("str1 is greater than str3");

    // assign a new string to str2
    str2 = "One Two Three One";

    idx = str2.indexOf("One");
    System.out.println("Index of first occurrence of One: " + idx);
    idx = str2.lastIndexOf("One");
    System.out.println("Index of last occurrence of One: " + idx);
  }
}
```

This program generates the following output:

```
Length of str1: 45
When it comes to Web programming, Java is #1.
str1 equals str2
str1 does not equal str3
str1 is greater than str3
Index of first occurrence of One: 0
Index of last occurrence of One: 14
```

You can *concatenate* (join together) two strings using the **+** operator. For example, this statement

```
String str1 = "One";
String str2 = "Two";
String str3 = "Three";
String str4 = str1 + str2 + str3;
```

initializes **str4** with the string "OneTwoThree".

Ask the Expert

Q: Why does String **define the** equals( ) **method? Can't I just use ==?**

A: The **equals( )** method compares the character sequences of two **String** objects for equality. Applying the **==** to two **String** references simply determines whether the two references refer to the same object.

Arrays of Strings

Like any other data type, strings can be assembled into arrays. For example:

```
// Demonstrate String arrays.
class StringArrays {
  public static void main(String[] args) {
    String[] strs = { "This", "is", "a", "test." };

    System.out.println("Original array: ");
    for(String s : strs)
      System.out.print(s + " ");
    System.out.println("\n");

    // change a string
    strs[1] = "was";
    strs[3] = "test, too!";

    System.out.println("Modified array: ");
    for(String s : strs)
      System.out.print(s + " ");
  }
}
```

Here is the output from this program:

```
Original array:
This is a test.

Modified array:
This was a test, too!
```

Strings Are Immutable

The contents of a **String** object are immutable. That is, once created, the character sequence that makes up the string cannot be altered. This restriction allows Java to implement strings more efficiently. Even though this probably sounds like a serious drawback, it isn't. When you need a string that is a variation on one that already exists, simply create a new string that contains the desired changes. Since unused **String** objects are automatically garbage collected, you don't even need to worry about what happens to the discarded strings. It must be made clear, however, that **String** reference variables may, of course, change the object to which they refer. It is just that the contents of a specific **String** object cannot be changed after it is created.

To fully understand why immutable strings are not a hindrance, we will use another of **String**'s methods: **substring( )**. The **substring( )** method returns a new string that contains a specified portion of the invoking string. Because a new **String** object is manufactured that contains the substring, the original string is unaltered, and the rule of immutability remains intact. The form of **substring( )** that we will be using is shown here:

String substring(int *startIndex*, int *endIndex*)

Ask the Expert

Q: You say that once created, String objects are immutable. I understand that, from a practical point of view, this is not a serious restriction, but what if I want to create a string that can be changed?

A: You're in luck. Java offers a class called **StringBuffer**, which creates string objects that can be changed. For example, in addition to the **charAt( )** method, which obtains the character at a specific location, **StringBuffer** defines **setCharAt( )**, which sets a character within the string. Java also supplies **StringBuilder**, which is related to **StringBuffer**, and also supports strings that can be changed. However, for most purposes you will want to use **String**, not **StringBuffer** or **StringBuilder**.

Here, *startIndex* specifies the beginning index, and *endIndex* specifies the stopping point. Here is a program that demonstrates **substring( )** and the principle of immutable strings:

```
// Use substring().
class SubStr {
  public static void main(String[] args) {
    String orgstr = "Java makes the Web move.";

    // construct a substring
    String substr = orgstr.substring(5, 18);

    System.out.println("orgstr: " + orgstr);
    System.out.println("substr: " + substr);
  }
}
```

This creates a new string that contains the desired substring.

Here is the output from the program:

```
orgstr: Java makes the Web move.
substr: makes the Web
```

As you can see, the original string **orgstr** is unchanged, and **substr** contains the substring.

Using a String to Control a switch Statement

As explained in Chapter 3, in the past a **switch** had to be controlled by an integer type, such as **int** or **char**. This precluded the use of a **switch** in situations in which one of several actions is selected based on the contents of a string. Instead, an **if-else-if** ladder was the typical solution.

Although an **if-else-if** ladder is semantically correct, a **switch** statement would be the more natural idiom for such a selection. Fortunately, this situation has been remedied. With a modern version of Java, you can use a **String** to control a switch. This results in more readable, streamlined code in many situations.

Here is an example that demonstrates controlling a **switch** with a **String**:

```
// Use a string to control a switch statement.

class StringSwitch {
  public static void main(String[] args) {

    String command = "cancel";

    switch(command) {
      case "connect":
        System.out.println("Connecting");
        break;
      case "cancel":
        System.out.println("Canceling");
        break;
      case "disconnect":
        System.out.println("Disconnecting");
        break;
      default:
        System.out.println("Command Error!");
        break;
    }
  }
}
```

As you would expect, the output from the program is

```
Canceling
```

The string contained in **command** (which is "cancel" in this program) is tested against the **case** constants. When a match is found (as it is in the second **case**), the code sequence associated with that sequence is executed.

Being able to use strings in a **switch** statement can be very convenient and can improve the readability of some code. For example, using a string-based **switch** is an improvement over using the equivalent sequence of **if/else** statements. However, switching on strings can be less efficient than switching on integers. Therefore, it is best to switch on strings only in cases in which the controlling data is already in string form. In other words, don't use strings in a **switch** unnecessarily.

Ask the Expert

Q: I have heard about another type of string literal called a *text block*. Can you tell me about it?

A: Yes, text blocks were added to Java by JDK 15. A text block is a new kind of string literal that is comprised of a sequence of characters that can occupy more than one line. A text block reduces the tedium programmers often face when creating multiline string literals because newline characters can be used in a text block without the need for the **\n** escape sequence. Furthermore, tab and double quote characters can also be entered directly, without using an escape sequence, and the indentation of a multiline string can be preserved. Thus, text blocks provide an elegant alternative to what can be a rather annoying process.

A text block is supported by a delimiter that consists of three double-quote characters: """. A block of text is created by enclosing a string within a set of these delimiters. Specifically, a text block begins immediately following the newline after the opening """. Thus, the opening delimiter must end with a newline. The text block begins on the next line. A text block ends at the first character of the closing """. It is important to emphasize that even though a text block uses the """ delimiter, it is still of type **String**. Thus, a text block can be used wherever any other string can.

Here is a simple example of a text block. It assigns a multiline text block to **str**.

```
String str = """
Text blocks make multiple lines easy because they eliminate
    the need to use \n escape sequences to indicate a newline.
As a result, text blocks make the programmer's life better!
""";
```

This example creates a string in which each line is separated from the next by a newline. It is not necessary to use the **\n** escape sequence to obtain the newline. Thus, the text block automatically preserves the newlines in the text. Notice that the second line is indented. When **str** is output using this statement:

```
System.out.println(str);
```

the following is displayed:

```
Text blocks make multiple lines easy because they eliminate
    the need to use \n escape sequences to indicate a newline.
As a result, text blocks make the programmer's life better!
```

As the output shows, the newlines and the indentation of the second line are preserved. These are key benefits of text blocks.

Text blocks have additional attributes, such as the ability to removed unwanted leading whitespace. It is a feature that you will want to look at more closely as you advance in your study of Java. Simply put, text blocks make what was often a difficult coding task easy.

Using Command-Line Arguments

Now that you know about the **String** class, you can understand the **args** parameter to **main( )** that has been in every program shown so far. Many programs accept what are called *command-line arguments.* A command-line argument is the information that directly follows the program's name on the command line when it is executed. To access the command-line arguments inside a Java program is quite easy—they are stored as strings in the **String** array passed to **main( )**. For example, the following program displays all of the command-line arguments that it is called with:

```
// Display all command-line information.
class CLDemo {
  public static void main(String[] args) {
    System.out.println("There are " + args.length +
                         " command-line arguments.");

    System.out.println("They are: ");
    for(int i=0; i<args.length; i++)
      System.out.println("arg[" + i + "]: " + args[i]);
  }
}
```

If **CLDemo** is executed like this,

```
java CLDemo one two three
```

you will see the following output:

```
There are 3 command-line arguments.
They are:
arg[0]: one
arg[1]: two
arg[2]: three
```

Notice that the first argument is stored at index 0, the second argument is stored at index 1, and so on.

To get a taste of the way command-line arguments can be used, consider the next program. It takes one command-line argument that specifies a person's name. It then searches through a two-dimensional array of strings for that name. If it finds a match, it displays that person's telephone number.

```
// A simple automated telephone directory.
class Phone {
  public static void main(String[] args) {
    String[][] numbers = {
      { "Tom", "555-3322" },
      { "Mary", "555-8976" },
      { "Jon", "555-1037" },
      { "Rachel", "555-1400" }
    };
```

```
    int i;

    if(args.length != 1)  ◄─────────────────────
      System.out.println("Usage: java Phone <name>");
    else {
      for(i=0; i<numbers.length; i++) {
        if(numbers[i][0].equals(args[0])) {
          System.out.println(numbers[i][0] + ": " +
                               numbers[i][1]);
          break;
        }
      }
      if(i == numbers.length)
        System.out.println("Name not found.");
    }
  }
}
```

To use the program, one command-line argument must be present.

Here is a sample run:

```
java Phone Mary
Mary: 555-8976
```

Using Type Inference with Local Variables

Not long ago, a feature called *local variable type inference* was added to the Java language. To begin, let's review two important aspects of variables. First, all variables in Java must be declared prior to their use. Second, a variable can be initialized with a value when it is declared. Furthermore, when a variable is initialized, the type of the initializer must be the same as (or convertible to) the declared type of the variable. Thus, in principle, it would not be necessary to specify an explicit type for an initialized variable because it could be inferred from the type of its initializer. Of course, in the past, Java did not support such inference and all variables required an explicitly declared type, whether they were initialized or not. Today, that situation has changed.

Beginning with JDK 10, it became possible to let the compiler infer the type of a local variable based on the type of its initializer, thus avoiding the need to explicitly specify the type. Local variable type inference offers a number of advantages. For example, it can streamline code by eliminating the need to redundantly specify a variable's type when it can be inferred from its initializer. It can simplify declarations in cases in which the type is quite lengthy, such as can be the case with some class names. It can also be helpful when a type is difficult to discern or cannot be denoted. (An example of a type that cannot be denoted is the type of an anonymous class, discussed in Chapter 17.) Furthermore, local variable type inference has become a common part of the contemporary programming environment. Its inclusion in Java helps keep Java up-to-date with evolving trends in language design. To support local variable type inference, the context-sensitive keyword **var** was added to Java.

To use local variable type inference, the variable must be declared with **var** as the type name and it must include an initializer. Let's begin with a simple example. Consider the following statement that declares a local **double** variable called **avg** that is initialized with the value 10.0:

```
double avg = 10.0;
```

Using type inference, this declaration can also be written like this:

```
var avg = 10.0;
```

In both cases, **avg** will be of type **double**. In the first case, its type is explicitly specified. In the second, its type is inferred as **double** because the initializer 10.0 is of type **double**.

As mentioned, **var** is context-sensitive. When it is used as the type name in the context of a local variable declaration, it tells the compiler to use type inference to determine the type of the variable being declared based on the type of the initializer. Thus, in a local variable declaration, **var** is a placeholder for the actual inferred type. However, when used in most other places, **var** is simply a user-defined identifier with no special meaning. For example, the following declaration is still valid:

```
int var = 1;  // In this case, var is simply a user-defined identifier.
```

In this case, the type is explicitly specified as **int** and **var** is the name of the variable being declared. Even though it is context-sensitive, there are a few places in which the use of **var** is illegal. It cannot be used as the name of a class, for example.

The following program puts the preceding discussion into action:

```
// A simple demonstration of local variable type inference.
class VarDemo {
  public static void main(String[] args) {

    // Use type inference to determine the type of the
    // variable named avg. In this case, double is inferred.
    var avg = 10.0;  ◄───────────────────────────────── Use var to infer
    System.out.println("Value of avg: " + avg);                type of avg.

    // In the following context, var is not a predefined identifier.
    // It is simply a user-defined variable name.
    int var = 1;
    System.out.println("Value of var: " + var);

    // Interestingly, in the following sequence, var is used
    // as both the type of the declaration and as a variable name
    // in the initializer.
    var k = -var;
    System.out.println("Value of k: " + k);
  }
}
```

Here is the output:

```
Value of avg: 10.0
Value of var: 1
Value of k: -1
```

The preceding example uses **var** to declare only simple variables, but you can also use **var** to declare an array. For example:

```
var myArray = new int[10]; // This is valid.
```

Notice that neither **var** nor **myArray** has brackets. Instead, the type of **myArray** is inferred to be **int[ ]**. Furthermore, you *cannot* use brackets on the left side of a **var** declaration. Thus, both of these declarations are invalid:

```
var[] myArray = new int[10]; // Wrong
var myArray[] = new int[10]; // Wrong
```

In the first line, an attempt is made to bracket **var**. In the second, an attempt is made to bracket **myArray**. In both cases, the use of the brackets is wrong because the type is inferred from the type of the initializer.

It is important to emphasize that **var** can be used to declare a variable only when that variable is initialized. Therefore, the following statement is wrong:

```
var counter; // Wrong! Initializer required.
```

Also, remember that **var** can be used only to declare local variables. It cannot be used when declaring instance variables, parameters, or return types, for example.

Local Variable Type Inference with Reference Types

The preceding examples introduced the fundamentals of local variable type inference using primitive types. However, it is with reference types, such as class types, that the full benefits of type inference become apparent. Moreover, local variable type inference with reference types constitutes a primary use of this feature.

Let's again begin with a simple example. The following declarations use type inference to declare two **String** variables called **myStr** and **mySubStr**:

```
var myStr = "This is a string";
var mySubStr = myStr.substring(5, 10);
```

Recall that a quoted string is an object of type **String**. Because a quoted string is used as an initializer, the type of **myStr** is inferred to be **String**. The type of **mySubStr** is also inferred to be **String** because the type of reference returned by the **substring( )** method is **String**.

Of course, you can also use local variable type inference with user-defined classes, as the following program illustrates. It creates a class called **MyClass** and then uses local variable type inference to declare and initialize an object of that class.

```
// Local variable type inference with a user-defined class type.
class MyClass {
  private int i;
```

```
  MyClass(int k) { i = k;}

  int geti() { return i; }
  void seti(int k) { if(k >= 0) i = k; }
}

class VarDemo2 {
  public static void main(String[] args) {
    var mc = new MyClass(10); // Notice the use of var here.

    System.out.println("Value of i in mc is " + mc.geti());
    mc.seti(19);
    System.out.println("Value of i in mc is now " + mc.geti());
  }
}
```

The output of the program is shown here:

```
Value of i in mc is 10
Value of i in mc is now 19
```

In the program, pay special attention to this line:

```
var mc = new MyClass(10); // Notice the use of var here.
```

Here, the type of **mc** will be inferred as **MyClass** because that is the type of the initializer, which is a new **MyClass** object.

As mentioned, one of the primary benefits of local variable type inference is its ability to streamline code, and it is with reference types where such streamlining is most apparent. As you advance in your study of Java, you will find that many class types have rather long names. For example, in Chapter 10 you will learn about the **FileInputStream** class, which is used to open a file for input operations. Without the use of type inference, you would declare and initialize a **FileInputStream** using a traditional declaration like the one shown here:

```
FileInputStream fin = new FileInputStream("test.txt");
```

With the use of **var**, it can now be written like this:

```
var fin = new FileInputStream("test.txt");
```

Here, **fin** is inferred to be of type **FileInputStream** because that is the type of its initializer. There is no need to explicitly repeat the type name. As a result, this declaration of **fin** is substantially shorter than writing it the traditional way. Thus, the use of **var** streamlines the declaration. In general, the streamlining attribute of local variable type inference helps lessen the tedium of entering long type names into your program. Of course, local variable type inference must be used carefully to avoid reducing the readability of your program and thus obscuring its meaning. In essence, it is a feature that you should use wisely.

Using Local Variable Type Inference in a for Loop

Another place that local variable type inference can be used is in a **for** loop when declaring
and initializing the loop control variable inside a traditional **for** loop, or when specifying the
iteration variable in a for-each **for**. The following program shows an example of each case:

```
// Use type inference in a for loop.
class VarDemo3 {
  public static void main(String[] args) {

    // Use type inference with the loop control variable.
    System.out.print("Values of x: ");
    for(var x = 2.5; x < 100.0; x = x * 2)          Use var in a for loop.
      System.out.print(x + " ");

    System.out.println();

    // Use type inference with the iteration variable.
    int[] nums = { 1, 2, 3, 4, 5, 6};
    System.out.print("Values in nums array: ");
    for(var v : nums)
      System.out.print(v + " ");

    System.out.println();
  }
}
```

The output is shown here:

```
Values of x: 2.5 5.0 10.0 20.0 40.0 80.0
Values in nums array: 1 2 3 4 5 6
```

In this example, loop control variable **x** in this line:

```
for(var x = 2.5; x < 100.0; x = x * 2)
```

is inferred to be type **double** because that is the type of its initializer. Iteration variable **v** in
this line:

```
for(var v : nums)
```

is inferred to be of type **int** because that is the element type of the array **nums**.

Some var Restrictions

In addition to those mentioned in the preceding discussion, several other restrictions apply
to the use of **var**. Only one variable can be declared at a time; a variable cannot use **null** as
an initializer; and the variable being declared cannot be used by the initializer expression.
Although you can declare an array type using **var**, you cannot use **var** with an array initializer.

For example, this is valid:

```
var myArray = new int[10]; // This is valid.
```

but this is not:

```
var myArray = { 1, 2, 3 }; // Wrong
```

As mentioned earlier, **var** cannot be used as the name of a class. It also cannot be used as the name of other reference types, including an interface, enumeration, or annotation, which are described later in this book. Here are two other restrictions that relate to Java features also described later, but mentioned here in the interest of completeness. Local variable type inference cannot be used to declare the exception type caught by a **catch** statement. Also, neither lambda expressions nor method references can be used as initializers.

NOTE

At the time of this writing, a number of readers will be using Java environments that predate JDK 10. So that as many of the code examples as possible can be compiled and run with older JDKs, local variable type inference will not be used by most of the programs in the remainder of this edition of the book. Using the full syntax also makes it very clear at a glance what type of variable is being created, which is important for example code. Of course, going forward, you should consider the use of local variable type inference where appropriate in your own code.

The Bitwise Operators

In Chapter 2 you learned about Java's arithmetic, relational, and logical operators. Although these are often the most commonly used, Java provides additional operators that expand the set of problems to which Java can be applied: the bitwise operators. The bitwise operators can be used on values of type **long**, **int**, **short**, **char**, or **byte**. Bitwise operations cannot be used on **boolean**, **float**, or **double**, or class types. They are called the bitwise operators because they are used to test, set, or shift the individual bits that make up a value. Bitwise operations are important to a wide variety of systems-level programming tasks in which status information from a device must be interrogated or constructed. Table 5-1 lists the bitwise operators.

Operator	Result
&	Bitwise AND
|	Bitwise OR
^	Bitwise exclusive OR
>>	Shift right
>>>	Unsigned shift right
<<	Shift left
~	One's complement (unary NOT)

Table 5-1 The Bitwise Operators

The Bitwise AND, OR, XOR, and NOT Operators

The bitwise operators AND, OR, XOR, and NOT are **&**, **|**, **^**, and **~**. They perform the same operations as their Boolean logical equivalents described in Chapter 2. The difference is that the bitwise operators work on a bit-by-bit basis. The following table shows the outcome of each operation using 1's and 0's:

p	q	p & q	p | q	p ^ q	~p
0	0	0	0	0	1
1	0	0	1	1	0
0	1	0	1	1	1
1	1	1	1	0	0

In terms of one common usage, you can think of the bitwise AND as a way to turn bits off. That is, any bit that is 0 in either operand will cause the corresponding bit in the outcome to be set to 0. For example:

```
  1 1 0 1  0 0 1 1
& 1 0 1 0  1 0 1 0
  1 0 0 0  0 0 1 0
```

The following program demonstrates the **&** by turning any lowercase letter into uppercase by resetting the 6th bit to 0. As the Unicode/ASCII character set is defined, the lowercase letters are the same as the uppercase ones except that the lowercase ones are greater in value by exactly 32. Therefore, to transform a lowercase letter to uppercase, just turn off the 6th bit, as this program illustrates:

```java
// Uppercase letters
class UpCase {
  public static void main(String[] args) {
    char ch;

    for(int i=0; i < 10; i++) {
      ch = (char) ('a' + i);
      System.out.print(ch);

      // This statement turns off the 6th bit.
      ch = (char) ((int) ch & 65503); // ch is now uppercase

      System.out.print(ch + " ");
    }
  }
}
```

The output from this program is shown here:

```
aA bB cC dD eE fF gG hH iI jJ
```

The value 65,503 used in the AND statement is the decimal representation of 1111 1111 1101 1111. Thus, the AND operation leaves all bits in **ch** unchanged except for the 6th one, which is set to 0.

The AND operator is also useful when you want to determine whether a bit is on or off. For example, this statement determines whether bit 4 in **status** is set:

```
if((status & 8) != 0) System.out.println("bit 4 is on");
```

The number 8 is used because it translates into a binary value that has only the 4th bit set. Therefore, the **if** statement can succeed only when bit 4 of **status** is also on. An interesting use of this concept is to show the bits of a **byte** value in binary format.

```
// Display the bits within a byte.
class ShowBits {
  public static void main(String[] args) {
    int t;
    byte val;

    val = 123;
    for(t=128; t > 0; t = t/2) {
      if((val & t) != 0) System.out.print("1 ");
      else System.out.print("0 ");
    }
  }
}
```

The output is shown here:

```
0 1 1 1 1 0 1 1
```

The **for** loop successively tests each bit in **val**, using the bitwise AND, to determine whether it is on or off. If the bit is on, the digit **1** is displayed; otherwise, **0** is displayed. In Try This 5-3, you will see how this basic concept can be expanded to create a class that will display the bits in any type of integer.

The bitwise OR, as the reverse of AND, can be used to turn bits on. Any bit that is set to 1 in either operand will cause the corresponding bit in the result to be set to 1. For example:

```
  1 1 0 1 0 0 1 1
| 1 0 1 0 1 0 1 0
  1 1 1 1 1 0 1 1
```

We can make use of the OR to change the uppercasing program into a lowercasing program, as shown here:

```
// Lowercase letters.
class LowCase {
  public static void main(String[] args) {
    char ch;

    for(int i=0; i < 10; i++) {
      ch = (char) ('A' + i);
      System.out.print(ch);

      // This statement turns on the 6th bit.
      ch = (char) ((int) ch | 32); // ch is now lowercase

      System.out.print(ch + " ");
    }
  }
}
```

The output from this program is shown here:

```
Aa Bb Cc Dd Ee Ff Gg Hh Ii Jj
```

The program works by ORing each character with the value 32, which is 0000 0000 0010 0000 in binary. Thus, 32 is the value that produces a value in binary in which only the 6th bit is set. When this value is ORed with any other value, it produces a result in which the 6th bit is set and all other bits remain unchanged. As explained, for characters this means that each uppercase letter is transformed into its lowercase equivalent.

An exclusive OR, usually abbreviated XOR, will result in a set bit if, and only if, the bits being compared are different, as illustrated here:

```
  0 1 1 1 1 1 1 1
^ 1 0 1 1 1 0 0 1
  1 1 0 0 0 1 1 0
```

The XOR operator has an interesting property that makes it a simple way to encode a message. When some value X is XORed with another value Y, and then that result is XORed with Y again, X is produced. That is, given the sequence

```
R1 = X ^ Y; R2 = R1 ^ Y;
```

then R2 is the same value as X. Thus, the outcome of a sequence of two XORs can produce the original value.

You can use this principle to create a simple cipher program in which some integer is the key that is used to both encode and decode a message by XORing the characters in that message. To encode, the XOR operation is applied the first time, yielding the cipher text. To decode, the XOR is applied a second time, yielding the plain text. Of course, such a cipher has no practical value, being trivially easy to break. It does, however, provide an interesting way to demonstrate the XOR. Here is a program that uses this approach to encode and decode a short message:

```
// Use XOR to encode and decode a message.
class Encode {
  public static void main(String[] args) {
    String msg = "This is a test";
    String encmsg = "";
    String decmsg = "";
    int key = 88;

    System.out.print("Original message: ");
    System.out.println(msg);

    // encode the message                              This constructs the encoded string.
    for(int i=0; i < msg.length(); i++)
      encmsg = encmsg + (char) (msg.charAt(i) ^ key);

    System.out.print("Encoded message: ");
    System.out.println(encmsg);

    // decode the message
    for(int i=0; i < msg.length(); i++)
      decmsg = decmsg + (char) (encmsg.charAt(i) ^ key);

    System.out.print("Decoded message: ");            This constructs the decoded string.
    System.out.println(decmsg);
  }
}
```

Here is the output:

```
Original message: This is a test
Encoded message: 01+x1+x9x,=+,
Decoded message: This is a test
```

As you can see, the result of two XORs using the same key produces the decoded message.

The unary one's complement (NOT) operator reverses the state of all the bits of the operand. For example, if some integer called **A** has the bit pattern 1001 0110, then ~**A** produces a result with the bit pattern 0110 1001.

The following program demonstrates the NOT operator by displaying a number and its complement in binary:

```
// Demonstrate the bitwise NOT.
class NotDemo {
  public static void main(String[] args) {
    byte b = -34;
```

```
    for(int t=128; t > 0; t = t/2) {
      if((b & t) != 0) System.out.print("1 ");
      else System.out.print("0 ");
    }
    System.out.println();

    // reverse all bits
    b = (byte) ~b;

    for(int t=128; t > 0; t = t/2) {
      if((b & t) != 0) System.out.print("1 ");
      else System.out.print("0 ");
    }
  }
}
```

Here is the output:

```
1 1 0 1 1 1 1 0
0 0 1 0 0 0 0 1
```

The Shift Operators

In Java it is possible to shift the bits that make up a value to the left or to the right by a specified amount. Java defines the three bit-shift operators shown here:

<<	Left shift
>>	Right shift
>>>	Unsigned right shift

The general forms for these operators are shown here:

value << *num-bits*
value >> *num-bits*
value >>> *num-bits*

Here, *value* is the value being shifted by the number of bit positions specified by *num-bits*.

Each left shift causes all bits within the specified value to be shifted left one position and a 0 bit to be brought in on the right. Each right shift shifts all bits to the right one position and preserves the sign bit. As you may know, negative numbers are usually represented by setting the high-order bit of an integer value to 1, and this is the approach used by Java. Thus, if the value being shifted is negative, each right shift brings in a 1 on the left. If the value is positive, each right shift brings in a 0 on the left.

In addition to the sign bit, there is something else to be aware of when right shifting. Java uses *two's complement* to represent negative values. In this approach negative values are stored by first reversing the bits in the value and then adding 1. Thus, the byte value for −1 in binary is 1111 1111. Right shifting this value will always produce −1!

If you don't want to preserve the sign bit when shifting right, you can use an unsigned right shift (>>>), which always brings in a 0 on the left. For this reason, the >>> is also called the *zero-fill* right shift. You will use the unsigned right shift when shifting bit patterns, such as status codes, that do not represent integers.

For all of the shifts, the bits shifted out are lost. Thus, a shift is not a rotate, and there is no way to retrieve a bit that has been shifted out.

Shown next is a program that graphically illustrates the effect of a left and right shift. Here, an integer is given an initial value of 1, which means that its low-order bit is set. Then, a series of eight shifts are performed on the integer. After each shift, the lower 8 bits of the value are shown. The process is then repeated, except that a 1 is put in the 8th bit position, and right shifts are performed.

```java
// Demonstrate the shift << and >> operators.
class ShiftDemo {
  public static void main(String[] args) {
    int val = 1;

    for(int i = 0; i < 8; i++) {
      for(int t=128; t > 0; t = t/2) {
        if((val & t) != 0) System.out.print("1 ");
        else System.out.print("0 ");
      }
      System.out.println();
      val = val << 1; // left shift
    }
    System.out.println();

    val = 128;
    for(int i = 0; i < 8; i++) {
      for(int t=128; t > 0; t = t/2) {
        if((val & t) != 0) System.out.print("1 ");
        else System.out.print("0 ");
      }
      System.out.println();
      val = val >> 1; // right shift
    }
  }
}
```

The output from the program is shown here:

```
0 0 0 0 0 0 0 1
0 0 0 0 0 0 1 0
0 0 0 0 0 1 0 0
```

```
0 0 0 0 1 0 0 0
0 0 0 1 0 0 0 0
0 0 1 0 0 0 0 0
0 1 0 0 0 0 0 0
1 0 0 0 0 0 0 0

1 0 0 0 0 0 0 0
0 1 0 0 0 0 0 0
0 0 1 0 0 0 0 0
0 0 0 1 0 0 0 0
0 0 0 0 1 0 0 0
0 0 0 0 0 1 0 0
0 0 0 0 0 0 1 0
0 0 0 0 0 0 0 1
```

You need to be careful when shifting **byte** and **short** values because Java will automatically promote these types to **int** when evaluating an expression. For example, if you right shift a **byte** value, it will first be promoted to **int** and then shifted. The result of the shift will also be of type **int**. Often this conversion is of no consequence. However, if you shift a negative **byte** or **short** value, it will be sign-extended when it is promoted to **int**. Thus, the high-order bits of the resulting integer value will be filled with ones. This is fine when performing a normal right shift. But when you perform a zero-fill right shift, there are 24 ones to be shifted before the byte value begins to see zeros.

Bitwise Shorthand Assignments

All of the binary bitwise operators have a shorthand form that combines an assignment with the bitwise operation. For example, the following two statements both assign to **x** the outcome of an XOR of **x** with the value 127.

```
x = x ^ 127;
x ^= 127;
```

Ask the Expert

Q: Since binary is based on powers of two, can the shift operators be used as a shortcut for multiplying or dividing an integer by two?

A: Yes. The bitwise shift operators can be used to perform very fast multiplication or division by two. A shift left doubles a value. A shift right halves it.

Try This 5-3 A ShowBits Class

ShowBitsDemo.java

This project creates a class called **ShowBits** that enables you to display in binary the bit pattern for any integer value. Such a class can be quite useful in programming. For example, if you are debugging device-driver code, then being able to monitor the data stream in binary is often a benefit.

1. Create a file called **ShowBitsDemo.java**.

2. Begin the **ShowBits** class as shown here:

```
class ShowBits {
  int numbits;

  ShowBits(int n) {
    numbits = n;
  }
```

ShowBits creates objects that display a specified number of bits. For example, to create an object that will display the low-order 8 bits of some value, use

```
ShowBits byteval = new ShowBits(8)
```

The number of bits to display is stored in **numbits**.

3. To actually display the bit pattern, **ShowBits** provides the method **show( )**, which is shown here:

```
void show(long val) {
  long mask = 1;

  // left-shift a 1 into the proper position
  mask <<= numbits-1;

  int spacer = 0;
  for(; mask != 0; mask >>>= 1) {
    if((val & mask) != 0) System.out.print("1");
    else System.out.print("0");
    spacer++;
    if((spacer % 8) == 0) {
      System.out.print(" ");
      spacer = 0;
    }
  }
  System.out.println();
}
```

Notice that **show( )** specifies one **long** parameter. This does not mean that you always have to pass **show( )** a **long** value, however. Because of Java's automatic type promotions, any integer type can be passed to **show( )**. The number of bits displayed is determined by the value stored in **numbits**. After each group of 8 bits, **show( )** outputs a space. This makes it easier to read the binary values of long bit patterns.

4. The **ShowBitsDemo** program is shown here:

```
/*
    Try This 5-3
    A class that displays the binary representation of a value.
*/

class ShowBits {
  int numbits;

  ShowBits(int n) {
    numbits = n;
  }

  void show(long val) {
    long mask = 1;

    // left-shift a 1 into the proper position
    mask <<= numbits-1;

    int spacer = 0;
    for(; mask != 0; mask >>>= 1) {
      if((val & mask) != 0) System.out.print("1");
      else System.out.print("0");
      spacer++;
      if((spacer % 8) == 0) {
        System.out.print(" ");
        spacer = 0;
      }
    }
    System.out.println();
  }
}

// Demonstrate ShowBits.
class ShowBitsDemo {
  public static void main(String[] args) {
    ShowBits b = new ShowBits(8);
    ShowBits i = new ShowBits(32);
    ShowBits li = new ShowBits(64);
```

(continued)

```
        System.out.println("123 in binary: ");
        b.show(123);

        System.out.println("\n87987 in binary: ");
        i.show(87987);

        System.out.println("\n237658768 in binary: ");
        li.show(237658768);


        // you can also show low-order bits of any integer
        System.out.println("\nLow order 8 bits of 87987 in binary: ");
        b.show(87987);
    }
}
```

5. The output from **ShowBitsDemo** is shown here:

```
123 in binary:
01111011

87987 in binary:
00000000 00000001 01010111 10110011

237658768 in binary:
00000000 00000000 00000000 00000000 00001110 00101010 01100010
10010000

Low order 8 bits of 87987 in binary:
10110011
```

The ? Operator

One of Java's most fascinating operators is the **?**. The **?** operator is often used to replace **if-else** statements of this general form:

if (*condition*)
  myVar = expression1;
else
  myVar = expression2;

Here, the value assigned to *myVar* depends upon the outcome of the condition controlling the **if**.

The **?** is called a *ternary operator* because it requires three operands. It takes the general form

Exp1 ? *Exp2* : *Exp3*;

where *Exp1* is a **boolean** expression, and *Exp2* and *Exp3* are expressions of any type other than **void**. The type of *Exp2* and *Exp3* must be the same (or compatible), though. Notice the use and placement of the colon.

The value of a **?** expression is determined like this: *Exp1* is evaluated. If it is true, then *Exp2* is evaluated and becomes the value of the entire **?** expression. If *Exp1* is false, then *Exp3* is evaluated and its value becomes the value of the expression. Consider this example, which assigns **absval** the absolute value of **val**:

```
absval = val < 0 ? -val : val; // get absolute value of val
```

Here, **absval** will be assigned the value of **val** if **val** is zero or greater. If **val** is negative, then **absval** will be assigned the negative of that value (which yields a positive value). The same code written using the **if-else** structure would look like this:

```
if(val < 0) absval = -val;
else absval = val;
```

Here is another example of the **?** operator. This program divides two numbers, but will not allow a division by zero.

```
// Prevent a division by zero using the ?.
class NoZeroDiv {
  public static void main(String[] args) {
    int result;

    for(int i = -5; i < 6; i++) {
      result = i != 0 ? 100 / i : 0;  ◄───────── This prevents a divide-by-zero.
      if(i != 0)
        System.out.println("100 / " + i + " is " + result);
    }
  }
}
```

The output from the program is shown here:

```
100 / -5 is -20
100 / -4 is -25
100 / -3 is -33
100 / -2 is -50
100 / -1 is -100
100 / 1 is 100
100 / 2 is 50
100 / 3 is 33
100 / 4 is 25
100 / 5 is 20
```

Pay special attention to this line from the program:

```
result = i != 0 ? 100 / i : 0;
```

Here, **result** is assigned the outcome of the division of 100 by **i**. However, this division takes place only if **i** is not zero. When **i** is zero, a placeholder value of zero is assigned to **result**.

You don't actually have to assign the value produced by the **?** to some variable. For example, you could use the value as an argument in a call to a method. Or, if the expressions are all of type **boolean**, the **?** can be used as the conditional expression in a loop or **if** statement. For example, here is the preceding program rewritten a bit more efficiently. It produces the same output as before.

```java
// Prevent a division by zero using the ?.
class NoZeroDiv2 {
  public static void main(String[] args) {

    for(int i = -5; i < 6; i++)
      if(i != 0 ? true : false)
        System.out.println("100 / " + i +
                            " is " + 100 / i);
  }
}
```

Notice the **if** statement. If **i** is zero, then the outcome of the **if** is false, the division by zero is prevented, and no result is displayed. Otherwise, the division takes place.

Chapter 5 Self Test

1. Show two ways to declare a one-dimensional array of 12 **double**s.

2. Show how to initialize a one-dimensional array of integers to the values 1 through 5.

3. Write a program that uses an array to find the average of 10 **double** values. Use any 10 values you like.

4. Change the sort in Try This 5-1 so that it sorts an array of strings. Demonstrate that it works.

5. What is the difference between the **String** methods **indexOf( )** and **lastIndexOf( )**?

6. Since all strings are objects of type **String**, show how you can call the **length( )** and **charAt( )** methods on this string literal: "I like Java".

7. Expanding on the **Encode** cipher class, modify it so that it uses an eight-character string as the key.

8. Can the bitwise operators be applied to the **double** type?

9. Show how this sequence can be rewritten using the **?** operator.

```
if(x < 0) y = 10;
else y = 20;
```

10. In the following fragment, is the **&** a bitwise or logical operator? Why?

```
boolean a, b;
// ...
if(a & b) ...
```

11. Is it an error to overrun the end of an array? Is it an error to index an array with a negative value?

12. What is the unsigned right-shift operator?

13. Rewrite the **MinMax** class shown earlier in this chapter so that it uses a for-each style **for** loop.

14. Can the **for** loops that perform sorting in the **Bubble** class shown in Try This 5-1 be converted into for-each style loops? If not, why not?

15. Can a **String** control a **switch** statement?

16. What keyword is reserved for use with local variable type inference?

17. Show how to use local variable type inference to declare a **boolean** variable called **done** that has an initial value of **false**.

18. Can **var** be the name of a variable? Can **var** be the name of a class?

19. Is the following declaration valid? If not, why not.

```
var[] avgTemps = new double[7];
```

20. Is the following declaration valid? If not, why not?

```
var alpha = 10, beta = 20;
```

21. In the **show( )** method of the **ShowBits** class developed in Try This 5-3, the local variable **mask** is declared as shown here:

```
long mask = 1;
```

Change this declaration so that it uses local variable type inference. When doing so, be sure that **mask** is of type **long** (as it is here), and not of type **int**.

Chapter 6

A Closer Look at Methods and Classes

Key Skills & Concepts

- Control access to members

- Pass objects to a method

- Return objects from a method

- Overload methods

- Overload constructors

- Use recursion

- Apply **static**

- Use inner classes

- Use varargs

This chapter resumes our examination of classes and methods. It begins by explaining how to control access to the members of a class. It then discusses the passing and returning of objects, method overloading, recursion, and the use of the keyword **static**. Also described are nested classes and variable-length arguments.

Controlling Access to Class Members

In its support for encapsulation, the class provides two major benefits. First, it links data with the code that manipulates it. You have been taking advantage of this aspect of the class since Chapter 4. Second, it provides the means by which access to members can be controlled. It is this feature that is examined here.

Although Java's approach is a bit more sophisticated, in essence, there are two basic types of class members: public and private. A public member can be freely accessed by code defined outside of its class. A private member can be accessed only by other methods defined by its class. It is through the use of private members that access is controlled.

Restricting access to a class' members is a fundamental part of object-oriented programming because it helps prevent the misuse of an object. By allowing access to private data only through a well-defined set of methods, you can prevent improper values from being assigned to that data—by performing a range check, for example. It is not possible for code outside the class to set the value of a private member directly. You can also control precisely how and when the data within an object is used. Thus, when correctly implemented, a class creates a "black box" that can be used, but the inner workings of which are not open to tampering.

Up to this point, you haven't had to worry about access control because Java provides a default access setting in which, for the types of programs shown earlier, the members of a class are freely available to the other code in the program. (Thus, for the preceding examples, the default access setting is essentially public.) Although convenient for simple classes (and example programs in books such as this one), this default setting is inadequate for many real-world situations. Here we introduce Java's other access control features.

Java's Access Modifiers

Member access control is achieved through the use of three *access modifiers*: **public**, **private**, and **protected**. As explained, if no access modifier is used, the default access setting is assumed. In this chapter, we will be concerned with **public** and **private**. The **protected** modifier applies only when inheritance is involved and is described in Chapter 8.

When a member of a class is modified by the **public** specifier, that member can be accessed by any other code in your program. This includes by methods defined inside other classes.

When a member of a class is specified as **private**, that member can be accessed only by other members of its class. Thus, methods in other classes cannot access a **private** member of another class.

The default access setting (in which no access modifier is used) is the same as **public** unless your program is broken down into packages. A *package* is, essentially, a grouping of classes. Packages are both an organizational and an access control feature, but a discussion of packages must wait until Chapter 8. For the types of programs shown in this and the preceding chapters, **public** access is the same as default access.

An access modifier precedes the rest of a member's type specification. That is, it must begin a member's declaration statement. Here are some examples:

```
public String errMsg;
private accountBalance bal;

private boolean isError(byte status) { // ...
```

To understand the effects of **public** and **private**, consider the following program:

```
// Public vs private access.
class MyClass {
  private int alpha; // private access
  public int beta; // public access
  int gamma; // default access

  /* Methods to access alpha. It is OK for a
     member of a class to access a private member
     of the same class.
  */
  void setAlpha(int a) {
    alpha = a;
  }
```

```
   int getAlpha() {
     return alpha;
   }
}

class AccessDemo {
  public static void main(String[] args) {
    MyClass ob = new MyClass();

    /* Access to alpha is allowed only through
       its accessor methods. */
    ob.setAlpha(-99);
    System.out.println("ob.alpha is " + ob.getAlpha());

    // You cannot access alpha like this:
//  ob.alpha = 10; // Wrong! alpha is private!   ◄──── Wrong—alpha is private!

    // These are OK because beta and gamma are public.
    ob.beta = 88;   ◄──── OK because these are public.
    ob.gamma = 99;
  }
}
```

As you can see, inside the **MyClass** class, **alpha** is specified as **private**, **beta** is explicitly specified as **public**, and **gamma** uses the default access, which for this example is the same as specifying **public**. Because **alpha** is private, it cannot be accessed by code outside of its class. Therefore, inside the **AccessDemo** class, **alpha** cannot be used directly. It must be accessed through its public accessor methods: **setAlpha( )** and **getAlpha( )**. If you were to remove the comment symbol from the beginning of the following line,

```
//  ob.alpha = 10; // Wrong! alpha is private!
```

you would not be able to compile this program because of the access violation. Although access to **alpha** by code outside of **MyClass** is not allowed, methods defined within **MyClass** can freely access it, as the **setAlpha( )** and **getAlpha( )** methods show.

The key point is this: A private member can be used freely by other members of its class, but it cannot be accessed by code outside its class.

To see how access control can be applied to a more practical example, consider the following program that implements a "fail-soft" **int** array, in which boundary errors are prevented, thus avoiding a run-time exception from being generated. This is accomplished by encapsulating the array as a private member of a class, allowing access to the array only through member methods. With this approach, any attempt to access the array beyond its boundaries can be prevented, with such an attempt failing gracefully (resulting in a "soft" landing rather than a "crash"). The fail-soft array is implemented by the **FailSoftArray** class, shown here:

```
/* This class implements a "fail-soft" array which prevents
   runtime errors.
*/
```

```
class FailSoftArray {
  private int[] a; // reference to array
  private int errval; // value to return if get() fails
  public int length; // length is public

  /* Construct array given its size and the value to
     return if get() fails. */
  public FailSoftArray(int size, int errv) {
    a = new int[size];
    errval = errv;
    length = size;
  }

  // Return value at given index.
  public int get(int index) {
    if(indexOK(index)) return a[index];      Trap on out-of-bounds index.
    return errval;
  }

  // Put a value at an index. Return false on failure.
  public boolean put(int index, int val) {
    if(indexOK(index)) {
      a[index] = val;
      return true;
    }
    return false;
  }

  // Return true if index is within bounds.
  private boolean indexOK(int index) {
    if(index >= 0 & index < length) return true;
    return false;
  }
}

// Demonstrate the fail-soft array.
class FSDemo {
  public static void main(String[] args) {
    FailSoftArray fs = new FailSoftArray(5, -1);
    int x;

    // show quiet failures
    System.out.println("Fail quietly.");
    for(int i=0; i < (fs.length * 2); i++)
      fs.put(i, i*10);       Access to array must be through its accessor methods.

    for(int i=0; i < (fs.length * 2); i++) {
      x = fs.get(i);
```

```
      if(x != -1) System.out.print(x + " ");
    }
    System.out.println("");

    // now, handle failures
    System.out.println("\nFail with error reports.");
    for(int i=0; i < (fs.length * 2); i++)
      if(!fs.put(i, i*10))
        System.out.println("Index " + i + " out-of-bounds");

    for(int i=0; i < (fs.length * 2); i++) {
      x = fs.get(i);
      if(x != -1) System.out.print(x + " ");
      else
        System.out.println("Index " + i + " out-of-bounds");
    }
  }
}
```

The output from the program is shown here:

```
Fail quietly.
0 10 20 30 40

Fail with error reports.
Index 5 out-of-bounds
Index 6 out-of-bounds
Index 7 out-of-bounds
Index 8 out-of-bounds
Index 9 out-of-bounds
0 10 20 30 40 Index 5 out-of-bounds
Index 6 out-of-bounds
Index 7 out-of-bounds
Index 8 out-of-bounds
Index 9 out-of-bounds
```

Let's look closely at this example. Inside **FailSoftArray** are defined three **private** members. The first is **a**, which stores a reference to the array that will actually hold information. The second is **errval**, which is the value that will be returned when a call to **get( )** fails. The third is the **private** method **indexOK( )**, which determines whether an index is within bounds. Thus, these three members can be used only by other members of the **FailSoftArray** class. Specifically, **a** and **errval** can be used only by other methods in the class, and **indexOK( )** can be called only by other members of **FailSoftArray**. The rest of the class members are **public** and can be called by any other code in a program that uses **FailSoftArray**.

When a **FailSoftArray** object is constructed, you must specify the size of the array and the value that you want to return if a call to **get( )** fails. The error value must be a value that would otherwise not be stored in the array. Once constructed, the actual array referred to by **a** and the error value stored in **errval** cannot be accessed by users of the **FailSoftArray** object. Thus, they are not open to misuse. For example, the user cannot try to index **a** directly, possibly exceeding its bounds. Access is available only through the **get( )** and **put( )** methods.

The **indexOK( )** method is **private** mostly for the sake of illustration. It would be harmless to make it **public** because it does not modify the object. However, since it is used internally by the **FailSoftArray** class, it can be **private**.

Notice that the **length** instance variable is **public**. This is in keeping with the way Java implements arrays. To obtain the length of a **FailSoftArray**, simply use its **length** member.

To use a **FailSoftArray** array, call **put( )** to store a value at the specified index. Call **get( )** to retrieve a value from a specified index. If the index is out-of-bounds, **put( )** returns **false** and **get( )** returns **errval**.

For the sake of convenience, the majority of the examples in this book will continue to use default access for most members. Remember, however, that in the real world, restricting access to members— especially instance variables—is an important part of successful object-oriented programming. As you will see in Chapter 7, access control is even more vital when inheritance is involved.

NOTE

The modules feature added by JDK 9 can also play a role in accessibility. Modules are discussed in Chapter 15.

Try This 6-1   Improving the Queue Class

Queue.java

You can use the **private** modifier to make a rather important improvement to the **Queue** class developed in Chapter 5, Try This 5-2. In that version, all members of the **Queue** class use the default access. This means that it would be possible for a program that uses a **Queue** to directly access the underlying array, possibly accessing its elements out of turn. Since the entire point of a queue is to provide a first-in, first-out list, allowing out-of-order access is not desirable. It would also be possible for a malicious programmer to alter the values stored in the **putloc** and **getloc** indices, thus corrupting the queue. Fortunately, these types of problems are easy to prevent by applying the **private** specifier.

1. Copy the original **Queue** class in Try This 5-2 to a new file called **Queue.java**.

2. In the **Queue** class, add the **private** modifier to the **q** array, and the indices **putloc** and **getloc**, as shown here:

```
// An improved queue class for characters.
class Queue {
  // these members are now private
  private char[] q; // this array holds the queue
  private int putloc, getloc; // the put and get indices

  Queue(int size) {
    q = new char[size]; // allocate memory for queue
    putloc = getloc = 0;
  }
```

(continued)

```
    // Put a character into the queue.
    void put(char ch) {
      if(putloc==q.length) {
        System.out.println(" - Queue is full.");
        return;
      }

      q[putloc++] = ch;
    }

    // Get a character from the queue.
    char get() {
      if(getloc == putloc) {
        System.out.println(" - Queue is empty.");
        return (char) 0;
      }

      return q[getloc++];
    }
  }
```

3. Changing **q**, **putloc**, and **getloc** from default access to private access has no effect on a program that properly uses **Queue**. For example, it still works fine with the **QDemo** class from Try This 5-2. However, it prevents the improper use of a **Queue**. For example, the following types of statements are illegal:

```
Queue test = new Queue(10);

test.q[0] = 99; // wrong!
test.putloc = -100; // won't work!
```

4. Now that **q**, **putloc**, and **getloc** are private, the **Queue** class strictly enforces the first-in, first-out attribute of a queue.

Pass Objects to Methods

Up to this point, the examples in this book have been using simple types as parameters to methods. However, it is both correct and common to pass objects to methods. For example, the following program defines a class called **Block** that stores the dimensions of a three-dimensional block:

```
// Objects can be passed to methods.
class Block {
  int a, b, c;
  int volume;
```

```
Block(int i, int j, int k) {
   a = i;
   b = j;
   c = k;
   volume = a * b * c;
}

// Return true if ob defines same block.
boolean sameBlock(Block ob) {                    ──────── Use object type for parameter.
   if((ob.a == a) & (ob.b == b) & (ob.c == c)) return true;
   else return false;
}

// Return true if ob has same volume.
boolean sameVolume(Block ob) {  ←
   if(ob.volume == volume) return true;
   else return false;
}
}

class PassOb {
  public static void main(String[] args) {
     Block ob1 = new Block(10, 2, 5);
     Block ob2 = new Block(10, 2, 5);
     Block ob3 = new Block(4, 5, 5);

     System.out.println("ob1 same dimensions as ob2: " +
                        ob1.sameBlock(ob2));  ←──────────── Pass an object.
     System.out.println("ob1 same dimensions as ob3: " +
                        ob1.sameBlock(ob3));  ←
     System.out.println("ob1 same volume as ob3: " +
                        ob1.sameVolume(ob3));  ←
  }
}
```

This program generates the following output:

```
ob1 same dimensions as ob2: true
ob1 same dimensions as ob3: false
ob1 same volume as ob3: true
```

The **sameBlock( )** and **sameVolume( )** methods compare the **Block** object passed as
a parameter to the invoking object. For **sameBlock( )**, the dimensions of the objects are
compared and **true** is returned only if the two blocks are the same. For **sameVolume( )**, the
two blocks are compared only to determine whether they have the same volume. In both cases,
notice that the parameter **ob** specifies **Block** as its type. Although **Block** is a class type created
by the program, it is used in the same way as Java's built-in types.

How Arguments Are Passed

As the preceding example demonstrated, passing an object to a method is a straightforward task. However, there are some nuances of passing an object that are not shown in the example. In certain cases, the effects of passing an object will be different from those experienced when passing non-object arguments. To see why, you need to understand in a general sense the two ways in which an argument can be passed to a subroutine.

The first way is *call-by-value*. This approach copies the *value* of an argument into the formal parameter of the subroutine. Therefore, changes made to the parameter of the subroutine have no effect on the argument in the call. The second way an argument can be passed is *call-by-reference*. In this approach, a reference to an argument (not the value of the argument) is passed to the parameter. Inside the subroutine, this reference is used to access the actual argument specified in the call. This means that changes made to the parameter *will* affect the argument used to call the subroutine. As you will see, although Java uses call-by-value to pass arguments, the precise effect differs between whether a primitive type or a reference type is passed.

When you pass a primitive type, such as **int** or **double**, to a method, it is passed by value. Thus, a copy of the argument is made, and what occurs to the parameter that receives the argument has no effect outside the method. For example, consider the following program:

```
// Primitive types are passed by value.
class Test {
  /* This method causes no change to the arguments
     used in the call. */
  void noChange(int i, int j) {
    i = i + j;
    j = -j;
  }
}

class CallByValue {
  public static void main(String[] args) {
    Test ob = new Test();

    int a = 15, b = 20;

    System.out.println("a and b before call: " +
                       a + " " + b);

    ob.noChange(a, b);

    System.out.println("a and b after call: " +
                       a + " " + b);
  }
}
```

The output from this program is shown here:

```
a and b before call: 15 20
a and b after call: 15 20
```

As you can see, the operations that occur inside **noChange( )** have no effect on the values of **a** and **b** used in the call.

When you pass an object to a method, the situation changes dramatically, because objects are implicitly passed by reference. Keep in mind that when you create a variable of a class type, you are creating a reference to an object. It is the reference, not the object itself, that is actually passed to the method. As a result, when you pass this reference to a method, the parameter that receives it will refer to the same object as that referred to by the argument. This effectively means that objects are passed to methods by use of call-by-reference. Changes to the object inside the method *do* affect the object used as an argument. For example, consider the following program:

```
// Objects are passed through their references.
class Test {
  int a, b;

  Test(int i, int j) {
    a = i;
    b = j;
  }
  /* Pass an object. Now, ob.a and ob.b in object
     used in the call will be changed. */
  void change(Test ob) {
    ob.a = ob.a + ob.b;
    ob.b = -ob.b;
  }
}

class PassObRef {
  public static void main(String[] args) {
    Test ob = new Test(15, 20);

    System.out.println("ob.a and ob.b before call: " +
                       ob.a + " " + ob.b);

    ob.change(ob);

    System.out.println("ob.a and ob.b after call: " +
                       ob.a + " " + ob.b);
  }
}
```

This program generates the following output:

```
ob.a and ob.b before call: 15 20
ob.a and ob.b after call: 35 -20
```

As you can see, in this case, the actions inside **change( )** have affected the object used as an argument.

Ask the Expert

Q: **Is there any way that I can pass a primitive type by reference?**

A: Not directly. However, Java defines a set of classes that *wrap* the primitive types in objects. These are **Double**, **Float**, **Byte**, **Short**, **Integer**, **Long**, and **Character**. In addition to allowing a primitive type to be passed by reference, these wrapper classes define several methods that enable you to manipulate their values. For example, the numeric type wrappers include methods that convert a numeric value from its binary form into its human-readable **String** form, and vice versa.

Remember, when an object reference is passed to a method, the reference itself is passed by use of call-by-value. However, since the value being passed refers to an object, the copy of that value will still refer to the same object referred to by its corresponding argument.

Returning Objects

A method can return any type of data, including class types. For example, the class **ErrorMsg** shown here could be used to report errors. Its method, **getErrorMsg( )**, returns a **String** object that contains a description of an error based upon the error code that it is passed.

```
// Return a String object.
class ErrorMsg {
  String[] msgs = {
    "Output Error",
    "Input Error",
    "Disk Full",
    "Index Out-Of-Bounds"
  };

  // Return the error message.
  String getErrorMsg(int i) {          ←——— Return an object of type String.
    if(i >=0 & i < msgs.length)
      return msgs[i];
    else
      return "Invalid Error Code";
  }
}

class ErrMsg {
  public static void main(String[] args) {
    ErrorMsg err = new ErrorMsg();
```

```
    System.out.println(err.getErrorMsg(2));
    System.out.println(err.getErrorMsg(19));
  }
}
```

Its output is shown here:

```
Disk Full
Invalid Error Code
```

You can, of course, also return objects of classes that you create. For example, here is a reworked version of the preceding program that creates two error classes. One is called **Err**, and it encapsulates an error message along with a severity code. The second is called **ErrorInfo**. It defines a method called **getErrorInfo( )**, which returns an **Err** object.

```
// Return a programmer-defined object.
class Err {
  String msg; // error message
  int severity; // code indicating severity of error

  Err(String m, int s) {
    msg = m;
    severity = s;
  }
}

class ErrorInfo {
  String[] msgs = {
    "Output Error",
    "Input Error",
    "Disk Full",
    "Index Out-Of-Bounds"
  };
  int[] howBad = { 3, 3, 2, 4 };

  Err getErrorInfo(int i) {  ◄─────── Return an object of type Err.
    if(i >= 0 & i < msgs.length)
      return new Err(msgs[i], howBad[i]);
    else
      return new Err("Invalid Error Code", 0);
  }
}

class ErrInfo {
  public static void main(String[] args) {
    ErrorInfo err = new ErrorInfo();
    Err e;
```

```
   e = err.getErrorInfo(2);
   System.out.println(e.msg + " severity: " + e.severity);

   e = err.getErrorInfo(19);
   System.out.println(e.msg + " severity: " + e.severity);
 }
}
```

Here is the output:

```
Disk Full severity: 2
Invalid Error Code severity: 0
```

Each time **getErrorInfo( )** is invoked, a new **Err** object is created, and a reference to it is returned to the calling routine. This object is then used within **main( )** to display the error message and severity code.

When an object is returned by a method, it remains in existence until there are no more references to it. At that point, it is subject to garbage collection. Thus, an object won't be destroyed just because the method that created it terminates.

Method Overloading

In this section, you will learn about one of Java's most exciting features: method overloading. In Java, two or more methods within the same class can share the same name, as long as their parameter declarations are different. When this is the case, the methods are said to be *overloaded,* and the process is referred to as *method overloading.* Method overloading is one of the ways that Java implements polymorphism.

In general, to overload a method, simply declare different versions of it. The compiler takes care of the rest. You must observe one important restriction: the type and/or number of the parameters of each overloaded method must differ. It is not sufficient for two methods to differ only in their return types. (Return types do not provide sufficient information in all cases for Java to decide which method to use.) Of course, overloaded methods *may* differ in their return types, too. When an overloaded method is called, the version of the method whose parameters match the arguments is executed.

Here is a simple example that illustrates method overloading:

```
// Demonstrate method overloading.
class Overload {
  void ovlDemo() {  ◄──────────────── First version
    System.out.println("No parameters");
  }

  // Overload ovlDemo for one integer parameter.
  void ovlDemo(int a) {  ◄──────────────── Second version
    System.out.println("One parameter: " + a);
  }
```

```
  // Overload ovlDemo for two integer parameters.
  int ovlDemo(int a, int b) {  ◄─────────────────── Third version
    System.out.println("Two parameters: " + a + " " + b);
    return a + b;
  }

  // Overload ovlDemo for two double parameters.
  double ovlDemo(double a, double b) {  ◄─────────── Fourth version
    System.out.println("Two double parameters: " +
                       a + " " + b);
    return a + b;
  }
}

class OverloadDemo {
  public static void main(String[] args) {
    Overload ob = new Overload();
    int resI;
    double resD;

    // call all versions of ovlDemo()
    ob.ovlDemo();
    System.out.println();

    ob.ovlDemo(2);
    System.out.println();

    resI = ob.ovlDemo(4, 6);
    System.out.println("Result of ob.ovlDemo(4, 6): " +
                       resI);
    System.out.println();

    resD = ob.ovlDemo(1.1, 2.32);
    System.out.println("Result of ob.ovlDemo(1.1, 2.32): " +
                       resD);
  }
}
```

This program generates the following output:

```
No parameters

One parameter: 2

Two parameters: 4 6
Result of ob.ovlDemo(4, 6): 10

Two double parameters: 1.1 2.32
Result of ob.ovlDemo(1.1, 2.32): 3.42
```

As you can see, **ovlDemo( )** is overloaded four times. The first version takes no parameters, the second takes one integer parameter, the third takes two integer parameters, and the fourth takes two **double** parameters. Notice that the first two versions of **ovlDemo( )** return **void**, and the second two return a value. This is perfectly valid, but as explained, overloading is not affected one way or the other by the return type of a method. Thus, attempting to use the following two versions of **ovlDemo( )** will cause an error:

```
// One ovlDemo(int) is OK.
void ovlDemo(int a) {                    ◄──────────────── Return types cannot be used to
  System.out.println("One parameter: " + a);               differentiate overloaded methods.
}

/* Error! Two ovlDemo(int)s are not OK even though
   return types differ.
*/
int ovlDemo(int a) {  ◄──────
  System.out.println("One parameter: " + a);
  return a * a;
}
```

As the comments suggest, the difference in their return types is insufficient for the purposes of overloading.

As you will recall from Chapter 2, Java provides certain automatic type conversions. These conversions also apply to parameters of overloaded methods. For example, consider the following:

```
/* Automatic type conversions can affect
   overloaded method resolution.
*/
class Overload2 {
  void f(int x) {
    System.out.println("Inside f(int): " + x);
  }

  void f(double x) {
    System.out.println("Inside f(double): " + x);
  }
}

class TypeConv {
  public static void main(String[] args) {
    Overload2 ob = new Overload2();

    int i = 10;
    double d = 10.1;

    byte b = 99;
    short s = 10;
    float f = 11.5F;
```

```
  ob.f(i); // calls ob.f(int)
  ob.f(d); // calls ob.f(double)

  ob.f(b); // calls ob.f(int) - type conversion
  ob.f(s); // calls ob.f(int) - type conversion
  ob.f(f); // calls ob.f(double) - type conversion
  }
}
```

The output from the program is shown here:

```
Inside f(int): 10
Inside f(double): 10.1
Inside f(int): 99
Inside f(int): 10
Inside f(double): 11.5
```

In this example, only two versions of **f( )** are defined: one that has an **int** parameter and one that has a **double** parameter. However, it is possible to pass **f( )** a **byte**, **short**, or **float** value. In the case of **byte** and **short**, Java automatically converts them to **int**. Thus, **f(int)** is invoked. In the case of **float**, the value is converted to **double** and **f(double)** is called.

It is important to understand, however, that the automatic conversions apply only if there is no direct match between a parameter and an argument. For example, here is the preceding program with the addition of a version of **f( )** that specifies a **byte** parameter:

```
// Add f(byte).
class Overload2 {
  void f(byte x) {         ◄──────────────────────────    This version specifies
    System.out.println("Inside f(byte): " + x);           a byte parameter.
  }

  void f(int x) {
    System.out.println("Inside f(int): " + x);
  }

  void f(double x) {
    System.out.println("Inside f(double): " + x);
  }
}

class TypeConv {
  public static void main(String[] args) {
    Overload2 ob = new Overload2();

    int i = 10;
    double d = 10.1;

    byte b = 99;
    short s = 10;
    float f = 11.5F;
```

```
ob.f(i); // calls ob.f(int)
ob.f(d); // calls ob.f(double)

ob.f(b); // calls ob.f(byte) - now, no type conversion

ob.f(s); // calls ob.f(int) - type conversion
ob.f(f); // calls ob.f(double) - type conversion
  }
}
```

Now when the program is run, the following output is produced:

```
Inside f(int): 10
Inside f(double): 10.1
Inside f(byte): 99
Inside f(int): 10
Inside f(double): 11.5
```

In this version, since there is a version of **f( )** that takes a **byte** argument, when **f( )** is called with a **byte** argument, **f(byte)** is invoked and the automatic conversion to **int** does not occur.

Method overloading supports polymorphism because it is one way that Java implements the "one interface, multiple methods" paradigm. To understand how, consider the following: In languages that do not support method overloading, each method must be given a unique name. However, frequently you will want to implement essentially the same method for different types of data. Consider the absolute value function. In languages that do not support overloading, there are usually three or more versions of this function, each with a slightly different name. For instance, in C, the function **abs( )** returns the absolute value of an integer, **labs( )** returns the absolute value of a long integer, and **fabs( )** returns the absolute value of a floating-point value. Since C does not support overloading, each function has to have its own name, even though all three functions do essentially the same thing. This makes the situation more complex, conceptually, than it actually is. Although the underlying concept of each function is the same, you still have three names to remember. This situation does not occur in Java, because each absolute value method can use the same name. Indeed, Java's standard class library includes an absolute value method, called **abs( )**. This method is overloaded by Java's **Math** class to handle all of the numeric types. Java determines which version of **abs( )** to call based upon the type of argument.

The value of overloading is that it allows related methods to be accessed by use of a common name. Thus, the name **abs** represents the *general action* that is being performed. It is left to the compiler to choose the correct *specific* version for a particular circumstance. You, the programmer, need only remember the general operation being performed. Through the application of polymorphism, several names have been reduced to one. Although this example is fairly simple, if you expand the concept, you can see how overloading can help manage greater complexity.

When you overload a method, each version of that method can perform any activity you desire. There is no rule stating that overloaded methods must relate to one another. However, from a stylistic point of view, method overloading implies a relationship. Thus, while you can

Ask the Expert

Q: I've heard the term *signature* used by Java programmers. What is it?

A: As it applies to Java, a signature is the name of a method plus its parameter list. Thus, for the purposes of overloading, no two methods within the same class can have the same signature. Notice that a signature does not include the return type, since it is not used by Java for overload resolution.

use the same name to overload unrelated methods, you should not. For example, you could use the name **sqr** to create methods that return the *square* of an integer and the *square root* of a floating-point value. But these two operations are fundamentally different. Applying method overloading in this manner defeats its original purpose. In practice, you should overload only closely related operations.

Overloading Constructors

Like methods, constructors can also be overloaded. Doing so allows you to construct objects in a variety of ways. For example, consider the following program:

```java
// Demonstrate an overloaded constructor.
class MyClass {
  int x;

  MyClass() {                                          // Construct objects in a variety of ways.
    System.out.println("Inside MyClass().");
    x = 0;
  }

  MyClass(int i) {
    System.out.println("Inside MyClass(int).");
    x = i;
  }

  MyClass(double d) {
    System.out.println("Inside MyClass(double).");
    x = (int) d;
  }

  MyClass(int i, int j) {
    System.out.println("Inside MyClass(int, int).");
    x = i * j;
  }
}
```

```
class OverloadConsDemo {
  public static void main(String[] args) {
    MyClass t1 = new MyClass();
    MyClass t2 = new MyClass(88);
    MyClass t3 = new MyClass(17.23);
    MyClass t4 = new MyClass(2, 4);

    System.out.println("t1.x: " + t1.x);
    System.out.println("t2.x: " + t2.x);
    System.out.println("t3.x: " + t3.x);
    System.out.println("t4.x: " + t4.x);
  }
}
```

The output from the program is shown here:

```
Inside MyClass().
Inside MyClass(int).
Inside MyClass(double).
Inside MyClass(int, int).
t1.x: 0
t2.x: 88
t3.x: 17
t4.x: 8
```

MyClass( ) is overloaded four ways, each constructing an object differently. The proper constructor is called based upon the parameters specified when **new** is executed. By overloading a class' constructor, you give the user of your class flexibility in the way objects are constructed.

One of the most common reasons that constructors are overloaded is to allow one object to initialize another. For example, consider this program that uses the **Summation** class to compute the summation of an integer value:

```
// Initialize one object with another.
class Summation {
  int sum;

  // Construct from an int.
  Summation(int num) {
    sum = 0;
    for(int i=1; i <= num; i++)
      sum += i;
  }

  // Construct from another object.
  Summation(Summation ob) {  ←——————— Construct one object from another.
    sum = ob.sum;
  }
}
```

```
class SumDemo {
  public static void main(String[] args) {
    Summation s1 = new Summation(5);
    Summation s2 = new Summation(s1);

    System.out.println("s1.sum: " + s1.sum);
    System.out.println("s2.sum: " + s2.sum);
  }
}
```

The output is shown here:

```
s1.sum: 15
s2.sum: 15
```

Often, as this example shows, an advantage of providing a constructor that uses one object to initialize another is efficiency. In this case, when **s2** is constructed, it is not necessary to recompute the summation. Of course, even in cases when efficiency is not an issue, it is often useful to provide a constructor that makes a copy of an object.

Try This 6-2   Overloading the Queue Constructor

QDemo2.java

In this project, you will enhance the **Queue** class by giving it two additional constructors. The first will construct a new queue from another queue. The second will construct a queue, giving it initial values. As you will see, adding these constructors enhances the usability of **Queue** substantially.

1. Create a file called **QDemo2.java** and copy the updated **Queue** class from Try This 6-1 into it.

2. First, add the following constructor, which constructs a queue from a queue.

```
// Construct a Queue from a Queue.
Queue(Queue ob) {
  putloc = ob.putloc;
  getloc = ob.getloc;
  q = new char[ob.q.length];

  // copy elements
  for(int i=getloc; i < putloc; i++)
    q[i] = ob.q[i];
}
```

Look closely at this constructor. It initializes **putloc** and **getloc** to the values contained in the **ob** parameter. It then allocates a new array to hold the queue and copies the elements from **ob** into that array. Once constructed, the new queue will be an identical copy of the original, but both will be completely separate objects.

(continued)

3. Now add the constructor that initializes the queue from a character array, as shown here:

```
// Construct a Queue with initial values.
Queue(char[] a) {
  putloc = 0;
  getloc = 0;
  q = new char[a.length];

  for(int i = 0; i < a.length; i++) put(a[i]);
}
```

This constructor creates a queue large enough to hold the characters in **a** and then stores those characters in the queue.

4. Here is the complete updated **Queue** class along with the **QDemo2** class, which demonstrates it:

```
// A queue class for characters.
class Queue {
  private char[] q; // this array holds the queue
  private int putloc, getloc; // the put and get indices

  // Construct an empty Queue given its size.
  Queue(int size) {
    q = new char[size]; // allocate memory for queue
    putloc = getloc = 0;
  }

  // Construct a Queue from a Queue.
  Queue(Queue ob) {
    putloc = ob.putloc;
    getloc = ob.getloc;
    q = new char[ob.q.length];

    // copy elements
    for(int i=getloc; i < putloc; i++)
      q[i] = ob.q[i];
  }

  // Construct a Queue with initial values.
  Queue(char[] a) {
    putloc = 0;
    getloc = 0;
    q = new char[a.length];

    for(int i = 0; i < a.length; i++) put(a[i]);
  }
```

```java
    // Put a character into the queue.
    void put(char ch) {
      if(putloc==q.length) {
        System.out.println(" - Queue is full.");
        return;
      }

      q[putloc++] = ch;
    }

    // Get a character from the queue.
    char get() {
      if(getloc == putloc) {
        System.out.println(" - Queue is empty.");
        return (char) 0;
      }

      return q[getloc++];
    }
  }

  // Demonstrate the Queue class.
  class QDemo2 {
    public static void main(String[] args) {
      // construct 10-element empty queue
      Queue q1 = new Queue(10);

      char[] name = {'T', 'o', 'm'};
      // construct queue from array
      Queue q2 = new Queue(name);

      char ch;
      int i;

      // put some characters into q1
      for(i=0; i < 10; i++)
        q1.put((char) ('A' + i));

      // construct queue from another queue
      Queue q3 = new Queue(q1);

      // Show the queues.
      System.out.print("Contents of q1: ");
      for(i=0; i < 10; i++) {
        ch = q1.get();
```

(continued)

```
        System.out.print(ch);
      }

      System.out.println("\n");

      System.out.print("Contents of q2: ");
      for(i=0; i < 3; i++) {
        ch = q2.get();
        System.out.print(ch);
      }

      System.out.println("\n");

      System.out.print("Contents of q3: ");
      for(i=0; i < 10; i++) {
        ch = q3.get();
        System.out.print(ch);
      }
    }
  }
}
```

The output from the program is shown here:

```
Contents of q1: ABCDEFGHIJ

Contents of q2: Tom

Contents of q3: ABCDEFGHIJ
```

Recursion

In Java, a method can call itself. This process is called *recursion,* and a method that calls itself is said to be *recursive.* In general, recursion is the process of defining something in terms of itself and is somewhat similar to a circular definition. The key component of a recursive method is a statement that executes a call to itself. Recursion is a powerful control mechanism.

The classic example of recursion is the computation of the factorial of a number. The *factorial* of a number N is the product of all the whole numbers between 1 and N. For example, 3 factorial is $1 \times 2 \times 3$, or 6. The following program shows a recursive way to compute the factorial of a number. For comparison purposes, a nonrecursive equivalent is also included.

```
// A simple example of recursion.
class Factorial {
  // This is a recursive function.
  int factR(int n) {
    int result;
```

```
      if(n==1) return 1;
      result = factR(n-1) * n;
      return result;
    }                         Execute the recursive call to factR( ).

  // This is an iterative equivalent.
  int factI(int n) {
    int t, result;

    result = 1;
    for(t=1; t <= n; t++) result *= t;
    return result;
  }
}

class Recursion {
  public static void main(String[] args) {
    Factorial f = new Factorial();

    System.out.println("Factorials using recursive method.");
    System.out.println("Factorial of 3 is " + f.factR(3));
    System.out.println("Factorial of 4 is " + f.factR(4));
    System.out.println("Factorial of 5 is " + f.factR(5));
    System.out.println();

    System.out.println("Factorials using iterative method.");
    System.out.println("Factorial of 3 is " + f.factI(3));
    System.out.println("Factorial of 4 is " + f.factI(4));
    System.out.println("Factorial of 5 is " + f.factI(5));
  }
}
```

The output from this program is shown here:

```
Factorials using recursive method.
Factorial of 3 is 6
Factorial of 4 is 24
Factorial of 5 is 120

Factorials using iterative method.
Factorial of 3 is 6
Factorial of 4 is 24
Factorial of 5 is 120
```

The operation of the nonrecursive method **factI( )** should be clear. It uses a loop starting at 1 and progressively multiplies each number by the moving product.

The operation of the recursive **factR( )** is a bit more complex. When **factR( )** is called with an argument of 1, the method returns 1; otherwise, it returns the product of **factR(n–1)*n**. To evaluate this expression, **factR( )** is called with **n–1**. This process repeats until **n** equals 1 and the calls to the method begin returning. For example, when the factorial of 2 is calculated, the first call to **factR( )** will cause a second call to be made with an argument of 1. This call will return 1, which is then multiplied by 2 (the original value of **n**). The answer is then 2. You might find it interesting to insert **println( )** statements into **factR( )** that show at what level each call is, and what the intermediate results are.

When a method calls itself, new local variables and parameters are allocated storage on the stack, and the method code is executed with these new variables from the start. A recursive call does not make a new copy of the method. Only the arguments are new. As each recursive call returns, the old local variables and parameters are removed from the stack, and execution resumes at the point of the call inside the method. Recursive methods could be said to "telescope" out and back.

Recursive versions of many routines may execute a bit more slowly than their iterative equivalents because of the added overhead of the additional method calls. Too many recursive calls to a method could cause a stack overrun. Because storage for parameters and local variables is on the stack and each new call creates a new copy of these variables, it is possible that the stack could be exhausted. If this occurs, the Java run-time system will cause an exception. However, you probably will not encounter this unless a recursive routine runs wild. The main advantage to recursion is that some types of algorithms can be implemented more clearly and simply recursively than they can be iteratively. For example, the Quicksort sorting algorithm is quite difficult to implement in an iterative way. Also, some problems, especially AI-related ones, seem to lend themselves to recursive solutions. When writing recursive methods, you must have a conditional statement, such as an **if**, somewhere to force the method to return without the recursive call being executed. If you don't do this, once you call the method, it will never return. This type of error is very common when working with recursion. Use **println( )** statements liberally so that you can watch what is going on and abort execution if you see that you have made a mistake.

Understanding static

There will be times when you will want to define a class member that will be used independently of any object of that class. Normally a class member must be accessed through an object of its class, but it is possible to create a member that can be used by itself, without reference to a specific instance. To create such a member, precede its declaration with the keyword **static**. When a member is declared **static**, it can be accessed before any objects of its class are created, and without reference to any object. You can declare both methods and variables to be **static**. The most common example of a **static** member is **main( )**. **main( )** is declared as **static** because it must be called by the JVM when your program begins. Outside the class, to use a **static** member, you need only specify the name of its class followed by the dot operator. No object needs to be created. For example, if you want to assign the value 10 to a **static** variable called **count** that is part of the **Timer** class, use this line:

```
Timer.count = 10;
```

This format is similar to that used to access normal instance variables through an object, except that the class name is used. A **static** method can be called in the same way—by use of the dot operator on the name of the class.

Variables declared as **static** are, essentially, global variables. When an object is declared, no copy of a **static** variable is made. Instead, all instances of the class share the same **static** variable. Here is an example that shows the differences between a **static** variable and an instance variable:

```
// Use a static variable.
class StaticDemo {
  int x; // a normal instance variable
  static int y; // a static variable          There is one copy of y
                                               for all objects to share.

  // Return the sum of the instance variable x
  // and the static variable y.
  int sum() {
    return x + y;
  }
}

class SDemo {
  public static void main(String[] args) {
    StaticDemo ob1 = new StaticDemo();
    StaticDemo ob2 = new StaticDemo();

    // Each object has its own copy of an instance variable.
    ob1.x = 10;
    ob2.x = 20;
    System.out.println("Of course, ob1.x and ob2.x " +
                        "are independent.");
    System.out.println("ob1.x: " + ob1.x +
                        "\nob2.x: " + ob2.x);
    System.out.println();

    // Each object shares one copy of a static variable.
    System.out.println("The static variable y is shared.");
    StaticDemo.y = 19;
    System.out.println("Set StaticDemo.y to 19.");

    System.out.println("ob1.sum(): " + ob1.sum());
    System.out.println("ob2.sum(): " + ob2.sum());
    System.out.println();

    StaticDemo.y = 100;
    System.out.println("Change StaticDemo.y to 100");

    System.out.println("ob1.sum(): " + ob1.sum());
    System.out.println("ob2.sum(): " + ob2.sum());
    System.out.println();   }
}
```

The output from the program is shown here:

```
Of course, ob1.x and ob2.x are independent.
ob1.x: 10
ob2.x: 20

The static variable y is shared.
Set StaticDemo.y to 19.
ob1.sum(): 29
ob2.sum(): 39

Change StaticDemo.y to 100
ob1.sum(): 110
ob2.sum(): 120
```

As you can see, the **static** variable **y** is shared by both **ob1** and **ob2**. Changing it affects the entire class, not just an instance.

The difference between a **static** method and a normal method is that the **static** method is called through its class name, without any object of that class being created. You have seen an example of this already: the **sqrt( )** method, which is a **static** method within Java's standard **Math** class. Here is an example that creates a **static** method:

```
// Use a static method.
class StaticMeth {
  static int val = 1024; // a static variable

  // a static method
  static int valDiv2() {  ◄─────── A static method.
    return val/2;
  }
}

class SDemo2 {
  public static void main(String[] args) {

    System.out.println("val is " + StaticMeth.val);
    System.out.println("StaticMeth.valDiv2(): " +
                       StaticMeth.valDiv2());

    StaticMeth.val = 4;
    System.out.println("val is " + StaticMeth.val);
    System.out.println("StaticMeth.valDiv2(): " +
                       StaticMeth.valDiv2());
  }
}
```

The output is shown here:

```
val is 1024
StaticMeth.valDiv2(): 512
val is 4
StaticMeth.valDiv2(): 2
```

Methods declared as **static** have several restrictions:

- They can directly call only other **static** methods in their class.

- They can directly access only **static** variables in their class.

- They do not have a **this** reference.

For example, in the following class, the **static** method **valDivDenom( )** is illegal:

```
class StaticError {
  int denom = 3; // a normal instance variable
  static int val = 1024; // a static variable

  /* Error! Can't access a non-static variable
     from within a static method. */
  static int valDivDenom() {
    return val/denom; // won't compile!
  }
}
```

Here, **denom** is a normal instance variable that cannot be accessed within a **static** method.

Static Blocks

Sometimes a class will require some type of initialization before it is ready to create objects. For example, it might need to establish a connection to a remote site. It also might need to initialize certain **static** variables before any of the class' **static** methods are used. To handle these types of situations, Java allows you to declare a **static** block. A **static** block is executed when the class is first loaded. Thus, it is executed before the class can be used for any other purpose. Here is an example of a **static** block:

```
// Use a static block
class StaticBlock {
  static double rootOf2;
  static double rootOf3;

  static {                                        ← This block is executed
    System.out.println("Inside static block.");     when the class is loaded.
    rootOf2 = Math.sqrt(2.0);
    rootOf3 = Math.sqrt(3.0);
  }
```

```
    StaticBlock(String msg) {
      System.out.println(msg);
    }
  }

  class SDemo3 {
    public static void main(String[] args) {
      StaticBlock ob = new StaticBlock("Inside Constructor");

      System.out.println("Square root of 2 is " +
                          StaticBlock.rootOf2);
      System.out.println("Square root of 3 is " +
                          StaticBlock.rootOf3);

    }
  }
```

The output is shown here:

```
Inside static block.
Inside Constructor
Square root of 2 is 1.4142135623730951
Square root of 3 is 1.7320508075688772
```

As you can see, the **static** block is executed before any objects are constructed.

Try This 6-3 The Quicksort

QSDemo.java

In Chapter 5 you were shown a simple sorting method called the Bubble sort. It was mentioned at the time that substantially better sorts exist. Here you will develop a version of one of the best: the Quicksort. The Quicksort, invented and named by C.A.R. Hoare, is arguably the best general-purpose sorting algorithm currently available. The reason it could not be shown in Chapter 5 is that the best implementations of the Quicksort rely on recursion. The version we will develop sorts a character array, but the logic can be adapted to sort any type of object you like.

The Quicksort is built on the idea of partitions. The general procedure is to select a value, called the *comparand,* and then to partition the array into two sections. All elements greater than or equal to the partition value are put on one side, and those less than the value are put on the other. This process is then repeated for each remaining section until the array is sorted. For example, given the array **fedacb** and using the value **d** as the comparand, the first pass of the Quicksort would rearrange the array as follows:

Initial	f e d a c b
Pass1	b c a d e f

This process is then repeated for each section—that is, **bca** and **def**. As you can see, the process is essentially recursive in nature, and indeed, the cleanest implementation of Quicksort is recursive.

Assuming that you have no information about the distribution of the data to be sorted, there are a number of ways you can select the comparand. Here are two. You can choose a value at random from within the data, or you can select it by averaging a small set of values taken from the data. For optimal sorting, you want a value that is precisely in the middle of the range of values. However, this is often not practical. In the worst case, the value chosen is at one extremity. Even in this case, however, Quicksort still performs correctly. The version of Quicksort that we will develop selects the middle element of the array as the comparand.

1. Create a file called **QSDemo.java**.

2. First, create the **Quicksort** class shown here:

```
// Try This 6-3: A simple version of the Quicksort.
class Quicksort {

  // Set up a call to the actual Quicksort method.
  static void qsort(char[] items) {
    qs(items, 0, items.length-1);
  }

  // A recursive version of Quicksort for characters.
  private static void qs(char[] items, int left, int right)
  {
    int i, j;
    char x, y;

    i = left; j = right;
    x = items[(left+right)/2];

    do {
      while((items[i] < x) && (i < right)) i++;
      while((x < items[j]) && (j > left)) j--;

      if(i <= j) {
        y = items[i];
        items[i] = items[j];
        items[j] = y;
        i++; j--;
      }
    } while(i <= j);

    if(left < j) qs(items, left, j);
    if(i < right) qs(items, i, right);
  }
}
```

(continued)

To keep the interface to the Quicksort simple, the **Quicksort** class provides the **qsort( )** method, which sets up a call to the actual Quicksort method, **qs( )**. This enables the Quicksort to be called with just the name of the array to be sorted, without having to provide an initial partition. Since **qs( )** is only used internally, it is specified as **private**.

3. To use the **Quicksort**, simply call **Quicksort.qsort( )**. Since **qsort( )** is specified as **static**, it can be called through its class rather than on an object. Thus, there is no need to create a **Quicksort** object. After the call returns, the array will be sorted. Remember, this version works only for character arrays, but you can adapt the logic to sort any type of arrays you want.

4. Here is a program that demonstrates **Quicksort**:

```
// Try This 6-3: A simple version of the Quicksort.
class Quicksort {

  // Set up a call to the actual Quicksort method.
  static void qsort(char[] items) {
    qs(items, 0, items.length-1);
  }

  // A recursive version of Quicksort for characters.
  private static void qs(char[] items, int left, int right)
  {
    int i, j;
    char x, y;

    i = left; j = right;
    x = items[(left+right)/2];

    do {
      while((items[i] < x) && (i < right)) i++;
      while((x < items[j]) && (j > left)) j--;

      if(i <= j) {
        y = items[i];
        items[i] = items[j];
        items[j] = y;
        i++; j--;
      }
    } while(i <= j);

    if(left < j) qs(items, left, j);
    if(i < right) qs(items, i, right);
  }
}

class QSDemo {
```

```
public static void main(String[] args) {
  char[] a = { 'd', 'x', 'a', 'r', 'p', 'j', 'i' };
  int i;

  System.out.print("Original array: ");
  for(i=0; i < a.length; i++)
    System.out.print(a[i]);

  System.out.println();

  // now, sort the array
  Quicksort.qsort(a);

  System.out.print("Sorted array: ");
  for(i=0; i < a.length; i++)
    System.out.print(a[i]);
  }
}
```

Introducing Nested and Inner Classes

In Java, you can define a *nested class*. This is a class that is declared within another class. Frankly, the nested class is a somewhat advanced topic. In fact, nested classes were not even allowed in the first version of Java. It was not until Java 1.1 that they were added. However, it is important that you know what they are and the mechanics of how they are used because they play an important role in many real-world programs.

A nested class does not exist independently of its enclosing class. Thus, the scope of a nested class is bounded by its outer class. A nested class that is declared directly within its enclosing class scope is a member of its enclosing class. It is also possible to declare a nested class that is local to a block.

There are two general types of nested classes: those that are preceded by the **static** modifier and those that are not. The only type that we are concerned about in this book is the non-static variety. This type of nested class is also called an *inner class*. It has access to all of the variables and methods of its outer class and may refer to them directly in the same way that other non-**static** members of the outer class do.

Sometimes an inner class is used to provide a set of services that is needed only by its enclosing class. Here is an example that uses an inner class to compute various values for its enclosing class:

```
// Use an inner class.
class Outer {
  int[] nums;
```

```
  Outer(int[] n) {
    nums = n;
  }

  void analyze() {
    Inner inOb = new Inner();

    System.out.println("Minimum: " + inOb.min());
    System.out.println("Maximum: " + inOb.max());
    System.out.println("Average: " + inOb.avg());
  }

  // This is an inner class.
  class Inner {        ◄─────── An inner class
    int min() {
      int m = nums[0];

      for(int i=1; i < nums.length; i++)
        if(nums[i] < m) m = nums[i];

      return m;
    }

    int max() {
      int m = nums[0];
      for(int i=1; i < nums.length; i++)
        if(nums[i] > m) m = nums[i];

      return m;
    }

    int avg() {
      int a = 0;
      for(int i=0; i < nums.length; i++)
        a += nums[i];

      return a / nums.length;
    }
  }
}

class NestedClassDemo {
  public static void main(String[] args) {
    int[] x = { 3, 2, 1, 5, 6, 9, 7, 8 };
    Outer outOb = new Outer(x);

    outOb.analyze();
  }
}
```

The output from the program is shown here:

```
Minimum: 1
Maximum: 9
Average: 5
```

In this example, the inner class **Inner** computes various values from the array **nums**, which is a member of **Outer**. As explained, an inner class has access to the members of its enclosing class, so it is perfectly acceptable for **Inner** to access the **nums** array directly. Of course, the opposite is not true. For example, it would not be possible for **analyze( )** to invoke the **min( )** method directly, without creating an **Inner** object.

As mentioned, it is possible to nest a class within a block scope. Doing so simply creates a localized class that is not known outside its block. The following example adapts the **ShowBits** class developed in Try This 5 3 for use as a local class.

```java
// Use ShowBits as a local class.
class LocalClassDemo {
  public static void main(String[] args) {

    // An inner class version of ShowBits.
    class ShowBits {  ◄─────────────────────── A local class nested within a method
      int numbits;

      ShowBits(int n) {
        numbits = n;
      }

      void show(long val) {
        long mask = 1;

        // left-shift a 1 into the proper position
        mask <<= numbits-1;

        int spacer = 0;
        for(; mask != 0; mask >>>= 1) {
          if((val & mask) != 0) System.out.print("1");
          else System.out.print("0");
          spacer++;
          if((spacer % 8) == 0) {
            System.out.print(" ");
            spacer = 0;
          }
        }
        System.out.println();
      }
    }

    for(byte b = 0; b < 10; b++) {
      ShowBits byteval = new ShowBits(8);
```

```
        System.out.print(b + " in binary: ");
        byteval.show(b);
      }
    }
}
```

The output from this version of the program is shown here:

```
0 in binary: 00000000
1 in binary: 00000001
2 in binary: 00000010
3 in binary: 00000011
4 in binary: 00000100
5 in binary: 00000101
6 in binary: 00000110
7 in binary: 00000111
8 in binary: 00001000
9 in binary: 00001001
```

In this example, the **ShowBits** class is not known outside of **main( )**, and any attempt to access it by any method other than **main( )** will result in an error.

One last point: You can create an inner class that does not have a name. This is called an *anonymous inner class*. An object of an anonymous inner class is instantiated when the class is declared, using **new**. Anonymous inner classes are discussed further in Chapter 17.

Varargs: Variable-Length Arguments

Sometimes you will want to create a method that takes a variable number of arguments, based on its precise usage. For example, a method that opens an Internet connection might take a user name, password, file name, protocol, and so on, but supply defaults if some of this information is not provided. In this situation, it would be convenient to pass only the arguments to which the defaults did not apply. To create such a method implies that there must be some way to create a list of arguments that is variable in length, rather than fixed.

In the early days of Java, methods that required a variable-length argument list could be handled two ways, neither of which was particularly pleasing. First, if the maximum number of arguments was small and known, then you could create overloaded versions of the method,

Ask the Expert

Q: **What makes a** static **nested class different from a non-**static **one?**

A: A **static** nested class is one that has the **static** modifier applied. Because it is **static**, it can access only other **static** members of the enclosing class directly. It must access other members of its outer class through an object reference.

one for each way the method could be called. Although this works and is suitable for some situations, it applies to only a narrow class of situations. In cases where the maximum number of potential arguments is larger, or unknowable, a second approach was used in which the arguments were put into an array, and then the array was passed to the method. Frankly, both of these approaches often resulted in clumsy solutions, and it was widely acknowledged that a better approach was needed.

Fortunately, today, Java includes a feature that greatly simplifies the creation of methods that require a variable number of arguments. This feature is called *varargs,* which is short for variable-length arguments. A method that takes a variable number of arguments is called a *variable-arity method,* or simply a *varargs method.* The parameter list for a varargs method is not fixed, but rather variable in length. Thus, a varargs method can take a variable number of arguments.

Varargs Basics

A variable-length argument is specified by three periods (**...**). For example, here is how to write a method called **vaTest( )** that takes a variable number of arguments:

```
// vaTest() uses a vararg.                      Declare a variable-length argument list.
static void vaTest(int ... v) {  ◄─────────────────────────────────────┐
  System.out.println("Number of args: " + v.length);
  System.out.println("Contents: ");

  for(int i=0; i < v.length; i++)
    System.out.println(" arg " + i + ": " + v[i]);

  System.out.println();
}
```

Notice that **v** is declared as shown here:

```
int ... v
```

This syntax tells the compiler that **vaTest( )** can be called with zero or more arguments. Furthermore, it causes **v** to be implicitly declared as an array of type **int[ ]**. Thus, inside **vaTest( )**, **v** is accessed using the normal array syntax.

Here is a complete program that demonstrates **vaTest( )**:

```
// Demonstrate variable-length arguments.
class VarArgs {

  // vaTest() uses a vararg.
  static void vaTest(int ... v) {
    System.out.println("Number of args: " + v.length);
    System.out.println("Contents: ");

    for(int i=0; i < v.length; i++)
      System.out.println("  arg " + i + ": " + v[i]);
```

```
    System.out.println();
  }

  public static void main(String[] args)
  {

    // Notice how vaTest() can be called with a
    // variable number of arguments.
    vaTest(10);       // 1 arg
    vaTest(1, 2, 3); // 3 args         Call with different numbers
    vaTest();         // no args        of arguments.
  }
}
```

The output from the program is shown here:

```
Number of args: 1
Contents:
  arg 0: 10

Number of args: 3
Contents:
  arg 0: 1
  arg 1: 2
  arg 2: 3

Number of args: 0
Contents:
```

There are two important things to notice about this program. First, as explained, inside **vaTest( )**, **v** is operated on as an array. This is because **v** *is an array*. The **...** syntax simply tells the compiler that a variable number of arguments will be used, and that these arguments will be stored in the array referred to by **v**. Second, in **main( )**, **vaTest( )** is called with different numbers of arguments, including no arguments at all. The arguments are automatically put in an array and passed to **v**. In the case of no arguments, the length of the array is zero.

A method can have "normal" parameters along with a variable-length parameter. However, the variable-length parameter must be the last parameter declared by the method. For example, this method declaration is perfectly acceptable:

```
int doIt(int a, int b, double c, int ... vals) {
```

In this case, the first three arguments used in a call to **doIt( )** are matched to the first three parameters. Then, any remaining arguments are assumed to belong to **vals**.

Here is a reworked version of the **vaTest( )** method that takes a regular argument and a variable-length argument:

```
// Use varargs with standard arguments.
class VarArgs2 {
```

```
// Here, msg is a normal parameter and v is a
// varargs parameter.
static void vaTest(String msg, int ... v) {        ◄──────── A "normal" and
  System.out.println(msg + v.length);                        vararg parameter
  System.out.println("Contents: ");

  for(int i=0; i < v.length; i++)
    System.out.println("  arg " + i + ": " + v[i]);

  System.out.println();
}

public static void main(String[] args)
{
  vaTest("One vararg: ", 10);
  vaTest("Three varargs: ", 1, 2, 3);
  vaTest("No varargs: ");
}
}
```

The output from this program is shown here:

```
One vararg: 1
Contents:
  arg 0: 10

Three varargs: 3
Contents:
  arg 0: 1
  arg 1: 2
  arg 2: 3

No varargs: 0
Contents:
```

Remember, the varargs parameter must be last. For example, the following declaration is incorrect:

```
int doIt(int a, int b, double c, int ... vals, boolean stopFlag) { // Error!
```

Here, there is an attempt to declare a regular parameter after the varargs parameter, which is illegal. There is one more restriction to be aware of: there must be only one varargs parameter. For example, this declaration is also invalid:

```
int doIt(int a, int b, double c, int ... vals, double ... morevals) { // Error!
```

The attempt to declare the second varargs parameter is illegal.

Overloading Varargs Methods

You can overload a method that takes a variable-length argument. For example, the following program overloads **vaTest( )** three times:

```
// Varargs and overloading.
class VarArgs3 {
                                      First version of vaTest( )
  static void vaTest(int ... v) {
    System.out.println("vaTest(int ...): " +
                       "Number of args: " + v.length);
    System.out.println("Contents: ");

    for(int i=0; i < v.length; i++)
      System.out.println("  arg " + i + ": " + v[i]);

    System.out.println();
  }
                                       Second version of vaTest( )
  static void vaTest(boolean ... v) {
    System.out.println("vaTest(boolean ...): " +
                       "Number of args: " + v.length);
    System.out.println("Contents: ");

    for(int i=0; i < v.length; i++)
      System.out.println("  arg " + i + ": " + v[i]);

    System.out.println();
  }
                                         Third version of vaTest( )
  static void vaTest(String msg, int ... v) {
    System.out.println("vaTest(String, int ...): " +
                       msg + v.length);
    System.out.println("Contents: ");

    for(int i=0; i < v.length; i++)
      System.out.println("  arg " + i + ": " + v[i]);

    System.out.println();
  }

  public static void main(String[] args)
  {
    vaTest(1, 2, 3);
    vaTest("Testing: ", 10, 20);
    vaTest(true, false, false);
  }
}
```

The output produced by this program is shown here:

```
vaTest(int ...): Number of args: 3
Contents:
  arg 0: 1
  arg 1: 2
  arg 2: 3

vaTest(String, int ...): Testing: 2
Contents:
  arg 0: 10
  arg 1: 20

vaTest(boolean ...): Number of args: 3
Contents:
  arg 0: true
  arg 1: false
  arg 2: false
```

This program illustrates both ways that a varargs method can be overloaded. First, the types of its vararg parameter can differ. This is the case for **vaTest(int ...)** and **vaTest(boolean ...)**. Remember, the **...** causes the parameter to be treated as an array of the specified type. Therefore, just as you can overload methods by using different types of array parameters, you can overload varargs methods by using different types of varargs. In this case, Java uses the type difference to determine which overloaded method to call.

The second way to overload a varargs method is to add one or more normal parameters. This is what was done with **vaTest(String, int ...)**. In this case, Java uses both the number of arguments and the type of the arguments to determine which method to call.

Varargs and Ambiguity

Somewhat unexpected errors can result when overloading a method that takes a variable-length argument. These errors involve ambiguity because it is possible to create an ambiguous call to an overloaded varargs method. For example, consider the following program:

```
// Varargs, overloading, and ambiguity.
//
// This program contains an error and will
// not compile!
class VarArgs4 {

  // Use an int vararg parameter.
  static void vaTest(int ... v) {          An int vararg
    // ...
  }

  // Use a boolean vararg parameter.
  static void vaTest(boolean ... v) {          A boolean vararg
    // ...
  }
```

```
public static void main(String[] args)
{
  vaTest(1, 2, 3); // OK
  vaTest(true, false, false); // OK

  vaTest(); // Error: Ambiguous!  ◄─────── Ambiguous!
}
}
```

In this program, the overloading of **vaTest( )** is perfectly correct. However, this program will not compile because of the following call:

```
vaTest(); // Error: Ambiguous!
```

Because the vararg parameter can be empty, this call could be translated into a call to **vaTest(int ...)** or to **vaTest(boolean ...)**. Both are equally valid. Thus, the call is inherently ambiguous.

Here is another example of ambiguity. The following overloaded versions of **vaTest( )** are inherently ambiguous even though one takes a normal parameter:

```
static void vaTest(int ... v) { // ...
```

```
static void vaTest(int n, int ... v) { // ...
```

Although the parameter lists of **vaTest( )** differ, there is no way for the compiler to resolve the following call:

vaTest(1)

Does this translate into a call to **vaTest(int ...)**, with one varargs argument, or into a call to **vaTest(int, int ...)** with no varargs arguments? There is no way for the compiler to answer this question. Thus, the situation is ambiguous.

Because of ambiguity errors like those just shown, sometimes you will need to forego overloading and simply use two different method names. Also, in some cases, ambiguity errors expose a conceptual flaw in your code, which you can remedy by more carefully crafting a solution.

Chapter 6 Self Test

1. Given this fragment,

```
class X {
  private int count;
```

is the following fragment correct?

```
class Y {
  public static void main(String[] args) {
    X ob = new X();

    ob.count = 10;
```

2. An access modifier must _____ a member's declaration.

3. The complement of a queue is a stack. It uses first-in, last-out accessing and is often likened to a stack of plates. The first plate put on the table is the last plate used. Create a stack class called **Stack** that can hold characters. Call the methods that access the stack **push( )** and **pop( )**. Allow the user to specify the size of the stack when it is created. Keep all other members of the **Stack** class private. (Hint: You can use the **Queue** class as a model; just change the way the data is accessed.)

4. Given this class,

```
class Test {
  int a;
  Test(int i) { a = i; }
}
```

write a method called **swap( )** that exchanges the contents of the objects referred to by two **Test** object references.

5. Is the following fragment correct?

```
class X {
  int meth(int a, int b) { ... }
  String meth(int a, int b) { ... }
```

6. Write a recursive method that displays the contents of a string backwards.

7. If all objects of a class need to share the same variable, how must you declare that variable?

8. Why might you need to use a **static** block?

9. What is an inner class?

10. To make a member accessible by only other members of its class, what access modifier must be used?

11. The name of a method plus its parameter list constitutes the method's _____.

12. An **int** argument is passed to a method by using call-by-_____.

13. Create a varargs method called **sum( )** that sums the **int** values passed to it. Have it return the result. Demonstrate its use.

14. Can a varargs method be overloaded?

15. Show an example of an overloaded varargs method that is ambiguous.

Chapter 7

Inheritance

Key Skills & Concepts

- Understand inheritance basics

- Call superclass constructors

- Use **super** to access superclass members

- Create a multilevel class hierarchy

- Know when constructors are called

- Understand superclass references to subclass objects

- Override methods

- Use overridden methods to achieve dynamic method dispatch

- Use abstract classes

- Use **final**

- Know the **Object** class

Inheritance is one of the three foundation principles of object-oriented programming because it allows the creation of hierarchical classifications. Using inheritance, you can create a general class that defines traits common to a set of related items. This class can then be inherited by other, more specific classes, each adding those things that are unique to it.

In the language of Java, a class that is inherited is called a *superclass*. The class that does the inheriting is called a *subclass*. Therefore, a subclass is a specialized version of a superclass. It inherits all of the variables and methods defined by the superclass and adds its own, unique elements.

Inheritance Basics

Java supports inheritance by allowing one class to incorporate another class into its declaration. This is done by using the **extends** keyword. Thus, the subclass adds to (extends) the superclass.

Let's begin with a short example that illustrates several of the key features of inheritance. The following program creates a superclass called **TwoDShape**, which stores the width and height of a two-dimensional object, and a subclass called **Triangle**. Notice how the keyword **extends** is used to create a subclass.

```
// A simple class hierarchy.

// A class for two-dimensional objects.
class TwoDShape {
```

```
  double width;
  double height;

  void showDim() {
    System.out.println("Width and height are " +
                       width + " and " + height);
  }
}

// A subclass of TwoDShape for triangles.
class Triangle extends TwoDShape {
  String style;                              Triangle inherits TwoDShape.

  double area() {
    return width * height / 2;              Triangle can refer to the members of TwoDShape
  }                                         as if they were declared by Triangle.

  void showStyle() {
    System.out.println("Triangle is " + style);
  }
}


class Shapes {
  public static void main(String[] args) {
    Triangle t1 = new Triangle();
    Triangle t2 = new Triangle();

    t1.width = 4.0;
    t1.height = 4.0;                    All members of Triangle are available to Triangle
    t1.style = "filled";               objects, even those inherited from TwoDShape.

    t2.width = 8.0;
    t2.height = 12.0;
    t2.style = "outlined";

    System.out.println("Info for t1: ");
    t1.showStyle();
    t1.showDim();
    System.out.println("Area is " + t1.area());

    System.out.println();

    System.out.println("Info for t2: ");
    t2.showStyle();
    t2.showDim();
    System.out.println("Area is " + t2.area());
  }
}
```

The output from this program is shown here:

```
Info for t1:
Triangle is filled
Width and height are 4.0 and 4.0
Area is 8.0

Info for t2:
Triangle is outlined
Width and height are 8.0 and 12.0
Area is 48.0
```

Here, **TwoDShape** defines the attributes of a "generic" two-dimensional shape, such as a square, rectangle, triangle, and so on. The **Triangle** class creates a specific type of **TwoDShape**, in this case, a triangle. The **Triangle** class includes all of **TwoDShape** and adds the field **style**, the method **area( )**, and the method **showStyle( )**. The triangle's style is stored in **style**. This can be any string that describes the triangle, such as "filled", "outlined", "transparent", or even something like "warning symbol", "isosceles", or "rounded". The **area( )** method computes and returns the area of the triangle, and **showStyle( )** displays the triangle style.

Because **Triangle** includes all of the members of its superclass, **TwoDShape**, it can access **width** and **height** inside **area( )**. Also, inside **main( )**, objects **t1** and **t2** can refer to **width** and **height** directly, as if they were declared by **Triangle**. Figure 7-1 depicts conceptually how **TwoDShape** is incorporated into **Triangle**.

Even though **TwoDShape** is a superclass for **Triangle**, it is also a completely independent, stand-alone class. Being a superclass for a subclass does not mean that the superclass cannot be used by itself. For example, the following is perfectly valid:

```
TwoDShape shape = new TwoDShape();

shape.width = 10;
shape.height = 20;

shape.showDim();
```

Of course, an object of **TwoDShape** has no knowledge of or access to any subclasses of **TwoDShape**.

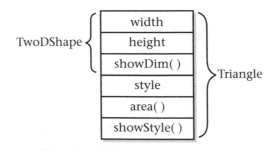

Figure 7-1 A conceptual depiction of the **Triangle** class

The general form of a **class** declaration that inherits a superclass is shown here:

class *subclass-name* extends *superclass-name* {
  // body of class
}

You can specify only one superclass for any subclass that you create. Java does not support the inheritance of multiple superclasses into a single subclass. (This differs from C++, in which you can inherit multiple base classes. Be aware of this when converting C++ code to Java.) You can, however, create a hierarchy of inheritance in which a subclass becomes a superclass of another subclass. Of course, no class can be a superclass of itself.

A major advantage of inheritance is that once you have created a superclass that defines the attributes common to a set of objects, it can be used to create any number of more specific subclasses. Each subclass can precisely tailor its own classification. For example, here is another subclass of **TwoDShape** that encapsulates rectangles:

```
// A subclass of TwoDShape for rectangles.
class Rectangle extends TwoDShape {
  boolean isSquare() {
    if(width == height) return true;
    return false;
  }

  double area() {
    return width * height;
  }
}
```

The **Rectangle** class includes **TwoDShape** and adds the methods **isSquare( )**, which determines if the rectangle is square, and **area( )**, which computes the area of a rectangle.

Member Access and Inheritance

As you learned in Chapter 6, often an instance variable of a class will be declared **private** to prevent its unauthorized use or tampering. Inheriting a class *does not* overrule the **private** access restriction. Thus, even though a subclass includes all of the members of its superclass, it cannot access those members of the superclass that have been declared **private**. For example, if, as shown here, **width** and **height** are made private in **TwoDShape**, then **Triangle** will not be able to access them:

```
// Private members are not inherited.

// This example will not compile.

// A class for two-dimensional objects.
class TwoDShape {
```

```
    private double width; // these are
    private double height; // now private

    void showDim() {
      System.out.println("Width and height are " +
                          width + " and " + height);
    }
}

// A subclass of TwoDShape for triangles.
class Triangle extends TwoDShape {
  String style;

  double area() {
    return width * height / 2; // Error! can't access
  }

  void showStyle() {
    System.out.println("Triangle is " + style);
  }
}
```

Can't access a **private** member of a superclass.

The **Triangle** class will not compile because the reference to **width** and **height** inside the **area( )** method causes an access violation. Since **width** and **height** are declared **private**, they are accessible only by other members of their own class. Subclasses have no access to them.

Remember that a class member that has been declared **private** will remain private to its class. It is not accessible by any code outside its class, including subclasses.

At first, you might think that the fact that subclasses do not have access to the private members of superclasses is a serious restriction that would prevent the use of private members in many situations. However, this is not true. As explained in Chapter 6, Java programmers typically use accessor methods to provide access to the private members of a class. Here is a rewrite of the **TwoDShape** and **Triangle** classes that uses methods to access the private instance variables **width** and **height**:

```
// Use accessor methods to set and get private members.

// A class for two-dimensional objects.
class TwoDShape {
  private double width; // these are
  private double height; // now private

  // Accessor methods for width and height.
  double getWidth() { return width; }
  double getHeight() { return height; }
  void setWidth(double w) { width = w; }
  void setHeight(double h) { height = h; }
```

Accessor methods for **width** and **height**

```
  void showDim() {
    System.out.println("Width and height are " +
                        width + " and " + height);
  }
}

// A subclass of TwoDShape for triangles.
class Triangle extends TwoDShape {
  String style;

  double area() {
    return getWidth() * getHeight() / 2;
  }

  void showStyle() {
    System.out.println("Triangle is " + style);
  }
}

class Shapes2 {
  public static void main(String[] args) {
    Triangle t1 = new Triangle();
    Triangle t2 = new Triangle();

    t1.setWidth(4.0);
    t1.setHeight(4.0);
    t1.style = "filled";

    t2.setWidth(8.0);
    t2.setHeight(12.0);
    t2.style = "outlined";

    System.out.println("Info for t1: ");
    t1.showStyle();
    t1.showDim();
    System.out.println("Area is " + t1.area());

    System.out.println();

    System.out.println("Info for t2: ");
    t2.showStyle();
    t2.showDim();
    System.out.println("Area is " + t2.area());
  }
}
```

Use accessor methods provided by superclass.

Ask the Expert

Q: **When should I make an instance variable private?**

A: There are no hard and fast rules, but here are two general principles. If an instance variable is to be used only by methods defined within its class, then it should be made private. If an instance variable must be within certain bounds, then it should be private and made available only through accessor methods. This way, you can prevent invalid values from being assigned.

Constructors and Inheritance

In a hierarchy, it is possible for both superclasses and subclasses to have their own constructors. This raises an important question: What constructor is responsible for building an object of the subclass—the one in the superclass, the one in the subclass, or both? The answer is this: The constructor for the superclass constructs the superclass portion of the object, and the constructor for the subclass constructs the subclass part. This makes sense because the superclass has no knowledge of or access to any element in a subclass. Thus, their construction must be separate. The preceding examples have relied upon the default constructors created automatically by Java, so this was not an issue. However, in practice, most classes will have explicit constructors. Here you will see how to handle this situation.

When only the subclass defines a constructor, the process is straightforward: simply construct the subclass object. The superclass portion of the object is constructed automatically using its default constructor. For example, here is a reworked version of **Triangle** that defines a constructor. It also makes **style** private, since it is now set by the constructor.

```
// Add a constructor to Triangle.

// A class for two-dimensional objects.
class TwoDShape {
  private double width; // these are
  private double height; // now private

  // Accessor methods for width and height.
  double getWidth() { return width; }
  double getHeight() { return height; }
  void setWidth(double w) { width = w; }
  void setHeight(double h) { height = h; }

  void showDim() {
    System.out.println("Width and height are " +
                       width + " and " + height);
  }
}
```

```
// A subclass of TwoDShape for triangles.
class Triangle extends TwoDShape {
  private String style;

  // Constructor
  Triangle(String s, double w, double h) {
    setWidth(w);
    setHeight(h);                             Initialize TwoDShape
                                              portion of object.
    style = s;
  }

  double area() {
    return getWidth() * getHeight() / 2;
  }

  void showStyle() {
    System.out.println("Triangle is " + style);
  }
}

class Shapes3 {
  public static void main(String[] args) {
    Triangle t1 = new Triangle("filled", 4.0, 4.0);
    Triangle t2 = new Triangle("outlined", 8.0, 12.0);

    System.out.println("Info for t1: ");
    t1.showStyle();
    t1.showDim();
    System.out.println("Area is " + t1.area());

    System.out.println();

    System.out.println("Info for t2: ");
    t2.showStyle();
    t2.showDim();
    System.out.println("Area is " + t2.area());
  }
}
```

Here, **Triangle**'s constructor initializes the members of **TwoDClass** that it inherits along with its own **style** field.

When both the superclass and the subclass define constructors, the process is a bit more complicated because both the superclass and subclass constructors must be executed. In this case, you must use another of Java's keywords, **super**, which has two general forms. The first calls a superclass constructor. The second is used to access a member of the superclass that has been hidden by a member of a subclass. Here, we will look at its first use.

Using super to Call Superclass Constructors

A subclass can call a constructor defined by its superclass by use of the following form of **super**:

super(*parameter-list*);

Here, *parameter-list* specifies any parameters needed by the constructor in the superclass. **super( )** must always be the first statement executed inside a subclass constructor. To see how **super( )** is used, consider the version of **TwoDShape** in the following program. It defines a constructor that initializes **width** and **height**.

```
// Add constructors to TwoDShape.
class TwoDShape {
  private double width;
  private double height;

  // Parameterized constructor.
  TwoDShape(double w, double h) {  ◄─────── A constructor for TwoDShape
    width = w;
    height = h;
  }

  // Accessor methods for width and height.
  double getWidth() { return width; }
  double getHeight() { return height; }
  void setWidth(double w) { width = w; }
  void setHeight(double h) { height = h; }

  void showDim() {
    System.out.println("Width and height are " +
                        width + " and " + height);
  }
}

// A subclass of TwoDShape for triangles.
class Triangle extends TwoDShape {
  private String style;

  Triangle(String s, double w, double h) {
    super(w, h); // call superclass constructor

    style = s;
  }                            Use super( ) to execute the
                               TwoDShape constructor.

  double area() {
    return getWidth() * getHeight() / 2;
  }
```

```
      void showStyle() {
        System.out.println("Triangle is " + style);
      }
    }

    class Shapes4 {
      public static void main(String[] args) {
        Triangle t1 = new Triangle("filled", 4.0, 4.0);
        Triangle t2 = new Triangle("outlined", 8.0, 12.0);

        System.out.println("Info for t1: ");
        t1.showStyle();
        t1.showDim();
        System.out.println("Area is " + t1.area());

        System.out.println();

        System.out.println("Info for t2: ");
        t2.showStyle();
        t2.showDim();
        System.out.println("Area is " + t2.area());
      }
    }
```

Here, **Triangle( )** calls **super( )** with the parameters **w** and **h**. This causes the **TwoDShape( )**
constructor to be called, which initializes **width** and **height** using these values. **Triangle**
no longer initializes these values itself. It need only initialize the value unique to it: **style**.
This leaves **TwoDShape** free to construct its subobject in any manner that it so chooses.
Furthermore, **TwoDShape** can add functionality about which existing subclasses have no
knowledge, thus preventing existing code from breaking.

Any form of constructor defined by the superclass can be called by **super( )**. The constructor
executed will be the one that matches the arguments. For example, here are expanded versions
of both **TwoDShape** and **Triangle** that include default constructors and constructors that take
one argument:

```
// Add more constructors to TwoDShape.
class TwoDShape {
  private double width;
  private double height;

  // A default constructor.
  TwoDShape() {
    width = height = 0.0;
  }

  // Parameterized constructor.
  TwoDShape(double w, double h) {
    width = w;
    height = h;
  }
```

```
   // Construct object with equal width and height.
   TwoDShape(double x) {
     width = height = x;
   }

   // Accessor methods for width and height.
   double getWidth() { return width; }
   double getHeight() { return height; }
   void setWidth(double w) { width = w; }
   void setHeight(double h) { height = h; }

   void showDim() {
     System.out.println("Width and height are " +
                        width + " and " + height);
   }
 }

// A subclass of TwoDShape for triangles.
class Triangle extends TwoDShape {
  private String style;

  // A default constructor.
  Triangle() {
    super();  ◄─────────────────────────────┐
    style = "none";                          │
  }                                          │
                                             │
  // Constructor                             │
  Triangle(String s, double w, double h) {   │
    super(w, h); // call superclass constructor ◄──┤
                                                   │
    style = s;                               
  }                          Use super( ) to call the
                            various forms of the
  // One argument constructor.  TwoDShape constructor.
  Triangle(double x) {                       │
    super(x); // call superclass constructor ◄──┘

    style = "filled";
  }

  double area() {
    return getWidth() * getHeight() / 2;
  }

  void showStyle() {
    System.out.println("Triangle is " + style);
  }
}
```

```
class Shapes5 {
  public static void main(String[] args) {
    Triangle t1 = new Triangle();
    Triangle t2 = new Triangle("outlined", 8.0, 12.0);
    Triangle t3 = new Triangle(4.0);

    t1 = t2;

    System.out.println("Info for t1: ");
    t1.showStyle();
    t1.showDim();
    System.out.println("Area is " + t1.area());

    System.out.println();

    System.out.println("Info for t2: ");
    t2.showStyle();
    t2.showDim();
    System.out.println("Area is " + t2.area());

    System.out.println();

    System.out.println("Info for t3: ");
    t3.showStyle();
    t3.showDim();
    System.out.println("Area is " + t3.area());

    System.out.println();
  }
}
```

Here is the output from this version:

```
Info for t1:
Triangle is outlined
Width and height are 8.0 and 12.0
Area is 48.0

Info for t2:
Triangle is outlined
Width and height are 8.0 and 12.0
Area is 48.0

Info for t3:
Triangle is filled
Width and height are 4.0 and 4.0
Area is 8.0
```

Let's review the key concepts behind **super( )**. When a subclass calls **super( )**, it is calling the constructor of its immediate superclass. Thus, **super( )** always refers to the superclass immediately above the calling class. This is true even in a multilevel hierarchy. Also, **super( )** must always be the first statement executed inside a subclass constructor.

Using super to Access Superclass Members

There is a second form of **super** that acts somewhat like **this**, except that it always refers to the superclass of the subclass in which it is used. This usage has the following general form:

super.*member*

Here, *member* can be either a method or an instance variable.

This form of **super** is most applicable to situations in which member names of a subclass hide members by the same name in the superclass. Consider this simple class hierarchy:

```java
// Using super to overcome name hiding.
class A {
  int i;
}

// Create a subclass by extending class A.
class B extends A {
  int i; // this i hides the i in A

  B(int a, int b) {
    super.i = a; // i in A          Here, super.i refers
    i = b; // i in B                to the i in A.
  }

  void show() {
    System.out.println("i in superclass: " + super.i);
    System.out.println("i in subclass: " + i);
  }
}

class UseSuper {
  public static void main(String[] args) {
    B subOb = new B(1, 2);

    subOb.show();
  }
}
```

This program displays the following:

```
i in superclass: 1
i in subclass: 2
```

Although the instance variable **i** in **B** hides the **i** in **A**, **super** allows access to the **i** defined in the superclass. **super** can also be used to call methods that are hidden by a subclass.

Try This 7-1   Extending the Vehicle Class

TruckDemo.java   To illustrate the power of inheritance, we will extend the **Vehicle** class first developed in Chapter 4. As you should recall, **Vehicle** encapsulates information about vehicles, including the number of passengers they can carry, their fuel capacity, and their fuel consumption rate. We can use the **Vehicle** class as a starting point from which more specialized classes are developed. For example, one type of vehicle is a truck. An important attribute of a truck is its cargo capacity. Thus, to create a **Truck** class, you can extend **Vehicle**, adding an instance variable that stores the carrying capacity. Here is a version of **Truck** that does this. In the process, the instance variables in **Vehicle** will be made **private**, and accessor methods are provided to get and set their values.

1. Create a file called **TruckDemo.java** and copy the last implementation of **Vehicle** from Chapter 4 into the file:

2. Create the **Truck** class as shown here:

```
// Extend Vehicle to create a Truck specialization.
class Truck extends Vehicle {
  private int cargocap; // cargo capacity in pounds

  // This is a constructor for Truck.
  Truck(int p, int f, int m, int c) {
    /* Initialize Vehicle members using
       Vehicle's constructor. */
    super(p, f, m);

    cargocap = c;
  }

  // Accessor methods for cargocap.
  int getCargo() { return cargocap; }
  void putCargo(int c) { cargocap = c; }
}
```

Here, **Truck** inherits **Vehicle**, adding **cargocap**, **getCargo( )**, and **putCargo( )**. Thus, **Truck** includes all of the general vehicle attributes defined by **Vehicle**. It need add only those items that are unique to its own class.

3. Next, make the instance variables of **Vehicle** private, as shown here:

```
private int passengers; // number of passengers
private int fuelcap;    // fuel capacity in gallons
private int mpg;        // fuel consumption in miles per gallon
```

4. Here is an entire program that demonstrates the **Truck** class:

```
// Try This 7-1
//
// Build a subclass of Vehicle for trucks.
```

(continued)

```java
class Vehicle {
  private int passengers; // number of passengers
  private int fuelcap;    // fuel capacity in gallons
  private int mpg;        // fuel consumption in miles per gallon

  // This is a constructor for Vehicle.
  Vehicle(int p, int f, int m) {
    passengers = p;
    fuelcap = f;
    mpg = m;
  }

  // Return the range.
  int range() {
    return mpg * fuelcap;
  }

  // Compute fuel needed for a given distance.
  double fuelNeeded(int miles) {
    return (double) miles / mpg;
  }

  // Accessor methods for instance variables.
  int getPassengers() { return passengers; }
  void setPassengers(int p) { passengers = p; }
  int getFuelcap() { return fuelcap; }
  void setFuelcap(int f) { fuelcap = f; }
  int getMpg() { return mpg; }
  void setMpg(int m) { mpg = m; }

}

// Extend Vehicle to create a Truck specialization.
class Truck extends Vehicle {
  private int cargocap; // cargo capacity in pounds

  // This is a constructor for Truck.
  Truck(int p, int f, int m, int c) {
    /* Initialize Vehicle members using
       Vehicle's constructor. */
    super(p, f, m);

    cargocap = c;
  }

  // Accessor methods for cargocap.
  int getCargo() { return cargocap; }
  void putCargo(int c) { cargocap = c; }
}
```

```
class TruckDemo {
  public static void main(String[] args) {

    // construct some trucks
    Truck semi = new Truck(2, 200, 7, 44000);
    Truck pickup = new Truck(3, 28, 15, 2000);
    double gallons;
    int dist = 252;

    gallons = semi.fuelNeeded(dist);

    System.out.println("Semi can carry " + semi.getCargo() +
                       " pounds.");
    System.out.println("To go " + dist + " miles semi needs " +
                       gallons + " gallons of fuel.\n");

    gallons = pickup.fuelNeeded(dist);

    System.out.println("Pickup can carry " + pickup.getCargo() +
                       " pounds.");
    System.out.println("To go " + dist + " miles pickup needs " +
                       gallons + " gallons of fuel.");
  }
}
```

5. The output from this program is shown here:

```
Semi can carry 44000 pounds.
To go 252 miles semi needs 36.0 gallons of fuel.

Pickup can carry 2000 pounds.
To go 252 miles pickup needs 16.8 gallons of fuel.
```

6. Many other types of classes can be derived from **Vehicle**. For example, the following skeleton creates an off-road class that stores the ground clearance of the vehicle.

```
// Create an off-road vehicle class
class OffRoad extends Vehicle {
  private int groundClearance; // ground clearance in inches

  // ...
}
```

The key point is that once you have created a superclass that defines the general aspects of an object, that superclass can be inherited to form specialized classes. Each subclass simply adds its own, unique attributes. This is the essence of inheritance.

Creating a Multilevel Hierarchy

Up to this point, we have been using simple class hierarchies that consist of only a superclass and a subclass. However, you can build hierarchies that contain as many layers of inheritance as you like. As mentioned, it is perfectly acceptable to use a subclass as a superclass of another. For example, given three classes called **A**, **B**, and **C**, **C** can be a subclass of **B**, which is a subclass of **A**. When this type of situation occurs, each subclass inherits all of the traits found in all of its superclasses. In this case, **C** inherits all aspects of **B** and **A**.

To see how a multilevel hierarchy can be useful, consider the following program. In it, the subclass **Triangle** is used as a superclass to create the subclass called **ColorTriangle**. **ColorTriangle** inherits all of the traits of **Triangle** and **TwoDShape** and adds a field called **color**, which holds the color of the triangle.

```
// A multilevel hierarchy.
class TwoDShape {
  private double width;
  private double height;

  // A default constructor.
  TwoDShape() {
    width = height = 0.0;
  }

  // Parameterized constructor.
  TwoDShape(double w, double h) {
    width = w;
    height = h;
  }

  // Construct object with equal width and height.
  TwoDShape(double x) {
    width = height = x;
  }

  // Accessor methods for width and height.
  double getWidth() { return width; }
  double getHeight() { return height; }
  void setWidth(double w) { width = w; }
  void setHeight(double h) { height = h; }

  void showDim() {
    System.out.println("Width and height are " +
                       width + " and " + height);
  }
}

// Extend TwoDShape.
class Triangle extends TwoDShape {
  private String style;
```

```java
    // A default constructor.
    Triangle() {
      super();
      style = "none";
    }

    Triangle(String s, double w, double h) {
      super(w, h); // call superclass constructor

      style = s;
    }

    // One argument constructor.
    Triangle(double x) {
      super(x); // call superclass constructor

      style = "filled";
    }

    double area() {
      return getWidth() * getHeight() / 2;
    }

    void showStyle() {
      System.out.println("Triangle is " + style);
    }
}

// Extend Triangle.
class ColorTriangle extends Triangle {
  private String color;

  ColorTriangle(String c, String s,
                double w, double h) {
    super(s, w, h);

    color = c;
  }

  String getColor() { return color; }

  void showColor() {
    System.out.println("Color is " + color);
  }
}

class Shapes6 {
  public static void main(String[] args) {
    ColorTriangle t1 =
        new ColorTriangle("Blue", "outlined", 8.0, 12.0);
```

ColorTriangle inherits **Triangle**, which is descended from **TwoDShape**, so **ColorTriangle** includes all members of **Triangle** and **TwoDShape**.

```
        ColorTriangle t2 =
            new ColorTriangle("Red", "filled", 2.0, 2.0);

    System.out.println("Info for t1: ");
    t1.showStyle();
    t1.showDim();
    t1.showColor();
    System.out.println("Area is " + t1.area());

    System.out.println();

    System.out.println("Info for t2: ");
    t2.showStyle();
    t2.showDim();           ◄─────────────────────── A ColorTriangle object can call methods
    t2.showColor();                                   defined by itself and its superclasses.
    System.out.println("Area is " + t2.area());
  }
}
```

The output of this program is shown here:

```
Info for t1:
Triangle is outlined
Width and height are 8.0 and 12.0
Color is Blue
Area is 48.0

Info for t2:
Triangle is filled
Width and height are 2.0 and 2.0
Color is Red
Area is 2.0
```

Because of inheritance, **ColorTriangle** can make use of the previously defined classes of **Triangle** and **TwoDShape**, adding only the extra information it needs for its own, specific application. This is part of the value of inheritance; it allows the reuse of code.

This example illustrates one other important point: **super( )** always refers to the constructor in the closest superclass. The **super( )** in **ColorTriangle** calls the constructor in **Triangle**. The **super( )** in **Triangle** calls the constructor in **TwoDShape**. In a class hierarchy, if a superclass constructor requires parameters, then all subclasses must pass those parameters "up the line." This is true whether or not a subclass needs parameters of its own.

When Are Constructors Executed?

In the foregoing discussion of inheritance and class hierarchies, an important question may have occurred to you: When a subclass object is created, whose constructor is executed first, the one in the subclass or the one defined by the superclass? For example, given a subclass called

B and a superclass called **A**, is **A**'s constructor executed before **B**'s, or vice versa? The answer is that in a class hierarchy, constructors complete their execution in order of derivation, from superclass to subclass. Further, since **super( )** must be the first statement executed in a subclass' constructor, this order is the same whether or not **super( )** is used. If **super( )** is not used, then the default (parameterless) constructor of each superclass will be executed. The following program illustrates when constructors are executed:

```java
// Demonstrate when constructors are executed.

// Create a super class.
class A {
  A() {
    System.out.println("Constructing A.");
  }
}

// Create a subclass by extending class A.
class B extends A {
  B() {
    System.out.println("Constructing B.");
  }
}

// Create another subclass by extending B.
class C extends B {
  C() {
    System.out.println("Constructing C.");
  }
}

class OrderOfConstruction {
  public static void main(String[] args) {
    C c = new C();
  }
}
```

The output from this program is shown here:

```
Constructing A.
Constructing B.
Constructing C.
```

As you can see, the constructors are executed in order of derivation.

If you think about it, it makes sense that constructors are executed in order of derivation. Because a superclass has no knowledge of any subclass, any initialization it needs to perform is separate from and possibly prerequisite to any initialization performed by the subclass. Therefore, it must complete its execution first.

Superclass References and Subclass Objects

As you know, Java is a strongly typed language. Aside from the standard conversions and automatic promotions that apply to its primitive types, type compatibility is strictly enforced. Therefore, a reference variable for one class type cannot normally refer to an object of another class type. For example, consider the following program:

```
// This will not compile.
class X {
  int a;

  X(int i) { a = i; }
}

class Y {
  int a;

  Y(int i) { a = i; }
}

class IncompatibleRef {
  public static void main(String[] args) {
    X x = new X(10);
    X x2;
    Y y = new Y(5);

    x2 = x; // OK, both of same type

    x2 = y; // Error, not of same type
  }
}
```

Here, even though class **X** and class **Y** are structurally the same, it is not possible to assign an **X** reference to a **Y** object because they have different types. In general, an object reference variable can refer only to objects of its type.

There is, however, an important exception to Java's strict type enforcement. A reference variable of a superclass can be assigned a reference to an object of any subclass derived from that superclass. In other words, a superclass reference can refer to a subclass object. Here is an example:

```
// A superclass reference can refer to a subclass object.
class X {
  int a;

  X(int i) { a = i; }
}

class Y extends X {
  int b;
```

```
    Y(int i, int j) {
      super(j);
      b = i;
    }
}

class SupSubRef {
  public static void main(String[] args) {
    X x = new X(10);
    X x2;
    Y y = new Y(5, 6);

    x2 = x; // OK, both of same type
    System.out.println("x2.a: " + x2.a);

    x2 = y; // still Ok because Y is derived from X
    System.out.println("x2.a: " + x2.a);

    // X references know only about X members
    x2.a = 19; // OK
//  x2.b = 27; // Error, X doesn't have a b member
  }
}
```

OK because **Y** is a subclass of **X**; thus **x2** can refer to **y**.

Here, **Y** is now derived from **X**; thus, it is permissible for **x2** to be assigned a reference to a **Y** object.

It is important to understand that it is the type of the reference variable—not the type of the object that it refers to—that determines what members can be accessed. That is, when a reference to a subclass object is assigned to a superclass reference variable, you will have access only to those parts of the object defined by the superclass. This is why **x2** can't access **b** even when it refers to a **Y** object. If you think about it, this makes sense, because the superclass has no knowledge of what a subclass adds to it. This is why the last line of code in the program is commented out.

Although the preceding discussion may seem a bit esoteric, it has some important practical applications. One is described here. The other is discussed later in this chapter, when method overriding is covered.

An important place where subclass references are assigned to superclass variables is when constructors are called in a class hierarchy. As you know, it is common for a class to define a constructor that takes an object of the class as a parameter. This allows the class to construct a copy of an object. Subclasses of such a class can take advantage of this feature. For example, consider the following versions of **TwoDShape** and **Triangle**. Both add constructors that take an object as a parameter.

```
class TwoDShape {
  private double width;
  private double height;

  // A default constructor.
  TwoDShape() {
```

```
    width = height = 0.0;
  }

  // Parameterized constructor.
  TwoDShape(double w, double h) {
    width = w;
    height = h;
  }

  // Construct an object with equal width and height.
  TwoDShape(double x) {
    width = height = x;
  }

  // Construct an object from an object.
  TwoDShape(TwoDShape ob) {  ◄─────────────── Construct object from an object.
    width = ob.width;
    height = ob.height;
  }

  // Accessor methods for width and height.
  double getWidth() { return width; }
  double getHeight() { return height; }
  void setWidth(double w) { width = w; }
  void setHeight(double h) { height = h; }

  void showDim() {
    System.out.println("Width and height are " +
                        width + " and " + height);
  }
}

// A subclass of TwoDShape for triangles.
class Triangle extends TwoDShape {
  private String style;

  // A default constructor.
  Triangle() {
    super();
    style = "none";
  }

  // Constructor for Triangle.
  Triangle(String s, double w, double h) {
    super(w, h); // call superclass constructor

    style = s;
  }
```

```
   // One argument constructor.
   Triangle(double x) {
     super(x); // call superclass constructor

     style = "filled";
   }

   // Construct an object from an object.
   Triangle(Triangle ob) {
     super(ob); // pass object to TwoDShape constructor
     style = ob.style;
   }

   double area() {
     return getWidth() * getHeight() / 2;
   }

   void showStyle() {
     System.out.println("Triangle is " + style);
   }
}

class Shapes7 {
   public static void main(String[] args) {
     Triangle t1 =
         new Triangle("outlined", 8.0, 12.0);

     // make a copy of t1
     Triangle t2 = new Triangle(t1);

     System.out.println("Info for t1: ");
     t1.showStyle();
     t1.showDim();
     System.out.println("Area is " + t1.area());

     System.out.println();

     System.out.println("Info for t2: ");
     t2.showStyle();
     t2.showDim();
     System.out.println("Area is " + t2.area());
   }
}
```

Pass a **Triangle** reference to **TwoDShape**'s constructor.

In this program, **t2** is constructed from **t1** and is, thus, identical. The output is shown here:

```
Info for t1:
Triangle is outlined
Width and height are 8.0 and 12.0
Area is 48.0
```

```
Info for t2:
Triangle is outlined
Width and height are 8.0 and 12.0
Area is 48.0
```

Pay special attention to this **Triangle** constructor:

```
// Construct an object from an object.
Triangle(Triangle ob) {
  super(ob); // pass object to TwoDShape constructor
  style = ob.style;
}
```

It receives an object of type **Triangle** and it passes that object (through **super**) to this **TwoDShape** constructor:

```
// Construct an object from an object.
TwoDShape(TwoDShape ob) {
  width = ob.width;
  height = ob.height;
}
```

The key point is that **TwoDshape( )** is expecting a **TwoDShape** object. However, **Triangle( )** passes it a **Triangle** object. The reason this works is because, as explained, a superclass reference can refer to a subclass object. Thus, it is perfectly acceptable to pass **TwoDShape( )** a reference to an object of a class derived from **TwoDShape**. Because the **TwoDShape( )** constructor is initializing only those portions of the subclass object that are members of **TwoDShape**, it doesn't matter that the object might also contain other members added by derived classes.

Method Overriding

In a class hierarchy, when a method in a subclass has the same return type and signature as a method in its superclass, then the method in the subclass is said to *override* the method in the superclass. When an overridden method is called from within a subclass, it will always refer to the version of that method defined by the subclass. The version of the method defined by the superclass will be hidden. Consider the following:

```
// Method overriding.
class A {
  int i, j;
  A(int a, int b) {
    i = a;
    j = b;
  }

  // display i and j
  void show() {
```

```
      System.out.println("i and j: " + i + " " + j);
  }
}

class B extends A {
  int k;

  B(int a, int b, int c) {
    super(a, b);
    k = c;
  }

  // display k - this overrides show() in A
  void show() {   ◄──────────────────────────
    System.out.println("k: " + k);
  }
}

class Override {
  public static void main(String[] args) {
    B subOb = new B(1, 2, 3);

    subOb.show(); // this calls show() in B
  }
}
```

This **show( )** in **B** overrides
the one defined by **A**.

The output produced by this program is shown here:

```
k: 3
```

When **show( )** is invoked on an object of type **B**, the version of **show( )** defined within **B** is used. That is, the version of **show( )** inside **B** overrides the version declared in **A**.

If you want to access the superclass version of an overridden method, you can do so by using **super**. For example, in this version of **B**, the superclass version of **show( )** is invoked within the subclass' version. This allows all instance variables to be displayed.

```
class B extends A {
  int k;

  B(int a, int b, int c) {
    super(a, b);
    k = c;
  }

  void show() {
    super.show(); // this calls A's show()
    System.out.println("k: " + k);
  }
}
```

Use **super** to call the version of
show( ) defined by superclass **A**.

If you substitute this version of **show( )** into the previous program, you will see the following output:

```
i and j: 1 2
k: 3
```

Here, **super.show( )** calls the superclass version of **show( )**.

Method overriding occurs only when the signatures of the two methods are identical. If they are not, then the two methods are simply overloaded. For example, consider this modified version of the preceding example:

```
/* Methods with differing signatures are
   overloaded and not overridden. */
class A {
  int i, j;

  A(int a, int b) {
    i = a;
    j = b;
  }

  // display i and j
  void show() {
    System.out.println("i and j: " + i + " " + j);
  }
}

// Create a subclass by extending class A.
class B extends A {
  int k;

  B(int a, int b, int c) {
    super(a, b);
    k = c;
  }

  // overload show()
  void show(String msg) {
    System.out.println(msg + k);
  }
}

class Overload {
  public static void main(String[] args) {
    B subOb = new B(1, 2, 3);
```

Because signatures differ, this **show( )** simply overloads **show( )** in superclass **A**.

```
    subOb.show("This is k: "); // this calls show() in B
    subOb.show(); // this calls show() in A
  }
}
```

The output produced by this program is shown here:

```
This is k: 3
i and j: 1 2
```

The version of **show( )** in **B** takes a string parameter. This makes its signature different from the one in **A**, which takes no parameters. Therefore, no overriding (or name hiding) takes place.

Overridden Methods Support Polymorphism

While the examples in the preceding section demonstrate the mechanics of method overriding, they do not show its power. Indeed, if there were nothing more to method overriding than a namespace convention, then it would be, at best, an interesting curiosity but of little real value. However, this is not the case. Method overriding forms the basis for one of Java's most powerful concepts: *dynamic method dispatch*. Dynamic method dispatch is the mechanism by which a call to an overridden method is resolved at run time rather than compile time. Dynamic method dispatch is important because this is how Java implements run-time polymorphism.

Let's begin by restating an important principle: a superclass reference variable can refer to a subclass object. Java uses this fact to resolve calls to overridden methods at run time. Here's how. When an overridden method is called through a superclass reference, Java determines which version of that method to execute based upon the type of the object being referred to at the time the call occurs. Thus, this determination is made at run time. When different types of objects are referred to, different versions of an overridden method will be called. In other words, *it is the type of the object being referred to* (not the type of the reference variable) that determines which version of an overridden method will be executed. Therefore, if a superclass contains a method that is overridden by a subclass, then when different types of objects are referred to through a superclass reference variable, different versions of the method are executed.

Here is an example that illustrates dynamic method dispatch:

```
// Demonstrate dynamic method dispatch.

class Sup {
  void who() {
    System.out.println("who() in Sup");
  }
}

class Sub1 extends Sup {
  void who() {
    System.out.println("who() in Sub1");
  }
}
```

```
class Sub2 extends Sup {
  void who() {
    System.out.println("who() in Sub2");
  }
}

class DynDispDemo {
  public static void main(String[] args) {
    Sup superOb = new Sup();
    Sub1 subOb1 = new Sub1();
    Sub2 subOb2 = new Sub2();

    Sup supRef;

    supRef = superOb;
    supRef.who();              In each case,
                               the version of
                               who( ) to call
    supRef = subOb1;           is determined
    supRef.who();              at run time by
                               the type of
                               object being
    supRef = subOb2;           referred to.
    supRef.who();
  }
}
```

The output from the program is shown here:

```
who() in Sup
who() in Sub1
who() in Sub2
```

This program creates a superclass called **Sup** and two subclasses of it, called **Sub1** and **Sub2**. **Sup** declares a method called **who( )**, and the subclasses override it. Inside the **main( )** method, objects of type **Sup**, **Sub1**, and **Sub2** are declared. Also, a reference of type **Sup**, called **supRef**, is declared. The program then assigns a reference to each type of object to **supRef** and uses that reference to call **who( )**. As the output shows, the version of **who( )** executed is determined by the type of object being referred to at the time of the call, not by the class type of **supRef**.

Ask the Expert

Q: Overridden methods in Java look a lot like virtual functions in C++. Is there a similarity?

A: Yes. Readers familiar with C++ will recognize that overridden methods in Java are equivalent in purpose and similar in operation to virtual functions in C++.

Why Overridden Methods?

As stated earlier, overridden methods allow Java to support run-time polymorphism. Polymorphism is essential to object-oriented programming for one reason: it allows a general class to specify methods that will be common to all of its derivatives, while allowing subclasses to define the specific implementation of some or all of those methods. Overridden methods are another way that Java implements the "one interface, multiple methods" aspect of polymorphism. Part of the key to successfully applying polymorphism is understanding that the superclasses and subclasses form a hierarchy that moves from lesser to greater specialization. Used correctly, the superclass provides all elements that a subclass can use directly. It also defines those methods that the derived class must implement on its own. This allows the subclass the flexibility to define its own methods, yet still enforces a consistent interface. Thus, by combining inheritance with overridden methods, a superclass can define the general form of the methods that will be used by all of its subclasses.

Applying Method Overriding to TwoDShape

To better understand the power of method overriding, we will apply it to the **TwoDShape** class. In the preceding examples, each class derived from **TwoDShape** defines a method called **area( )**. This suggests that it might be better to make **area( )** part of the **TwoDShape** class, allowing each subclass to override it, defining how the area is calculated for the type of shape that the class encapsulates. The following program does this. For convenience, it also adds a name field to **TwoDShape**. (This makes it easier to write demonstration programs.)

```
// Use dynamic method dispatch.
class TwoDShape {
  private double width;
  private double height;
  private String name;

  // A default constructor.
  TwoDShape() {
    width = height = 0.0;
    name = "none";
  }

  // Parameterized constructor.
  TwoDShape(double w, double h, String n) {
    width = w;
    height = h;
    name = n;
  }

  // Construct object with equal width and height.
  TwoDShape(double x, String n) {
    width = height = x;
    name = n;
  }
```

```
    // Construct an object from an object.
    TwoDShape(TwoDShape ob) {
      width = ob.width;
      height = ob.height;
      name = ob.name;
    }

    // Accessor methods for width and height.
    double getWidth() { return width; }
    double getHeight() { return height; }
    void setWidth(double w) { width = w; }
    void setHeight(double h) { height = h; }

    String getName() { return name; }

    void showDim() {
      System.out.println("Width and height are " +
                          width + " and " + height);
    }
```

The **area( )** method defined by **TwoDShape**

```
    double area() {
      System.out.println("area() must be overridden");
      return 0.0;
    }
}

// A subclass of TwoDShape for triangles.
class Triangle extends TwoDShape {
  private String style;

  // A default constructor.
  Triangle() {
    super();
    style = "none";
  }

  // Constructor for Triangle.
  Triangle(String s, double w, double h) {
    super(w, h, "triangle");

    style = s;
  }

  // One argument constructor.
  Triangle(double x) {
    super(x, "triangle"); // call superclass constructor

    style = "filled";
  }
```

```
  // Construct an object from an object.
  Triangle(Triangle ob) {
    super(ob); // pass object to TwoDShape constructor
    style = ob.style;
  }

  // Override area() for Triangle.
  double area() {                                    Override area( ) for Triangle
    return getWidth() * getHeight() / 2;
  }

  void showStyle() {
    System.out.println("Triangle is " + style);
  }
}

// A subclass of TwoDShape for rectangles.
class Rectangle extends TwoDShape {
  // A default constructor.
  Rectangle() {
    super();
  }

  // Constructor for Rectangle.
  Rectangle(double w, double h) {
    super(w, h, "rectangle"); // call superclass constructor
  }

  // Construct a square.
  Rectangle(double x) {
    super(x, "rectangle"); // call superclass constructor
  }

  // Construct an object from an object.
  Rectangle(Rectangle ob) {
    super(ob); // pass object to TwoDShape constructor
  }

  boolean isSquare() {
    if(getWidth() == getHeight()) return true;
    return false;
  }

  // Override area() for Rectangle.
  double area() {                                    Override area( ) for Rectangle
    return getWidth() * getHeight();
  }
}
```

```
class DynShapes {
  public static void main(String[] args) {
    TwoDShape[] shapes = new TwoDShape[5];

    shapes[0] = new Triangle("outlined", 8.0, 12.0);
    shapes[1] = new Rectangle(10);
    shapes[2] = new Rectangle(10, 4);
    shapes[3] = new Triangle(7.0);
    shapes[4] = new TwoDShape(10, 20, "generic");

    for(int i=0; i < shapes.length; i++) {
      System.out.println("object is " + shapes[i].getName());
      System.out.println("Area is " + shapes[i].area());
      System.out.println();
    }
  }
}
```

The proper version of **area( )** is called for each shape.

The output from the program is shown here:

```
object is triangle
Area is 48.0

object is rectangle
Area is 100.0

object is rectangle
Area is 40.0

object is triangle
Area is 24.5

object is generic
area() must be overridden
Area is 0.0
```

Let's examine this program closely. First, as explained, **area( )** is now part of the **TwoDShape** class and is overridden by **Triangle** and **Rectangle**. Inside **TwoDShape**, **area( )** is given a placeholder implementation that simply informs the user that this method must be overridden by a subclass. Each override of **area( )** supplies an implementation that is suitable for the type of object encapsulated by the subclass. Thus, if you were to implement an ellipse class, for example, then **area( )** would need to compute the **area( )** of an ellipse.

There is one other important feature in the preceding program. Notice in **main( )** that **shapes** is declared as an array of **TwoDShape** objects. However, the elements of this array are assigned **Triangle**, **Rectangle**, and **TwoDShape** references. This is valid because, as explained, a superclass reference can refer to a subclass object. The program then cycles through the array, displaying information about each object. Although quite simple, this illustrates the power of both inheritance and method overriding. The type of object referred to by a superclass reference variable is determined at run time and acted on accordingly.

If an object is derived from **TwoDShape**, then its area can be obtained by calling **area( )**. The interface to this operation is the same no matter what type of shape is being used.

Using Abstract Classes

Sometimes you will want to create a superclass that defines only a generalized form that will be shared by all of its subclasses, leaving it to each subclass to fill in the details. Such a class determines the nature of the methods that the subclasses must implement but does not, itself, provide an implementation of one or more of these methods. One way this situation can occur is when a superclass is unable to create a meaningful implementation for a method. This is the case with the version of **TwoDShape** used in the preceding example. The definition of **area( )** is simply a placeholder. It will not compute and display the area of any type of object.

As you will see as you create your own class libraries, it is not uncommon for a method to have no meaningful definition in the context of its superclass. You can handle this situation in two ways. One way, as shown in the previous example, is to simply have it report a warning message. While this approach can be useful in certain situations—such as debugging—it is not usually appropriate. You may have methods which must be overridden by the subclass in order for the subclass to have any meaning. Consider the class **Triangle**. It is incomplete if **area( )** is not defined. In this case, you want some way to ensure that a subclass does, indeed, override all necessary methods. Java's solution to this problem is the *abstract method*.

An abstract method is created by specifying the **abstract** type modifier. An abstract method contains no body and is, therefore, not implemented by the superclass. Thus, a subclass must override it—it cannot simply use the version defined in the superclass. To declare an abstract method, use this general form:

abstract *type name*(*parameter-list*);

As you can see, no method body is present. The **abstract** modifier can be used only on instance methods. It cannot be applied to **static** methods or to constructors.

A class that contains one or more abstract methods must also be declared as abstract by preceding its **class** declaration with the **abstract** modifier. Since an abstract class does not define a complete implementation, there can be no objects of an abstract class. Thus, attempting to create an object of an abstract class by using **new** will result in a compile-time error.

When a subclass inherits an abstract class, it must implement all of the abstract methods in the superclass. If it doesn't, then the subclass must also be specified as **abstract**. Thus, the **abstract** attribute is inherited until such time as a complete implementation is achieved.

Using an abstract class, you can improve the **TwoDShape** class. Since there is no meaningful concept of area for an undefined two-dimensional figure, the following version of the preceding program declares **area( )** as **abstract** inside **TwoDShape**, and **TwoDShape** as **abstract**. This, of course, means that all classes derived from **TwoDShape** must override **area( )**.

```
// Create an abstract class.
abstract class TwoDShape {  ◄────────── TwoDShape is now abstract.
  private double width;
  private double height;
  private String name;
```

```
  // A default constructor.
  TwoDShape() {
    width = height = 0.0;
    name = "none";
  }

  // Parameterized constructor.
  TwoDShape(double w, double h, String n) {
    width = w;
    height = h;
    name = n;
  }

  // Construct object with equal width and height.
  TwoDShape(double x, String n) {
    width = height = x;
    name = n;
  }

  // Construct an object from an object.
  TwoDShape(TwoDShape ob) {
    width = ob.width;
    height = ob.height;
    name = ob.name;
  }

  // Accessor methods for width and height.
  double getWidth() { return width; }
  double getHeight() { return height; }
  void setWidth(double w) { width = w; }
  void setHeight(double h) { height = h; }

  String getName() { return name; }

  void showDim() {
    System.out.println("Width and height are " +
                        width + " and " + height);
  }

  // Now, area() is abstract.
  abstract double area();    ◄──────────── Make area( ) into an
}                                          abstract method.

// A subclass of TwoDShape for triangles.
class Triangle extends TwoDShape {
  private String style;
```

```java
    // A default constructor.
    Triangle() {
      super();
      style = "none";
    }

    // Constructor for Triangle.
    Triangle(String s, double w, double h) {
      super(w, h, "triangle");

      style = s;
    }

    // One argument constructor.
    Triangle(double x) {
      super(x, "triangle"); // call superclass constructor

      style = "filled";
    }

    // Construct an object from an object.
    Triangle(Triangle ob) {
      super(ob); // pass object to TwoDShape constructor
      style = ob.style;
    }

    double area() {
      return getWidth() * getHeight() / 2;
    }

    void showStyle() {
      System.out.println("Triangle is " + style);
    }
}

// A subclass of TwoDShape for rectangles.
class Rectangle extends TwoDShape {
  // A default constructor.
  Rectangle() {
    super();
  }

  // Constructor for Rectangle.
  Rectangle(double w, double h) {
    super(w, h, "rectangle"); // call superclass constructor
  }
```

```
    // Construct a square.
    Rectangle(double x) {
      super(x, "rectangle"); // call superclass constructor
    }

    // Construct an object from an object.
    Rectangle(Rectangle ob) {
      super(ob); // pass object to TwoDShape constructor
    }

    boolean isSquare() {
      if(getWidth() == getHeight()) return true;
      return false;
    }

    double area() {
      return getWidth() * getHeight();
    }
  }

class AbsShape {
  public static void main(String[] args) {
    TwoDShape[] shapes = new TwoDShape[4];

    shapes[0] = new Triangle("outlined", 8.0, 12.0);
    shapes[1] = new Rectangle(10);
    shapes[2] = new Rectangle(10, 4);
    shapes[3] = new Triangle(7.0);

    for(int i=0; i < shapes.length; i++) {
      System.out.println("object is " +
                         shapes[i].getName());
      System.out.println("Area is " + shapes[i].area());

      System.out.println();
    }
  }
}
```

As the program illustrates, all subclasses of **TwoDShape** *must* override **area( )**. To prove this to yourself, try creating a subclass that does not override **area( )**. You will receive a compile-time error. Of course, it is still possible to create an object reference of type **TwoDShape**, which the program does. However, it is no longer possible to declare objects of type **TwoDShape**. Because of this, in **main( )** the **shapes** array has been shortened to 4, and a **TwoDShape** object is no longer created.

One last point: Notice that **TwoDShape** still includes the **showDim( )** and **getName( )** methods and that these are not modified by **abstract**. It is perfectly acceptable—indeed, quite common—for an abstract class to contain concrete methods which a subclass is free to use as is. Only those methods declared as **abstract** need be overridden by subclasses.

Using final

As powerful and useful as method overriding and inheritance are, sometimes you will want to prevent them. For example, you might have a class that encapsulates control of some hardware device. Further, this class might offer the user the ability to initialize the device, making use of private, proprietary information. In this case, you don't want users of your class to be able to override the initialization method. Whatever the reason, in Java it is easy to prevent a method from being overridden or a class from being inherited by using the keyword **final**.

final Prevents Overriding

To prevent a method from being overridden, specify **final** as a modifier at the start of its declaration. Methods declared as **final** cannot be overridden. The following fragment illustrates **final**:

```
class A {
  final void meth() {
    System.out.println("This is a final method.");
  }
}

class B extends A {
  void meth() { // ERROR! Can't override.
    System.out.println("Illegal!");
  }
}
```

Because **meth( )** is declared as **final**, it cannot be overridden in **B**. If you attempt to do so, a compile-time error will result.

final Prevents Inheritance

You can prevent a class from being inherited by preceding its declaration with **final**. Declaring a class as **final** implicitly declares all of its methods as **final**, too. As you might expect, it is illegal to declare a class as both **abstract** and **final** since an abstract class is incomplete by itself and relies upon its subclasses to provide complete implementations.

Here is an example of a **final** class:

```
final class A {
  // ...
}

// The following class is illegal.
class B extends A { // ERROR! Can't subclass A
  // ...
}
```

As the comments imply, it is illegal for **B** to inherit **A** since **A** is declared as **final**.

NOTE

Beginning with JDK 17, the ability to *seal* a class was added to Java. Sealing offers fine-grained control over inheritance. Sealing is described in Chapter 16.

Using final with Data Members

In addition to the uses of **final** just shown, **final** can also be applied to member variables to create what amounts to named constants. If you precede an instance variable's name with **final**, its value cannot be changed throughout the lifetime of your program. You can, of course, give that variable an initial value. For example, in Chapter 6 a simple error-management class called **ErrorMsg** was shown. That class mapped a human-readable string to an error code. Here, that original class is improved by the addition of **final** constants which stand for the errors. Now, instead of passing **getErrorMsg( )** a number such as 2, you can pass the named integer constant **DISKERR**.

```java
// Return a String object.
class ErrorMsg {
  // Error codes.
  final int OUTERR    = 0;
  final int INERR     = 1;  // ←——— Declare final constants.
  final int DISKERR   = 2;
  final int INDEXERR  = 3;

  String[] msgs = {
    "Output Error",
    "Input Error",
    "Disk Full",
    "Index Out-Of-Bounds"
  };

  // Return the error message.
  String getErrorMsg(int i) {
    if(i >=0 & i < msgs.length)
      return msgs[i];
    else
      return "Invalid Error Code";
  }
}

class FinalD {
  public static void main(String[] args) {   // ←——— Use final constants.
    ErrorMsg err = new ErrorMsg();

    System.out.println(err.getErrorMsg(err.OUTERR));
    System.out.println(err.getErrorMsg(err.DISKERR));
  }
}
```

Notice how the **final** constants are used in **main( )**. Since they are members of the **ErrorMsg** class, they must be accessed via an object of that class. Of course, they can also be inherited by subclasses and accessed directly inside those subclasses.

Ask the Expert

Q: Can final **member variables be made** static? **Can** final **be used on method parameters and local variables?**

A: The answer to both is Yes. Making a **final** member variable **static** lets you refer to the constant through its class name rather than through an object. For example, if the constants in **ErrorMsg** were modified by **static**, then the **println( )** statements in **main( )** could look like this:

```
System.out.println(err.getErrorMsg(ErrorMsg.OUTERR));
System.out.println(err.getErrorMsg(ErrorMsg.DISKERR));
```

Declaring a parameter **final** prevents it from being changed within the method. Declaring a local variable **final** prevents it from being assigned a value more than once.

As a point of style, many Java programmers use uppercase identifiers for **final** constants, as does the preceding example. But this is not a hard and fast rule.

The Object Class

Java defines one special class called **Object** that is an implicit superclass of all other classes. In other words, all other classes are subclasses of **Object**. This means that a reference variable of type **Object** can refer to an object of any other class. Also, since arrays are implemented as classes, a variable of type **Object** can also refer to any array.

Object defines the following methods, which means that they are available in every object:

Method	Purpose
Object clone( )	Creates a new object that is the same as the object being cloned.
boolean equals(Object *object*)	Determines whether one object is equal to another.
void finalize( )	Called before an unused object is recycled. (Deprecated by JDK 9.)
Class<?> getClass( )	Obtains the class of an object at run time.
int hashCode( )	Returns the hash code associated with the invoking object.
void notify( )	Resumes execution of a thread waiting on the invoking object.
void notifyAll( )	Resumes execution of all threads waiting on the invoking object.
String toString( )	Returns a string that describes the object.
void wait( ) void wait(long *milliseconds*) void wait(long *milliseconds*, int *nanoseconds*)	Waits on another thread of execution.

The methods **getClass( )**, **notify( )**, **notifyAll( )**, and **wait( )** are declared as **final**. You can override the others. Several of these methods are described later in this book. However, notice two methods now: **equals( )** and **toString( )**. The **equals( )** method compares two objects. It returns **true** if the objects are equal, and **false** otherwise. The precise definition of equality can vary, depending on the type of objects to be compared. The **toString( )** method returns a string that contains a description of the object on which it is called. Also, this method is automatically called when an object is output using **println( )**. Many classes override this method. Doing so allows them to tailor a description specifically for the types of objects that they create.

One last point: Notice the unusual syntax in the return type for **getClass( )**. This relates to Java's *generics* feature. Generics allow the type of data used by a class or method to be specified as a parameter. Generics are discussed in Chapter 13.

Chapter 7 Self Test

1. Does a superclass have access to the members of a subclass? Does a subclass have access to the members of a superclass?

2. Create a subclass of **TwoDShape** called **Circle**. Include an **area( )** method that computes the area of the circle and a constructor that uses **super** to initialize the **TwoDShape** portion.

3. How do you prevent a subclass from having access to a member of a superclass?

4. Describe the purpose and use of the two versions of **super** described in this chapter.

5. Given the following hierarchy:

   ```
   class Alpha { ...

   class Beta extends Alpha { ...

   Class Gamma extends Beta { ...
   ```

   In what order do the constructors for these classes complete their execution when a **Gamma** object is instantiated?

6. A superclass reference can refer to a subclass object. Explain why this is important as it relates to method overriding.

7. What is an abstract class?

8. How do you prevent a method from being overridden? How do you prevent a class from being inherited?

9. Explain how inheritance, method overriding, and abstract classes are used to support polymorphism.

10. What class is a superclass of every other class?

11. A class that contains at least one abstract method must, itself, be declared abstract. True or False?

12. What keyword is used to create a named constant?

13. Assume that class **B** inherits class **A**. Further, assume a method called **makeObj( )** that is declared as shown here:

```
A makeObj(int which ) {
   if(which == 0) return new A();
   else return new B();
}
```

Notice that **makeObj( )** returns a reference to an object of either type **A** or **B**, depending on the value of **which**. Notice, however, that the return type of **makeObj( )** is **A**. (Recall that a superclass reference can refer to a subclass object.) Given this situation and assuming that you are using JDK 10 or later, what is the type of **myRef** in the following declaration and why?

```
var myRef = makeObj(1);
```

14. Assuming the situation described in Question 13, what will the type of **myRef** be given this statement?

```
var myRef = (B) makeObj(1);
```

Chapter 8

Packages
and Interfaces

Key Skills & Concepts

- Use packages

- Understand how packages affect access

- Apply the **protected** access modifier

- Import packages

- Know Java's standard packages

- Understand interface fundamentals

- Implement an interface

- Apply interface references

- Understand interface variables

- Extend interfaces

- Create default, static, and private interface methods

This chapter examines two of Java's most innovative features: packages and interfaces. *Packages* are groups of related classes. Packages help organize your code and provide another layer of encapsulation. As you will see in Chapter 15, packages also play an important role with modules. An *interface* defines a set of methods that will be implemented by a class. Thus, an interface gives you a way to specify what a class will do, but not how it will do it. Packages and interfaces give you greater control over the organization of your program.

Packages

In programming, it is often helpful to group related pieces of a program together. In Java, this can be accomplished by using a package. A package serves two purposes. First, it provides a mechanism by which related pieces of a program can be organized as a unit. Classes defined within a package must be accessed through their package name. Thus, a package provides a way to name a collection of classes. Second, a package participates in Java's access control mechanism. Classes defined within a package can be made private to that package and not accessible by code outside the package. Thus, the package provides a means by which classes can be encapsulated. Let's examine each feature a bit more closely.

In general, when you name a class, you are allocating a name from the *namespace*. A namespace defines a declarative region. In Java, no two classes can use the same name from the same namespace. Thus, within a given namespace, each class name must be unique. The examples shown in the preceding chapters have all used the default namespace.

While this is fine for short sample programs, it becomes a problem as programs grow and the default namespace becomes crowded. In large programs, finding unique names for each class can be difficult. Furthermore, you must avoid name collisions with code created by other programmers working on the same project, and with Java's library. The solution to these problems is the package because it gives you a way to partition the namespace. When a class is defined within a package, the name of that package is attached to each class, thus avoiding name collisions with other classes that have the same name, but are in other packages.

Since a package usually contains related classes, Java defines special access rights to code within a package. In a package, you can define code that is accessible by other code within the same package but not by code outside the package. This enables you to create self-contained groups of related classes that keep their operation private.

Defining a Package

All classes in Java belong to some package. When no **package** statement is specified, the default package is used. Furthermore, the default package has no name, which makes the default package transparent. This is why you haven't had to worry about packages before now. While the default package is fine for short, sample programs, it is inadequate for real applications. Most of the time, you will define one or more packages for your code.

To create a package, put a **package** command at the top of a Java source file. The classes declared within that file will then belong to the specified package. Since a package defines a namespace, the names of the classes that you put into the file become part of that package's namespace.

This is the general form of the **package** statement:

package *pkg*;

Here, *pkg* is the name of the package. For example, the following statement creates a package called **mypack**:

```
package mypack;
```

Typically, Java uses the file system to manage packages, with each package stored in its own directory, and this is the approach assumed by the discussions and examples of packages in this book. For example, the **.class** files for any classes you declare to be part of **mypack** must be stored in a directory called **mypack**.

Like the rest of Java, package names are case sensitive. This means that the directory in which a package is stored must be precisely the same as the package name. If you have trouble trying the examples in this chapter, remember to check your package and directory names carefully. Lowercase is often used for package names.

More than one file can include the same **package** statement. The **package** statement simply specifies to which package the classes defined in a file belong. It does not exclude other classes in other files from being part of that same package. Most real-world packages are spread across many files.

You can create a hierarchy of packages. To do so, simply separate each package name from the one above it by use of a period. The general form of a multileveled package statement is shown here:

package *pack1.pack2.pack3...packN*;

Of course, you must create directories that support the package hierarchy that you create. For example,

```
package alpha.beta.gamma;
```

must be stored in .../alpha/beta/gamma, where ... specifies the path to the specified directories.

Finding Packages and CLASSPATH

As just explained, packages are typically mirrored by directories. This raises an important question: How does the Java run-time system know where to look for packages that you create? As it relates to the examples in this chapter, the answer has three parts. First, by default, the Java run-time system uses the current working directory as its starting point. Thus, if your package is in a subdirectory of the current directory, it will be found. Second, you can specify a directory path or paths by setting the **CLASSPATH** environmental variable. Third, you can use the **-classpath** option with **java** and **javac** to specify the path to your classes. It is useful to point out that, beginning with JDK 9, a package can be part of a module, and thus found on the module path. However, a discussion of modules and module paths is deferred until Chapter 15. For now, we will use only class paths.

For example, assuming the following package specification:

```
package mypack
```

In order for a program to find **mypack,** the program can be executed from a directory immediately above **mypack**, or **CLASSPATH** must be set to include the path to **mypack**, or the **-classpath** option must specify the path to **mypack** when the program is run via **java**.

The easiest way to try the examples shown in this chapter is to simply create the package directories below your current development directory, put the **.class** files into the appropriate directories, and then execute the programs from the development directory. This is the approach used by the following examples.

One last point: To avoid problems, it is best to keep all **.java** and **.class** files associated with a package in that package's directory. Also, compile each file from the directory above the package directory.

A Short Package Example

Keeping the preceding discussion in mind, try this short package example. It creates a simple book database that is contained within a package called **bookpack**.

```
// A short package demonstration.
package bookpack;  ◄──────────────── This file is part of the bookpack package.

class Book {  ◄──────────────── Thus, Book is part of bookpack.
  private String title;
  private String author;
  private int pubDate;

  Book(String t, String a, int d) {
    title = t;
```

```
      author = a;
      pubDate = d;
  }

  void show() {
    System.out.println(title);
    System.out.println(author);
    System.out.println(pubDate);
    System.out.println();
  }
}
```

BookDemo is also part of **bookpack**.

```
class BookDemo {
  public static void main(String[] args) {
    Book[] books = new Book[5];

    books[0] = new Book("Java: A Beginner's Guide",
                        "Schildt", 2022);
    books[1] = new Book("Java: The Complete Reference",
                        "Schildt", 2022);
    books[2] = new Book("1984",
                        "Orwell", 1949);
    books[3] = new Book("Red Storm Rising",
                        "Clancy", 1986);
    books[4] = new Book("On the Road",
                        "Kerouac", 1955);

    for(int i=0; i < books.length; i++) books[i].show();
  }
}
```

Call this file **BookDemo.java** and put it in a directory called **bookpack**.
   Next, compile the file. You can do this by specifying

```
javac bookpack/BookDemo.java
```

from the directory directly above **bookpack**. Then try executing the class, using the following command line:

```
java bookpack.BookDemo
```

Remember, you will need to be in the directory above **bookpack** when you execute this command. (Or, use one of the other two options described in the preceding section to specify the path to **bookpack**.)
   As explained, **BookDemo** and **Book** are now part of the package **bookpack**. This means that **BookDemo** cannot be executed by itself. That is, you cannot use this command line:

```
java BookDemo
```

Instead, **BookDemo** must be qualified with its package name.

Packages and Member Access

The preceding chapters have introduced the fundamentals of access control, including the **private** and **public** modifiers, but they have not told the entire story. One reason for this is that packages also participate in Java's access control mechanism, and this aspect of access control had to wait until packages were covered. Before we continue, it is important to note that the modules feature also offers another dimension to accessibility, but here we focus strictly on the interplay between packages and classes.

The visibility of an element is affected by its access specification—**private**, **public**, **protected**, or default—and the package in which it resides. Thus, as it relates to classes and packages, the visibility of an element is determined by its visibility within a class and its visibility within a package. This multilayered approach to access control supports a rich assortment of access privileges. Table 8-1 summarizes the various access levels. Let's examine each access option individually.

If a member of a class has no explicit access modifier, then it is visible within its package but not outside its package. Therefore, you will use the default access specification for elements that you want to keep private to a package but public within that package.

Members explicitly declared **public** are the most visible, and can be accessed from different classes and different packages. A **private** member is accessible only to the other members of its class. A **private** member is unaffected by its membership in a package. A member specified as **protected** is accessible within its package and to subclasses in other packages.

Table 8-1 applies only to members of classes. A top-level class has only two possible access levels: default and public. When a class is declared as **public**, it is accessible outside its package. If a class has default access, it can be accessed only by other code within its same package. Also, a class that is declared **public** must reside in a file by the same name.

	Private Member	Default Member	Protected Member	Public Member
Visible within same class	Yes	Yes	Yes	Yes
Visible within same package by subclass	No	Yes	Yes	Yes
Visible within same package by non-subclass	No	Yes	Yes	Yes
Visible within different package by subclass	No	No	Yes	Yes
Visible within different package by non-subclass	No	No	No	Yes

Table 8-1 Class Member Access

NOTE

Remember, the modules feature can also affect accessibility. Modules are discussed in Chapter 15.

A Package Access Example

In the **package** example shown earlier, both **Book** and **BookDemo** were in the same package, so there was no problem with **BookDemo** using **Book** because the default access privilege grants all members of the same package access. However, if **Book** were in one package and **BookDemo** were in another, the situation would be different. In this case, access to **Book** would be denied. To make **Book** available to other packages, you must make three changes. First, **Book** needs to be declared **public**. This makes **Book** visible outside of **bookpack**. Second, its constructor must be made **public**, and finally, its **show( )** method needs to be **public**. This allows them to be visible outside of **bookpack**, too. Thus, to make **Book** usable by other packages, it must be recoded as shown here:

```
// Book recoded for public access.
package bookpack;

public class Book {  ◄──────── Book and its members must be public
  private String title;          in order to be used by other packages.
  private String author;
  private int pubDate;

  // Now public.
  public Book(String t, String a, int d) {
    title = t;
    author = a;
    pubDate = d;
  }

  // Now public.
  public void show() {
    System.out.println(title);
    System.out.println(author);
    System.out.println(pubDate);
    System.out.println();
  }
}
```

To use **Book** from another package, either you must use the **import** statement described in the next section, or you must fully qualify its name to include its full package specification. For example, here is a class called **UseBook**, which is contained in the **bookpackext** package. It fully qualifies **Book** in order to use it.

```
// This class is in package bookpackext.
package bookpackext;
```

```
// Use the Book class from bookpack.
class UseBook {
  public static void main(String[] args) {
    bookpack.Book[] books = new bookpack.Book[5];

    books[0] = new bookpack.Book("Java: A Beginner's Guide",
                      "Schildt", 2022);
    books[1] = new bookpack.Book("Java: The Complete Reference",
                      "Schildt", 2022);
    books[2] = new bookpack.Book("1984",
                      "Orwell", 1949);
    books[3] = new bookpack.Book("Red Storm Rising",
                      "Clancy", 1986);
    books[4] = new bookpack.Book("On the Road",
                      "Kerouac", 1955);

    for(int i=0; i < books.length; i++) books[i].show();
  }
}
```

Qualify **Book** with its package name: **bookpack**.

Notice how every use of **Book** is preceded with the **bookpack** qualifier. Without this specification, **Book** would not be found when you tried to compile **UseBook**.

Understanding Protected Members

Newcomers to Java are sometimes confused by the meaning and use of **protected**. As explained, the **protected** modifier creates a member that is accessible within its package and to subclasses in other packages. Thus, a **protected** member is available for all subclasses to use but is still protected from arbitrary access by code outside its package.

To better understand the effects of **protected**, let's work through an example. First, change the **Book** class so that its instance variables are **protected**, as shown here:

```
// Make the instance variables in Book protected.
package bookpack;

public class Book {
  // these are now protected
  protected String title;
  protected String author;      —— These are now protected.
  protected int pubDate;

  public Book(String t, String a, int d) {
    title = t;
    author = a;
    pubDate = d;
  }
```

```
  public void show() {
    System.out.println(title);
    System.out.println(author);
    System.out.println(pubDate);
    System.out.println();
  }
}
```

Next, create a subclass of **Book**, called **ExtBook**, and a class called **ProtectDemo** that uses **ExtBook**. **ExtBook** adds a field that stores the name of the publisher and several accessor methods. Both of these classes will be in their own package called **bookpackext**. They are shown here:

```
// Demonstrate protected.
package bookpackext;

class ExtBook extends bookpack.Book {
  private String publisher;

  public ExtBook(String t, String a, int d, String p) {
    super(t, a, d);
    publisher = p;
  }

  public void show() {
    super.show();
    System.out.println(publisher);
    System.out.println();
  }

  public String getPublisher() { return publisher; }
  public void setPublisher(String p) { publisher = p; }

  /* These are OK because subclass can access
     a protected member. */
  public String getTitle() { return title; }
  public void setTitle(String t) { title = t; }
  public String getAuthor() { return author; }       ◄——— Access to **Book**'s members
  public void setAuthor(String a) { author = a; }          is allowed for subclasses.
  public int getPubDate() { return pubDate; }
  public void setPubDate(int d) { pubDate = d; }
}

class ProtectDemo {
  public static void main(String[] args) {
    ExtBook[] books = new ExtBook[5];
```

```
        books[0] = new ExtBook("Java: A Beginner's Guide",
                       "Schildt", 2022, "McGraw Hill");
        books[1] = new ExtBook("Java: The Complete Reference",
                       "Schildt", 2022, "McGraw Hill");
        books[2] = new ExtBook("1984",
                       "Orwell", 1949,
                       "Harcourt Brace Jovanovich");
        books[3] = new ExtBook("Red Storm Rising",
                       "Clancy", 1986, "Putnam");
        books[4] = new ExtBook("On the Road",
                       "Kerouac", 1955, "Viking");

        for(int i=0; i < books.length; i++) books[i].show();

        // Find books by author
        System.out.println("Showing all books by Schildt.");
        for(int i=0; i < books.length; i++)
          if(books[i].getAuthor() == "Schildt")
            System.out.println(books[i].getTitle());

//      books[0].title = "test title"; // Error - not accessible
    }
}
```

—— Access to **protected** field not allowed by non-subclass.

Look first at the code inside **ExtBook**. Because **ExtBook** extends **Book**, it has access to the **protected** members of **Book**, even though **ExtBook** is in a different package. Thus, it can access **title**, **author**, and **pubDate** directly, as it does in the accessor methods it creates for those variables. However, in **ProtectDemo**, access to these variables is denied because **ProtectDemo** is not a subclass of **Book**. For example, if you remove the comment symbol from the following line, the program will not compile.

```
//      books[0].title = "test title"; // Error - not accessible
```

Importing Packages

When you use a class from another package, you can fully qualify the name of the class with the name of its package, as the preceding examples have done. However, such an approach could easily become tiresome and awkward, especially if the classes you are qualifying are deeply nested in a package hierarchy. Since Java was invented by programmers for programmers—and programmers don't like tedious constructs—it should come as no surprise that a more convenient method exists for using the contents of packages: the **import** statement. Using **import** you can bring one or more members of a package into view. This allows you to use those members directly, without explicit package qualification.

Here is the general form of the **import** statement:

import *pkg.classname*;

Here, *pkg* is the name of the package, which can include its full path, and *classname* is the name of the class being imported. If you want to import the entire contents of a package, use an asterisk (*) for the class name. Here are examples of both forms:

```
import mypack.MyClass
import mypack.*;
```

In the first case, the **MyClass** class is imported from **mypack**. In the second, all of the classes in **mypack** are imported. In a Java source file, **import** statements occur immediately following the **package** statement (if it exists) and before any class definitions.

You can use **import** to bring the **bookpack** package into view so that the **Book** class can be used without qualification. To do so, simply add this **import** statement to the top of any file that uses **Book**.

```
import bookpack.*;
```

For example, here is the **UseBook** class recoded to use **import**:

```
// Demonstrate import.
package bookpackext;
import bookpack.*;   ◄——————  Import bookpack.

// Use the Book class from bookpack.
class UseBook {
  public static void main(String[] args) {
    Book[] books = new Book[5];   ◄———— Now, you can refer to Book
                                         directly, without qualification.

    books[0] = new Book("Java: A Beginner's Guide",
                        "Schildt", 2022);
    books[1] = new Book("Java: The Complete Reference",
                        "Schildt", 2022);
    books[2] = new Book("1984",
                        "Orwell", 1949);
    books[3] = new Book("Red Storm Rising",
                        "Clancy", 1986);
    books[4] = new Book("On the Road",
                        "Kerouac", 1955);

    for(int i=0; i < books.length; i++) books[i].show();
  }
}
```

Notice that you no longer need to qualify **Book** with its package name.

Java's Class Library Is Contained in Packages

As explained earlier in this book, Java defines a large number of standard classes that are available to all programs. This class library is often referred to as the Java API (Application Programming Interface). The Java API is stored in packages. At the top of the package hierarchy is **java**. Descending from **java** are several subpackages. Here are a few examples:

Subpackage	Description
java.lang	Contains a large number of general-purpose classes
java.io	Contains I/O classes
java.net	Contains classes that support networking
java.util	Contains a large number of utility classes, including the Collections Framework
java.awt	Contains classes that support the Abstract Window Toolkit

Since the beginning of this book, you have been using **java.lang**. It contains, among several others, the **System** class, which you have been using when performing output using **println( )**. The **java.lang** package is unique because it is imported automatically into every Java program. This is why you did not have to import **java.lang** in the preceding sample programs. However, you must explicitly import the other packages. We will be examining several packages in subsequent chapters.

Interfaces

In object-oriented programming, it is sometimes helpful to define what a class must do but not how it will do it. You have already seen an example of this: the abstract method. An abstract method defines the signature for a method but provides no implementation. A subclass must provide its own implementation of each abstract method defined by its superclass. Thus, an abstract method specifies the *interface* to the method but not the *implementation*. While abstract classes and methods are useful, it is possible to take this concept a step further. In Java, you can fully separate a class' interface from its implementation by using the keyword **interface**.

An **interface** is syntactically similar to an abstract class, in that you can specify one or more methods that have no body. Those methods must be implemented by a class in order for their actions to be defined. Thus, an interface specifies what must be done, but not how to do it. Once an interface is defined, any number of classes can implement it. Also, one class can implement any number of interfaces.

To implement an interface, a class must provide bodies (implementations) for the methods described by the interface. Each class is free to determine the details of its own implementation. Two classes might implement the same interface in different ways, but each class still supports the same set of methods. Thus, code that has knowledge of the interface can use objects of either class since the interface to those objects is the same. By providing the **interface** keyword, Java allows you to fully utilize the "one interface, multiple methods" aspect of polymorphism.

Before continuing an important point needs to be made. JDK 8 added a feature to **interface** that made a significant change to its capabilities. Prior to JDK 8, an interface could not define any implementation whatsoever. Thus, prior to JDK 8, an interface could define only what, but not how, as just described. JDK 8 changed this. Today, it is possible to add a *default implementation* to an interface method. Furthermore, static interface methods are now supported, and beginning with JDK 9, an interface can also include private methods. Thus, it is now possible for **interface** to specify some behavior. However, such methods constitute what are, in essence, special-use features, and the original intent behind **interface** still remains. Therefore, as a general rule, you will still often create and use interfaces in which no use is made of these new features. For this reason, we will begin by discussing the interface in its traditional form. The expanded interface features are described at the end of this chapter.

Here is a simplified general form of a traditional interface:

access interface *name* {
  ret-type method-name1(param-list);
  ret-type method-name2(param-list);
  type var1 = value;
  type var2 = value;
  // ...
  ret-type method-nameN(param-list);
  type varN = value;
}

For a top-level interface, *access* is either **public** or not used. When no access modifier is included, then default access results, and the interface is available only to other members of its package. When it is declared as **public**, the interface can be used by any other code. (When an **interface** is declared **public**, it must be in a file of the same name.) *name* is the name of the interface and can be any valid identifier.

In the traditional form of an interface, methods are declared using only their return type and signature. They are, essentially, abstract methods. Thus, each class that includes such an **interface** must implement all of its methods. In an interface, methods are implicitly **public**.

Variables declared in an **interface** are not instance variables. Instead, they are implicitly **public**, **final**, and **static** and must be initialized. Thus, they are essentially constants.

Here is an example of an **interface** definition. It specifies the interface to a class that generates a series of numbers.

```
public interface Series {
  int getNext(); // return next number in series
  void reset(); // restart
  void setStart(int x); // set starting value
}
```

This interface is declared **public** so that it can be implemented by code in any package.

Implementing Interfaces

Once an **interface** has been defined, one or more classes can implement that interface. To implement an interface, include the **implements** clause in a class definition and then create the methods required by the interface. The general form of a class that includes the **implements** clause looks like this:

class *classname* extends *superclass* implements *interface* {
  // class-body
}

To implement more than one interface, the interfaces are separated with a comma. Of course, the **extends** clause is optional.

The methods that implement an interface must be declared **public**. Also, the type signature of the implementing method must match exactly the type signature specified in the **interface** definition.

Here is an example that implements the **Series** interface shown earlier. It creates a class called **ByTwos**, which generates a series of numbers, each two greater than the previous one.

```
// Implement Series.
class ByTwos implements Series {
  int start;
  int val;                    Implement the Series interface.

  ByTwos() {
    start = 0;
    val = 0;
  }

  public int getNext() {
    val += 2;
    return val;
  }

  public void reset() {
    val = start;
  }

  public void setStart(int x) {
    start = x;
    val = x;
  }
}
```

Notice that the methods **getNext( )**, **reset( )**, and **setStart( )** are declared using the **public** access specifier. This is necessary. Whenever you implement a method defined by an interface, it must be implemented as **public** because all members of an interface are implicitly **public**.

Here is a class that demonstrates **ByTwos**:

```
class SeriesDemo {
  public static void main(String[] args) {
    ByTwos ob = new ByTwos();

    for(int i=0; i < 5; i++)
      System.out.println("Next value is " +
                          ob.getNext());

    System.out.println("\nResetting");
    ob.reset();
    for(int i=0; i < 5; i++)
      System.out.println("Next value is " +
```

```
                                        ob.getNext());

      System.out.println("\nStarting at 100");
      ob.setStart(100);
      for(int i=0; i < 5; i++)
        System.out.println("Next value is " +
                                ob.getNext());
  }
}
```

The output from this program is shown here:

```
Next value is 2
Next value is 4
Next value is 6
Next value is 8
Next value is 10

Resetting
Next value is 2
Next value is 4
Next value is 6
Next value is 8
Next value is 10

Starting at 100
Next value is 102
Next value is 104
Next value is 106
Next value is 108
Next value is 110
```

It is both permissible and common for classes that implement interfaces to define additional members of their own. For example, the following version of **ByTwos** adds the method **getPrevious( )**, which returns the previous value:

```
// Implement Series and add getPrevious().
class ByTwos implements Series {
  int start;
  int val;
  int prev;

  ByTwos() {
    start = 0;
    val = 0;
    prev = -2;
  }

  public int getNext() {
    prev = val;
    val += 2;
```

```
      return val;
    }

  public void reset() {
    val = start;
    prev = start - 2;
  }

  public void setStart(int x) {
    start = x;
    val = x;
    prev = x - 2;
  }

  int getPrevious() {  ◄──────── Add a method not defined by Series.
    return prev;
  }
}
```

Notice that the addition of **getPrevious( )** required a change to the implementations of the methods defined by **Series**. However, since the interface to those methods stays the same, the change is seamless and does not break preexisting code. This is one of the advantages of interfaces.

As explained, any number of classes can implement an **interface**. For example, here is a class called **ByThrees** that generates a series that consists of multiples of three:

```
// Implement Series.
class ByThrees implements Series {  ◄──────── Implement Series a different way.
  int start;
  int val;

  ByThrees() {
    start = 0;
    val = 0;
  }

  public int getNext() {
    val += 3;
    return val;
  }

  public void reset() {
    val = start;
  }

  public void setStart(int x) {
    start = x;
    val = x;
  }
}
```

One more point: If a class includes an interface but does not fully implement the methods defined by that interface, then that class must be declared **abstract**. No objects of such a class can be created, but it can be used as an abstract superclass, allowing subclasses to provide the complete implementation.

Using Interface References

You might be somewhat surprised to learn that you can declare a reference variable of an interface type. In other words, you can create an interface reference variable. Such a variable can refer to any object that implements its interface. When you call a method on an object through an interface reference, it is the version of the method implemented by the object that is executed. This process is similar to using a superclass reference to access a subclass object, as described in Chapter 7.

The following example illustrates this process. It uses the same interface reference variable to call methods on objects of both **ByTwos** and **ByThrees**.

```
// Demonstrate interface references.

class ByTwos implements Series {
  int start;
  int val;

  ByTwos() {
    start = 0;
    val = 0;
  }

  public int getNext() {
    val += 2;
    return val;
  }

  public void reset() {
    val = start;
  }

  public void setStart(int x) {
    start = x;
    val = x;
  }
}

class ByThrees implements Series {
  int start;
  int val;
```

```
   ByThrees() {
     start = 0;
     val = 0;
   }

   public int getNext() {
     val += 3;
     return val;
   }

   public void reset() {
     val = start;
   }

   public void setStart(int x) {
     start = x;
     val = x;
   }
 }

 class SeriesDemo2 {
   public static void main(String[] args) {
     ByTwos twoOb = new ByTwos();
     ByThrees threeOb = new ByThrees();
     Series ob;

     for(int i=0; i < 5; i++) {
       ob = twoOb;
       System.out.println("Next ByTwos value is " +
                         ob.getNext());
       ob = threeOb;
       System.out.println("Next ByThrees value is " +
                         ob.getNext());
     }
   }
 }
```

— Access an object via an interface reference.

In **main( )**, **ob** is declared to be a reference to a **Series** interface. This means that it can be used to store references to any object that implements **Series**. In this case, it is used to refer to **twoOb** and **threeOb**, which are objects of type **ByTwos** and **ByThrees**, respectively, which both implement **Series**. An interface reference variable has knowledge only of the methods declared by its **interface** declaration. Thus, **ob** could not be used to access any other variables or methods that might be supported by the object.

Try This 8-1   Creating a Queue Interface

ICharQ.java
IQDemo.java

To see the power of interfaces in action, we will look at a practical example. In earlier chapters, you developed a class called **Queue** that implemented a simple fixed-size queue for characters. However, there are many ways to implement a queue. For example, the queue can be of a fixed size or it can be "growable." The queue can be linear, in which case it can be used up, or it can be circular, in which case elements can be put in as long as elements are being taken off. The queue can also be held in an array, a linked list, a binary tree, and so on. No matter how the queue is implemented, the interface to the queue remains the same, and the methods **put( )** and **get( )** define the interface to the queue independently of the details of the implementation. Because the interface to a queue is separate from its implementation, it is easy to define a queue interface, leaving it to each implementation to define the specifics.

In this project, you will create an interface for a character queue and three implementations. All three implementations will use an array to store the characters. One queue will be the fixed-size, linear queue developed earlier. Another will be a circular queue. In a circular queue, when the end of the underlying array is encountered, the get and put indices automatically loop back to the start. Thus, any number of items can be stored in a circular queue as long as items are also being taken out. The final implementation creates a dynamic queue, which grows as necessary when its size is exceeded.

1. Create a file called **ICharQ.java** and put into that file the following interface definition:

```
// A character queue interface.
public interface ICharQ {
  // Put a character into the queue.
  void put(char ch);

  // Get a character from the queue.
  char get();
}
```

As you can see, this interface is very simple, consisting of only two methods. Each class that implements **ICharQ** will need to implement these methods.

2. Create a file called **IQDemo.java**.

3. Begin creating **IQDemo.java** by adding the **FixedQueue** class shown here:

```
// A fixed-size queue class for characters.
class FixedQueue implements ICharQ {
  private char[] q; // this array holds the queue
  private int putloc, getloc; // the put and get indices

  // Construct an empty queue given its size.
  public FixedQueue(int size) {
    q = new char[size]; // allocate memory for queue
```

(continued)

```
      putloc = getloc = 0;
    }

    // Put a character into the queue.
    public void put(char ch) {
      if(putloc==q.length) {
        System.out.println(" - Queue is full.");
        return;
      }

      q[putloc++] = ch;
    }

    // Get a character from the queue.
    public char get() {
      if(getloc == putloc) {
        System.out.println(" - Queue is empty.");
        return (char) 0;
      }

      return q[getloc++];
    }
}
```

This implementation of **ICharQ** is adapted from the **Queue** class shown in Chapter 5 and should already be familiar to you.

4. To **IQDemo.java** add the **CircularQueue** class shown here. It implements a circular queue for characters.

```
// A circular queue.
class CircularQueue implements ICharQ {
  private char[] q; // this array holds the queue
  private int putloc, getloc; // the put and get indices

  // Construct an empty queue given its size.
  public CircularQueue(int size) {
    q = new char[size+1]; // allocate memory for queue
    putloc = getloc = 0;
  }

  // Put a character into the queue.
  public void put(char ch) {
    /* Queue is full if either putloc is one less than
       getloc, or if putloc is at the end of the array
       and getloc is at the beginning. */
    if(putloc+1==getloc |
```

```
      ((putloc==q.length-1) & (getloc==0))) {
        System.out.println(" - Queue is full.");
        return;
      }

    q[putloc++] = ch;
    if(putloc==q.length) putloc = 0; // loop back
  }

  // Get a character from the queue.
  public char get() {
    if(getloc == putloc) {
      System.out.println(" - Queue is empty.");
      return (char) 0;
    }

    char ch = q[getloc++];
    if(getloc==q.length) getloc = 0; // loop back
    return ch;
  }
}
```

The circular queue works by reusing space in the array that is freed when elements are retrieved. Thus, it can store an unlimited number of elements as long as elements are also being removed. While conceptually simple—just reset the appropriate index to zero when the end of the array is reached—the boundary conditions are a bit confusing at first. In a circular queue, the queue is full not when the end of the underlying array is reached, but rather when storing an item would cause an unretrieved item to be overwritten. Thus, **put( )** must check several conditions in order to determine if the queue is full. As the comments suggest, the queue is full when either **putloc** is one less than **getloc**, or if **putloc** is at the end of the array and **getloc** is at the beginning. As before, the queue is empty when **getloc** and **putloc** are equal. To make these checks easier, the underlying array is created one size larger than the queue size.

5. Put into **IQDemo.java** the **DynQueue** class shown next. It implements a "growable" queue that expands its size when space is exhausted.

```
// A dynamic queue.
class DynQueue implements ICharQ {
  private char[] q; // this array holds the queue
  private int putloc, getloc; // the put and get indices

  // Construct an empty queue given its size.
  public DynQueue(int size) {
    q = new char[size]; // allocate memory for queue
    putloc = getloc = 0;
  }
```

(continued)

```
   // Put a character into the queue.
   public void put(char ch) {
     if(putloc==q.length) {
       // increase queue size
       char[] t = new char[q.length * 2];

       // copy elements into new queue
       for(int i=0; i < q.length; i++)
         t[i] = q[i];

       q = t;
     }

     q[putloc++] = ch;
   }

   // Get a character from the queue.
   public char get() {
     if(getloc == putloc) {
       System.out.println(" - Queue is empty.");
       return (char) 0;
     }

     return q[getloc++];
   }
 }
```

In this queue implementation, when the queue is full, an attempt to store another element causes a new underlying array to be allocated that is twice as large as the original, the current contents of the queue are copied into this array, and a reference to the new array is stored in **q**.

6. To demonstrate the three **ICharQ** implementations, enter the following class into **IQDemo.java**. It uses an **ICharQ** reference to access all three queues.

```
// Demonstrate the ICharQ interface.
class IQDemo {
  public static void main(String[] args) {
    FixedQueue q1 = new FixedQueue(10);
    DynQueue q2 = new DynQueue(5);
    CircularQueue q3 = new CircularQueue(10);

    ICharQ iQ;

    char ch;
    int i;
```

```
iQ = q1;
// Put some characters into fixed queue.
for(i=0; i < 10; i++)
  iQ.put((char) ('A' + i));

// Show the queue.
System.out.print("Contents of fixed queue: ");
for(i=0; i < 10; i++) {
  ch = iQ.get();
  System.out.print(ch);
}
System.out.println();

iQ = q2;
// Put some characters into dynamic queue.
for(i=0; i < 10; i++)
  iQ.put((char) ('Z' - i));

// Show the queue.
System.out.print("Contents of dynamic queue: ");
for(i=0; i < 10; i++) {
  ch = iQ.get();
  System.out.print(ch);
}

System.out.println();

iQ = q3;
// Put some characters into circular queue.
for(i=0; i < 10; i++)
  iQ.put((char) ('A' + i));

// Show the queue.
System.out.print("Contents of circular queue: ");
for(i=0; i < 10; i++) {
  ch = iQ.get();
  System.out.print(ch);
}

System.out.println();

// Put more characters into circular queue.
for(i=10; i < 20; i++)
  iQ.put((char) ('A' + i));

// Show the queue.
System.out.print("Contents of circular queue: ");
for(i=0; i < 10; i++) {
```

(continued)

```
      ch = iQ.get();
      System.out.print(ch);
    }

    System.out.println("\nStore and consume from" +
                       " circular queue.");

    // Store in and consume from circular queue.
    for(i=0; i < 20; i++) {
      iQ.put((char) ('A' + i));
      ch = iQ.get();
      System.out.print(ch);
    }
  }
}
```

7. The output from this program is shown here:

```
Contents of fixed queue: ABCDEFGHIJ
Contents of dynamic queue: ZYXWVUTSRQ
Contents of circular queue: ABCDEFGHIJ
Contents of circular queue: KLMNOPQRST
Store and consume from circular queue.
ABCDEFGHIJKLMNOPQRST
```

8. Here are some things to try on your own. Create a circular version of **DynQueue**. Add a **reset( )** method to **ICharQ**, which resets the queue. Create a **static** method that copies the contents of one type of queue into another.

Variables in Interfaces

As mentioned, variables can be declared in an interface, but they are implicitly **public**, **static**, and **final**. At first glance, you might think that there would be very limited use for such variables, but the opposite is true. Large programs typically make use of several constant values that describe such things as array size, various limits, special values, and the like. Since a large program is typically held in a number of separate source files, there needs to be a convenient way to make these constants available to each file. In Java, interface variables offer one solution.

To define a set of shared constants, create an **interface** that contains only these constants, without any methods. Each file that needs access to the constants simply "implements" the interface. This brings the constants into view. Here is an example:

```
// An interface that contains constants.
interface IConst {
  int MIN = 0;
  int MAX = 10;                               These are constants.
  String ERRORMSG = "Boundary Error";
}
```

```
class IConstD implements IConst {
  public static void main(String[] args) {
    int[] nums = new int[MAX];

    for(int i=MIN; i < 11; i++) {
      if(i >= MAX) System.out.println(ERRORMSG);
      else {
        nums[i] = i;
        System.out.print(nums[i] + " ");
      }
    }
  }
}
```

NOTE

The technique of using an **interface** to define shared constants is controversial. It is described here for completeness.

Interfaces Can Be Extended

One interface can inherit another by use of the keyword **extends**. The syntax is the same as for inheriting classes. When a class implements an interface that inherits another interface, it must provide implementations for all methods required by the interface inheritance chain. Following is an example:

```
// One interface can extend another.
interface A {
  void meth1();
  void meth2();
}

// B now includes meth1() and meth2() - it adds meth3().
interface B extends A {
  void meth3();
}
```
B inherits A.

```
// This class must implement all of A and B
class MyClass implements B {
  public void meth1() {
    System.out.println("Implement meth1().");
  }
```

```
    public void meth2() {
      System.out.println("Implement meth2().");
    }

    public void meth3() {
      System.out.println("Implement meth3().");
    }
}

class IFExtend {
  public static void main(String[] args) {
    MyClass ob = new MyClass();

    ob.meth1();
    ob.meth2();
    ob.meth3();
  }
}
```

As an experiment, you might try removing the implementation for **meth1( )** in **MyClass**. This will cause a compile-time error. As stated earlier, any class that implements an interface must implement all methods required by that interface, including any that are inherited from other interfaces.

Default Interface Methods

As explained earlier, prior to JDK 8, an interface could not define any implementation whatsoever. This meant that for all previous versions of Java, the methods specified by an interface were abstract, containing no body. This is the traditional form of an interface and is the type of interface that the preceding discussions have used. The release of JDK 8 changed this by adding a new capability to **interface** called the *default method*. A default method lets you define a default implementation for an interface method. In other words, by use of a default method, it is possible for an interface method to provide a body, rather than being abstract. During its development, the default method was also referred to as an *extension method*, and you will likely see both terms used.

A primary motivation for the default method was to provide a means by which interfaces could be expanded without breaking existing code. Recall that there must be implementations for all methods defined by an interface. In the past, if a new method were added to a popular, widely used interface, then the addition of that method would break existing code because no implementation would be found for that method. The default method solves this problem by supplying an implementation that will be used if no other implementation is explicitly provided. Thus, the addition of a default method will not cause preexisting code to break.

Another motivation for the default method was the desire to specify methods in an interface that are, essentially, optional, depending on how the interface is used. For example, an interface might define a group of methods that act on a sequence of elements. One of these methods might be called **remove( )**, and its purpose is to remove an element from the sequence. However, if the interface is intended to support both modifiable and non-modifiable sequences, then **remove( )** is essentially optional because it won't be used by non-modifiable sequences. In the past, a class that implemented a non-modifiable sequence would have had to define an empty implementation of **remove( )**, even though it was not needed. Today, a default implementation for **remove( )** can be specified in the interface that either does nothing or reports an error. Providing this default prevents a class used for non-modifiable sequences from having to define its own, placeholder version of **remove( )**. Thus, by providing a default, the interface makes the implementation of **remove( )** by a class optional.

It is important to point out that the addition of default methods does not change a key aspect of **interface**: an interface still cannot have instance variables. Thus, the defining difference between an interface and a class is that a class can maintain state information, but an interface cannot. Furthermore, it is still not possible to create an instance of an interface by itself. It must be implemented by a class. Therefore, even though modern versions of Java allow an interface to define default methods, the interface must still be implemented by a class if an instance is to be created.

One last point: As a general rule, default methods constitute a special-purpose feature. Interfaces that you create will still be used primarily to specify what and not how. However, the inclusion of the default method gives you added flexibility.

Default Method Fundamentals

An interface default method is defined similar to the way a method is defined by a **class**. The primary difference is that the declaration is preceded by the keyword **default**. For example, consider this simple interface:

```
public interface MyIF {
  // This is a "normal" interface method declaration.
  // It does NOT define a default implementation.
  int getUserID();

  // This is a default method. Notice that it provides
  // a default implementation.
  default int getAdminID() {
    return 1;
  }
}
```

MyIF declares two methods. The first, **getUserID( )**, is a standard interface method declaration. It defines no implementation whatsoever. The second method is **getAdminID( )**, and it does include a default implementation. In this case, it simply returns 1. Pay special attention to the way **getAdminID( )** is declared. Its declaration is preceded by the **default** modifier. This syntax can be generalized. To define a default method, precede its declaration with **default**.

Because **getAdminID( )** includes a default implementation, it is not necessary for an implementing class to override it. In other words, if an implementing class does not provide its own implementation, the default is used. For example, the **MyIFImp** class shown next is perfectly valid:

```
// Implement MyIF.
class MyIFImp implements MyIF {
  // Only getUserID() defined by MyIF needs to be implemented.
  // getAdminID() can be allowed to default.
  public int getUserID() {
    return 100;
  }
}
```

The following code creates an instance of **MyIFImp** and uses it to call both **getUserID( )** and **getAdminID( )**.

```
// Use the default method.
class DefaultMethodDemo {
  public static void main(String[] args) {

    MyIFImp obj = new MyIFImp();

    // Can call getUserID(), because it is explicitly
    // implemented by MyIFImp:
    System.out.println("User ID is " + obj.getUserID());

    // Can also call getAdminID(), because of default
    // implementation:
    System.out.println("Administrator ID is " + obj.getAdminID());
  }
}
```

The output is shown here:

```
User ID is 100
Administrator ID is 1
```

As you can see, the default implementation of **getAdminID( )** was automatically used. It was not necessary for **MyIFImp** to define it. Thus, for **getAdminID( )**, implementation by a class is optional. (Of course, its implementation by a class will be *required* if the class needs to return a different ID.)

It is both possible and common for an implementing class to define its own implementation of a default method. For example, **MyIFImp2** overrides **getAdminID( )**, as shown here:

```
class MyIFImp2 implements MyIF {
  // Here, implementations for both getUserID( ) and getAdminID( ) are
  // provided.
  public int getUserID() {
    return 100;
  }
```

```
   public int getAdminID() {
     return 42;
   }
}
```

Now, when **getAdminID( )** is called, a value other than its default is returned.

A More Practical Example of a Default Method

Although the preceding shows the mechanics of using default methods, it doesn't illustrate their usefulness in a more practical setting. To do this, let's return to the **Series** interface shown earlier in this chapter. For the sake of discussion, assume that **Series** is widely used and many programs rely on it. Further assume that through an analysis of usage patterns, it was discovered that many implementations of **Series** were adding a method that returned an array that contained the next *n* elements in the series. Given this situation, you decide to enhance **Series** so that it includes such a method, calling the new method **getNextArray( )** and declaring it as shown here:

```
int[] getNextArray(int n)
```

Here, **n** specifies the number of elements to retrieve. Prior to default methods, adding this method to **Series** would have broken preexisting code because existing implementations would not have defined the method. However, by providing a default for this new method, it can be added to **Series** without causing harm. Let's work through the process.

In some cases, when a default method is added to an existing interface, its implementation simply reports an error if an attempt is made to use the default. This approach is necessary in the case of default methods for which no implementation can be provided that will work in all cases. These types of default methods define what is, essentially, optional code. However, in some cases, you can define a default method that will work in all cases. This is the situation for **getNextArray( )**. Because **Series** already requires that a class implement **getNext( )**, the default version of **getNextArray( )** can use it. Thus, here is one way to implement the new version of **Series** that includes the default **getNextArray( )** method:

```
// An enhanced version of Series that includes a default
// method called getNextArray().
public interface Series {
  int getNext(); // return next number in series

  // Return an array that contains the next n elements
  // in the series beyond the current element.
  default int[] getNextArray(int n) {
    int[] vals = new int[n];

    for(int i=0; i < n; i++) vals[i] = getNext();
    return vals;
  }

  void reset(); // restart
  void setStart(int x); // set starting value
}
```

Pay special attention to the way that the default method **getNextArray( )** is implemented. Because **getNext( )** was part of the original specification for **Series**, any class that implements **Series** will provide that method. Thus, it can be used inside **getNextArray( )** to obtain the next *n* elements in the series. As a result, any class that implements the enhanced version of **Series** will be able to use **getNextArray( )** as is, and no class is required to override it. Therefore, no preexisting code is broken. Of course, it is still possible for a class to provide its own implementation of **getNextArray( )**, if you choose.

As the preceding example shows, the default method provides two major benefits:

- It gives you a way to gracefully evolve interfaces over time without breaking existing code.

- It provides optional functionality without requiring that a class provide a placeholder implementation when that functionality is not needed.

In the case of **getNextArray( )**, the second point is especially important. If an implementation of **Series** does not require the capability offered by **getNextArray( )**, it need not provide its own placeholder implementation. This allows cleaner code to be created.

Multiple Inheritance Issues

As explained earlier in this book, Java does not support the multiple inheritance of classes. Now that an interface can include default methods, you might be wondering if an interface can provide a way around this restriction. The answer is, essentially, no. Recall that there is still a key difference between a class and an interface: a class can maintain state information (through the use of instance variables), but an interface cannot.

The preceding notwithstanding, default methods do offer a bit of what one would normally associate with the concept of multiple inheritance. For example, you might have a class that implements two interfaces. If each of these interfaces provides default methods, then some behavior is inherited from both. Thus, to a limited extent, default methods do support multiple inheritance of behavior. As you might guess, in such a situation, it is possible that a name conflict will occur.

For example, assume that two interfaces called **Alpha** and **Beta** are implemented by a class called **MyClass**. What happens if both **Alpha** and **Beta** provide a method called **reset( )** for which both declare a default implementation? Is the version by **Alpha** or the version by **Beta** used by **MyClass**? Or, consider a situation in which **Beta** extends **Alpha**. Which version of the default method is used? Or, what if **MyClass** provides its own implementation of the method? To handle these and other similar types of situations, Java defines a set of rules that resolve such conflicts.

First, in all cases a class implementation takes priority over an interface default implementation. Thus, if **MyClass** provides an override of the **reset( )** default method, **MyClass**'s version is used. This is the case even if **MyClass** implements both **Alpha** and **Beta**. In this case, both defaults are overridden by **MyClass**'s implementation.

Second, in cases in which a class inherits two interfaces that both have the same default method, if the class does not override that method, then an error will result. Continuing with the example, if **MyClass** inherits both **Alpha** and **Beta**, but does not override **reset( )**, then an error will occur.

In cases in which one interface inherits another, with both defining a common default method, the inheriting interface's version of the method takes precedence. Therefore, continuing the example, if **Beta** extends **Alpha**, then **Beta**'s version of **reset( )** will be used.

It is possible to refer explicitly to a default implementation by using a new form of **super**. Its general form is shown here:

InterfaceName.super.*methodName( )*

For example, if **Beta** wants to refer to **Alpha**'s default for **reset( )**, it can use this statement:

```
Alpha.super.reset();
```

Use static Methods in an Interface

JDK 8 added another new capability to **interface**: the ability to define one or more **static** methods. Like **static** methods in a class, a **static** method defined by an interface can be called independently of any object. Thus, no implementation of the interface is necessary, and no instance of the interface is required in order to call a **static** method. Instead, a **static** method is called by specifying the interface name, followed by a period, followed by the method name. Here is the general form.

InterfaceName.*staticMethodName*

Notice that this is similar to the way that a **static** method in a class is called.

The following shows an example of a **static** method in an interface by adding one to **MyIF**, shown earlier. The **static** method is **getUniversalID( )**. It returns zero.

```
public interface MyIF {
  // This is a "normal" interface method declaration.
  // It does NOT define a default implementation.
  int getUserID();

  // This is a default method. Notice that it provides
  // a default implementation.
  default int getAdminID() {
    return 1;
  }

  // This is a static interface method.
  static int getUniversalID() {
    return 0;
  }
}
```

The **getUniversalID( )** method can be called, as shown here:

```
int uID = MyIF.getUniversalID();
```

As mentioned, no implementation or instance of **MyIF** is required to call **getUniversalID( )** because it is **static**.

One last point: **static** interface methods are not inherited by either an implementing class or a subinterface.

Private Interface Methods

Beginning with JDK 9, an interface can include a private method. A private interface method can be called only by a default method or another private method defined by the same interface. Because a private interface method is specified **private**, it cannot be used by code outside the interface in which it is defined. This restriction includes subinterfaces because a private interface method is not inherited by a subinterface.

The key benefit of a private interface method is that it lets two or more default methods use a common piece of code, thus avoiding code duplication. For example, here is a further enhanced version of the **Series** interface that adds a second default method called **skipAndGetNextArray( )**. It skips a specified number of elements and then returns an array that contains the subsequent elements. It uses a private method called **getArray( )** to obtain an element array of a specified size.

```java
// A further enhanced version of Series that includes two
// default methods that use a private method called getArray();
public interface Series {
  int getNext(); // return next number in series

  // Return an array that contains the next n elements
  // in the series beyond the current element.
  default int[] getNextArray(int n) {
    return getArray(n);
  }

  // Return an array that contains the next n elements
  // in the series, after skipping elements.
  default int[] skipAndGetNextArray(int skip, int n) {

    // Skip the specified number of elements.
    getArray(skip);

    return getArray(n);
  }

  // A private method that returns an array containing
  // the next n elements.
  private int[] getArray(int n) {
    int[] vals = new int[n];
```

```
      for(int i=0; i < n; i++) vals[i] = getNext();
      return vals;
   }

   void reset(); // restart
   void setStart(int x); // set starting value
}
```

Notice that both **getNextArray( )** and **skipAndGetNextArray( )** use the private **getArray( )** method to obtain the array to return. This prevents both methods from having to duplicate the same code sequence. Keep in mind that because **getArray( )** is private, it cannot be called by code outside **Series**. Thus, its use is limited to the default methods inside **Series**.

Although the private interface method is a feature that you will seldom need, in those cases in which you *do* need it, you will find it quite useful.

Final Thoughts on Packages and Interfaces

Although the examples we've included in this book do not make frequent use of packages or interfaces, both of these tools are an important part of the Java programming environment. Virtually all real programs that you write in Java will be contained within packages. A number will probably implement interfaces as well. As you will see in Chapter 15, packages play an important role in the modules feature. It is important, therefore, that you be comfortable with their usage.

Chapter 8 Self Test

1. Using the code from Try This 8-1, put the **ICharQ** interface and its three implementations into a package called **qpack**. Keeping the queue demonstration class **IQDemo** in the default package, show how to import and use the classes in **qpack**.

2. What is a namespace? Why is it important that Java allows you to partition the namespace?

3. Typically, packages are stored in _____.

4. Explain the difference between **protected** and default access.

5. Explain the two ways that the members of a package can be used by other packages.

6. "One interface, multiple methods" is a key tenet of Java. What feature best exemplifies it?

7. How many classes can implement an interface? How many interfaces can a class implement?

8. Can interfaces be extended?

9. Create an interface for the **Vehicle** class from Chapter 7. Call the interface **IVehicle**.

10. Variables declared in an interface are implicitly **static** and **final**. Can they be shared with other parts of a program?

11. A package is, in essence, a container for classes. True or False?

12. What standard Java package is automatically imported into a program?

13. What keyword is used to declare a default **interface** method?

14. Is it possible to define a **static** method in an **interface**?

15. Assume that the **ICharQ** interface shown in Try This 8-1 has been in widespread use for several years. Now, you want to add a method to it called **reset( )**, which will be used to reset the queue to its empty, starting condition. How can this be accomplished without breaking preexisting code?

16. How is a **static** method in an interface called?

17. Can an **interface** have a private method?

Chapter 9

Exception Handling

Key Skills & Concepts

- Know the exception hierarchy

- Use **try** and **catch**

- Understand the effects of an uncaught exception

- Use multiple **catch** statements

- Catch subclass exceptions

- Nest **try** blocks

- Throw an exception

- Know the members of **Throwable**

- Use **finally**

- Use **throws**

- Know Java's built-in exceptions

- Create custom exception classes

This chapter discusses exception handling. An exception is an error that occurs at run time. Using Java's exception handling subsystem you can, in a structured and controlled manner, handle run-time errors. Although most modern programming languages offer some form of exception handling, Java's support for it is both easy-to-use and flexible.

A principal advantage of exception handling is that it automates much of the error handling code that previously had to be entered "by hand" into any large program. For example, in some older computer languages, error codes are returned when a method fails, and these values must be checked manually, each time the method is called. This approach is both tedious and error-prone. Exception handling streamlines error handling by allowing your program to define a block of code, called an *exception handler,* that is executed automatically when an error occurs. It is not necessary to manually check the success or failure of each specific operation or method call. If an error occurs, it will be processed by the exception handler.

Another reason that exception handling is important is that Java defines standard exceptions for common program errors, such as divide-by-zero or file-not-found. To respond to these errors, your program must watch for and handle these exceptions. Also, Java's API library makes extensive use of exceptions.

In the final analysis, to be a successful Java programmer means that you are fully capable of navigating Java's exception handling subsystem.

The Exception Hierarchy

In Java, all exceptions are represented by classes. All exception classes are derived from a class called **Throwable**. Thus, when an exception occurs in a program, an object of some type of exception class is generated. There are two direct subclasses of **Throwable**: **Exception** and **Error**. Exceptions of type **Error** are related to errors that occur in the Java Virtual Machine itself, and not in your program. These types of exceptions are beyond your control, and your program will not usually deal with them. Thus, these types of exceptions are not described here.

Errors that result from program activity are represented by subclasses of **Exception**. For example, divide-by-zero, array boundary, and file errors fall into this category. In general, your program should handle exceptions of these types. An important subclass of **Exception** is **RuntimeException**, which is used to represent various common types of run-time errors.

Exception Handling Fundamentals

Java exception handling is managed via five keywords: **try**, **catch**, **throw**, **throws**, and **finally**. They form an interrelated subsystem in which the use of one implies the use of another. Throughout the course of this chapter, each keyword is examined in detail. However, it is useful at the outset to have a general understanding of the role each plays in exception handling. Briefly, here is how they work.

Program statements that you want to monitor for exceptions are contained within a **try** block. If an exception occurs within the **try** block, it is *thrown*. Your code can catch this exception using **catch** and handle it in some rational manner. System-generated exceptions are automatically thrown by the Java run-time system. To manually throw an exception, use the keyword **throw**. In some cases, an exception that is thrown out of a method must be specified as such by a **throws** clause. Any code that absolutely must be executed upon exiting from a **try** block is put in a **finally** block.

Ask the Expert

Q: Just to be sure, could you review the conditions that cause an exception to be generated?

A: Exceptions are generated in three different ways. First, the Java Virtual Machine can generate an exception in response to some internal error which is beyond your control. Normally, your program won't handle these types of exceptions. Second, standard exceptions, such as those corresponding to divide-by-zero or array index out-of-bounds, are generated by errors in program code. You need to handle these exceptions. Third, you can manually generate an exception by using the **throw** statement. No matter how an exception is generated, it is handled in the same way.

Using try and catch

At the core of exception handling are **try** and **catch**. These keywords work together; you can't have a **catch** without a **try**. Here is the general form of the **try/catch** exception handling blocks:

```
try {
  // block of code to monitor for errors
}
catch (ExcepType1 exOb) {
  // handler for ExcepType1
}
catch (ExcepType2 exOb) {
  // handler for ExcepType2
}
.
.
.
```

Here, *ExcepType* is the type of exception that has occurred. When an exception is thrown, it is caught by its corresponding **catch** statement, which then processes the exception. As the general form shows, there can be more than one **catch** statement associated with a **try**. The type of the exception determines which **catch** statement is executed. That is, if the exception type specified by a **catch** statement matches that of the exception, then that **catch** statement is executed (and all others are bypassed). When an exception is caught, *exOb* will receive its value.

Here is an important point: If no exception is thrown, then a **try** block ends normally, and all of its **catch** statements are bypassed. Execution resumes with the first statement following the last **catch**. Thus, **catch** statements are executed only if an exception is thrown.

NOTE

There is another form of the **try** statement that supports *automatic resource management*. This form of **try** is called *try-with-resources*. It is described in Chapter 10, in the context of managing I/O streams (such as those connected to a file) because I/O streams are some of the most commonly used resources.

A Simple Exception Example

Here is a simple example that illustrates how to watch for and catch an exception. As you know, it is an error to attempt to index an array beyond its boundaries. When this occurs, the JVM throws an **ArrayIndexOutOfBoundsException**. The following program purposely generates such an exception and then catches it:

```
// Demonstrate exception handling.
class ExcDemo1 {
  public static void main(String[] args) {
    int[] nums = new int[4];
```

```
try {  ◄───────── Create a try block.
    System.out.println("Before exception is generated.");

    // Generate an index out-of-bounds exception.
    nums[7] = 10;  ◄─────────────────────────────    Attempt to index past
    System.out.println("this won't be displayed");    nums boundary.
}
catch (ArrayIndexOutOfBoundsException exc) {  ◄─────── Catch array boundary
    // catch the exception                            errors.
    System.out.println("Index out-of-bounds!");
}
System.out.println("After catch statement.");
  }
}
```

This program displays the following output:

```
Before exception is generated.
Index out-of-bounds!
After catch statement.
```

Although quite short, the preceding program illustrates several key points about exception handling. First, the code that you want to monitor for errors is contained within a **try** block. Second, when an exception occurs (in this case, because of the attempt to index **nums** beyond its bounds), the exception is thrown out of the **try** block and caught by the **catch** statement. At this point, control passes to the **catch**, and the **try** block is terminated. That is, **catch** is *not* called. Rather, program execution is transferred to it. Thus, the **println( )** statement following the out-of-bounds index will never execute. After the **catch** statement executes, program control continues with the statements following the **catch**. Thus, it is the job of your exception handler to remedy the problem that caused the exception so that program execution can continue normally.

Remember, if no exception is thrown by a **try** block, no **catch** statements will be executed and program control resumes after the **catch** statement. To confirm this, in the preceding program, change the line

```
nums[7] = 10;
```

to

```
nums[0] = 10;
```

Now, no exception is generated, and the **catch** block is not executed.

It is important to understand that all code within a **try** block is monitored for exceptions. This includes exceptions that might be generated by a method called from within the **try** block. An exception thrown by a method called from within a **try** block can be caught by the **catch** statements associated with that **try** block—assuming, of course, that the method did not catch the exception itself. For example, this is a valid program:

```
/* An exception can be generated by one
   method and caught by another. */
```

```
class ExcTest {
  // Generate an exception.
  static void genException() {
    int[] nums = new int[4];

    System.out.println("Before exception is generated.");

    // generate an index out-of-bounds exception
    nums[7] = 10;                                              Exception generated here.
    System.out.println("this won't be displayed");
  }
}

class ExcDemo2 {
  public static void main(String[] args) {

    try {                                                      Exception caught here.
      ExcTest.genException();
    } catch (ArrayIndexOutOfBoundsException exc) {
      // catch the exception
      System.out.println("Index out-of-bounds!");
    }
    System.out.println("After catch statement.");
  }
}
```

This program produces the following output, which is the same as that produced by the first version of the program shown earlier:

```
Before exception is generated.
Index out-of-bounds!
After catch statement.
```

Since **genException( )** is called from within a **try** block, the exception that it generates (and does not catch) is caught by the **catch** in **main( )**. Understand, however, that if **genException( )** had caught the exception itself, it never would have been passed back to **main( )**.

The Consequences of an Uncaught Exception

Catching one of Java's standard exceptions, as the preceding program does, has a side benefit: It prevents abnormal program termination. When an exception is thrown, it must be caught by some piece of code, somewhere. In general, if your program does not catch an exception, then it will be caught by the JVM. The trouble is that the JVM's default exception handler terminates execution and displays a stack trace and error message. For example, in this version of the preceding example, the index out-of-bounds exception is not caught by the program.

```
// Let JVM handle the error.
class NotHandled {
  public static void main(String[] args) {
```

```
    int[] nums = new int[4];

    System.out.println("Before exception is generated.");

    // generate an index out-of-bounds exception
    nums[7] = 10;
  }
}
```

When the array index error occurs, execution is halted, and the following error message is displayed. (The exact output you see may vary because of differences between JDKs.)

```
Exception in thread "main" java.lang.ArrayIndexOutOfBoundsException:
      Index 7 out of bounds for length 4
      at NotHandled.main(NotHandled.java:9)
```

While such a message is useful for you while debugging, it would not be something that you would want others to see, to say the least! This is why it is important for your program to handle exceptions itself, rather than rely upon the JVM.

As mentioned earlier, the type of the exception must match the type specified in a **catch** statement. If it doesn't, the exception won't be caught. For example, the following program tries to catch an array boundary error with a **catch** statement for an **ArithmeticException** (another of Java's built-in exceptions). When the array boundary is overrun, an **ArrayIndexOutOfBoundsException** is generated, but it won't be caught by the **catch** statement. This results in abnormal program termination.

```
// This won't work!
class ExcTypeMismatch {
  public static void main(String[] args) {
    int[] nums = new int[4];                        This throws an
                                                    ArrayIndexOutOfBoundsException.
    try {
      System.out.println("Before exception is generated.");

      //generate an index out-of-bounds exception
      nums[7] = 10;  ◄───────────────────────────────┘
      System.out.println("this won't be displayed");
    }

    /* Can't catch an array boundary error with an
       ArithmeticException. */
    catch (ArithmeticException exc) {  ◄────── This tries to catch it with an
      // catch the exception                    ArithmeticException.
      System.out.println("Index out-of-bounds!");
    }
    System.out.println("After catch statement.");
  }
}
```

The output is shown here. (Again, your output may vary based on differences between JDKs.)

```
Before exception is generated.
Exception in thread "main" java.lang.ArrayIndexOutOfBoundsException:
        Index 7 out of bounds for length 4
        at ExcTypeMismatch.main(ExcTypeMismatch.java:10)
```

As the output demonstrates, a **catch** for **ArithmeticException** won't catch an **ArrayIndexOutOfBoundsException**.

Exceptions Enable You to Handle Errors Gracefully

One of the key benefits of exception handling is that it enables your program to respond to an error and then continue running. For example, consider the following example that divides the elements of one array by the elements of another. If a division by zero occurs, an **ArithmeticException** is generated. In the program, this exception is handled by reporting the error and then continuing with execution. Thus, attempting to divide by zero does not cause an abrupt run-time error resulting in the termination of the program. Instead, it is handled gracefully, allowing program execution to continue.

```java
// Handle error gracefully and continue.
class ExcDemo3 {
  public static void main(String[] args) {
    int[] numer = { 4, 8, 16, 32, 64, 128 };
    int[] denum = { 2, 0, 4, 4, 0, 8 };

    for(int i=0; i<numer.length; i++) {
      try {
        System.out.println(numer[i] + " / " +
                           denom[i] + " is " +
                           numer[i]/denom[i]);
      }
      catch (ArithmeticException exc) {
        // catch the exception
        System.out.println("Can't divide by Zero!");
      }
    }
  }
}
```

The output from the program is shown here:

```
4 / 2 is 2
Can't divide by Zero!
16 / 4 is 4
32 / 4 is 8
Can't divide by Zero!
128 / 8 is 16
```

This example makes another important point: Once an exception has been handled, it is removed from the system. Therefore, in the program, each pass through the loop enters the **try** block anew; any prior exceptions have been handled. This enables your program to handle repeated errors.

Using Multiple catch Statements

As stated earlier, you can associate more than one **catch** statement with a **try**. In fact, it is common to do so. However, each **catch** must catch a different type of exception. For example, the program shown here catches both array boundary and divide-by-zero errors:

```
// Use multiple catch statements.
class ExcDemo4 {
  public static void main(String[] args) {
    // Here, numer is longer than denom.
    int[] numer = { 4, 8, 16, 32, 64, 128, 256, 512 };
    int[] denom = { 2, 0, 4, 4, 0, 8 };

    for(int i=0; i<numer.length; i++) {
      try {
        System.out.println(numer[i] + " / " +
                           denom[i] + " is " +
                           numer[i]/denom[i]);
      }
      catch (ArithmeticException exc) {          Multiple catch statements
        // catch the exception
        System.out.println("Can't divide by Zero!");
      }
      catch (ArrayIndexOutOfBoundsException exc) {
        // catch the exception
        System.out.println("No matching element found.");
      }
    }
  }
}
```

This program produces the following output:

```
4 / 2 is 2
Can't divide by Zero!
16 / 4 is 4
32 / 4 is 8
Can't divide by Zero!
128 / 8 is 16
No matching element found.
No matching element found.
```

As the output confirms, each **catch** statement responds only to its own type of exception.

In general, **catch** expressions are checked in the order in which they occur in a program. Only a matching statement is executed. All other **catch** blocks are ignored.

Catching Subclass Exceptions

There is one important point about multiple **catch** statements that relates to subclasses. A **catch** clause for a superclass will also match any of its subclasses. For example, since the superclass of all exceptions is **Throwable**, to catch all possible exceptions, catch **Throwable**. If you want to catch exceptions of both a superclass type and a subclass type, put the subclass first in the **catch** sequence. If you don't, then the superclass **catch** will also catch all derived classes. This rule is self-enforcing because putting the superclass first causes unreachable code to be created, since the subclass **catch** clause can never execute. In Java, unreachable code is an error.

For example, consider the following program:

```
// Subclasses must precede superclasses in catch statements.
class ExcDemo5 {
  public static void main(String[] args) {
    // Here, numer is longer than denom.
    int[] numer = { 4, 8, 16, 32, 64, 128, 256, 512 };
    int[] denom = { 2, 0, 4, 4, 0, 8 };

    for(int i=0; i<numer.length; i++) {
      try {
        System.out.println(numer[i] + " / " +
                           denom[i] + " is " +
                           numer[i]/denom[i]);
      }
      catch (ArrayIndexOutOfBoundsException exc) {    ◄—— Catch subclass
        // catch the exception
        System.out.println("No matching element found.");
      }
      catch (Throwable exc) {  ◄————————————————————————— Catch superclass
        System.out.println("Some exception occurred.");
      }
    }
  }
}
```

The output from the program is shown here:

```
4 / 2 is 2
Some exception occurred.
16 / 4 is 4
32 / 4 is 8
Some exception occurred.
128 / 8 is 16
No matching element found.
No matching element found.
```

Ask the Expert

Q: Why would I want to catch superclass exceptions?

A: There are, of course, a variety of reasons. Here are a couple. First, if you add a **catch** clause that catches exceptions of type **Exception**, then you have effectively added a "catch all" clause to your exception handler that deals with all program-related exceptions. Such a "catch all" clause might be useful in a situation in which abnormal program termination must be avoided no matter what occurs. Second, in some situations, an entire category of exceptions can be handled by the same clause. Catching the superclass of these exceptions allows you to handle all without duplicated code.

In this case, **catch(Throwable)** catches all exceptions except for **ArrayIndexOutOfBounds-Exception**. The issue of catching subclass exceptions becomes more important when you create exceptions of your own.

Try Blocks Can Be Nested

One **try** block can be nested within another. An exception generated within the inner **try** block that is not caught by a **catch** associated with that **try** is propagated to the outer **try** block. For example, here the **ArrayIndexOutOfBoundsException** is not caught by the inner **catch**, but by the outer **catch**:

```
// Use a nested try block.
class NestTrys {
  public static void main(String[] args) {
    // Here, numer is longer than denom.
    int[] numer = { 4, 8, 16, 32, 64, 128, 256, 512 };
    int[] denom = { 2, 0, 4, 4, 0, 8 };

    try { // outer try  ◀─────────────────────── Nested try blocks
      for(int i=0; i<numer.length; i++) {
        try { // nested try  ◀──────────────────┘
          System.out.println(numer[i] + " / " +
                             denom[i] + " is " +
                             numer[i]/denom[i]);
        }
        catch (ArithmeticException exc) {
          // catch the exception
          System.out.println("Can't divide by Zero!");
        }
      }
    }
  }
```

```
      catch (ArrayIndexOutOfBoundsException exc) {
        // catch the exception
        System.out.println("No matching element found.");
        System.out.println("Fatal error - program terminated.");
      }
    }
  }
}
```

The output from the program is shown here:

```
4 / 2 is 2
Can't divide by Zero!
16 / 4 is 4
32 / 4 is 8
Can't divide by Zero!
128 / 8 is 16
No matching element found.
Fatal error - program terminated.
```

In this example, an exception that can be handled by the inner **try**—in this case, a divide-by-zero error—allows the program to continue. However, an array boundary error is caught by the outer **try**, which causes the program to terminate.

Although certainly not the only reason for nested **try** statements, the preceding program makes an important point that can be generalized. Often nested **try** blocks are used to allow different categories of errors to be handled in different ways. Some types of errors are catastrophic and cannot be fixed. Some are minor and can be handled immediately. You might use an outer **try** block to catch the most severe errors, allowing inner **try** blocks to handle less serious ones.

Throwing an Exception

The preceding examples have been catching exceptions generated automatically by the JVM. However, it is possible to manually throw an exception by using the **throw** statement. Its general form is shown here:

throw *exceptOb*;

Here, *exceptOb* must be an object of an exception class derived from **Throwable**.

Here is an example that illustrates the **throw** statement by manually throwing an **ArithmeticException**:

```
// Manually throw an exception.
class ThrowDemo {
  public static void main(String[] args) {
    try {
      System.out.println("Before throw.");
      throw new ArithmeticException();  ◄——————— Throw an exception.
```

```
    }
    catch (ArithmeticException exc) {
      // catch the exception
      System.out.println("Exception caught.");
    }
    System.out.println("After try/catch block.");
  }
}
```

The output from the program is shown here:

```
Before throw.
Exception caught.
After try/catch block.
```

Notice how the **ArithmeticException** was created using **new** in the **throw** statement. Remember, **throw** throws an object. Thus, you must create an object for it to throw. That is, you can't just throw a type.

Rethrowing an Exception

An exception caught by one **catch** statement can be rethrown so that it can be caught by an outer **catch**. The most likely reason for rethrowing this way is to allow multiple handlers access to the exception. For example, perhaps one exception handler manages one aspect of an exception, and a second handler copes with another aspect. Remember, when you rethrow an exception, it will not be recaught by the same **catch** statement. It will propagate to the next **catch** statement. The following program illustrates rethrowing an exception:

```
// Rethrow an exception.
class Rethrow {
  public static void genException() {
    // here, numer is longer than denom
    int[] numer = { 4, 8, 16, 32, 64, 128, 256, 512 };
    int[] denom = { 2, 0, 4, 4, 0, 8 };
```

Ask the Expert

Q: Why would I want to manually throw an exception?

A: Most often, the exceptions that you will throw will be instances of exception classes that you created. As you will see later in this chapter, creating your own exception classes allows you to handle errors in your code as part of your program's overall exception handling strategy.

```
      for(int i=0; i<numer.length; i++) {
        try {
          System.out.println(numer[i] + " / " +
                              denom[i] + " is " +
                              numer[i]/denom[i]);
        }
        catch (ArithmeticException exc) {
          // catch the exception
          System.out.println("Can't divide by Zero!");
        }
        catch (ArrayIndexOutOfBoundsException exc) {
          // catch the exception
          System.out.println("No matching element found.");
          throw exc; // rethrow the exception
        }
      }
    }
  }

class RethrowDemo {
  public static void main(String[] args) {
    try {
      Rethrow.genException();
    }
    catch(ArrayIndexOutOfBoundsException exc) {
      // recatch exception
      System.out.println("Fatal error - " +
                          "program terminated.");
    }
  }
}
```

— Rethrow the exception.

— Catch rethrown exception.

In this program, divide-by-zero errors are handled locally, by **genException( )**, but an array boundary error is rethrown. In this case, it is caught by **main( )**.

A Closer Look at Throwable

Up to this point, we have been catching exceptions, but we haven't been doing anything with the exception object itself. As the preceding examples all show, a **catch** clause specifies an exception type and a parameter. The parameter receives the exception object. Since all exceptions are subclasses of **Throwable**, all exceptions support the methods defined by **Throwable**. Several commonly used ones are shown in Table 9-1.

Of the methods defined by **Throwable**, two of the most interesting are **printStackTrace( )** and **toString( )**. You can display the standard error message plus a record of the method calls that lead up to the exception by calling **printStackTrace( )**. You can use **toString( )** to retrieve

Method	Description
Throwable fillInStackTrace( )	Returns a **Throwable** object that contains a completed stack trace. This object can be rethrown.
String getLocalizedMessage( )	Returns a localized description of the exception.
String getMessage( )	Returns a description of the exception.
void printStackTrace( )	Displays the stack trace.
void printStackTrace(PrintStream *stream*)	Sends the stack trace to the specified stream.
void printStackTrace(PrintWriter *stream*)	Sends the stack trace to the specified stream.
String toString( )	Returns a **String** object containing a complete description of the exception. This method is called by **println( )** when outputting a **Throwable** object.

Table 9-1 Commonly Used Methods Defined by **Throwable**

the standard error message. The **toString( )** method is also called when an exception is used as an argument to **println( )**. The following program demonstrates these methods:

```
// Using the Throwable methods.

class ExcTest {
  static void genException() {
    int[] nums = new int[4];

    System.out.println("Before exception is generated.");

    // generate an index out-of-bounds exception
    nums[7] = 10;
    System.out.println("this won't be displayed");
  }
}

class UseThrowableMethods {
  public static void main(String[] args) {

    try {
      ExcTest.genException();
    }
    catch (ArrayIndexOutOfBoundsException exc) {
      // catch the exception
      System.out.println("Standard message is: ");
      System.out.println(exc);
      System.out.println("\nStack trace: ");
```

```
    exc.printStackTrace();
  }
  System.out.println("After catch statement.");
 }
}
```

The output from this program is shown here. (Your output may vary because of differences between JDKs.)

```
Before exception is generated.
Standard message is:
java.lang.ArrayIndexOutOfBoundsException: Index 7 out of bounds for length 4

Stack trace:
java.lang.ArrayIndexOutOfBoundsException: Index 7 out of bounds for length 4
    at ExcTest.genException(UseThrowableMethods.java:10)
    at UseThrowableMethods.main(UseThrowableMethods.java:19)
After catch statement.
```

Using finally

Sometimes you will want to define a block of code that will execute when a **try/catch** block is left. For example, an exception might cause an error that terminates the current method, causing its premature return. However, that method may have opened a file or a network connection that needs to be closed. Such types of circumstances are common in programming, and Java provides a convenient way to handle them: **finally**.

To specify a block of code to execute when a **try/catch** block is exited, include a **finally** block at the end of a **try/catch** sequence. The general form of a **try/catch** that includes **finally** is shown here.

```
try {
 // block of code to monitor for errors
}
catch (ExcepType1 exOb) {
 // handler for ExcepType1
}
catch (ExcepType2 exOb) {
 // handler for ExcepType2
}
//...
finally {
 // finally code
}
```

The **finally** block will be executed whenever execution leaves a **try/catch** block, no matter what conditions cause it. That is, whether the **try** block ends normally, or because of an exception, the last code executed is that defined by **finally**. The **finally** block is also executed if any code within the **try** block or any of its **catch** statements return from the method.

Here is an example of **finally**:

```
// Use finally.
class UseFinally {
  public static void genException(int what) {
    int t;
    int[] nums = new int[2];

    System.out.println("Receiving " + what);
    try {
      switch(what) {
        case 0:
          t = 10 / what; // generate div-by-zero error
          break;
        case 1:
          nums[4] = 4; // generate array index error.
          break;
        case 2:
          return; // return from try block
      }
    }
    catch (ArithmeticException exc) {
      // catch the exception
      System.out.println("Can't divide by Zero!");
      return; // return from catch
    }
    catch (ArrayIndexOutOfBoundsException exc) {
      // catch the exception
      System.out.println("No matching element found.");
    }
    finally {  ◄────────────────────────────────────
      System.out.println("Leaving try.");
    }
  }
}

class FinallyDemo {
  public static void main(String[] args) {
```

This is executed on the way out of **try/catch** blocks.

```
        for(int i=0; i < 3; i++) {
          UseFinally.genException(i);
          System.out.println();
        }
      }
  }
```

Here is the output produced by the program:

```
Receiving 0
Can't divide by Zero!
Leaving try.

Receiving 1
No matching element found.
Leaving try.

Receiving 2
Leaving try.
```

As the output shows, no matter how the **try** block is exited, the **finally** block is executed.

Using throws

In some cases, if a method generates an exception that it does not handle, it must declare that exception in a **throws** clause. Here is the general form of a method that includes a **throws** clause:

ret-type methName(param-list) throws *except-list* {
  // body
}

Here, *except-list* is a comma-separated list of exceptions that the method might throw outside of itself.

You might be wondering why you did not need to specify a **throws** clause for some of the preceding examples, which threw exceptions outside of methods. The answer is that exceptions that are subclasses of **Error** or **RuntimeException** don't need to be specified in a **throws** list. Java simply assumes that a method may throw one. All other types of exceptions *do* need to be declared. Failure to do so causes a compile-time error.

Actually, you saw an example of a **throws** clause earlier in this book. As you will recall, when performing keyboard input, you needed to add the clause

```
throws java.io.IOException
```

to **main( )**. Now you can understand why. An input statement might generate an **IOException**, and at that time, we weren't able to handle that exception. Thus, such an exception would be thrown out of **main( )** and needed to be specified as such. Now that you know about exceptions, you can easily handle **IOException**.

Let's look at an example that handles **IOException**. It creates a method called **prompt( )**, which displays a prompting message and then reads a character from the keyboard. Since input is being performed, an **IOException** might occur. However, the **prompt( )** method does not handle **IOException** itself. Instead, it uses a **throws** clause, which means that the calling method must handle it. In this example, the calling method is **main( )**, and it deals with the error.

```
// Use throws.
class ThrowsDemo {
  public static char prompt(String str)
    throws java.io.IOException {          ◄──────── Notice the throws clause.

    System.out.print(str + ": ");
    return (char) System.in.read();
  }

  public static void main(String[] args) {
    char ch;

    try {
      ch = prompt("Enter a letter");  ◄──────  Since prompt( ) might throw an
    }                                          exception, a call to it must be
    catch(java.io.IOException exc) {           enclosed within a try block.
      System.out.println("I/O exception occurred.");
      ch = 'X';
    }

    System.out.println("You pressed " + ch);
  }
}
```

On a related point, notice that **IOException** is fully qualified by its package name **java.io**. As you will learn in Chapter 10, Java's I/O system is contained in the **java.io** package. Thus, the **IOException** is also contained there. It would also have been possible to import **java.io** and then refer to **IOException** directly.

Three Additional Exception Features

In addition to the exception handling features already discussed, modern versions of Java include three more. The first supports *automatic resource management,* which automates the process of releasing a resource, such as a file, when it is no longer needed. It is based on an expanded form of **try**, called the *try-with-resources* statement, and it is described in Chapter 10, when files are discussed. The second feature is called *multi-catch,* and the third is sometimes called *final rethrow* or *more precise rethrow.* These two features are described here.

Multi-catch allows two or more exceptions to be caught by the same **catch** clause. As you learned earlier, it is possible (indeed, common) for a **try** to be followed by two or more **catch** clauses. Although each **catch** clause often supplies its own unique code sequence, it is not uncommon to have situations in which two or more **catch** clauses execute *the same code sequence* even though they catch different exceptions. Instead of having to catch each exception type individually, you can use a single **catch** clause to handle the exceptions without code duplication.

To create a multi-catch, specify a list of exceptions within a single **catch** clause. You do this by separating each exception type in the list with the OR operator. Each multi-catch parameter is implicitly **final**. (You can explicitly specify **final**, if desired, but it is not necessary.) Because each multi-catch parameter is implicitly **final**, it can't be assigned a new value.

Here is how you can use the multi-catch feature to catch both **ArithmeticException** and **ArrayIndexOutOfBoundsException** with a single **catch** clause:

```
catch(ArithmeticException | ArrayIndexOutOfBoundsException e) {
```

Here is a simple program that demonstrates the use of this multi-catch:

```
// Use the multi-catch feature.  Note: This code requires JDK 7 or
// later to compile.
class MultiCatch {
  public static void main(String[] args) {
    int a=88, b=0;
    int result;
    char[] chrs = { 'A', 'B', 'C' };

    for(int i=0; i < 2; i++) {
      try {
        if(i == 0)
          result = a / b; // generate an ArithmeticException
        else
          chrs[5] = 'X'; // generate an ArrayIndexOutOfBoundsException

      // This catch clause catches both exceptions.
      }
      catch(ArithmeticException | ArrayIndexOutOfBoundsException e) {
        System.out.println("Exception caught: " + e);
      }
    }

    System.out.println("After multi-catch.");
  }
}
```

The program will generate an **ArithmeticException** when the division by zero is attempted. It will generate an **ArrayIndexOutOfBoundsException** when the attempt is made to access outside the bounds of **chrs**. Both exceptions are caught by the single **catch** statement.

The more precise rethrow feature restricts the type of exceptions that can be rethrown to only those checked exceptions that the associated **try** block throws, that are not handled by a preceding **catch** clause, and that are a subtype or supertype of the parameter. While this capability might not be needed often, it is now available for use. For the final rethrow feature to be in force, the **catch** parameter must be effectively **final**. This means that it must not be assigned a new value inside the **catch** block. It can also be explicitly specified as **final**, but this is not necessary.

Java's Built-in Exceptions

Inside the standard package **java.lang**, Java defines several exception classes. A few have been used by the preceding examples. The most general of these exceptions are subclasses of the standard type **RuntimeException**. Since **java.lang** is implicitly imported into all Java programs, many exceptions derived from **RuntimeException** are automatically available. Furthermore, they need not be included in any method's **throws** list. In the language of Java, these are called *unchecked exceptions* because the compiler does not check to see if a method handles or throws these exceptions. The unchecked exceptions defined in **java.lang** are listed in Table 9-2. Table 9-3 lists those exceptions defined by **java.lang** that must be included in a method's **throws** list if that method can generate one of these exceptions and does not handle it itself. These are called *checked exceptions*. In addition to the exceptions in **java.lang**, Java defines several other types of exceptions that relate to other packages, such as **IOException** mentioned earlier.

Exception	Meaning
ArithmeticException	Arithmetic error, such as integer divide-by-zero.
ArrayIndexOutOfBoundsException	Array index is out-of-bounds.
ArrayStoreException	Assignment to an array element of an incompatible type.
ClassCastException	Invalid cast.
EnumConstantNotPresentException	An attempt is made to use an undefined enumeration value.
IllegalArgumentException	Illegal argument used to invoke a method.
IllegalCallerException	A method cannot be legally executed by the calling code.
IllegalMonitorStateException	Illegal monitor operation, such as waiting on an unlocked thread.
IllegalStateException	Environment or application is in incorrect state.
IllegalThreadStateException	Requested operation not compatible with current thread state.
IndexOutOfBoundsException	Some type of index is out-of-bounds.
LayerInstantiationException	A module layer cannot be created.
NegativeArraySizeException	Array created with a negative size.
NullPointerException	Invalid use of a null reference.
NumberFormatException	Invalid conversion of a string to a numeric format.
SecurityException	Attempt to violate security.
StringIndexOutOfBoundsException	Attempt to index outside the bounds of a string.
TypeNotPresentException	Type not found.
UnsupportedOperationException	An unsupported operation was encountered.

Table 9-2 The Unchecked Exceptions Defined in **java.lang**

Exception	Meaning
ClassNotFoundException	Class not found.
CloneNotSupportedException	Attempt to clone an object that does not implement the **Cloneable** interface.
IllegalAccessException	Access to a class is denied.
InstantiationException	Attempt to create an object of an abstract class or interface.
InterruptedException	One thread has been interrupted by another thread.
NoSuchFieldException	A requested field does not exist.
NoSuchMethodException	A requested method does not exist.
ReflectiveOperationException	Superclass of reflection-related exceptions.

Table 9-3 The Checked Exceptions Defined in **java.lang**

Ask the Expert

Q: I have heard that Java supports something called *chained exceptions*. What are they?

A: A number of years ago, chained exceptions were incorporated into Java. The chained exception feature allows you to specify one exception as the underlying cause of another. For example, imagine a situation in which a method throws an **ArithmeticException** because of an attempt to divide by zero. However, the actual cause of the problem was that an I/O error occurred, which caused the divisor to be set improperly. Although the method must certainly throw an **ArithmeticException**, since that is the error that occurred, you might also want to let the calling code know that the underlying cause was an I/O error. Chained exceptions let you handle this, and any other situation, in which layers of exceptions exist.

To allow chained exceptions, two constructors and two methods were added to **Throwable**. The constructors are shown here:

Throwable(Throwable *causeExc*)
Throwable(String *msg*, Throwable *causeExc*)

In the first form, *causeExc* is the exception that causes the current exception. That is, *causeExc* is the underlying reason that an exception occurred. The second form allows you to specify a description at the same time that you specify a cause exception. These two constructors were also added to the **Error**, **Exception**, and **RuntimeException** classes.

(continued)

The chained exception methods added to **Throwable** are **getCause( )** and **initCause( )**. These methods are shown here:

Throwable getCause( )
Throwable initCause(Throwable *causeExc*)

The **getCause( )** method returns the exception that underlies the current exception. If there is no underlying exception, **null** is returned. The **initCause( )** method associates *causeExc* with the invoking exception and returns a reference to the exception. Thus, you can associate a cause with an exception after the exception has been created. In general, **initCause( )** is used to set a cause for legacy exception classes that don't support the two additional constructors described earlier.

Chained exceptions are not something that every program will need. However, in cases in which knowledge of an underlying cause is useful, they offer an elegant solution.

Creating Exception Subclasses

Although Java's built-in exceptions handle most common errors, Java's exception handling mechanism is not limited to these errors. In fact, part of the power of Java's approach to exceptions is its ability to handle exception types that you create. Through the use of custom exceptions, you can manage errors that relate specifically to your application. Creating an exception class is easy. Just define a subclass of **Exception** (which is, of course, a subclass of **Throwable**). Your subclasses don't need to actually implement anything—it is their existence in the type system that allows you to use them as exceptions.

The **Exception** class does not define any methods of its own. It does, of course, inherit those methods provided by **Throwable**. Thus, all exceptions, including those that you create, have the methods defined by **Throwable** available to them. Of course, you can override one or more of these methods in exception subclasses that you create.

Here is an example that creates an exception called **NonIntResultException**, which is generated when the result of dividing two integer values produces a result with a fractional component. **NonIntResultException** has two fields which hold the integer values; a constructor; and an override of the **toString( )** method, allowing the description of the exception to be displayed using **println( )**.

```
// Use a custom exception.

// Create an exception.
class NonIntResultException extends Exception {
  int n;
  int d;

  NonIntResultException(int i, int j) {
    n = i;
    d = j;
  }
}
```

```java
  public String toString() {
    return "Result of " + n + " / " + d +
          " is non-integer.";
  }
}


class CustomExceptDemo {
  public static void main(String[] args) {

    // Here, numer contains some odd values.
    int[] numer = { 4, 8, 15, 32, 64, 127, 256, 512 };
    int[] denom = { 2, 0, 4, 4, 0, 8 };

    for(int i=0; i<numer.length; i++) {
      try {
        if((numer[i]%2) != 0)
          throw new
            NonIntResultException(numer[i], denom[i]);

        System.out.println(numer[i] + " / " +
                           denom[i] + " is " +
                           numer[i]/denom[i]);
      }
      catch (ArithmeticException exc) {
        // catch the exception
        System.out.println("Can't divide by Zero!");
      }
      catch (ArrayIndexOutOfBoundsException exc) {
        // catch the exception
        System.out.println("No matching element found.");
      }
      catch (NonIntResultException exc) {
        System.out.println(exc);
      }
    }
  }
}
```

The output from the program is shown here:

```
4 / 2 is 2
Can't divide by Zero!
Result of 15 / 4 is non-integer.
32 / 4 is 8
Can't divide by Zero!
Result of 127 / 8 is non-integer.
No matching element found.
No matching element found.
```

Ask the Expert

Q: When should I use exception handling in a program? When should I create my own custom exception classes?

A: Since the Java API makes extensive use of exceptions to report errors, nearly all real-world programs will make use of exception handling. This is the part of exception handling that most new Java programmers find easy. It is harder to decide when and how to use your own custom-made exceptions. In general, errors can be reported in two ways: return values and exceptions. When is one approach better than the other? Simply put, in Java, exception handling should be the norm. Certainly, returning an error code is a valid alternative in some cases, but exceptions provide a more powerful, structured way to handle errors. They are the way professional Java programmers handle errors in their code.

Try This 9-1 Adding Exceptions to the Queue Class

```
QueueFullException.java
QueueEmptyException.java
FixedQueue.java
QExcDemo.java
```

In this project, you will create two exception classes that can be used by the queue classes developed by Project 8-1. They will indicate the queue-full and queue-empty error conditions. These exceptions can be thrown by the **put( )** and **get( )** methods, respectively. For the sake of simplicity, this project will add these exceptions to the **FixedQueue** class, but you can easily incorporate them into the other queue classes from Project 8-1.

1. You will create two files that will hold the queue exception classes. Call the first file **QueueFullException.java** and enter into it the following:

```
// An exception for queue-full errors.
public class QueueFullException extends Exception {
  int size;

  QueueFullException(int s) { size = s; }

  public String toString() {
   return "\nQueue is full. Maximum size is " +
        size;
  }
}
```

(continued)

A **QueueFullException** will be generated when an attempt is made to store an item in an already full queue.

2. Create the second file **QueueEmptyException.java** and enter into it the following:

```
// An exception for queue-empty errors.
public class QueueEmptyException extends Exception {

  public String toString() {
   return "\nQueue is empty.";
  }
}
```

A **QueueEmptyException** will be generated when an attempt is made to remove an element from an empty queue.

3. Modify the **FixedQueue** class so that it throws exceptions when an error occurs, as shown here. Put it in a file called **FixedQueue.java**.

```
// A fixed-size queue class for characters that uses exceptions.
class FixedQueue implements ICharQ {
  private char[] q; // this array holds the queue
  private int putloc, getloc; // the put and get indices

  // Construct an empty queue given its size.
  public FixedQueue(int size) {
    q = new char[size]; // allocate memory for queue
    putloc = getloc = 0;
  }

  // Put a character into the queue.
  public void put(char ch)
    throws QueueFullException {

    if(putloc==q.length)
      throw new QueueFullException(q.length);

    q[putloc++] = ch;
  }

  // Get a character from the queue.
  public char get()
    throws QueueEmptyException {

    if(getloc == putloc)
      throw new QueueEmptyException();

    return q[getloc++];
  }
}
```

Notice that two steps are required to add exceptions to **FixedQueue**. First, **get( )** and **put( )** must have a **throws** clause added to their declarations. Second, when an error occurs, these methods throw an exception. Using exceptions allows the calling code to handle the error in a rational fashion. You might recall that the previous versions simply reported the error. Throwing an exception is a much better approach.

4. To try the updated **FixedQueue** class, use the **QExcDemo** class shown here. Put it into a file called **QExcDemo.java**:

```
// Demonstrate the queue exceptions.
class QExcDemo {
  public static void main(String[] args) {
    FixedQueue q = new FixedQueue(10);
    char ch;
    int i;

    try {
      // overrun the queue
      for(i=0; i < 11; i++) {
        System.out.print("Attempting to store : " +
                          (char) ('A' + i));
        q.put((char) ('A' + i));
        System.out.println(" - OK");
      }
      System.out.println();
    }
    catch (QueueFullException exc) {
      System.out.println(exc);
    }
    System.out.println();

    try {
      // over-empty the queue
      for(i=0; i < 11; i++) {
        System.out.print("Getting next char: ");
        ch = q.get();
        System.out.println(ch);
      }
    }
    catch (QueueEmptyException exc) {
      System.out.println(exc);
    }
  }
}
```

(continued)

5. Since **FixedQueue** implements the **ICharQ** interface, which defines the two queue methods **get( )** and **put( )**, **ICharQ** will need to be changed to reflect the **throws** clause. Here is the updated **ICharQ** interface. Remember, this must be in a file by itself called **ICharQ.java**.

```java
// A character queue interface that throws exceptions.
public interface ICharQ {
    // Put a character into the queue.
    void put(char ch) throws QueueFullException;

    // Get a character from the queue.
    char get() throws QueueEmptyException;
}
```

6. Now, compile the updated **ICharQ.java** file. Then, compile **FixedQueue.java**, **QueueFullException.java**, **QueueEmptyException.java**, and **QExcDemo.java**. Finally, run **QExcDemo**. You will see the following output:

```
Attempting to store : A - OK
Attempting to store : B - OK
Attempting to store : C - OK
Attempting to store : D - OK
Attempting to store : E - OK
Attempting to store : F - OK
Attempting to store : G - OK
Attempting to store : H - OK
Attempting to store : I - OK
Attempting to store : J - OK
Attempting to store : K
Queue is full. Maximum size is 10

Getting next char: A
Getting next char: B
Getting next char: C
Getting next char: D
Getting next char: E
Getting next char: F
Getting next char: G
Getting next char: H
Getting next char: I
Getting next char: J
Getting next char:
Queue is empty.
```

Chapter 9 Self Test

1. What class is at the top of the exception hierarchy?

2. Briefly explain how to use **try** and **catch**.

3. What is wrong with this fragment?

```
// ...
vals[18] = 10;
catch (ArrayIndexOutOfBoundsException exc) {
  // handle error
}
```

4. What happens if an exception is not caught?

5. What is wrong with this fragment?

```
class A extends Exception { ...

class B extends A { ...

// ...

try {
  // ...
}
catch (A exc) { ... }
catch (B exc) { ... }
```

6. Can an inner **catch** rethrow an exception to an outer **catch**?

7. The **finally** block is the last bit of code executed before your program ends. True or False? Explain your answer.

8. What type of exceptions must be explicitly declared in a **throws** clause of a method?

9. What is wrong with this fragment?

```
class MyClass { // ... }
// ...
throw new MyClass();
```

10. In question 3 of the Chapter 6 Self Test, you created a **Stack** class. Add custom exceptions to your class that report stack full and stack empty conditions.

11. What are the three ways that an exception can be generated?

12. What are the two direct subclasses of **Throwable**?

13. What is the multi-catch feature?

14. Should your code typically catch exceptions of type **Error**?

Chapter 10

Using I/O

Key Skills & Concepts

- Understand the stream

- Know the difference between byte and character streams

- Know Java's byte stream classes

- Know Java's character stream classes

- Know the predefined streams

- Use byte streams

- Use byte streams for file I/O

- Automatically close a file by using **try**-with-resources

- Read and write binary data

- Use random-access files

- Use character streams

- Use character streams for file I/O

- Apply Java's type wrappers to convert numeric strings

Since the beginning of this book, you have been using parts of the Java I/O system, such as **println( )**. However, you have been doing so without much formal explanation. Because the Java I/O system is based upon a hierarchy of classes, it was not possible to present its theory and details without first discussing classes, inheritance, and exceptions. Now it is time to examine Java's approach to I/O in detail.

Be forewarned, Java's I/O system is quite large, containing many classes, interfaces, and methods. Part of the reason for its size is that Java defines two complete I/O systems: one for byte I/O and the other for character I/O. It won't be possible to discuss every aspect of Java's I/O here. (An entire book could easily be dedicated to Java's I/O system!) This chapter will, however, introduce you to many important and commonly used features. Fortunately, Java's I/O system is cohesive and consistent; once you understand its fundamentals, the rest of the I/O system is easy to master.

Before we begin, an important point needs to be made. The I/O classes described in this chapter support text-based console I/O and file I/O. They are not used to create graphical user interfaces (GUIs). Thus, you will not use them to create windowed applications, for example. However, Java *does* include substantial support for building graphical user interfaces. The basics of GUI programming are found in Chapter 17, which offers an introduction to Swing, Java's most widely used GUI toolkit.

Java's I/O Is Built upon Streams

Java programs perform I/O through streams. An I/O stream is an abstraction that either produces or consumes information. A stream is linked to a physical device by the Java I/O system. All streams behave in the same manner, even if the actual physical devices they are linked to differ. Thus, the same I/O classes and methods can be applied to different types of devices. For example, the same methods that you use to write to the console can also be used to write to a disk file. Java implements I/O streams within class hierarchies defined in the **java.io** package.

Byte Streams and Character Streams

Modern versions of Java define two types of I/O streams: byte and character. (The original version of Java defined only the byte stream, but character streams were quickly added.) Byte streams provide a convenient means for handling input and output of bytes. They are used, for example, when reading or writing binary data. They are especially helpful when working with files. Character streams are designed for handling the input and output of characters. They use Unicode and, therefore, can be internationalized. Also, in some cases, character streams are more efficient than byte streams.

The fact that Java defines two different types of streams makes the I/O system quite large because two separate sets of class hierarchies (one for bytes, one for characters) are needed. The sheer number of I/O classes can make the I/O system appear more intimidating than it actually is. Just remember, for the most part, the functionality of byte streams is paralleled by that of the character streams.

One other point: At the lowest level, all I/O is still byte-oriented. The character-based streams simply provide a convenient and efficient means for handling characters.

The Byte Stream Classes

Byte streams are defined by using two class hierarchies. At the top of these are two abstract classes: **InputStream** and **OutputStream**. **InputStream** defines the characteristics common to byte input streams and **OutputStream** describes the behavior of byte output streams.

From **InputStream** and **OutputStream** are created several concrete subclasses that offer varying functionality and handle the details of reading and writing to various devices, such as disk files. The non-deprecated byte stream classes in **java.io** are shown in Table 10-1. Don't be overwhelmed by the number of different classes. Once you can use one byte stream, the others are easy to master.

Byte Stream Class	Meaning
BufferedInputStream	Buffered input stream
BufferedOutputStream	Buffered output stream
ByteArrayInputStream	Input stream that reads from a byte array
ByteArrayOutputStream	Output stream that writes to a byte array
DataInputStream	An input stream that contains methods for reading the Java standard data types
DataOutputStream	An output stream that contains methods for writing the Java standard data types
FileInputStream	Input stream that reads from a file
FileOutputStream	Output stream that writes to a file
FilterInputStream	Implements **InputStream**
FilterOutputStream	Implements **OutputStream**
InputStream	Abstract class that describes stream input
ObjectInputStream	Input stream for objects
ObjectOutputStream	Output stream for objects
OutputStream	Abstract class that describes stream output
PipedInputStream	Input pipe
PipedOutputStream	Output pipe
PrintStream	Output stream that contains **print( )** and **println( )**
PushbackInputStream	Input stream that allows bytes to be returned to the stream
SequenceInputStream	Input stream that is a combination of two or more input streams that will be read sequentially, one after the other

Table 10-1 The Non-Deprecated Byte Stream Classes in **java.io**

The Character Stream Classes

Character streams are defined by using two class hierarchies topped by these two abstract classes: **Reader** and **Writer**. **Reader** is used for input, and **Writer** is used for output. Concrete classes derived from **Reader** and **Writer** operate on Unicode character streams.

From **Reader** and **Writer** are derived several concrete subclasses that handle various I/O situations. In general, the character-based classes parallel the byte-based classes. The character stream classes in **java.io** are shown in Table 10-2.

Character Stream Class	Meaning
BufferedReader	Buffered input character stream
BufferedWriter	Buffered output character stream
CharArrayReader	Input stream that reads from a character array
CharArrayWriter	Output stream that writes to a character array
FileReader	Input stream that reads from a file
FileWriter	Output stream that writes to a file
FilterReader	Filtered reader
FilterWriter	Filtered writer
InputStreamReader	Input stream that translates bytes to characters
LineNumberReader	Input stream that counts lines
OutputStreamWriter	Output stream that translates characters to bytes
PipedReader	Input pipe
PipedWriter	Output pipe
PrintWriter	Output stream that contains **print( )** and **println( )**
PushbackReader	Input stream that allows characters to be returned to the input stream
Reader	Abstract class that describes character stream input
StringReader	Input stream that reads from a string
StringWriter	Output stream that writes to a string
Writer	Abstract class that describes character stream output

Table 10-2   The Character Stream I/O Classes in **java.io**

The Predefined Streams

As you know, all Java programs automatically import the **java.lang** package. This package defines a class called **System**, which encapsulates several aspects of the run-time environment. Among other things, it contains three predefined stream variables, called **in**, **out**, and **err**. These fields are declared as **public**, **final**, and **static** within **System**. This means that they can be used by any other part of your program and without reference to a specific **System** object.

System.out refers to the standard output stream. By default, this is the console. **System.in** refers to standard input, which is by default the keyboard. **System.err** refers to the standard error stream, which is also the console by default. However, these streams can be redirected to any compatible I/O device.

System.in is an object of type **InputStream**; **System.out** and **System.err** are objects of type **PrintStream**. These are byte streams, even though they are typically used to read and write characters from and to the console. The reason they are byte and not character streams is that the predefined streams were part of the original specification for Java, which did not include the character streams. As you will see, it is possible to wrap these within character-based streams if desired.

Using the Byte Streams

We will begin our examination of Java's I/O with the byte streams. As explained, at the top of the byte stream hierarchy are the **InputStream** and **OutputStream** classes. Table 10-3 shows the methods in **InputStream**, and Table 10-4 shows the methods in **OutputStream**.

Method	Description
int available( )	Returns the number of bytes of input currently available for reading.
void close( )	Closes the input source. Subsequent read attempts will generate an **IOException**.
void mark(int *numBytes*)	Places a mark at the current point in the input stream that will remain valid until *numBytes* bytes are read.
boolean markSupported( )	Returns **true** if **mark( )**/**reset( )** are supported by the invoking stream.
static InputStream nullInputStream( )	Returns an open, but null stream, which is a stream that contains no data. Thus, the stream is always at the end of the stream and no input can be obtained. The stream can, however, be closed.
int read( )	Returns an integer representation of the next available byte of input. −1 is returned when an attempt is made to read at the end of the stream.
int read(byte[ ] *buffer*)	Attempts to read up to *buffer.length* bytes into *buffer* and returns the actual number of bytes that were successfully read. −1 is returned when an attempt is made to read at the end of the stream.
int read(byte[ ] *buffer*, int *offset*, int *numBytes*)	Attempts to read up to *numBytes* bytes into *buffer* starting at *buffer[offset]*, returning the number of bytes successfully read. −1 is returned when an attempt is made to read at the end of the stream.
byte[ ] readAllBytes( )	Reads and returns, in the form of an array of bytes, all bytes available in the stream. An attempt to read at the end of the stream results in an empty array.
byte[ ] readNBytes(int *numBytes*)	Attempts to read *numBytes* bytes, returning the result in a byte array. If the end of the stream is reached before *numBytes* bytes have been read, then the returned array will contain less than *numBytes* bytes.
int readNBytes(byte[ ] *buffer*, int *offset*, int *numBytes*)	Attempts to read up to *numBytes* bytes into *buffer* starting at *buffer[offset]*, returning the number of bytes successfully read. An attempt to read at the end of the stream results in zero bytes being read.
void reset( )	Resets the input pointer to the previously set mark.
long skip(long *numBytes*)	Ignores (that is, skips) *numBytes* bytes of input, returning the number of bytes actually ignored.
void skipNBytes(long *numBytes*)	Ignores (that is, skips) *numBytes* of input. Throws **EOFException** if the end of the stream is reached before *numBytes* are skipped, or **IOException** if an I/O error occurs.
long transferTo(OutputStream *outStrm*)	Copies the contents of the invoking stream to *outStrm*, returning the number of bytes copied.

Table 10-3 The Methods Defined by **InputStream**

Method	Description
void close( )	Closes the output stream. Subsequent write attempts will generate an **IOException**.
void flush( )	Causes any output that has been buffered to be sent to its destination. That is, it flushes the output buffer.
static OutputStream nullOutputStream( )	Returns an open, but null output stream, which is a stream to which no output is written. The stream can, however, be closed.
void write(int *b*)	Writes a single byte to an output stream. Note that the parameter is an **int**, which allows you to call **write( )** with expressions without having to cast them back to **byte**.
void write(byte[ ] *buffer*)	Writes a complete array of bytes to an output stream.
void write(byte[ ] *buffer*, int *offset*, int *numBytes*)	Writes a subrange of *numBytes* bytes from the array *buffer*, beginning at *buffer*[*offset*].

Table 10-4 The Methods Defined by **OutputStream**

In general, the methods in **InputStream** and **OutputStream** can throw an **IOException** on error. The methods defined by these two abstract classes are available to all of their subclasses. Thus, they form a minimal set of I/O functions that all byte streams will have.

Reading Console Input

Originally, the only way to perform console input was to use a byte stream, and much Java code still uses the byte streams exclusively. Today, you can use byte or character streams. For commercial code, the preferred method of reading console input is to use a character-oriented stream. Doing so makes your program easier to internationalize and easier to maintain. It is also more convenient to operate directly on characters rather than converting back and forth between characters and bytes. However, for sample programs, simple utility programs for your own use, and applications that deal with raw keyboard input, using the byte streams is acceptable. For this reason, console I/O using byte streams is examined here.

Because **System.in** is an instance of **InputStream**, you automatically have access to the methods defined by **InputStream**. This means that, for example, you can use the **read( )** method to read bytes from **System.in**. There are three versions of **read( )**, which are shown here:

int read( ) throws IOException
int read(byte[ ] *data*) throws IOException
int read(byte[ ] *data*, int *start*, int *max*) throws IOException

In Chapter 3, you saw how to use the first version of **read( )** to read a single character from the keyboard (from **System.in**). It returns –1 when an attempt is made to read at the end of the stream. The second version reads bytes from the input stream and puts them into *data* until either the array is full, the end of stream is reached, or an error occurs. It returns the number of bytes read, or –1 when an attempt is made to read at the end of the stream. The third version reads input into *data* beginning at the location specified by *start*. Up to *max* bytes are stored.

It returns the number of bytes read, or –1 when an attempt is made to read at the end of the stream. All throw an **IOException** when an error occurs.

Here is a program that demonstrates reading an array of bytes from **System.in**. Notice that any I/O exceptions that might be generated are simply thrown out of **main( )**. Such an approach is common when reading from the console, but you can handle these types of errors yourself, if you choose.

```java
// Read an array of bytes from the keyboard.

import java.io.*;

class ReadBytes {
  public static void main(String[] args)
    throws IOException {
      byte[] data = new byte[10];

      System.out.println("Enter some characters.");
      System.in.read(data);  ◄─────────────── Read an array of bytes
      System.out.print("You entered: ");        from the keyboard.
      for(int i=0; i < data.length; i++)
        System.out.print((char) data[i]);
  }
}
```

Here is a sample run:

```
Enter some characters.
Read Bytes
You entered: Read Bytes
```

Writing Console Output

As is the case with console input, Java originally provided only byte streams for console output. Java 1.1 added character streams. For the most portable code, character streams are recommended. Because **System.out** is a byte stream, however, byte-based console output is still widely used. In fact, all of the programs in this book up to this point have used it! Thus, it is examined here.

Console output is most easily accomplished with **print( )** and **println( )**, with which you are already familiar. These methods are defined by the class **PrintStream** (which is the type of the object referenced by **System.out**). Even though **System.out** is a byte stream, it is still acceptable to use this stream for simple console output.

Since **PrintStream** is an output stream derived from **OutputStream**, it also implements the low-level method **write( )**. Thus, it is possible to write to the console by using **write( )**. The simplest form of **write( )** defined by **PrintStream** is shown here:

void write(int *byteval*)

This method writes the byte specified by *byteval* to the file. Although *byteval* is declared as an integer, only the low-order 8 bits are written. Here is a short example that uses **write( )** to output the character X followed by a new line:

```
// Demonstrate System.out.write().
class WriteDemo {
  public static void main(String[] args) {
    int b;

    b = 'X';
    System.out.write(b);  ◄──────── Write a byte to the screen.
    System.out.write('\n');
  }
}
```

You will not often use **write( )** to perform console output (although it might be useful in some situations), since **print( )** and **println( )** are substantially easier to use.

PrintStream supplies two additional output methods: **printf( )** and **format( )**. Both give you detailed control over the precise format of data that you output. For example, you can specify the number of decimal places displayed, a minimum field width, or the format of a negative value. Although we won't be using these methods in the examples in this book, they are features that you will want to look into as you advance in your knowledge of Java.

Reading and Writing Files Using Byte Streams

Java provides a number of classes and methods that allow you to read and write files. Of course, the most common types of files are disk files. In Java, all files are byte-oriented, and Java provides methods to read and write bytes from and to a file. Thus, reading and writing files using byte streams is very common. However, Java allows you to wrap a byte-oriented file stream within a character-based object, which is shown later in this chapter.

To create a byte stream linked to a file, use **FileInputStream** or **FileOutputStream**. To open a file, simply create an object of one of these classes, specifying the name of the file as an argument to the constructor. Once the file is open, you can read from or write to it.

Inputting from a File

A file is opened for input by creating a **FileInputStream** object. Here is a commonly used constructor:

FileInputStream(String *fileName*) throws FileNotFoundException

Here, *fileName* specifies the name of the file you want to open. If the file does not exist, then **FileNotFoundException** is thrown. **FileNotFoundException** is a subclass of **IOException**.

To read from a file, you can use **read( )**. The version that we will use is shown here:

int read( ) throws IOException

Each time it is called, **read( )** reads a single byte from the file and returns it as an integer value. It returns –1 when the end of the file is encountered. It throws an **IOException** when an error occurs. Thus, this version of **read( )** is the same as the one used to read from the console.

When you are done with a file, you must close it by calling **close( )**. Its general form is shown here:

void close( ) throws IOException

Closing a file releases the system resources allocated to the file, allowing them to be used by another file. Failure to close a file can result in "memory leaks" because of unused resources remaining allocated.

The following program uses **read( )** to input and display the contents of a text file, the name of which is specified as a command-line argument. Notice how the **try/catch** blocks handle I/O errors that might occur.

```java
/* Display a text file.

   To use this program, specify the name
   of the file that you want to see.
   For example, to see a file called TEST.TXT,
   use the following command line.

   java ShowFile TEST.TXT
*/

import java.io.*;

class ShowFile {
  public static void main(String[] args)
  {
    int i;
    FileInputStream fin;

    // First make sure that a file has been specified.
    if(args.length != 1) {
      System.out.println("Usage: ShowFile File");
      return;
    }

    try {
      fin = new FileInputStream(args[0]);  ◄──────── Open the file.
    } catch(FileNotFoundException exc) {
      System.out.println("File Not Found");
      return;
    }

    try {
      // read bytes until EOF is encountered
      do {
        i = fin.read();  ◄──────── Read from the file.
        if(i != -1) System.out.print((char) i);
      } while(i != -1);  ◄────────────────────── When i equals –1, the end of
    } catch(IOException exc) {                    the file has been reached.
```

```
      System.out.println("Error reading file.");
    }

    try {
      fin.close();    ◄──────── Close the file.
    } catch(IOException exc) {
      System.out.println("Error closing file.");
    }
  }
}
```

Notice that the preceding example closes the file stream after the **try** block that reads the file has completed. Although this approach is occasionally useful, Java supports a variation that is often a better choice. The variation is to call **close( )** within a **finally** block. In this approach, all of the methods that access the file are contained within a **try** block, and the **finally** block is used to close the file. This way, no matter how the **try** block terminates, the file is closed. Assuming the preceding example, here is how the **try** block that reads the file can be recoded:

```
try {
  do {
    i = fin.read();
    if(i != -1) System.out.print((char) i);
  } while(i != -1);
} catch(IOException exc) {
  System.out.println("Error Reading File");
} finally {    ◄──────────────────────────────┐
  // Close file on the way out of the try block.│  Use a finally clause to
  try {                                          │  close the file.
    fin.close();   ◄────────────────────────────┘
  } catch(IOException exc) {
    System.out.println("Error Closing File");
  }
}
```

One advantage to this approach in general is that if the code that accesses a file terminates because of some non-I/O-related exception, the file is still closed by the **finally** block. Although not an issue in this example (or most other example programs) because the program simply ends if an unexpected exception occurs, this can be a major source of trouble in larger programs. Using **finally** avoids this trouble.

Sometimes it's easier to wrap the portions of a program that open the file and access the file within a single **try** block (rather than separating the two), and then use a **finally** block to close the file. For example, here is another way to write the **ShowFile** program:

```
/* This variation wraps the code that opens and
   accesses the file within a single try block.
   The file is closed by the finally block.
*/

import java.io.*;
```

```
class ShowFile {
  public static void main(String[] args)
  {
    int i;
    FileInputStream fin = null;    ◄──────── Here, fin is initialized to null.

    // First, confirm that a file name has been specified.
    if(args.length != 1) {
      System.out.println("Usage: ShowFile filename");
      return;
    }

    // The following code opens a file, reads characters until EOF
    // is encountered, and then closes the file via a finally block.
    try {
      fin = new FileInputStream(args[0]);

      do {
        i = fin.read();
        if(i != -1) System.out.print((char) i);
      } while(i != -1);

    } catch(FileNotFoundException exc) {
      System.out.println("File Not Found.");
    } catch(IOException exc) {
      System.out.println("An I/O Error Occurred");
    } finally {
      // Close file in all cases.
      try {
        if(fin != null) fin.close();   ◄──────── Close fin only if it is not null.
      } catch(IOException exc) {
        System.out.println("Error Closing File");
      }
    }
  }
}
```

In this approach, notice that **fin** is initialized to **null**. Then, in the **finally** block, the file is closed only if **fin** is not **null**. This works because **fin** will be non-**null** only if the file was successfully opened. Thus, **close( )** will not be called if an exception occurs while opening the file.

It is possible to make the **try/catch** sequence in the preceding example a bit more compact. Because **FileNotFoundException** is a subclass of **IOException**, it need not be caught separately. For example, this **catch** clause could be used to catch both exceptions, eliminating the need to catch **FileNotFoundException** separately. In this case, the standard exception message, which describes the error, is displayed.

```
...
} catch(IOException exc) {
  System.out.println("I/O Error: " + exc);
} finally {
...
```

Ask the Expert

Q: I noticed that read( ) **returns –1 when the end of the file has been reached, but that it does not have a special return value for a file error. Why not?**

A: In Java, errors are handled by exceptions. Thus, if **read( )**, or any other I/O method, returns a value, it means that no error has occurred. This is a much cleaner way than handling I/O errors by using special error codes.

In this approach, any error, including an error opening the file, will simply be handled by the single **catch** statement. Because of its compactness, this approach is used by most of the I/O examples in this book. Be aware, however, that it will not be appropriate in cases in which you want to deal separately with a failure to open a file, such as might be caused if a user mistypes a file name. In such a situation, you might want to prompt for the correct name, for example, before entering a **try** block that accesses the file.

Writing to a File

To open a file for output, create a **FileOutputStream** object. Here are two commonly used constructors:

FileOutputStream(String *fileName*) throws FileNotFoundException
FileOutputStream(String *fileName*, boolean *append*)
    throws FileNotFoundException

If the file cannot be created, then **FileNotFoundException** is thrown. In the first form, when an output file is opened, any preexisting file by the same name is destroyed. In the second form, if *append* is **true**, then output is appended to the end of the file. Otherwise, the file is overwritten.

To write to a file, you will use the **write( )** method. Its simplest form is shown here:

void write(int *byteval*) throws IOException

This method writes the byte specified by *byteval* to the file. Although *byteval* is declared as an integer, only the low-order 8 bits are written to the file. If an error occurs during writing, an **IOException** is thrown.

Once you are done with an output file, you must close it using **close( )**, shown here:

void close( ) throws IOException

Closing a file releases the system resources allocated to the file, allowing them to be used by another file. It also helps ensure that any output remaining in an output buffer is actually written to the physical device.

The following example copies a text file. The names of the source and destination files are specified on the command line.

```java
/* Copy a text file.
   To use this program, specify the name
   of the source file and the destination file.
   For example, to copy a file called FIRST.TXT
   to a file called SECOND.TXT, use the following
   command line.

   java CopyFile FIRST.TXT SECOND.TXT
*/

import java.io.*;

class CopyFile {
  public static void main(String[] args) throws IOException
  {
    int i;
    FileInputStream fin = null;
    FileOutputStream fout = null;

    // First, make sure that both files has been specified.
    if(args.length != 2) {
      System.out.println("Usage: CopyFile from to");
      return;
    }

    // Copy a File.
    try {
      // Attempt to open the files.
      fin = new FileInputStream(args[0]);
      fout = new FileOutputStream(args[1]);

      do {
        i = fin.read();                                    Read bytes from one file
        if(i != -1) fout.write(i);                         and write them to another.
      } while(i != -1);

    } catch(IOException exc) {
      System.out.println("I/O Error: " + exc);
    } finally {
      try {
        if(fin != null) fin.close();
      } catch(IOException exc) {
        System.out.println("Error Closing Input File");
      }
      try {
        if(fout != null) fout.close();
      } catch(IOException exc) {
        System.out.println("Error Closing Output File");
```

```
            }
          }
        }
      }
```

Automatically Closing a File

In the preceding section, the example programs have made explicit calls to **close( )** to close a file once it is no longer needed. This is the way files have been closed since Java was first created. As a result, this approach is widespread in existing code. Furthermore, this approach is still valid and useful. However, beginning with JDK 7, Java has included a feature that offers another, more streamlined way to manage resources, such as file streams, by automating the closing process. It is based on another version of the **try** statement called *try-with-resources*, and is sometimes referred to as *automatic resource management*. The principal advantage of **try**-with-resources is that it prevents situations in which a file (or other resource) is inadvertently not released after it is no longer needed. As explained, forgetting to close a file can result in memory leaks and could lead to other problems.

The **try**-with-resources statement has this general form:

```
try (resource specification) {
  // use the resource
}
```

Often, *resource specification* is a statement that declares and initializes a resource, such as a file. In this case, it consists of a variable declaration in which the variable is initialized with a reference to the object being managed. When the **try** block ends, the resource is automatically released. In the case of a file, this means that the file is automatically closed. (Thus, there is no need to call **close( )** explicitly.) A **try**-with-resources statement can also include **catch** and **finally** clauses

NOTE

Beginning with JDK 9, it is also possible for the resource specification of the **try** to consist of a variable that has been declared and initialized earlier in the program. However, that variable must be *effectively final*, which means that it has not been assigned a new value after being given its initial value.

The **try**-with-resources statement can be used only with those resources that implement the **AutoCloseable** interface defined by **java.lang**. This interface defines the **close( )** method. **AutoCloseable** is inherited by the **Closeable** interface defined in **java.io**. Both interfaces are implemented by the stream classes, including **FileInputStream** and **FileOutputStream**. Thus, **try**-with-resources can be used when working with I/O streams, including file streams.

As a first example of automatically closing a file, here is a reworked version of the **ShowFile** program that uses it:

```
/* This version of the ShowFile program uses a try-with-resources
   statement to automatically close a file when it is no longer needed.
*/

import java.io.*;
```

```
class ShowFile {
  public static void main(String[] args)
  {
    int i;

    // First, make sure that a file name has been specified.
    if(args.length != 1) {
      System.out.println("Usage: ShowFile filename");
      return;
    }

    // The following code uses try-with-resources to open a file
    // and then automatically close it when the try block is left.
    try(FileInputStream fin = new FileInputStream(args[0])) {

      do {
        i = fin.read();
        if(i != -1) System.out.print((char) i);
      } while(i != -1);

    } catch(IOException exc) {
      System.out.println("I/O Error: " + exc);
    }
  }
}
```

A **try**-with-resources block.

In the program, pay special attention to how the file is opened within the **try**-with-resources statement:

```
try(FileInputStream fin = new FileInputStream(args[0])) {
```

Notice how the resource-specification portion of the **try** declares a **FileInputStream** called **fin**, which is then assigned a reference to the file opened by its constructor. Thus, in this version of the program the variable **fin** is local to the **try** block, being created when the **try** is entered. When the **try** is exited, the file associated with **fin** is automatically closed by an implicit call to **close( )**. You don't need to call **close( )** explicitly, which means that you can't forget to close the file. This is a key advantage of automatic resource management.

It is important to understand that a resource declared in the **try** statement is implicitly **final**. This means that you can't assign to the resource after it has been created. Also, the scope of the resource is limited to the **try**-with-resources statement.

Before moving on, it is useful to mention that beginning with JDK 10, you can use local variable type inference to specify the type of the resource declared in a **try**-with-resources statement. To do so, specify the type as **var**. When this is done, the type of the resource is inferred from its initializer. For example, the **try** statement in the preceding program can now be written like this:

```
try(var fin = new FileInputStream(args[0])) {
```

Here, **fin** is inferred to be of type **FileInputStream** because that is the type of its initializer. To enable readers working in Java environments that predate JDK 10 to compile the examples,

try-with-resource statements in the remainder of this book will not make use of type inference. Of course, going forward, you should consider using it in your own code.

You can manage more than one resource within a single **try** statement. To do so, simply separate each resource specification with a semicolon. The following program shows an example. It reworks the **CopyFile** program shown earlier so that it uses a single **try**-with-resources statement to manage both **fin** and **fout**.

```
/* A version of CopyFile that uses try-with-resources.
   It demonstrates two resources (in this case files) being
   managed by a single try statement.

*/

import java.io.*;

class CopyFile {
  public static void main(String[] args) throws IOException
  {
    int i;

    // First, confirm that both files have been specified.
    if(args.length != 2) {
      System.out.println("Usage: CopyFile from to");
      return;
    }

    // Open and manage two files via the try statement.
    try (FileInputStream fin = new FileInputStream(args[0]);
         FileOutputStream fout = new FileOutputStream(args[1]))
    {

      do {
        i = fin.read();
        if(i != -1) fout.write(i);
      } while(i != -1);

    } catch(IOException exc) {
      System.out.println("I/O Error: " + exc);
    }
  }
}
```

Manage two resources.

In this program, notice how the input and output files are opened within the **try**:

```
try (FileInputStream fin = new FileInputStream(args[0]);
     FileOutputStream fout = new FileOutputStream(args[1]))
{
```

After this **try** block ends, both **fin** and **fout** will have been closed. If you compare this version of the program to the previous version, you will see that it is much shorter. The ability to streamline source code is a side-benefit of **try**-with-resources.

There is one other aspect to **try**-with-resources that needs to be mentioned. In general, when a **try** block executes, it is possible that an exception inside the **try** block will lead to another exception that occurs when the resource is closed in a **finally** clause. In the case of a "normal" **try** statement, the original exception is lost, being preempted by the second exception. However, with a **try**-with-resources statement, the second exception is *suppressed*. It is not, however, lost. Instead, it is added to the list of suppressed exceptions associated with the first exception. The list of suppressed exceptions can be obtained by use of the **getSuppressed( )** method defined by **Throwable**.

Because of its advantages, **try**-with-resources will be used by the remaining examples in this chapter. However, it is still very important that you are familiar with the traditional approach shown earlier in which **close( )** is called explicitly. There are several reasons for this. First, you may encounter legacy code that still relies on the traditional approach. It is important that all Java programmers be fully versed in and comfortable with the traditional approach when maintaining or updating this older code. Second, you might need to work in an environment that predates JDK 7. In such a situation, the **try**-with-resources statement will not be available and the traditional approach must be employed. Finally, there may be cases in which explicitly closing a resource is more appropriate than the automated approach. The foregoing notwithstanding, if you are using a modern version of Java, then you will usually want to use the automated approach to resource management. It offers a streamlined, robust alternative to the traditional approach.

Reading and Writing Binary Data

So far, we have just been reading and writing bytes containing ASCII characters, but it is possible—indeed, common—to read and write other types of data. For example, you might want to create a file that contains **int**s, **double**s, or **short**s. To read and write binary values of the Java primitive types, you will use **DataInputStream** and **DataOutputStream**.

DataOutputStream implements the **DataOutput** interface. This interface defines methods that write all of Java's primitive types to a file. It is important to understand that this data is written using its internal, binary format, not its human-readable text form. Several commonly used output methods for Java's primitive types are shown in Table 10-5. Each throws an **IOException** on failure.

Output Method	Purpose
void writeBoolean(boolean *val*)	Writes the **boolean** specified by *val*.
void writeByte(int *val*)	Writes the low-order byte specified by *val*.
void writeChar(int *val*)	Writes the value specified by *val* as a **char**.
void writeDouble(double *val*)	Writes the **double** specified by *val*.
void writeFloat(float *val*)	Writes the **float** specified by *val*.
void writeInt(int *val*)	Writes the **int** specified by *val*.
void writeLong(long *val*)	Writes the **long** specified by *val*.
void writeShort(int *val*)	Writes the value specified by *val* as a **short**.

Table 10-5 Commonly Used Output Methods Defined by **DataOutputStream**

Input Method	Purpose
boolean readBoolean( )	Reads a **boolean**.
byte readByte( )	Reads a **byte**.
char readChar( )	Reads a **char**.
double readDouble( )	Reads a **double**.
float readFloat( )	Reads a **float**.
int readInt( )	Reads an **int**.
long readLong( )	Reads a **long**.
short readShort( )	Reads a **short**.

Table 10-6 Commonly Used Input Methods Defined by **DataInputStream**

Here is the constructor for **DataOutputStream**. Notice that it is built upon an instance of **OutputStream**.

DataOutputStream(OutputStream *outputStream*)

Here, *outputStream* is the stream to which data is written. To write output to a file, you can use the object created by **FileOutputStream** for this parameter.

DataInputStream implements the **DataInput** interface, which provides methods for reading all of Java's primitive types. These methods are shown in Table 10-6, and each can throw an **IOException**. **DataInputStream** uses an **InputStream** instance as its foundation, overlaying it with methods that read the various Java data types. Remember that **DataInputStream** reads data in its binary format, not its human-readable form. The constructor for **DataInputStream** is shown here:

DataInputStream(InputStream *inputStream*)

Here, *inputStream* is the stream that is linked to the instance of **DataInputStream** being created. To read input from a file, you can use the object created by **FileInputStream** for this parameter.

Here is a program that demonstrates **DataOutputStream** and **DataInputStream**. It writes and then reads back various types of data to and from a file.

```java
// Write and then read back binary data.

import java.io.*;

class RWData {
  public static void main(String[] args)
  {
    int i = 10;
    double d = 1023.56;
    boolean b = true;
```

```
    // Write some values.
    try (DataOutputStream dataOut =
            new DataOutputStream(new FileOutputStream("testdata")))
    {
      System.out.println("Writing " + i);
      dataOut.writeInt(i); ◄─────────────────────────────┐
                                                          │
      System.out.println("Writing " + d);                │
      dataOut.writeDouble(d); ◄─────────────────────────┤
                                                          │
      System.out.println("Writing " + b);                ├──── Write binary data.
      dataOut.writeBoolean(b); ◄────────────────────────┤
                                                          │
      System.out.println("Writing " + 12.2 * 7.4);       │
      dataOut.writeDouble(12.2 * 7.4); ◄─────────────────┘
    }
    catch(IOException exc) {
      System.out.println("Write error.");
      return;
    }

    System.out.println();

    // Now, read them back.
    try (DataInputStream dataIn =
            new DataInputStream(new FileInputStream("testdata")))
    {
      i = dataIn.readInt(); ◄───────────────────────────┐
      System.out.println("Reading " + i);               │
                                                         │
      d = dataIn.readDouble(); ◄────────────────────────┤
      System.out.println("Reading " + d);               │
                                                         ├──── Read binary data.
      b = dataIn.readBoolean(); ◄───────────────────────┤
      System.out.println("Reading " + b);               │
                                                         │
      d = dataIn.readDouble(); ◄────────────────────────┘
      System.out.println("Reading " + d);
    }
    catch(IOException exc) {
      System.out.println("Read error.");
    }
  }
}
```

The output from the program is shown here.

```
Writing 10
Writing 1023.56
Writing true
Writing 90.28
```

```
Reading 10
Reading 1023.56
Reading true
Reading 90.28
```

Try This 10-1   A File Comparison Utility

CompFiles.java

This project develops a simple, yet useful file comparison utility. It works by opening both files to be compared and then reading and comparing each corresponding set of bytes. If a mismatch is found, the files differ. If the end of each file is reached at the same time and if no mismatches have been found, then the files are the same. Notice that it uses a **try**-with-resources statement to automatically close the files.

1. Create a file called **CompFiles.java**.

2. Into **CompFiles.java**, add the following program:

```java
/*
   Try This 10-1

   Compare two files.

   To use this program, specify the names
   of the files to be compared on the command line.

   java CompFile FIRST.TXT SECOND.TXT
*/

import java.io.*;

class CompFiles {
  public static void main(String[] args)
  {
    int i=0, j=0;

    // First make sure that both files have been specified.
    if(args.length !=2 ) {
      System.out.println("Usage: CompFiles f1 f2");
      return;
    }

    // Compare the files.
    try (FileInputStream f1 = new FileInputStream(args[0]);
         FileInputStream f2 = new FileInputStream(args[1]))
    {
      // Check the contents of each file.
```

(continued)

```
        do {
          i = f1.read();
          j = f2.read();
          if(i != j) break;
        } while(i != -1 && j != -1);

        if(i != j)
          System.out.println("Files differ.");
        else
          System.out.println("Files are the same.");
      } catch(IOException exc) {
        System.out.println("I/O Error: " + exc);
      }
    }
  }
```

3. To try **CompFiles**, first copy **CompFiles.java** to a file called **temp**. Then, try this command line:

```
java CompFiles CompFiles.java temp
```

The program will report that the files are the same. Next, compare **CompFiles.java** to **CopyFile.java** (shown earlier) using this command line:

```
java CompFiles CompFiles.java CopyFile.java
```

These files differ and **CompFiles** will report this fact.

4. On your own, try enhancing **CompFiles** with various options. For example, add an option that ignores the case of letters. Another idea is to have **CompFiles** display the position within the file where the files differ.

Random-Access Files

Up to this point, we have been using *sequential files*, which are files that are accessed in a strictly linear fashion, one byte after another. However, Java also allows you to access the contents of a file in random order. To do this, you will use **RandomAccessFile**, which encapsulates a random-access file. **RandomAccessFile** is not derived from **InputStream** or **OutputStream**. Instead, it implements the interfaces **DataInput** and **DataOutput**, which define the basic I/O methods. It also supports positioning requests—that is, you can position the *file pointer* within the file. The constructor that we will be using is shown here:

RandomAccessFile(String *fileName*, String *access*)
      throws FileNotFoundException

Here, the name of the file is passed in *fileName* and *access* determines what type of file access is permitted. If it is "r", the file can be read but not written. If it is "rw", the file is opened in read-write mode. (The *access* parameter also supports "rws" and "rwd", which (for local devices) ensure that changes to the file are immediately written to the physical device.)

The method **seek( )**, shown here, is used to set the current position of the file pointer within the file:

void seek(long *newPos*) throws IOException

Here, *newPos* specifies the new position, in bytes, of the file pointer from the beginning of the file. After a call to **seek( )**, the next read or write operation will occur at the new file position.

Because **RandomAccessFile** implements the **DataInput** and **DataOuput** interfaces, methods to read and write the primitive types, such as **readInt( )** and **writeDouble( )**, are available. The **read( )** and **write( )** methods are also supported.

Here is an example that demonstrates random-access I/O. It writes six **double**s to a file and then reads them back in nonsequential order.

```java
// Demonstrate random access files.

import java.io.*;

class RandomAccessDemo {
  public static void main(String[] args)
  {
    double[] data = { 19.4, 10.1, 123.54, 33.0, 87.9, 74.25 };
    double d;

    // Open and use a random access file.                    Open random-access file.
    try (RandomAccessFile raf = new RandomAccessFile("random.dat", "rw"))
    {
      // Write values to the file.
      for(int i=0; i < data.length; i++) {
        raf.writeDouble(data[i]);
      }

      // Now, read back specific values
      raf.seek(0); // seek to first double       ←──── Use seek( ) to set
      d = raf.readDouble();                              the file pointer.
      System.out.println("First value is " + d);

      raf.seek(8); // seek to second double
      d = raf.readDouble();
      System.out.println("Second value is " + d);

      raf.seek(8 * 3); // seek to fourth double
      d = raf.readDouble();
      System.out.println("Fourth value is " + d);

      System.out.println();

      // Now, read every other value.
      System.out.println("Here is every other value: ");
      for(int i=0; i < data.length; i+=2) {
        raf.seek(8 * i); // seek to ith double
        d = raf.readDouble();
        System.out.print(d + " ");
```

```
        }
      }
    catch(IOException exc) {
      System.out.println("I/O Error: " + exc);
    }
  }
}
```

The output from the program is shown here.

```
First value is 19.4
Second value is 10.1
Fourth value is 33.0

Here is every other value:
19.4 123.54 87.9
```

Notice how each value is located. Since each **double** value is 8 bytes long, each value starts on an 8-byte boundary. Thus, the first value is located at zero, the second begins at byte 8, the third starts at byte 16, and so on. Thus, to read the fourth value, the program seeks to location 24.

Using Java's Character-Based Streams

As the preceding sections have shown, Java's byte streams are both powerful and flexible. However, they are not the ideal way to handle character-based I/O. For this purpose, Java defines the character stream classes. At the top of the character stream hierarchy are the abstract classes **Reader** and **Writer**. Table 10-7 shows the methods in **Reader**, and Table 10-8 shows the methods in **Writer**. Most of the methods can throw an **IOException** on error. The methods defined by these two abstract classes are available to all of their subclasses. Thus, they form a minimal set of I/O functions that all character streams will have.

Method	Description
abstract void close( )	Closes the input source. Subsequent read attempts will generate an **IOException**.
void mark(int *numChars*)	Places a mark at the current point in the input stream that will remain valid until *numChars* characters are read.
boolean markSupported( )	Returns **true** if **mark( )/reset( )** are supported on this stream.
static Reader nullReader( )	Returns an open, but null reader, which is a reader that contains no data. Thus, the reader is always at the end of the stream and no input can be obtained. The reader can, however, be closed.
int read( )	Returns an integer representation of the next available character from the invoking input stream. −1 is returned when an attempt is made to read at the end of the stream.

Table 10-7 The Methods Defined by **Reader** *(continued)*

Method	Description
int read(char[ ] *buffer*)	Attempts to read up to *buffer.length* characters into *buffer* and returns the actual number of characters that were successfully read. −1 is returned when an attempt is made to read at the end of the stream.
abstract int read(char[ ] *buffer*, int *offset*, int *numChars*)	Attempts to read up to *numChars* characters into *buffer* starting at *buffer*[*offset*], returning the number of characters successfully read. −1 is returned when an attempt is made to read at the end of the stream.
int read(CharBuffer *buffer*)	Attempts to fill the buffer specified by *buffer*, returning the number of characters successfully read. −1 is returned when an attempt is made to read at the end of the stream. **CharBuffer** is a class that encapsulates a sequence of characters, such as a string.
boolean ready( )	Returns **true** if the next input request will not wait. Otherwise, it returns **false**.
void reset( )	Resets the input pointer to the previously set mark.
long skip(long *numChars*)	Skips over *numChars* characters of input, returning the number of characters actually skipped.
long transferTo(Writer *writer*)	Copies the contents of the invoking reader to *writer*, returning the number of characters copied.

Table 10-7 The Methods Defined by **Reader**

Method	Description
Writer append(char *ch*)	Appends *ch* to the end of the invoking output stream. Returns a reference to the invoking stream.
Writer append(CharSequence *chars*)	Appends *chars* to the end of the invoking output stream. Returns a reference to the invoking stream. **CharSequence** is an interface that defines read-only operations on a sequence of characters.
Writer append(CharSequence *chars*, int *begin*, int *end*)	Appends the sequence of *chars* starting at *begin* and stopping with *end* to the end of the invoking output stream. Returns a reference to the invoking stream. **CharSequence** is an interface that defines read-only operations on a sequence of characters.
abstract void close( )	Closes the output stream. Subsequent write attempts will generate an **IOException**.
abstract void flush( )	Causes any output that has been buffered to be sent to its destination. That is, it flushes the output buffer.
static Writer nullWriter( )	Returns an open, but null output writer, which is a writer to which no output is written. The writer can, however, be closed.

Table 10-8 The Methods Defined by **Writer** *(continued)*

Method	Description
void write(int *ch*)	Writes a single character to the invoking output stream. Note that the parameter is an **int**, which allows you to call **write( )** with expressions without having to cast them back to **char**.
void write(char[ ] *buffer*)	Writes a complete array of characters to the invoking output stream.
abstract void write(char[ ] *buffer*, int *offset*, int *numChars*)	Writes a subrange of *numChars* characters from the array *buffer*, beginning at *buffer*[*offset*] to the invoking output stream.
void write(String *str*)	Writes *str* to the invoking output stream.
void write(String *str*, int *offset*, int *numChars*)	Writes a subrange of *numChars* characters from the array *str*, beginning at the specified *offset*.

Table 10-8   The Methods Defined by **Writer**

Console Input Using Character Streams

For code that will be internationalized, inputting from the console using Java's character-based streams is a better, more convenient way to read characters from the keyboard than is using the byte streams. However, since **System.in** is a byte stream, you will need to wrap **System.in** inside some type of **Reader**. The best class for reading console input is **BufferedReader**, which supports a buffered input stream. A commonly used constructor is shown here:

BufferedReader(Reader *inputReader*)

Here, *inputReader* is the stream that is linked to the instance of **BufferedReader** that is being created. You cannot construct a **BufferedReader** directly from **System.in** because **System.in** is an **InputStream**, not a **Reader**. Instead, you must first convert it into a character stream. To do this, you will use **InputStreamReader**.

Beginning with JDK 17, the precise way you obtain an **InputStreamReader** linked to **System.in** has changed. In the past, it was common to use the following **InputStreamReader** constructor for this purpose:

InputStreamReader(InputStream *inputStream*)

Because **System.in** refers to an object of type **InputStream**, it can be used for *inputStream*. Thus, in the past, the following line of code shows a commonly used approach to creating a **BufferedReader** connected to the keyboard:

```
BufferedReader br = new BufferedReader(new InputStreamReader(System.in));
```

After this statement executes, **br** is a character-based stream that is linked to the console through **System.in**.

However, beginning with JDK 17, it is now recommended to explicitly specify the charset associated with the console when creating the **InputStreamReader**. A *charset* defines the way that bytes are mapped to characters. Normally, when a charset is not specified, the default charset of the JVM is used. However, in the case of the console, the charset used for console input may differ from this default charset. Thus, it is now recommended that this form of **InputStreamReader** constructor be used:

InputStreamReader(InputStream *inputStream,* Charset *charset*)

For *charset*, use the charset associated with the console. This charset is returned by calling **charset( )**, which is a new method added by JDK 17 to the **Console** class. You obtain a **Console** object by calling **System.console( )**. It returns reference to the console, or **null** if no console is present. Therefore, today the following sequence shows one way to wrap **System.in** in a **BufferedReader**:

```
Console con = System.console(); // get the console
if(con==null) return; // if no console present, return

BufferedReader br = new
    BufferedReader(new InputStreamReader(System.in, con.charset()));
```

Of course, in cases in which you know that a console will be present, the sequence can be shortened to:

```
BufferedReader br = new
    BufferedReader(new InputStreamReader(System.in,
                                         System.console().charset()));
```

Because a console is (obviously) required to run the examples in this book, this is the form we will use.

Reading Characters

Characters can be read from **System.in** using the **read( )** method defined by **BufferedReader** in much the same way as they were read using byte streams. Here are three versions of **read( )** supported by **BufferedReader**.

int read( ) throws IOException
int read(char[ ] *data*) throws IOException
int read(char[ ] *data*, int *start*, int *max*) throws IOException

The first version of **read( )** reads a single Unicode character. It returns –1 when an attempt is made to read at the end of the stream. The second version reads characters from the input stream and puts them into *data* until either the array is full, the end of stream is reached, or an error occurs. It returns the number of characters read or –1 when an attempt is made to read at the end of the stream. The third version reads input into *data* beginning at the location specified by *start*. Up to *max* characters are stored. It returns the number of characters read or –1 when an attempt is made to read at the end of the stream. All throw an **IOException** on error.

The following program demonstrates **read( )** by reading characters from the console until the user types a period. Notice that any I/O exceptions that might be generated are simply thrown out of **main( )**. As mentioned earlier in this chapter, such an approach is common when reading from the console. Of course, you can handle these types of errors under program control, if you choose.

```java
// Use a BufferedReader to read characters from the console.
import java.io.*;

class ReadChars {
  public static void main(String[] args)
    throws IOException
  {
    char c;

    BufferedReader br = new BufferedReader(new
      InputStreamReader(System.in, System.console().charset()));

    System.out.println("Enter characters, period to quit.");

    // read characters
    do {
      c = (char) br.read();
      System.out.println(c);
    } while(c != '.');
  }
}
```

Create **BufferedReader** linked to **System.in**.

Here is a sample run:

```
Enter characters, period to quit.
One Two.
O
n
e

T
w
o
.
```

Reading Strings

To read a string from the keyboard, use the version of **readLine( )** that is a member of the **BufferedReader** class. Its general form is shown here:

String readLine( ) throws IOException

It returns a **String** object that contains the characters read. It returns **null** if an attempt is made to read when at the end of the stream.

The following program demonstrates **BufferedReader** and the **readLine( )** method. The program reads and displays lines of text until you enter the word "stop".

```java
// Read a string from console using a BufferedReader.
import java.io.*;
```

```
class ReadLines {
  public static void main(String[] args)
    throws IOException
  {
    // create a BufferedReader using System.in
    BufferedReader br = new BufferedReader(new
      InputStreamReader(System.in, System.console().charset()));

    String str;

    System.out.println("Enter lines of text.");
    System.out.println("Enter 'stop' to quit.");
    do {
      str = br.readLine();    ◄──────── Use readLine( ) from BufferedReader
      System.out.println(str);          to read a line of text.
    } while(!str.equals("stop"));
  }
}
```

Ask the Expert

Q: In the preceding discussion, you mentioned the Console class. What else can you tell me about it?

A: The **Console** class was added a number of years ago (by JDK 6), and it is used to read from and write to the console. **Console** is primarily a convenience class because most of its functionality is available through **System.in** and **System.out**. However, its use can simplify some types of console interactions, especially when reading strings from the console.

Console supplies no constructors. As explained, a **Console** object is obtained by calling **System.console( )**. If a console is available, then a reference to it is returned. Otherwise, **null** is returned. A console may not be available in all cases, such as when a program runs as a background task. Therefore, if **null** is returned, no console I/O is possible.

Console offers a useful array of functionality that you will find interesting to explore. For example, it defines several methods that perform I/O, such as **readLine( )** and **printf( )**. It also defines a method called **readPassword( )**, which can be used to obtain a password. It lets your application read a password without echoing what is typed. As you have seen, beginning with JDK 17 **Console** provides the **charset( )** method, which obtains the charset used by the console. You can also obtain a reference to the **Reader** and the **Writer** that are attached to the console. Using the **Reader** obtained from **Console** offers an alternative to wrapping **System.in** in an **InputStreamReader**. However, this book uses the **InputStreamReader** approach because it explicitly demonstrates the way that byte streams and characters streams can interact.

Console Output Using Character Streams

While it is still permissible to use **System.out** to write to the console under Java, its use is recommended mostly for debugging purposes or for sample programs such as those found in this book. For real-world programs, the preferred method of writing to the console when using Java is through a **PrintWriter** stream. **PrintWriter** is one of the character-based classes. As explained, using a character-based class for console output makes it easier to internationalize your program.

PrintWriter defines several constructors. The one we will use is shown here:

PrintWriter(OutputStream *outputStream*, boolean *flushingOn*)

Here, *outputStream* is an object of type **OutputStream** and *flushingOn* controls whether Java flushes the output stream every time a **println( )** method (among others) is called. If *flushingOn* is **true**, flushing automatically takes place. If **false**, flushing is not automatic.

PrintWriter supports the **print( )** and **println( )** methods for all types including **Object**. Thus, you can use these methods in just the same way as they have been used with **System.out**. If an argument is not a primitive type, the **PrintWriter** methods will call the object's **toString( )** method and then print out the result.

To write to the console using a **PrintWriter**, specify **System.out** for the output stream and flush the stream after each call to **println( )**. For example, this line of code creates a **PrintWriter** that is connected to console output.

```
PrintWriter pw = new PrintWriter(System.out, true);
```

The following application illustrates using a **PrintWriter** to handle console output.

```
// Demonstrate PrintWriter.
import java.io.*;

public class PrintWriterDemo {
  public static void main(String[] args) {
    PrintWriter pw = new PrintWriter(System.out, true);
    int i = 10;
    double d = 123.65;

    pw.println("Using a PrintWriter.");
    pw.println(i);
    pw.println(d);

    pw.println(i + " + " + d + " is " + (i+d));
  }
}
```

Create a **PrintWriter** linked to **System.out**.

The output from this program is

```
Using a PrintWriter.
10
123.65
10 + 123.65 is 133.65
```

Remember that there is nothing wrong with using **System.out** to write simple text output to the console when you are learning Java or debugging your programs. However, using a **PrintWriter** will make your real-world applications easier to internationalize. Since no advantage is to be gained by using a **PrintWriter** in the sample programs shown in this book, for convenience we will continue to use **System.out** to write to the console.

File I/O Using Character Streams

Although byte-oriented file handling is often the most common, it is possible to use character-based streams for this purpose. The advantage to the character streams is that they operate directly on Unicode characters. Thus, if you want to store Unicode text, the character streams are certainly your best option. In general, to perform character-based file I/O, you will use the **FileReader** and **FileWriter** classes.

Using a FileWriter

FileWriter creates a **Writer** that you can use to write to a file. Two commonly used constructors are shown here:

FileWriter(String *fileName*) throws IOException
FileWriter(String *fileName*, boolean *append*) throws IOException

Here, *fileName* is the full path name of a file. If *append* is **true**, then output is appended to the end of the file. Otherwise, the file is overwritten. Either throws an **IOException** on failure. **FileWriter** is derived from **OutputStreamWriter** and **Writer**. Thus, it has access to the methods defined by these classes.

Here is a simple key-to-disk utility that reads lines of text entered at the keyboard and writes them to a file called "test.txt". Text is read until the user enters the word "stop". It uses a **FileWriter** to output to the file.

```
// A simple key-to-disk utility that demonstrates a FileWriter.
import java.io.*;

class KtoD {
  public static void main(String[] args)
  {

    String str;
    BufferedReader br = new BufferedReader(new
      InputStreamReader(System.in, System.console().charset()));

    System.out.println("Enter text ('stop' to quit).");
```

```
    try (FileWriter fw = new FileWriter("test.txt"))    ◄── Create a FileWriter.
    {
      do {
        System.out.print(": ");
        str = br.readLine();

        if(str.compareTo("stop") == 0) break;

        str = str + "\r\n"; // add newline
        fw.write(str);  ◄─────────────────────────── Write strings to the file.
      } while(str.compareTo("stop") != 0);
    } catch(IOException exc) {
      System.out.println("I/O Error: " + exc);
    }
  }
}
```

Using a FileReader

The **FileReader** class creates a **Reader** that you can use to read the contents of a file. A commonly used constructor is shown here:

FileReader(String *fileName*) throws FileNotFoundException

Here, *fileName* is the full path name of a file. It throws a **FileNotFoundException** if the file does not exist. **FileReader** is derived from **InputStreamReader** and **Reader**. Thus, it has access to the methods defined by these classes.

The following program creates a simple disk-to-screen utility that reads a text file called "test.txt" and displays its contents on the screen. Thus, it is the complement of the key-to-disk utility shown in the previous section.

```
// A simple disk-to-screen utilitiy that demonstrates a FileReader.

import java.io.*;

class DtoS {
  public static void main(String[] args) {
    String s;
                                                    Create a File Reader. ──┐
    // Create and use a FileReader wrapped in a BufferedReader.             │
    try (BufferedReader br = new BufferedReader(new FileReader("test.txt"))) ◄┘
```

```
    {
      while((s = br.readLine()) != null) {
        System.out.println(s);
      }
    } catch(IOException exc) {
      System.out.println("I/O Error: " + exc);
    }
  }
}
```
Read lines from the file and display them on the screen.

In this example, notice that the **FileReader** is wrapped in a **BufferedReader**. This gives it access to **readLine( )**. Also, closing the **BufferedReader**, **br** in this case, automatically closes the file.

Ask the Expert

Q: I have heard about another I/O package called NIO. Can you tell me about it?

A: Originally called *New I/O*, NIO was added to Java several years ago. It supports a channel-based approach to I/O operations. The NIO classes are contained in **java.nio** and its subordinate packages, such as **java.nio.channels** and **java.nio.charset**.

NIO is built on two foundational items: *buffers* and *channels*. A buffer holds data. A channel represents an open connection to an I/O device, such as a file or a socket. In general, to use the new I/O system, you obtain a channel to an I/O device and a buffer to hold data. You then operate on the buffer, inputting or outputting data as needed.

Two other entities used by NIO are charsets and selectors. A *charset* defines the way that bytes are mapped to characters. You can encode a sequence of characters into bytes using an *encoder*. You can decode a sequence of bytes into characters using a *decoder*. A *selector* supports key-based, non-blocking, multiplexed I/O. In other words, selectors enable you to perform I/O through multiple channels. Selectors are most applicable to socket-backed channels.

Beginning with JDK 7, NIO was substantially enhanced, so much so that the term *NIO.2* is often used. The improvements included three new packages (**java.nio.file**, **java.nio.file.attribute**, and **java.nio.file.spi**); several new classes, interfaces, and methods; and direct support for stream-based I/O. The additions greatly expanded the ways in which NIO can be used, especially with files.

It is important to understand that NIO does not replace the I/O classes found in **java.io**, which are discussed in this chapter. Instead, the NIO classes are designed to supplement the standard I/O system, offering an alternative approach, which can be beneficial in some circumstances.

Using Java's Type Wrappers to Convert Numeric Strings

Before leaving the topic of I/O, we will examine a technique useful when reading numeric strings. As you know, Java's **println( )** method provides a convenient way to output various types of data to the console, including numeric values of the built-in types, such as **int** and **double**. Thus, **println( )** automatically converts numeric values into their human-readable form. However, methods like **read( )** do not provide a parallel functionality that reads and converts a string containing a numeric value into its internal, binary format. For example, there is no version of **read( )** that reads a string such as "100" and then automatically converts it into its corresponding binary value that is able to be stored in an **int** variable. Instead, Java provides various other ways to accomplish this task. Perhaps the easiest is to use one of Java's *type wrappers*.

Java's type wrappers are classes that encapsulate, or *wrap*, the primitive types. Type wrappers are needed because the primitive types are not objects. This limits their use to some extent. For example, a primitive type cannot be passed by reference. To address this kind of need, Java provides classes that correspond to each of the primitive types.

The type wrappers are **Double**, **Float**, **Long**, **Integer**, **Short**, **Byte**, **Character**, and **Boolean**. These classes offer a wide array of methods that allow you to fully integrate the primitive types into Java's object hierarchy. As a side benefit, the numeric wrappers also define methods that convert a numeric string into its corresponding binary equivalent. Several of these conversion methods are shown here. Each returns a binary value that corresponds to the string.

Wrapper	Conversion Method
Double	static double parseDouble(String *str*) throws NumberFormatException
Float	static float parseFloat(String *str*) throws NumberFormatException
Long	static long parseLong(String *str*) throws NumberFormatException
Integer	static int parseInt(String *str*) throws NumberFormatException
Short	static short parseShort(String *str*) throws NumberFormatException
Byte	static byte parseByte(String *str*) throws NumberFormatException

The integer wrappers also offer a second parsing method that allows you to specify the radix.

The parsing methods give us an easy way to convert a numeric value, read as a string from the keyboard or a text file, into its proper internal format. For example, the following program demonstrates **parseInt( )** and **parseDouble( )**. It averages a list of numbers entered by the user. It first asks the user for the number of values to be averaged. It then reads that number using **readLine( )** and uses **parseInt( )** to convert the string into an integer. Next, it inputs the values, using **parseDouble( )** to convert the strings into their **double** equivalents.

```
// This program averages a list of numbers entered by the user.
import java.io.*;
```

```
class AvgNums {
  public static void main(String[] args)
    throws IOException
  {
    // create a BufferedReader using System.in
    BufferedReader br = new BufferedReader(new
      InputStreamReader(System.in, System.console().charset()));

    String str;
    int n;
    double sum = 0.0;
    double avg, t;

    System.out.print("How many numbers will you enter: ");
    str = br.readLine();
    try {
      n = Integer.parseInt(str);  ◄─────── Convert string to int.
    }
    catch(NumberFormatException exc) {
      System.out.println("Invalid format");
      n = 0;
    }

    System.out.println("Enter " + n + " values.");
    for(int i=0; i < n ; i++)  {
      System.out.print(": ");
      str = br.readLine();
      try {
        t = Double.parseDouble(str);  ◄─────── Convert string to double.
      } catch(NumberFormatException exc) {
        System.out.println("Invalid format");
        t = 0.0;
      }
      sum += t;
    }
    avg = sum / n;
    System.out.println("Average is " + avg);
  }
}
```

Here is a sample run:

```
How many numbers will you enter: 5
Enter 5 values.
: 1.1
: 2.2
: 3.3
: 4.4
: 5.5
Average is 3.3
```

Ask the Expert

Q: What else can the primitive type wrapper classes do?

A: The primitive type wrappers provide a number of methods that help integrate the primitive types into the object hierarchy. For example, various storage mechanisms provided by the Java library, including maps, lists, and sets, work only with objects. Thus, to store an **int**, for example, in a list, it must be wrapped in an object. Also, all type wrappers have a method called **compareTo( )**, which compares the value contained within the wrapper; **equals( )**, which tests two values for equality; and methods that return the value of the object in various forms. The topic of type wrappers is taken up again in Chapter 12, when autoboxing is discussed.

Try This 10-2 ## Creating a Disk-Based Help System

FileHelp.java

In Try This 4-1, you created a **Help** class that displayed information about Java's control statements. In that implementation, the help information was stored within the class itself, and the user selected help from a menu of numbered options.

Although this approach was fully functional, it is certainly not the ideal way of creating a Help system. For example, to add to or change the help information, the source code of the program needed to be modified. Also, the selection of the topic by number rather than by name is tedious, and is not suitable for long lists of topics. Here, we will remedy these shortcomings by creating a disk-based Help system.

The disk-based Help system stores help information in a help file. The help file is a standard text file that can be changed or expanded at will, without changing the Help program. The user obtains help about a topic by typing in its name. The Help system searches the help file for the topic. If it is found, information about the topic is displayed.

1. Create the help file that will be used by the Help system. The help file is a standard text file that is organized like this:

   #topic-name1
   topic info

   #topic-name2
   topic info

   .
   .
   .

   #topic-nameN
   topic info

The name of each topic must be preceded by a #, and the topic name must be on a line of its own. Preceding each topic name with a # allows the program to quickly find the start of each topic. After the topic name are any number of information lines about the topic. However, there must be a blank line between the end of one topic's information and the start of the next topic. Also, there must be no trailing spaces at the end of any help-topic lines.

Here is a simple help file that you can use to try the disk-based Help system. It stores information about Java's control statements.

```
#if
if(condition) statement;
else statement;

#switch
switch(expression) { // traditional form
  case constant:
    statement sequence
    break;
    // ...
  }

#for
for(init; condition; iteration) statement;

#while
while(condition) statement;

#do
do {
  statement;
} while (condition);

#break
break; or break label;

#continue
continue; or continue label;
```

Call this file **helpfile.txt**.

2. Create a file called **FileHelp.java**.

3. Begin creating the new **Help** class with these lines of code.

```
class Help {
  String helpfile; // name of help file

  Help(String fname) {
    helpfile = fname;
  }
```

(continued)

The name of the help file is passed to the **Help** constructor and stored in the instance variable **helpfile**. Since each instance of **Help** will have its own copy of **helpfile**, each instance can use a different file. Thus, you can create different sets of help files for different sets of topics.

4. Add the **helpOn( )** method shown here to the **Help** class. This method retrieves help on the specified topic.

```java
// Display help on a topic.
boolean helpOn(String what) {
  int ch;
  String topic, info;

  // Open the help file.
  try (BufferedReader helpRdr =
          new BufferedReader(new FileReader(helpfile)))
  {
    do {
      // read characters until a # is found
      ch = helpRdr.read();

      // now, see if topics match
      if(ch == '#') {
        topic = helpRdr.readLine();
        if(what.compareTo(topic) == 0) { // found topic
          do {
            info = helpRdr.readLine();
            if(info != null) System.out.println(info);
          } while((info != null) &&
                    (info.compareTo("") != 0));
          return true;
        }
      }
    } while(ch != -1);
  }
  catch(IOException exc) {
    System.out.println("Error accessing help file.");
    return false;
  }
  return false; // topic not found
}
```

The first thing to notice is that **helpOn( )** handles all possible I/O exceptions itself and does not include a **throws** clause. By handling its own exceptions, it prevents this burden from being passed on to all code that uses it. Thus, other code can simply call **helpOn( )** without having to wrap that call in a **try/catch** block.

The help file is opened using a **FileReader** that is wrapped in a **BufferedReader**. Since the help file contains text, using a character stream allows the Help system to be more efficiently internationalized.

The **helpOn( )** method works like this. A string containing the name of the topic is passed in the **what** parameter. The help file is then opened. Then, the file is searched, looking for a match between **what** and a topic in the file. Remember, in the file, each topic is preceded by a #, so the search loop scans the file for #s. When it finds one, it then checks to see if the topic following that # matches the one passed in **what**. If it does, the information associated with that topic is displayed. If a match is found, **helpOn( )** returns **true**. Otherwise, it returns **false**.

5. The **Help** class also provides a method called **getSelection( )**. It prompts the user for a topic and returns the topic string entered by the user.

```java
// Get a Help topic.
String getSelection() {
  String topic = "";

  BufferedReader br = new BufferedReader(
    new InputStreamReader(System.in, System.console().charset()));

  System.out.print("Enter topic: ");
  try {
    topic = br.readLine();
  }
  catch(IOException exc) {
    System.out.println("Error reading console.");
  }
  return topic;
}
```

This method creates a **BufferedReader** attached to **System.in**. It then prompts for the name of a topic, reads the topic, and returns it to the caller.

6. The entire disk-based Help system is shown here:

```java
/*
   Try This 10-2

   A help program that uses a disk file
   to store help information.
*/

import java.io.*;

/* The Help class opens a help file,
   searches for a topic, and then displays
   the information associated with that topic.
   Notice that it handles all I/O exceptions
   itself, avoiding the need for calling
   code to do so. */
class Help {
  String helpfile; // name of help file
```

(continued)

```
Help(String fname) {
  helpfile = fname;
}

// Display help on a topic.
boolean helpOn(String what) {
  int ch;
  String topic, info;

  // Open the help file.
  try (BufferedReader helpRdr =
          new BufferedReader(new FileReader(helpfile)))
  {
    do {
      // read characters until a # is found
      ch = helpRdr.read();

      // now, see if topics match
      if(ch == '#') {
        topic = helpRdr.readLine();
        if(what.compareTo(topic) == 0) { // found topic
          do {
            info = helpRdr.readLine();
            if(info != null) System.out.println(info);
          } while((info != null) &&
                  (info.compareTo("") != 0));
          return true;
        }
      }
    } while(ch != -1);
  }
  catch(IOException exc) {
    System.out.println("Error accessing help file.");
    return false;
  }
  return false; // topic not found
}

// Get a Help topic.
String getSelection() {
  String topic = "";

  BufferedReader br = new BufferedReader(
    new InputStreamReader(System.in, System.console().charset()));

  System.out.print("Enter topic: ");
  try {
    topic = br.readLine();
```

```
      }
      catch(IOException exc) {
        System.out.println("Error reading console.");
      }
      return topic;
    }
  }

  // Demonstrate the file-based Help system.
  class FileHelp {
    public static void main(String[] args) {
      Help hlpobj = new Help("helpfile.txt");
      String topic;

      System.out.println("Try the help system. " +
                         "Enter 'stop' to end.");
      do {
        topic = hlpobj.getSelection();

        if(!hlpobj.helpOn(topic))
          System.out.println("Topic not found.\n");

      } while(topic.compareTo("stop") != 0);
    }
  }
```

Ask the Expert

Q: In addition to the parse **methods defined by the primitive type wrappers, is there another easy way to convert a numeric string entered at the keyboard into its equivalent binary format?**

A: Yes! Another way to convert a numeric string into its internal, binary format is to use one of the methods defined by the **Scanner** class, packaged in **java.util**. **Scanner** reads formatted (that is, human-readable) input and converts it into its binary form. **Scanner** can be used to read input from a variety of sources, including the console and files. Therefore, you can use **Scanner** to read a numeric string entered at the keyboard and assign its value to a variable. Although **Scanner** contains far too many features to describe in detail, the following illustrates its basic usage.

(continued)

To use **Scanner** to read from the keyboard, you must first create a **Scanner** linked to console input. To do this, you will use the following constructor:

Scanner(InputStream *from*)

This creates a **Scanner** that uses the stream specified by *from* as a source for input. You can use this constructor to create a **Scanner** linked to console input, as shown here:

```
Scanner conin = new Scanner(System.in);
```

This works because **System.in** is an object of type **InputStream**. After this line executes, **conin** can be used to read input from the keyboard.

Once you have created a **Scanner**, it is a simple matter to use it to read numeric input. Here is the general procedure:

1. Determine if a specific type of input is available by calling one of **Scanner**'s **hasNext***X* methods, where *X* is the type of data desired.

2. If input is available, read it by calling one of **Scanner**'s **next***X* methods.

As the preceding indicates, **Scanner** defines two sets of methods that enable you to read input. The first are the **hasNext** methods. These include methods such as **hasNextInt( )** and **hasNextDouble( )**, for example. Each of the **hasNext** methods returns **true** if the desired data type is the next available item in the data stream, and **false** otherwise. For example, calling **hasNextInt( )** returns **true** only if the next item in the stream is the human-readable form of an integer. If the desired data is available, you can read it by calling one of **Scanner**'s **next** methods, such as **nextInt( )** or **nextDouble( )**. These methods convert the human-readable form of the data into its internal, binary representation and return the result. For example, to read an integer, call **nextInt( )**.

The following sequence shows how to read an integer from the keyboard.

```
Scanner conin = new Scanner(System.in);
int i;

if (conin.hasNextInt()) i = conin.nextInt();
```

Using this code, if you enter the number **123** on the keyboard, then **i** will contain the value 123.

Technically, you can call a **next** method without first calling a **hasNext** method. However, doing so is not usually a good idea. If a **next** method cannot find the type of data it is looking for, it throws an **InputMismatchException**. For this reason, it is best to first confirm that the desired type of data is available by calling a **hasNext** method before calling its corresponding **next** method.

Chapter 10 Self Test

1. Why does Java define both byte and character streams?

2. Even though console input and output is text-based, why does Java still use byte streams for this purpose?

3. Show how to open a file for reading bytes.

4. Show how to open a file for reading characters.

5. Show how to open a file for random-access I/O.

6. How can you convert a numeric string such as "123.23" into its binary equivalent?

7. Write a program that copies a text file. In the process, have it convert all spaces into hyphens. Use the byte stream file classes. Use the traditional approach to closing a file by explicitly calling **close( )**.

8. Rewrite the program described in question 7 so that it uses the character stream classes. This time, use the **try**-with-resources statement to automatically close the file.

9. What type of stream is **System.in**?

10. What does the **read( )** method of **InputStream** return when an attempt is made to read at the end of the stream?

11. What type of stream is used to read binary data?

12. **Reader** and **Writer** are at the top of the _____ class hierarchies.

13. The **try**-with-resources statement is used for _____ _____ _____.

14. If you are using the traditional method of closing a file, then closing a file within a **finally** block is generally a good approach. True or False?

15. Can local variable type inference be used when declaring the resource in a **try**-with-resources statement?

Chapter 11

Multithreaded Programming

Key Skills & Concepts

- Understand multithreading fundamentals
- Know the **Thread** class and the **Runnable** interface
- Create a thread
- Create multiple threads
- Determine when a thread ends
- Use thread priorities
- Understand thread synchronization
- Use synchronized methods
- Use synchronized blocks
- Communicate between threads
- Suspend, resume, and stop threads

Although Java contains many innovative features, one of its most exciting is its built-in support for *multithreaded programming*. A multithreaded program contains two or more parts that can run concurrently. Each part of such a program is called a *thread,* and each thread defines a separate path of execution. Thus, multithreading is a specialized form of multitasking.

Multithreading Fundamentals

There are two distinct types of multitasking: process-based and thread-based. It is important to understand the difference between the two. A process is, in essence, a program that is executing. Thus, *process-based* multitasking is the feature that allows your computer to run two or more programs concurrently. For example, it is process-based multitasking that allows you to run the Java compiler at the same time you are using a text editor or browsing the Internet. In process-based multitasking, a program is the smallest unit of code that can be dispatched by the scheduler.

In a *thread-based* multitasking environment, the thread is the smallest unit of dispatchable code. This means that a single program can perform two or more tasks at once. For instance, a text editor can be formatting text at the same time that it is printing, as long as these two actions are being performed by two separate threads. Although Java programs make use of process-based multitasking environments, process-based multitasking is not under the control of Java. Multithreaded multitasking is.

A principal advantage of multithreading is that it enables you to write very efficient programs because it lets you utilize the idle time that is present in most programs. As you probably know, most I/O devices, whether they be network ports, disk drives, or the keyboard, are much slower than the CPU. Thus, a program will often spend a majority of its execution time waiting to send or receive information to or from a device. By using multithreading, your program can execute another task during this idle time. For example, while one part of your program is sending a file over the Internet, another part can be reading keyboard input, and still another can be buffering the next block of data to send.

As you probably know, over the past few years, multiprocessor and multicore systems have become commonplace. Of course, single-processor systems are still in widespread use. It is important to understand that Java's multithreading features work in both types of systems. In a single-core system, concurrently executing threads share the CPU, with each thread receiving a slice of CPU time. Therefore, in a single-core system, two or more threads do not actually run at the same time, but idle CPU time is utilized. However, in multiprocessor/multicore systems, it is possible for two or more threads to actually execute simultaneously. In many cases, this can further improve program efficiency and increase the speed of certain operations.

A thread can be in one of several states. It can be *running*. It can be *ready to run* as soon as it gets CPU time. A running thread can be *suspended,* which is a temporary halt to its execution. It can later be *resumed.* A thread can be *blocked* when waiting for a resource. A thread can be *terminated,* in which case its execution ends and cannot be resumed.

Along with thread-based multitasking comes the need for a special type of feature called *synchronization,* which allows the execution of threads to be coordinated in certain well-defined ways. Java has a complete subsystem devoted to synchronization, and its key features are also described here.

If you have programmed for operating systems such as Windows, then you are already familiar with multithreaded programming. However, the fact that Java manages threads through language elements makes multithreading especially convenient. Many of the details are handled for you.

The Thread Class and Runnable Interface

Java's multithreading system is built upon the **Thread** class and its companion interface, **Runnable**. Both are packaged in **java.lang**. **Thread** encapsulates a thread of execution. To create a new thread, your program will either extend **Thread** or implement the **Runnable** interface.

The **Thread** class defines several methods that help manage threads. Here are some of the more commonly used ones (we will be looking at these more closely as they are used):

Method	Meaning
final String getName( )	Obtains a thread's name.
final int getPriority( )	Obtains a thread's priority.
final boolean isAlive( )	Determines whether a thread is still running.
final void join( )	Waits for a thread to terminate.
void run( )	Entry point for the thread.
static void sleep(long *milliseconds*)	Suspends a thread for a specified period of milliseconds.
void start( )	Starts a thread by calling its **run( )** method.

All processes have at least one thread of execution, which is usually called the *main thread,* because it is the one that is executed when your program begins. Thus, the main thread is the thread that all of the preceding example programs in the book have been using. From the main thread, you can create other threads.

Creating a Thread

You create a thread by instantiating an object of type **Thread**. The **Thread** class encapsulates an object that is runnable. As mentioned, Java defines two ways in which you can create a runnable object:

● You can implement the **Runnable** interface.

● You can extend the **Thread** class.

Most of the examples in this chapter will use the approach that implements **Runnable**. However, Try This 11-1 shows how to implement a thread by extending **Thread**. Remember: Both approaches still use the **Thread** class to instantiate, access, and control the thread. The only difference is how a thread-enabled class is created.

The **Runnable** interface abstracts a unit of executable code. You can construct a thread on any object that implements the **Runnable** interface. **Runnable** defines only one method called **run( )**, which is declared like this:

```
public void run( )
```

Inside **run( )**, you will define the code that constitutes the new thread. It is important to understand that **run( )** can call other methods, use other classes, and declare variables just like the main thread. The only difference is that **run( )** establishes the entry point for another, concurrent thread of execution within your program. This thread will end when **run( )** returns.

After you have created a class that implements **Runnable**, you will instantiate an object of type **Thread** on an object of that class. **Thread** defines several constructors. The one that we will use first is shown here:

Thread(Runnable *threadOb*)

In this constructor, *threadOb* is an instance of a class that implements the **Runnable** interface. This defines where execution of the thread will begin.

Once created, the new thread will not start running until you call its **start( )** method, which is declared within **Thread**. In essence, **start( )** executes a call to **run( )**. The **start( )** method is shown here:

void start( )

Here is an example that creates a new thread and starts it running:

```
// Create a thread by implementing Runnable.

class MyThread implements Runnable {
  String thrdName;
```

Objects of **MyThread** can be run in their own threads because **MyThread** implements **Runnable**.

```
    MyThread(String name) {
      thrdName = name;
    }

    // Entry point of thread.
    public void run() {  ◄──────────────────────── Threads start executing here.
      System.out.println(thrdName + " starting.");
      try {
        for(int count=0; count < 10; count++) {
          Thread.sleep(400);
          System.out.println("In " + thrdName +
                              ", count is " + count);
        }
      }
      catch(InterruptedException exc) {
        System.out.println(thrdName + " interrupted.");
      }
      System.out.println(thrdName + " terminating.");
    }
  }

  class UseThreads {
    public static void main(String[] args) {
      System.out.println("Main thread starting.");

      // First, construct a MyThread object.
      MyThread mt = new MyThread("Child #1");  ◄─────── Create a runnable object.

      // Next, construct a thread from that object.
      Thread newThrd = new Thread(mt);  ◄────────── Construct a thread on that object.

      // Finally, start execution of the thread.
      newThrd.start();  ◄─────────────────────── Start running the thread.

      for(int i=0; i<50; i++) {
        System.out.print(".");
        try {
          Thread.sleep(100);
        }
        catch(InterruptedException exc) {
          System.out.println("Main thread interrupted.");
        }
      }

      System.out.println("Main thread ending.");
    }
  }
```

Let's look closely at this program. First, **MyThread** implements **Runnable**. This means that an object of type **MyThread** is suitable for use as a thread and can be passed to the **Thread** constructor.

Inside **run( )**, a loop is established that counts from 0 to 9. Notice the call to **sleep( )**. The **sleep( )** method causes the thread from which it is called to suspend execution for the specified period of milliseconds. Its general form is shown here:

static void sleep(long *milliseconds*) throws InterruptedException

The number of milliseconds to suspend is specified in *milliseconds*. This method can throw an **InterruptedException**. Thus, calls to it must be wrapped in a **try** block. The **sleep( )** method also has a second form, which allows you to specify the period in terms of milliseconds and nanoseconds if you need that level of precision. In **run( )**, **sleep( )** pauses the thread for 400 milliseconds each time through the loop. This lets the thread run slow enough for you to watch it execute.

Inside **main( )**, a new **Thread** object is created by the following sequence of statements:

```
// First, construct a MyThread object.
MyThread mt = new MyThread("Child #1");

// Next, construct a thread from that object.
Thread newThrd = new Thread(mt);

// Finally, start execution of the thread.
newThrd.start();
```

As the comments suggest, first an object of **MyThread** is created. This object is then used to construct a **Thread** object. This is possible because **MyThread** implements **Runnable**. Finally, execution of the new thread is started by calling **start( )**. This causes the child thread's **run( )** method to begin. After calling **start( )**, execution returns to **main( )**, and it enters **main( )**'s **for** loop. Notice that this loop iterates 50 times, pausing 100 milliseconds each time through the loop. Both threads continue running, sharing the CPU in single-CPU systems, until their loops finish. The output produced by this program is as follows. Because of differences between computing environments, the precise output that you see may differ slightly from that shown here:

```
Main thread starting.
.Child #1 starting.
...In Child #1, count is 0
....In Child #1, count is 1
....In Child #1, count is 2
...In Child #1, count is 3
....In Child #1, count is 4
....In Child #1, count is 5
....In Child #1, count is 6
...In Child #1, count is 7
....In Child #1, count is 8
....In Child #1, count is 9
Child #1 terminating.
...........Main thread ending.
```

There is another point of interest to notice in this first threading example. To illustrate the fact that the main thread and **mt** execute concurrently, it is necessary to keep **main( )** from terminating until **mt** is finished. Here, this is done through the timing differences between the two threads. Because the calls to **sleep( )** inside **main( )**'s **for** loop cause a total delay of 5 seconds (50 iterations times 100 milliseconds), but the total delay within **run( )**'s loop is only 4 seconds (10 iterations times 400 milliseconds), **run( )** will finish approximately 1 second before **main( )**. As a result, both the main thread and **mt** will execute concurrently until **mt** ends. Then, about 1 second later **main( )** ends.

Although this use of timing differences to ensure that **main( )** finishes last is sufficient for this simple example, it is not something that you would normally use in practice. Java provides much better ways of waiting for a thread to end. It is, however, sufficient for the next few programs. Later in this chapter, you will see a better way for one thread to wait until another completes.

One other point. In a multithreaded program, you often will want the main thread to be the last thread to finish running. As a general rule, a program continues to run until all of its threads have ended. Thus, having the main thread finish last is not a requirement. It is, however, often a good practice to follow—especially when you are first learning about threads.

One Improvement and Two Simple Variations

The preceding program demonstrates the fundamentals of creating a **Thread** based on a **Runnable** and then starting the thread. The approach shown in that program is perfectly valid and is often exactly what you will want. However, two simple variations can make **MyThread** more flexible and easier to use in some cases. Furthermore, you may find that these variations are helpful when you create your own **Runnable** classes. It is also possible to make one significant improvement to **MyThread** that takes advantage of another feature of the **Thread** class. Let's begin with the improvement.

In the preceding program, notice that an instance variable called **thrdName** is defined by **MyThread** and is used to hold the name of the thread. However, there is no need for **MyThread** to store the name of the thread since it is possible to give a name to a thread when it is created. To do so, use this version of **Thread**'s constructor:

Thread(Runnable *threadOb*, String *name*)

Ask the Expert

Q: You state that in a multithreaded program, one will often want the main thread to finish last. Can you explain?

A: The main thread is a convenient place to perform the orderly shutdown of your program, such as the closing of files. It also provides a well-defined exit point for your program. Therefore, it often makes sense for it to finish last. Fortunately, as you will soon see, it is trivially easy for the main thread to wait until the child threads have completed.

Here, *name* becomes the name of the thread. You can obtain the name of the thread by calling **getName( )** defined by **Thread**. Its general form is shown here:

final String getName( )

Giving a thread a name when it is created provides two advantages. First, there is no need for you to use a separate variable to hold the name because **Thread** already provides this capability. Second, the name of the thread will be available to any code that holds a reference to the thread. One other point: although not needed by this example, you can set the name of a thread after it is created by using **setName( )**, which is shown here:

final void setName(String *threadName*)

Here, *threadName* specifies the new name of the thread.

As mentioned, there are two variations that can, depending on the situation, make **MyThread** more convenient to use. First, it is possible for the **MyThread** constructor to create a **Thread** object for the thread, storing a reference to that thread in an instance variable. With this approach, the thread is ready to start as soon as the **MyThread** constructor returns. You simply call **start( )** on the **Thread** instance encapsulated by **MyThread**.

The second variation offers a way to have a thread begin execution as soon as it is created. This approach is useful in cases in which there is no need to separate thread creation from thread execution. One way to accomplish this for **MyThread** is to provide a static *factory method* that:

1. creates a new **MyThread** instance,

2. calls **start( )** on the thread associated with that instance,

3. and then returns a reference to the newly created **MyThread** object.

With this approach, it becomes possible to create and start a thread through a single method call. This can streamline the use of **MyThread**, especially in cases in which several threads must be created and started.

The following version of the preceding program incorporates the changes just described:

```
// MyThread variations. This version of MyThread
// creates a Thread when its constructor is called and
// stores it in an instance variable called thrd.
// It also sets the name of the thread and provides
// a factory method to create and start a thread.

class MyThread implements Runnable {
    Thread thrd;  ◄─────────────── A reference to the thread is stored in thrd.

    // Construct a new thread using this Runnable and give
    // it a name.
```

```
  MyThread(String name) {
    thrd = new Thread(this, name);      ◄──── The thread is named when it is created.
  }

  // A factory method that creates and starts a thread.
  public static MyThread createAndStart(String name) {
    MyThread myThrd = new MyThread(name);

    myThrd.thrd.start(); // start the thread ◄────────── Begin executing the thread.
    return myThrd;
  }

  // Entry point of thread.
  public void run() {
    System.out.println(thrd.getName() + " starting.");
    try {
      for(int count=0; count<10; count++) {
        Thread.sleep(400);
        System.out.println("In " + thrd.getName() +
                           ", count is " + count);
      }
    }
    catch(InterruptedException exc) {
      System.out.println(thrd.getName() + " interrupted.");
    }
    System.out.println(thrd.getName() + " terminating.");
  }
}

class ThreadVariations {
  public static void main(String[] args) {
    System.out.println("Main thread starting.");

    // Create and start a thread.
    MyThread mt = MyThread.createAndStart("Child #1");

    for(int i=0; i < 50; i++) {               Now the thread starts when it is created.
      System.out.print(".");
      try {
        Thread.sleep(100);
      }
      catch(InterruptedException exc) {
        System.out.println("Main thread interrupted.");
      }
    }

    System.out.println("Main thread ending.");
  }
}
```

This version produces the same output as before. However, notice that now **MyThread** no longer contains the name of the thread. Instead, it provides an instance variable called **thrd** that holds a reference to the **Thread** object created by **MyThread**'s constructor, shown here:

```
MyThread(String name) {
  thrd = new Thread(this, name);
}
```

Thus, after **MyThread**'s constructor executes, **thrd** will contain a reference to the newly created thread. To start the thread, you will simply call **start( )** on **thrd**.

Next, pay special attention to the **createAndStart( )** factory method, shown here:

```
// A factory method that creates and starts a thread.
public static MyThread createAndStart(String name) {
  MyThread myThrd = new MyThread(name);

  myThrd.thrd.start(); // start the thread
  return myThrd;
}
```

When this method is called, it creates a new instance of **MyThread** called **myThrd**. It then calls **start( )** on **myThrd**'s copy of **thrd**. Finally, it returns a reference to the newly created **MyThread** instance. Thus, once the call to **createAndStart( )** returns, the thread will already have been started. Therefore, in **main( )**, this line creates and begins the execution of a thread in a single call:

```
MyThread mt = MyThread.createAndStart("Child #1");
```

Because of the convenience that **createAndStart( )** offers, it will be used by several of the examples in this chapter. Furthermore, you may find it helpful to adapt such a method for use in thread-based applications of your own. Of course, in cases in which you want a thread's execution to be separate from its creation, you can simply create a **MyThread** object and then call **start( )** later.

Ask the Expert

Q: Earlier, you used the term *factory method* and showed one example in the method called createAndStart( ). Can you give me a more general definition?

A: Yes. In general, a factory method is a method that returns an object of a class. Typically, factory methods are **static** methods of a class. Factory methods are useful in a variety of situations. Here are some examples. As you just saw in the case of **createAndStart( )**, a factory method enables an object to be constructed and then set to some specified state prior to being returned to the caller. Another type of factory method is used to provide an easy-to-remember name that indicates the variety of object that is being constructed.

(continued)

For example, assuming a class called **Line**, you might have factory methods that create lines of specific colors, such as **createRedLine( )** or **createBlueLine( )**. Instead of having to remember a potentially complex call to a constructor, you can simply use the factory method whose name indicates the type of line you want. In some cases it is also possible for a factory method to reuse an object, rather than constructing a new one. As you will see as you advance in your study of Java, factory methods are common in the Java API library.

Try This 11-1   Extending Thread

```
ExtendThread.java
```

Implementing **Runnable** is one way to create a class that can instantiate thread objects. Extending **Thread** is the other. In this project, you will see how to extend **Thread** by creating a program functionally similar to the **UseThreads** program shown at the start of this chapter.

When a class extends **Thread**, it must override the **run( )** method, which is the entry point for the new thread. It must also call **start( )** to begin execution of the new thread. It is possible to override other **Thread** methods, but doing so is not required.

1. Create a file called **ExtendThread.java**. Begin this file with the following lines:

```
/*
   Try This 11-1

   Extend Thread.
*/
class MyThread extends Thread {
```

Notice that **MyThread** now extends **Thread** instead of implementing **Runnable**.

2. Add the following **MyThread** constructor:

```
// Construct a new thread.
MyThread(String name) {
  super(name); // name thread
}
```

Here, **super** is used to call this version of **Thread**'s constructor:

Thread(String *threadName*)

Here, *threadName* specifies the name of the thread. As explained previously, **Thread** provides the ability to hold a thread's name. Thus, no instance variable is required by **MyThread** to store the name.

3. Conclude **MyThread** by adding the following **run( )** method:

```
// Entry point of thread.
public void run() {
```

(continued)

```
System.out.println(getName() + " starting.");
try {
  for(int count=0; count < 10; count++) {
    Thread.sleep(400);
    System.out.println("In " + getName() +
                         ", count is " + count);
  }
}
catch(InterruptedException exc) {
  System.out.println(getName() + " interrupted.");
}

System.out.println(getName() + " terminating.");
  }
}
```

Notice the calls to **getName( )**. Because **ExtendThread** extends **Thread**, it can directly call all of **Thread**'s methods, including the **getName( )** method.

4. Next, add the **ExtendThread** class shown here:

```
class ExtendThread {
  public static void main(String[] args) {
    System.out.println("Main thread starting.");

    MyThread mt = new MyThread("Child #1");

    mt.start();

    for(int i=0; i < 50; i++) {
      System.out.print(".");
      try {
        Thread.sleep(100);
      }
      catch(InterruptedException exc) {
        System.out.println("Main thread interrupted.");
      }
    }

    System.out.println("Main thread ending.");
  }
}
```

In **main( )**, notice how an instance of **MyThread** is created and then started with these two lines:

```
MyThread mt = new MyThread("Child #1");
mt.start();
```

Because **MyThread** now implements **Thread**, **start( )** is called directly on the **MyThread** instance, **mt**.

5. Here is the complete program. Its output is the same as the **UseThreads** example, but in this case, **Thread** is extended rather than **Runnable** being implemented.

```
/*
   Try This 11-1

   Extend Thread.
*/
class MyThread extends Thread {

  // Construct a new thread.
  MyThread(String name) {
    super(name); // name thread
  }

  // Entry point of thread.
  public void run() {
    System.out.println(getName() + " starting.");
    try {
      for(int count=0; count < 10; count++) {
        Thread.sleep(400);
        System.out.println("In " + getName() +
                           ", count is " + count);
      }
    }
    catch(InterruptedException exc) {
      System.out.println(getName() + " interrupted.");
    }

    System.out.println(getName() + " terminating.");
  }
}

class ExtendThread {
  public static void main(String[] args) {
    System.out.println("Main thread starting.");

    MyThread mt = new MyThread("Child #1");

    mt.start();

    for(int i=0; i < 50; i++) {
      System.out.print(".");
      try {
```

(continued)

```
      Thread.sleep(100);
    }
    catch(InterruptedException exc) {
      System.out.println("Main thread interrupted.");
    }
  }

  System.out.println("Main thread ending.");
  }
}
```

6. When extending **Thread**, it is also possible to include the ability to create and start a thread in one step by using a **static** factory method, similar to that used by the **ThreadVariations** program shown earlier. To try this, add the following method to **MyThread**:

```
public static MyThread createAndStart(String name) {
  MyThread myThrd = new MyThread(name);
  myThrd.start();
  return myThrd;
}
```

As you can see, this method creates a new **MyThread** instance with the specified name, calls **start( )** on that thread, and returns a reference to the thread. To try **createAndStart( )**, replace these two lines in **main( )**:

```
System.out.println("Main thread starting.");
MyThread mt = new MyThread("Child #1");
```

with this line:

```
MyThread mt = MyThread.createAndStart("Child #1");
```

After making these changes, the program will run the same as before, but you will be creating and starting the thread using a single method call.

Creating Multiple Threads

The preceding examples have created only one child thread. However, your program can spawn as many threads as it needs. For example, the following program creates three child threads:

```
// Create multiple threads.

class MyThread implements Runnable {
  Thread thrd;

  // Construct a new thread.
  MyThread(String name) {
    thrd = new Thread(this, name);
  }
```

```java
    // A factory method that creates and starts a thread.
    public static MyThread createAndStart(String name) {
      MyThread myThrd = new MyThread(name);

      myThrd.thrd.start(); // start the thread
      return myThrd;
    }

    // Entry point of thread.
    public void run() {
      System.out.println(thrd.getName() + " starting.");
      try {
        for(int count=0; count < 10; count++) {
          Thread.sleep(400);
          System.out.println("In " + thrd.getName() +
                             ", count is " + count);
        }
      }
      catch(InterruptedException exc) {
        System.out.println(thrd.getName() + " interrupted.");
      }
      System.out.println(thrd.getName() + " terminating.");
    }
}

class MoreThreads {
  public static void main(String[] args) {
    System.out.println("Main thread starting.");

    MyThread mt1 = MyThread.createAndStart("Child #1");
    MyThread mt2 = MyThread.createAndStart("Child #2");    ◄──── Create and start
    MyThread mt3 = MyThread.createAndStart("Child #3");          executing three threads.

    for(int i=0; i < 50; i++) {
      System.out.print(".");
      try {
        Thread.sleep(100);
      }
      catch(InterruptedException exc) {
        System.out.println("Main thread interrupted.");
      }
    }

    System.out.println("Main thread ending.");
  }
}
```

Ask the Expert

Q: Why does Java have two ways to create child threads (by extending Thread or implementing Runnable) and which approach is better?

A: The **Thread** class defines several methods that can be overridden by a derived class. Of these methods, the only one that must be overridden is **run( )**. This is, of course, the same method required when you implement **Runnable**. Some Java programmers feel that classes should be extended only when they are being expanded or customized in some way. So, if you will not be overriding any of **Thread**'s other methods, it is probably best to simply implement **Runnable**. Also, by implementing **Runnable**, you enable your thread to inherit a class other than **Thread**.

Sample output from this program follows:

```
Main thread starting.
Child #1 starting.
.Child #2 starting.
Child #3 starting.
...In Child #3, count is 0
In Child #2, count is 0
In Child #1, count is 0
....In Child #1, count is 1
In Child #2, count is 1
In Child #3, count is 1
....In Child #2, count is 2
In Child #3, count is 2
In Child #1, count is 2
...In Child #1, count is 3
In Child #2, count is 3
In Child #3, count is 3
....In Child #1, count is 4
In Child #3, count is 4
In Child #2, count is 4
....In Child #1, count is 5
In Child #3, count is 5
In Child #2, count is 5
...In Child #3, count is 6
.In Child #2, count is 6
In Child #1, count is 6
...In Child #3, count is 7
In Child #1, count is 7
In Child #2, count is 7
....In Child #2, count is 8
```

```
In Child #1, count is 8
In Child #3, count is 8
....In Child #1, count is 9
Child #1 terminating.
In Child #2, count is 9
Child #2 terminating.
In Child #3, count is 9
Child #3 terminating.
...........Main thread ending.
```

As you can see, once started, all three child threads share the CPU. Notice that in this run the threads are started in the order in which they are created. However, this may not always be the case. Java is free to schedule the execution of threads in its own way. Of course, because of differences in timing or environment, the precise output from the program may differ, so don't be surprised if you see slightly different results when you try the program.

Determining When a Thread Ends

It is often useful to know when a thread has ended. For example, in the preceding examples, for the sake of illustration it was helpful to keep the main thread alive until the other threads ended. In those examples, this was accomplished by having the main thread sleep longer than the child threads that it spawned. This is, of course, hardly a satisfactory or generalizable solution!

Fortunately, **Thread** provides two means by which you can determine if a thread has ended. First, you can call **isAlive( )** on the thread. Its general form is shown here:

final boolean isAlive( )

The **isAlive( )** method returns **true** if the thread upon which it is called is still running. It returns **false** otherwise. To try **isAlive( )**, substitute this version of **MoreThreads** for the one shown in the preceding program:

```
// Use isAlive().
class MoreThreads {
  public static void main(String[] args) {
    System.out.println("Main thread starting.");

    MyThread mt1 = MyThread.createAndStart("Child #1");
    MyThread mt2 = MyThread.createAndStart("Child #2");
    MyThread mt3 = MyThread.createAndStart("Child #3");

    do {
      System.out.print(".");
      try {
        Thread.sleep(100);
      }
      catch(InterruptedException exc) {
        System.out.println("Main thread interrupted.");
      }
```

```
      } while (mt1.thrd.isAlive() ||
               mt2.thrd.isAlive() ||        ◄——— This waits until all threads terminate.
               mt3.thrd.isAlive());

    System.out.println("Main thread ending.");
  }
}
```

This version produces output that is similar to the previous version, except that **main( )** ends as soon as the other threads finish. The difference is that it uses **isAlive( )** to wait for the child threads to terminate. Another way to wait for a thread to finish is to call **join( )**, shown here:

final void join( ) throws InterruptedException

This method waits until the thread on which it is called terminates. Its name comes from the concept of the calling thread waiting until the specified thread *joins* it. Additional forms of **join( )** allow you to specify a maximum amount of time that you want to wait for the specified thread to terminate.

Here is a program that uses **join( )** to ensure that the main thread is the last to stop:

```
// Use join().

class MyThread implements Runnable {
  Thread thrd;

  // Construct a new thread.
  MyThread(String name) {
    thrd = new Thread(this, name);
  }

  // A factory method that creates and starts a thread.
  public static MyThread createAndStart(String name) {
    MyThread myThrd = new MyThread(name);

    myThrd.thrd.start(); // start the thread
    return myThrd;
  }

  // Entry point of thread.
  public void run() {
    System.out.println(thrd.getName() + " starting.");
    try {
      for(int count=0; count < 10; count++) {
        Thread.sleep(400);
        System.out.println("In " + thrd.getName() +
                           ", count is " + count);
      }
    }
  }
}
```

```
    catch(InterruptedException exc) {
      System.out.println(thrd.getName() + " interrupted.");
    }
    System.out.println(thrd.getName() + " terminating.");
  }
}

class JoinThreads {
  public static void main(String[] args) {
    System.out.println("Main thread starting.");

    MyThread mt1 = MyThread.createAndStart("Child #1");
    MyThread mt2 = MyThread.createAndStart("Child #2");
    MyThread mt3 = MyThread.createAndStart("Child #3");

    try {
      mt1.thrd.join();   ◄─────────────────────────┐
      System.out.println("Child #1 joined.");       │
      mt2.thrd.join();   ◄────────────────────────  │ Wait until the specified
      System.out.println("Child #2 joined.");       │ thread ends.
      mt3.thrd.join();   ◄────────────────────────  │
      System.out.println("Child #3 joined.");
    }
    catch(InterruptedException exc) {
      System.out.println("Main thread interrupted.");
    }
    System.out.println("Main thread ending.");
  }
}
```

Sample output from this program is shown here. Remember that when you try the program, your precise output may vary slightly.

```
Main thread starting.
Child #1 starting.
Child #2 starting.
Child #3 starting.
In Child #2, count is 0
In Child #1, count is 0
In Child #3, count is 0
In Child #2, count is 1
In Child #3, count is 1
In Child #1, count is 1
In Child #2, count is 2
In Child #1, count is 2
In Child #3, count is 2
In Child #2, count is 3
```

```
In Child #3, count is 3
In Child #1, count is 3
In Child #3, count is 4
In Child #2, count is 4
In Child #1, count is 4
In Child #3, count is 5
In Child #1, count is 5
In Child #2, count is 5
In Child #3, count is 6
In Child #2, count is 6
In Child #1, count is 6
In Child #3, count is 7
In Child #1, count is 7
In Child #2, count is 7
In Child #3, count is 8
In Child #2, count is 8
In Child #1, count is 8
In Child #3, count is 9
Child #3 terminating.
In Child #2, count is 9
Child #2 terminating.
In Child #1, count is 9
Child #1 terminating.
Child #1 joined.
Child #2 joined.
Child #3 joined.
Main thread ending.
```

As you can see, after the calls to **join( )** return, the threads have stopped executing.

Thread Priorities

Each thread has associated with it a priority setting. A thread's priority determines, in part, how much CPU time a thread receives relative to the other active threads. In general, over a given period of time, low-priority threads receive little. High-priority threads receive a lot. As you might expect, how much CPU time a thread receives has profound impact on its execution characteristics and its interaction with other threads currently executing in the system.

It is important to understand that factors other than a thread's priority also affect how much CPU time a thread receives. For example, if a high-priority thread is waiting on some resource, perhaps for keyboard input, then it will be blocked, and a lower-priority thread will run. However, when that high-priority thread gains access to the resource, it can preempt the low-priority thread and resume execution. Another factor that affects the scheduling of threads is the way the operating system implements multitasking. (See "Ask the Expert" at the end of this section.) Thus, just because you give one thread a high priority and another a low priority does not necessarily mean that one thread will run faster or more often than the other. It's just that the high-priority thread has greater potential access to the CPU.

When a child thread is started, its priority setting is equal to that of its parent thread. You can change a thread's priority by calling **setPriority( )**, which is a member of **Thread**. This is its general form:

final void setPriority(int *level*)

Here, *level* specifies the new priority setting for the calling thread. The value of *level* must be within the range **MIN_PRIORITY** and **MAX_PRIORITY**. Currently, these values are 1 and 10, respectively. To return a thread to default priority, specify **NORM_PRIORITY**, which is currently 5. These priorities are defined as **static final** variables within **Thread**.

You can obtain the current priority setting by calling the **getPriority( )** method of **Thread**, shown here:

final int getPriority( )

The following example demonstrates threads at different priorities. The threads are created as instances of **Priority**. The **run( )** method contains a loop that counts the number of iterations. The loop stops when either the count reaches 10,000,000 or the static variable **stop** is **true**. Initially, **stop** is set to **false**, but the first thread to finish counting sets **stop** to **true**. This causes each other thread to terminate with its next time slice. Each time through the loop the string in **currentName** is checked against the name of the executing thread. If they don't match, it means that a task-switch occurred. Each time a task-switch happens, the name of the new thread is displayed, and **currentName** is given the name of the new thread. Displaying each thread switch allows you to watch (in a very imprecise way) when the threads gain access to the CPU. After the threads stop, the number of iterations for each loop is displayed.

```
// Demonstrate thread priorities.

class Priority implements Runnable {
  int count;
  Thread thrd;

  static boolean stop = false;
  static String currentName;

  // Construct a new thread.
  Priority(String name) {
    thrd = new Thread(this, name);
    count = 0;
    currentName = name;
  }
```

```
      // Entry point of thread.
    public void run() {
      System.out.println(thrd.getName() + " starting.");
      do {
        count++;

        if(currentName.compareTo(thrd.getName()) != 0) {
          currentName = thrd.getName();
          System.out.println("In " + currentName);
        }

      } while(stop == false && count < 10000000);
      stop = true;

      System.out.println("\n" + thrd.getName() +
                         " terminating.");
    }
  }
}

class PriorityDemo {
  public static void main(String[] args) {
    Priority mt1 = new Priority("High Priority");
    Priority mt2 = new Priority("Low Priority");
    Priority mt3 = new Priority("Normal Priority #1");
    Priority mt4 = new Priority("Normal Priority #2");
    Priority mt5 = new Priority("Normal Priority #3");

    // set the priorities
    mt1.thrd.setPriority(Thread.NORM_PRIORITY+2);
    mt2.thrd.setPriority(Thread.NORM_PRIORITY-2);
    // Leave mt3, mt4, and mt5 at the default, normal priority level

    // start the threads
    mt1.thrd.start();
    mt2.thrd.start();
    mt3.thrd.start();
    mt4.thrd.start();
    mt5.thrd.start();

    try {
      mt1.thrd.join();
      mt2.thrd.join();
      mt3.thrd.join();
      mt4.thrd.join();
      mt5.thrd.join();
    }
```

The first thread to 10,000,000 stops all threads.

Give **mt1** a higher priority than **mt2**.

```
catch(InterruptedException exc) {
  System.out.println("Main thread interrupted.");
}

System.out.println("\nHigh priority thread counted to " +
                   mt1.count);
System.out.println("Low priority thread counted to " +
                   mt2.count);
System.out.println("1st Normal priority thread counted to " +
                   mt3.count);
System.out.println("2nd Normal priority thread counted to " +
                   mt4.count);
System.out.println("3rd Normal priority thread counted to " +
                   mt5.count);

  }
}
```

Here are the results of a sample run:

```
High priority thread counted to 10000000
Low priority thread counted to 3477862
1st Normal priority thread counted to 7000045
2nd Normal priority thread counted to 6576054
3rd Normal priority thread counted to 7373846
```

In this run, the high-priority thread got the greatest amount of the CPU time. Of course, the exact output produced by this program will depend upon a number of factors, including the speed of your CPU, the number of CPUs in your system, the operating system you are using, and the number and nature of other tasks running in the system. Thus, for any given run, it is actually possible for the low priority thread to get the most CPU time if the circumstances are right.

Ask the Expert

Q: Does the operating system's implementation of multitasking affect how much CPU time a thread receives?

A: Aside from a thread's priority setting, the most important factor affecting thread execution is the way the operating system implements multitasking and scheduling. Some operating systems use preemptive multitasking in which each thread receives a time slice, at least occasionally. Other systems use nonpreemptive scheduling in which one thread must yield execution before another thread will execute. In nonpreemptive systems, it is easy for one thread to dominate, preventing others from running.

Synchronization

When using multiple threads, it is sometimes necessary to coordinate the activities of two or more. The process by which this is achieved is called *synchronization.* The most common reason for synchronization is when two or more threads need access to a shared resource that can be used by only one thread at a time. For example, when one thread is writing to a file, a second thread must be prevented from doing so at the same time. Another reason for synchronization is when one thread is waiting for an event that is caused by another thread. In this case, there must be some means by which the first thread is held in a suspended state until the event has occurred. Then, the waiting thread must resume execution.

Key to synchronization in Java is the concept of the *monitor,* which controls access to an object. A monitor works by implementing the concept of a *lock.* When an object is locked by one thread, no other thread can gain access to the object. When the thread exits, the object is unlocked and is available for use by another thread.

All objects in Java have a monitor. This feature is built into the Java language itself. Thus, all objects can be synchronized. Synchronization is supported by the keyword **synchronized** and a few well-defined methods that all objects have. Since synchronization was designed into Java from the start, it is much easier to use than you might first expect. In fact, for many programs, the synchronization of objects is almost transparent.

There are two ways that you can synchronize your code. Both involve the use of the **synchronized** keyword, and both are examined here.

Using Synchronized Methods

You can synchronize access to a method by modifying it with the **synchronized** keyword. When that method is called, the calling thread enters the object's monitor, which then locks the object. While locked, no other thread can enter the method, or enter any other synchronized method defined by the object's class. When the thread returns from the method, the monitor unlocks the object, allowing it to be used by the next thread. Thus, synchronization is achieved with virtually no programming effort on your part.

The following program demonstrates synchronization by controlling access to a method called **sumArray( )**, which sums the elements of an integer array.

```
// Use synchronize to control access.

class SumArray {
  private int sum;

  synchronized int sumArray(int[] nums) {        sumArray( ) is synchronized.
    sum = 0; // reset sum

    for(int i=0; i<nums.length; i++) {
      sum += nums[i];
      System.out.println("Running total for " +
              Thread.currentThread().getName() +
              " is " + sum);
      try {
```

```
          Thread.sleep(10); // allow task-switch
        }
        catch(InterruptedException exc) {
          System.out.println("Thread interrupted.");
        }
      }
      return sum;
    }
}

class MyThread implements Runnable {
  Thread thrd;
  static SumArray sa = new SumArray();
  int[] a;
  int answer;

  // Construct a new thread.
  MyThread(String name, int[] nums) {
    thrd = new Thread(this, name);
    a = nums;
  }

  // A factory method that creates and starts a thread.
  public static MyThread createAndStart(String name, int[] nums) {
    MyThread myThrd = new MyThread(name, nums);

    myThrd.thrd.start(); // start the thread
    return myThrd;
  }

  // Entry point of thread.
  public void run() {
    int sum;

    System.out.println(thrd.getName() + " starting.");

    answer = sa.sumArray(a);
    System.out.println("Sum for " + thrd.getName() +
                       " is " + answer);

    System.out.println(thrd.getName() + " terminating.");
  }
}

class Sync {
  public static void main(String[] args) {
    int[] a = {1, 2, 3, 4, 5};

    MyThread mt1 = MyThread.createAndStart("Child #1", a);
    MyThread mt2 = MyThread.createAndStart("Child #2", a);
```

```
      try {
        mt1.thrd.join();
        mt2.thrd.join();
      }
      catch(InterruptedException exc) {
        System.out.println("Main thread interrupted.");
      }

    }
}
```

The output from the program is shown here. (The precise output may differ on your computer.)

```
Child #1 starting.
Running total for Child #1 is 1
Child #2 starting.
Running total for Child #1 is 3
Running total for Child #1 is 6
Running total for Child #1 is 10
Running total for Child #1 is 15
Sum for Child #1 is 15
Child #1 terminating.
Running total for Child #2 is 1
Running total for Child #2 is 3
Running total for Child #2 is 6
Running total for Child #2 is 10
Running total for Child #2 is 15
Sum for Child #2 is 15
Child #2 terminating.
```

Let's examine this program in detail. The program creates three classes. The first is **SumArray**. It contains the method **sumArray( )**, which sums an integer array. The second class is **MyThread**, which uses a **static** object of type **SumArray** to obtain the sum of an integer array. This object is called **sa** and because it is **static**, there is only one copy of it that is shared by all instances of **MyThread**. Finally, the class **Sync** creates two threads and has each compute the sum of an integer array.

Inside **sumArray( )**, **sleep( )** is called to purposely allow a task switch to occur, if one can—but it can't. Because **sumArray( )** is synchronized, it can be used by only one thread at a time. Thus, when the second child thread begins execution, it does not enter **sumArray( )** until after the first child thread is done with it. This ensures that the correct result is produced.

To fully understand the effects of **synchronized**, try removing it from the declaration of **sumArray( )**. After doing this, **sumArray( )** is no longer synchronized, and any number of threads may use it concurrently. The problem with this is that the running total is stored in **sum**, which will be changed by each thread that calls **sumArray( )** through the **static** object **sa**. Thus, when two threads call **sa.sumArray( )** at the same time, incorrect results are produced because **sum** reflects the summation of both threads, mixed together. For example, here is sample output from the program after **synchronized** has been removed from **sumArray( )**'s declaration. (The precise output may differ on your computer.)

```
Child #1 starting.
Running total for Child #1 is 1
Child #2 starting.
Running total for Child #2 is 1
Running total for Child #1 is 3
Running total for Child #2 is 5
Running total for Child #2 is 8
Running total for Child #1 is 11
Running total for Child #2 is 15
Running total for Child #1 is 19
Running total for Child #2 is 24
Sum for Child #2 is 24
Child #2 terminating.
Running total for Child #1 is 29
Sum for Child #1 is 29
Child #1 terminating.
```

As the output shows, both child threads are calling **sa.sumArray( )** concurrently, and the value of **sum** is corrupted. Before moving on, let's review the key points of a synchronized method:

- A synchronized method is created by preceding its declaration with **synchronized**.

- For any given object, once a synchronized method has been called, the object is locked and no synchronized methods on the same object can be used by another thread of execution.

- Other threads trying to call an in-use synchronized object will enter a wait state until the object is unlocked.

- When a thread leaves the synchronized method, the object is unlocked.

The synchronized Statement

Although creating **synchronized** methods within classes that you create is an easy and effective means of achieving synchronization, it will not work in all cases. For example, you might want to synchronize access to some method that is not modified by **synchronized**. This can occur because you want to use a class that was not created by you but by a third party, and you do not have access to the source code. Thus, it is not possible for you to add **synchronized** to the appropriate methods within the class. How can access to an object of this class be synchronized? Fortunately, the solution to this problem is quite easy: You simply put calls to the methods defined by this class inside a **synchronized** block.

This is the general form of a **synchronized** block:

```
synchronized(objref) {
  // statements to be synchronized
}
```

Here, *objref* is a reference to the object being synchronized. Once a synchronized block has been entered, no other thread can call a synchronized method on the object referred to by *objref* until the block has been exited.

For example, another way to synchronize calls to **sumArray( )** is to call it from within a synchronized block, as shown in this version of the program:

```
// Use a synchronized block to control access to SumArray.
class SumArray {
  private int sum;

  int sumArray(int[] nums) {          ←————— Here, sumArray()
    sum = 0; // reset sum                            is not synchronized.

    for(int i=0; i<nums.length; i++) {
      sum += nums[i];
      System.out.println("Running total for " +
              Thread.currentThread().getName() +
              " is " + sum);
      try {
        Thread.sleep(10); // allow task-switch
      }
      catch(InterruptedException exc) {
        System.out.println("Thread interrupted.");
      }
    }
    return sum;
  }
}

class MyThread implements Runnable {
  Thread thrd;
  static SumArray sa = new SumArray();
  int[] a;
  int answer;

  // Construct a new thread.
  MyThread(String name, int[] nums) {
    thrd = new Thread(this, name);
    a = nums;
  }

  // A factory method that creates and starts a thread.
  public static MyThread createAndStart(String name, int[] nums) {
    MyThread myThrd = new MyThread(name, nums);

    myThrd.thrd.start(); // start the thread
    return myThrd;
  }
```

```java
    // Entry point of thread.
    public void run() {
      int sum;

      System.out.println(thrd.getName() + " starting.");

      // synchronize calls to sumArray()
      synchronized(sa) {  ◄──────────────────  Here, calls to sumArray()
        answer = sa.sumArray(a);                on sa are synchronized.
      }
      System.out.println("Sum for " + thrd.getName() +
                      " is " + answer);

      System.out.println(thrd.getName() + " terminating.");
    }
  }

class Sync {
  public static void main(String[] args) {
    int[] a = {1, 2, 3, 4, 5};

    MyThread mt1 = MyThread.createAndStart("Child #1", a);
    MyThread mt2 = MyThread.createAndStart("Child #2", a);

    try {
      mt1.thrd.join();
      mt2.thrd.join();
    } catch(InterruptedException exc) {
      System.out.println("Main thread interrupted.");
    }
  }
}
```

This version produces the same, correct output as the one shown earlier that uses a synchronized method.

Ask the Expert

Q: I have heard of something called the "concurrency utilities." What are these? Also, what is the Fork/Join Framework?

A: The concurrency utilities, which are packaged in **java.util.concurrent** (and its subpackages), support concurrent programming. Among several other items, they offer synchronizers, thread pools, execution managers, and locks that expand your control over thread execution. One of the most exciting features of the concurrent API is the Fork/Join Framework.

(continued)

The Fork/Join Framework supports what is often termed *parallel programming*. This is the name commonly given to the techniques that take advantage of computers that contain two or more processors (including multicore systems) by subdividing a task into subtasks, with each subtask executing on its own processor. As you can imagine, such an approach can lead to significantly higher throughput and performance. The key advantage of the Fork/Join Framework is ease of use; it streamlines the development of multithreaded code that automatically scales to utilize the number of processors in a system. Thus, it facilitates the creation of concurrent solutions to some common programming tasks, such as performing operations on the elements of an array. The concurrency utilities in general, and the Fork/Join Framework specifically, are features that you will want to explore after you have become more experienced with multithreading.

Thread Communication Using notify( ), wait( ), and notifyAll( )

Consider the following situation. A thread called T is executing inside a synchronized method and needs access to a resource called R that is temporarily unavailable. What should T do? If T enters some form of polling loop that waits for R, T ties up the object, preventing other threads' access to it. This is a less than optimal solution because it partially defeats the advantages of programming for a multithreaded environment. A better solution is to have T temporarily relinquish control of the object, allowing another thread to run. When R becomes available, T can be notified and resume execution. Such an approach relies upon some form of interthread communication in which one thread can notify another that it is blocked and be notified that it can resume execution. Java supports interthread communication with the **wait( )**, **notify( )**, and **notifyAll( )** methods.

The **wait( )**, **notify( )**, and **notifyAll( )** methods are part of all objects because they are implemented by the **Object** class. These methods should be called only from within a **synchronized** context. Here is how they are used. When a thread is temporarily blocked from running, it calls **wait( )**. This causes the thread to go to sleep and the monitor for that object to be released, allowing another thread to use the object. At a later point, the sleeping thread is awakened when some other thread enters the same monitor and calls **notify( )**, or **notifyAll( )**.

Following are the various forms of **wait( )** defined by **Object**:

final void wait( ) throws InterruptedException

final void wait(long *millis*) throws InterruptedException

final void wait(long *millis*, int *nanos*) throws InterruptedException

The first form waits until notified. The second form waits until notified or until the specified period of milliseconds has expired. The third form allows you to specify the wait period in terms of nanoseconds.

Here are the general forms for **notify( )** and **notifyAll( )**:

final void notify( )

final void notifyAll( )

A call to **notify( )** resumes one waiting thread. A call to **notifyAll( )** notifies all threads, with the scheduler determining which thread gains access to the object.

Before looking at an example that uses **wait( )**, an important point needs to be made. Although **wait( )** normally waits until **notify( )** or **notifyAll( )** is called, there is a possibility that in very rare cases the waiting thread could be awakened due to a *spurious wakeup*. The conditions that lead to a spurious wakeup are complex and beyond the scope of this book. However, the Java API documentation recommends that because of the remote possibility of a spurious wakeup, calls to **wait( )** should take place within a loop that checks the condition on which the thread is waiting. The following example shows this technique.

An Example That Uses wait( ) and notify( )

To understand the need for and the application of **wait( )** and **notify( )**, we will create a program that simulates the ticking of a clock by displaying the words Tick and Tock on the screen. To accomplish this, we will create a class called **TickTock** that contains two methods: **tick( )** and **tock( )**. The **tick( )** method displays the word "Tick", and **tock( )** displays "Tock". To run the clock, two threads are created, one that calls **tick( )** and one that calls **tock( )**. The goal is to make the two threads execute in a way that the output from the program displays a consistent "Tick Tock"—that is, a repeated pattern of one tick followed by one tock.

```
// Use wait() and notify() to create a ticking clock.

class TickTock {

  String state; // contains the state of the clock

  synchronized void tick(boolean running) {
    if (!running) { // stop the clock
      state = "ticked";
      notify(); // notify any waiting threads
      return;
    }

    System.out.print("Tick ");

    state = "ticked"; // set the current state to ticked

    notify(); // let tock() run  ◄──────── tick( ) notifies tock( ).
    try {
      while (!state.equals("tocked"))
        wait(); // wait for tock() to complete  ◄──────── tick( ) waits for tock( ).
    }
```

```
            catch(InterruptedException exc) {
              System.out.println("Thread interrupted.");
            }
          }

          synchronized void tock(boolean running) {
            if(!running) { // stop the clock
              state = "tocked";
              notify(); // notify any waiting threads
              return;
            }

            System.out.println("Tock");

            state = "tocked"; // set the current state to tocked

            notify(); // let tick() run ◄─────── tock( ) notifies tick( ).
            try {
              while(!state.equals("ticked"))
                wait(); // wait for tick to complete ◄─────── tock( ) waits for tick( ).
            }
            catch(InterruptedException exc) {
              System.out.println("Thread interrupted.");
            }
          }
        }

        class MyThread implements Runnable {
          Thread thrd;
          TickTock ttOb;

          // Construct a new thread.
          MyThread(String name, TickTock tt) {
            thrd = new Thread(this, name);
            ttOb = tt;
          }

          // A factory method that creates and starts a thread.
          public static MyThread createAndStart(String name, TickTock tt) {
            MyThread myThrd = new MyThread(name, tt);

            myThrd.thrd.start(); // start the thread
            return myThrd;
          }

          // Entry point of thread.
          public void run() {
```

```
      if(thrd.getName().compareTo("Tick") == 0) {
        for(int i=0; i<5; i++) ttOb.tick(true);
        ttOb.tick(false);
      }
      else {
        for(int i=0; i<5; i++) ttOb.tock(true);
        ttOb.tock(false);
      }
    }
  }

class ThreadCom {
  public static void main(String[] args) {
    TickTock tt = new TickTock();
    MyThread mt1 = MyThread.createAndStart("Tick", tt);
    MyThread mt2 = MyThread.createAndStart("Tock", tt);

    try {
      mt1.thrd.join();
      mt2.thrd.join();
    } catch(InterruptedException exc) {
      System.out.println("Main thread interrupted.");
    }
  }
}
```

Here is the output produced by the program:

```
Tick Tock
Tick Tock
Tick Tock
Tick Tock
Tick Tock
```

Let's take a close look at this program. The heart of the clock is the **TickTock** class. It contains two methods, **tick( )** and **tock( )**, which communicate with each other to ensure that a Tick is always followed by a Tock, which is always followed by a Tick, and so on. Notice the **state** field. When the clock is running, **state** will hold either the string "ticked" or "tocked", which indicates the current state of the clock. In **main( )**, a **TickTock** object called **tt** is created, and this object is used to start two threads of execution.

The threads are based on objects of type **MyThread**. Both the **MyThread** constructor and the **createAndStart( )** method are passed two arguments. The first becomes the name of the thread. This will be either "Tick" or "Tock". The second is a reference to the **TickTock** object, which is **tt** in this case. Inside the **run( )** method of **MyThread**, if the name of the thread is "Tick", then calls to **tick( )** are made. If the name of the thread is "Tock", then the **tock( )** method is called. Five calls that pass **true** as an argument are made to each method. The clock runs as long as **true** is passed. A final call that passes **false** to each method stops the clock.

The most important part of the program is found in the **tick( )** and **tock( )** methods of **TickTock**. We will begin with the **tick( )** method, which, for convenience, is shown here:

```
synchronized void tick(boolean running) {
  if(!running) { // stop the clock
    state = "ticked";
    notify(); // notify any waiting threads
    return;
  }

  System.out.print("Tick ");

  state = "ticked"; // set the current state to ticked

  notify(); // let tock() run
  try {
    while(!state.equals("tocked"))
      wait(); // wait for tock() to complete
  }
  catch(InterruptedException exc) {
    System.out.println("Thread interrupted.");
  }
}
```

First, notice that **tick( )** is modified by **synchronized**. Remember, **wait( )** and **notify( )** apply only to synchronized methods. The method begins by checking the value of the **running** parameter. This parameter is used to provide a clean shutdown of the clock. If it is **false**, then the clock has been stopped. If this is the case, **state** is set to "ticked" and a call to **notify( )** is made to enable any waiting thread to run. We will return to this point in a moment.

Assuming that the clock is running when **tick( )** executes, the word "Tick" is displayed, **state** is set to "ticked", and then a call to **notify( )** takes place. The call to **notify( )** allows a thread waiting on the same object to run. Next, **wait( )** is called within a **while** loop. The call to **wait( )** causes **tick( )** to suspend until another thread calls **notify( )**. Therefore, the loop will not iterate until another thread calls **notify( )** on the same object. As a result, when **tick( )** is called, it displays one "Tick", lets another thread run, and then suspends.

The **while** loop that calls **wait( )** checks the value of **state**, waiting for it to equal "tocked", which will be the case only after the **tock( )** method executes. As explained, using a **while** loop to check this condition prevents a spurious wakeup from incorrectly restarting the thread. If **state** does not equal "tocked" when **wait( )** returns, it means that a spurious wakeup occurred, and **wait( )** is simply called again.

The **tock( )** method is an exact copy of **tick( )** except that it displays "Tock" and sets **state** to "tocked". Thus, when entered, it displays "Tock", calls **notify( )**, and then waits. When viewed as a pair, a call to **tick( )** can only be followed by a call to **tock( )**, which can only be followed by a call to **tick( )**, and so on. Therefore, the two methods are mutually synchronized.

The reason for the call to **notify( )** when the clock is stopped is to allow a final call to **wait( )** to succeed. Remember, both **tick( )** and **tock( )** execute a call to **wait( )** after displaying their message. The problem is that when the clock is stopped, one of the methods will still be waiting. Thus, a final call to **notify( )** is required in order for the waiting method to run. As an experiment, try removing this call to **notify( )** and watch what happens. As you will see, the program will "hang," and you will need to press CTRL-C to exit. The reason for this is that when the final call to **tock( )** calls **wait( )**, there is no corresponding call to **notify( )** that lets **tock( )** conclude. Thus, **tock( )** just sits there, waiting forever.

Before moving on, if you have any doubt that the calls to **wait( )** and **notify( )** are actually needed to make the "clock" run right, substitute this version of **TickTock** into the preceding program. It has all calls to **wait( )** and **notify( )** removed.

```
// No calls to wait() or notify().
class TickTock {

  String state; // contains the state of the clock

  synchronized void tick(boolean running) {
    if(!running) { // stop the clock
      state = "ticked";
      return;
    }

    System.out.print("Tick ");

    state = "ticked"; // set the current state to ticked
  }

  synchronized void tock(boolean running) {
    if(!running) { // stop the clock
      state = "tocked";
      return;
    }

    System.out.println("Tock");

    state = "tocked"; // set the current state to tocked
  }
}
```

After the substitution, the output produced by the program will look like this:

```
Tick Tick Tick Tick Tick Tock
Tock
Tock
Tock
Tock
```

Clearly, the **tick( )** and **tock( )** methods are no longer working together!

Ask the Expert

Q: I have heard the term *deadlock* applied to misbehaving multithreaded programs. What is it, and how can I avoid it? Also, what is a *race condition*, and how can I avoid that, too?

A: Deadlock is, as the name implies, a situation in which one thread is waiting for another thread to do something, but that other thread is waiting on the first. Thus, both threads are suspended, waiting on each other, and neither executes. This situation is analogous to two overly polite people, both insisting that the other step through a door first!

Avoiding deadlock seems easy, but it's not. For example, deadlock can occur in roundabout ways. The cause of the deadlock often is not readily understood just by looking at the source code to the program because concurrently executing threads can interact in complex ways at run time. To avoid deadlock, careful programming and thorough testing is required. Remember, if a multithreaded program occasionally "hangs," deadlock is the likely cause.

A race condition occurs when two (or more) threads attempt to access a shared resource at the same time, without proper synchronization. For example, one thread may be writing a new value to a variable while another thread is incrementing the variable's current value. Without synchronization, the new value of the variable will depend upon the order in which the threads execute. (Does the second thread increment the original value or the new value written by the first thread?) In situations like this, the two threads are said to be "racing each other," with the final outcome determined by which thread finishes first. Like deadlock, a race condition can occur in difficult-to-discover ways. The solution is prevention: careful programming that properly synchronizes access to shared resources.

Suspending, Resuming, and Stopping Threads

It is sometimes useful to suspend execution of a thread. For example, a separate thread can be used to display the time of day. If the user does not desire a clock, then its thread can be suspended. Whatever the case, it is a simple matter to suspend a thread. Once suspended, it is also a simple matter to restart the thread.

The mechanisms to suspend, stop, and resume threads differ between early versions of Java and more modern versions, beginning with Java 2. Prior to Java 2, a program used **suspend( )**, **resume( )**, and **stop( )**, which are methods defined by **Thread**, to pause, restart, and stop the execution of a thread. They have the following forms:

final void resume( )

final void suspend( )

final void stop( )

While these methods seem to be a perfectly reasonable and convenient approach to managing the execution of threads, they must no longer be used. Here's why. The **suspend( )** method of the **Thread** class was deprecated by Java 2. This was done because **suspend( )** can sometimes cause serious problems that involve deadlock. The **resume( )** method is also deprecated. It does not cause problems but cannot be used without the **suspend( )** method as its counterpart. The **stop( )** method of the **Thread** class was also deprecated by Java 2. This was done because this method too can sometimes cause serious problems.

Since you cannot now use the **suspend( )**, **resume( )**, or **stop( )** methods to control a thread, you might at first be thinking that there is no way to pause, restart, or terminate a thread. But, fortunately, this is not true. Instead, a thread must be designed so that the **run( )** method periodically checks to determine if that thread should suspend, resume, or stop its own execution. Typically, this is accomplished by establishing two flag variables: one for suspend and resume, and one for stop. For suspend and resume, as long as the flag is set to "running," the **run( )** method must continue to let the thread execute. If this variable is set to "suspend," the thread must pause. For the stop flag, if it is set to "stop," the thread must terminate.

The following example shows one way to implement your own versions of **suspend( )**, **resume( )**, and **stop( )**:

```
// Suspending, resuming, and stopping a thread.

class MyThread implements Runnable {
  Thread thrd;
  boolean suspended;  ◄──────── Suspends thread when true.
  boolean stopped;  ◄──────── Stops thread when true.

  MyThread(String name) {
    thrd = new Thread(this, name);
    suspended = false;
    stopped = false;
  }

  // A factory method that creates and starts a thread.
  public static MyThread createAndStart(String name) {
    MyThread myThrd = new MyThread(name);

    myThrd.thrd.start(); // start the thread
    return myThrd;
  }

  // Entry point of thread.
  public void run() {
    System.out.println(thrd.getName() + " starting.");
    try {
      for(int i = 1; i < 1000; i++) {
        System.out.print(i + " ");
        if((i%10)==0) {
          System.out.println();
          Thread.sleep(250);
        }
```

```
         // Use synchronized block to check suspended and stopped.
         synchronized(this) {
           while(suspended) {
             wait();
           }
           if(stopped) break;
         }
       }
     } catch (InterruptedException exc) {
       System.out.println(thrd.getName() + " interrupted.");
     }
     System.out.println(thrd.getName() + " exiting.");
   }

   // Stop the thread.
   synchronized void mystop() {
     stopped = true;

    // The following ensures that a suspended thread can be stopped.
     suspended = false;
     notify();
   }

   // Suspend the thread.
   synchronized void mysuspend() {
     suspended = true;
   }

   // Resume the thread.
   synchronized void myresume() {
     suspended = false;
     notify();
   }
 }

class Suspend {
  public static void main(String[] args) {
    MyThread mt1 = MyThread.createAndStart("My Thread");

    try {
      Thread.sleep(1000); // let ob1 thread start executing

      mt1.mysuspend();
      System.out.println("Suspending thread.");
      Thread.sleep(1000);

      mt1.myresume();
      System.out.println("Resuming thread.");
      Thread.sleep(1000);
```

This synchronized block checks **suspended** and **stopped**.

```
      mt1.mysuspend();
      System.out.println("Suspending thread.");
      Thread.sleep(1000);

      mt1.myresume();
      System.out.println("Resuming thread.");
      Thread.sleep(1000);

      mt1.mysuspend();
      System.out.println("Stopping thread.");
      mt1.mystop();

    } catch (InterruptedException e) {
      System.out.println("Main thread Interrupted");
    }

    // wait for thread to finish
    try {
      mt1.thrd.join();
    } catch (InterruptedException e) {
      System.out.println("Main thread Interrupted");
    }

    System.out.println("Main thread exiting.");
  }
}
```

Sample output from this program is shown here. (Your output may differ slightly.)

```
My Thread starting.
1 2 3 4 5 6 7 8 9 10
11 12 13 14 15 16 17 18 19 20
21 22 23 24 25 26 27 28 29 30
31 32 33 34 35 36 37 38 39 40
Suspending thread.
Resuming thread.
41 42 43 44 45 46 47 48 49 50
51 52 53 54 55 56 57 58 59 60
61 62 63 64 65 66 67 68 69 70
71 72 73 74 75 76 77 78 79 80
Suspending thread.
Resuming thread.
81 82 83 84 85 86 87 88 89 90
91 92 93 94 95 96 97 98 99 100
101 102 103 104 105 106 107 108 109 110
111 112 113 114 115 116 117 118 119 120
Stopping thread.
My Thread exiting.
Main thread exiting.
```

Ask the Expert

Q: **Multithreading seems like a great way to improve the efficiency of my programs. Can you give me any tips on effectively using it?**

A: The key to effectively utilizing multithreading is to think concurrently rather than serially. For example, when you have two subsystems within a program that are fully independent of each other, consider making them into individual threads. A word of caution is in order, however. If you create too many threads, you can actually degrade the performance of your program rather than enhance it. Remember, overhead is associated with context switching. If you create too many threads, more CPU time will be spent changing contexts than in executing your program!

Here is how the program works. The thread class **MyThread** defines two Boolean variables, **suspended** and **stopped**, which govern the suspension and termination of a thread. Both are initialized to **false** by the constructor. The **run( )** method contains a **synchronized** statement block that checks **suspended**. If that variable is **true**, the **wait( )** method is invoked to suspend the execution of the thread. To suspend execution of the thread, call **mysuspend( )**, which sets **suspended** to **true**. To resume execution, call **myresume( )**, which sets **suspended** to **false** and invokes **notify( )** to restart the thread.

To stop the thread, call **mystop( )**, which sets **stopped** to **true**. In addition, **mystop( )** sets **suspended** to **false** and then calls **notify( )**. These steps are necessary to stop a suspended thread.

Try This 11-2   Using the Main Thread

```
UseMain.java
```

All Java programs have at least one thread of execution, called the *main thread,* which is given to the program automatically when it begins running. So far, we have been taking the main thread for granted. In this project, you will see that the main thread can be handled just like all other threads.

1. Create a file called **UseMain.java**.

2. To access the main thread, you must obtain a **Thread** object that refers to it. You do this by calling the **currentThread( )** method, which is a **static** member of **Thread**. Its general form is shown here:

   static Thread currentThread( )

   This method returns a reference to the thread in which it is called. Therefore, if you call **currentThread( )** while execution is inside the main thread, you will obtain a reference to the main thread. Once you have this reference, you can control the main thread just like any other thread.

3. Enter the following program into the file. It obtains a reference to the main thread, and then gets and sets the main thread's name and priority.

```
/*
   Try This 11-2

   Controlling the main thread.
*/

class UseMain {
  public static void main(String[] args) {
    Thread thrd;

    // Get the main thread.
    thrd = Thread.currentThread();

    // Display main thread's name.
    System.out.println("Main thread is called: " +
                       thrd.getName());

    // Display main thread's priority.
    System.out.println("Priority: " +
                       thrd.getPriority());

    System.out.println();

    // Set the name and priority.
    System.out.println("Setting name and priority.\n");
    thrd.setName("Thread #1");
    thrd.setPriority(Thread.NORM_PRIORITY+3);

    System.out.println("Main thread is now called: " +
                       thrd.getName());

    System.out.println("Priority is now: " +
                       thrd.getPriority());
  }
}
```

4. The output from the program is shown here:

```
Main thread is called: main
Priority: 5

Setting name and priority.

Main thread is now called: Thread #1
Priority is now: 8
```

(continued)

5. You need to be careful about what operations you perform on the main thread. For example, if you add the following code to the end of **main( )**, the program will never terminate because it will be waiting for the main thread to end!

```
try {
   thrd.join();
} catch(InterruptedException exc) {
   System.out.println("Interrupted");
}
```

Chapter 11 Self Test

1. How does Java's multithreading capability enable you to write more efficient programs?

2. Multithreading is supported by the _____ class and the _____ interface.

3. When creating a runnable object, why might you want to extend **Thread** rather than implement **Runnable**?

4. Show how to use **join( )** to wait for a thread object called **MyThrd** to end.

5. Show how to set a thread called **MyThrd** to three levels above normal priority.

6. What is the effect of adding the **synchronized** keyword to a method?

7. The **wait( )** and **notify( )** methods are used to perform _____.

8. Change the **TickTock** class so that it actually keeps time. That is, have each tick take one half second, and each tock take one half second. Thus, each tick-tock will take one second. (Don't worry about the time it takes to switch tasks, etc.)

9. Why can't you use **suspend( )**, **resume( )**, and **stop( )** for new programs?

10. What method defined by **Thread** obtains the name of a thread?

11. What does **isAlive( )** return?

12. On your own, try adding synchronization to the **Queue** class developed in previous chapters so that it is safe for multithreaded use.

Chapter 12

Enumerations, Autoboxing, Annotations, and More

Key Skills & Concepts

- Understand enumeration fundamentals
- Use the class-based features of enumerations
- Apply the **values( )** and **valueof( )** methods to enumerations
- Create enumerations that have constructors, instance variables, and methods
- Employ the **ordinal( )** and **compareTo( )** methods that enumerations inherit from **Enum**
- Use Java's type wrappers
- Know the basics of autoboxing and auto-unboxing
- Use autoboxing with methods
- Understand how autoboxing works with expressions
- Apply static import
- Gain an overview of annotations
- Use the **instanceof** operator

This chapter discusses a number of important Java features. Interestingly, four—enumerations, autoboxing, static import, and annotations—were not part of the original definition of Java, each having been added by JDK 5. However, each significantly enhanced the power and usability of the language. In the case of enumerations and autoboxing, both addressed what was, at the time, long-standing needs. Static import streamlined the use of static members. Annotations expanded the kinds of information that can be embedded within a source file. Collectively, these features offered a better way to solve common programming problems. Frankly, today, it is difficult to imagine Java without them. They have become that important. Also discussed in this chapter are Java's type wrappers, which provide a bridge between the primitive types and object types. Finally, the **instanceof** operator is introduced. It lets you check the type of an object at run time.

Enumerations

In its simplest form, an *enumeration* is a list of named constants that define a new data type. An object of an enumeration type can hold only the values that are defined by the list. Thus, an enumeration gives you a way to precisely define a new type of data that has a fixed number of valid values.

Enumerations are common in everyday life. For example, an enumeration of the coins used in the United States is penny, nickel, dime, quarter, half-dollar, and dollar. An enumeration of the months in the year consists of the names January through December. An enumeration of the days of the week is Sunday, Monday, Tuesday, Wednesday, Thursday, Friday, and Saturday.

From a programming perspective, enumerations are useful whenever you need to define a set of values that represent a collection of items. For example, you might use an enumeration to represent a set of status codes, such as success, waiting, failed, and retrying, which indicate the progress of some action. In the past, such values were defined as **final** variables, but enumerations offer a more structured approach.

Enumeration Fundamentals

An enumeration is created using the **enum** keyword. For example, here is a simple enumeration that lists various forms of transportation:

```
// An enumeration of transportation.
enum Transport {
  CAR, TRUCK, AIRPLANE, TRAIN, BOAT
}
```

The identifiers **CAR, TRUCK,** and so on, are called *enumeration constants*. Each is implicitly declared as a public, static member of **Transport**. Furthermore, the enumeration constants' type is the type of the enumeration in which the constants are declared, which is **Transport** in this case. Thus, in the language of Java, these constants are called *self-typed,* where "self" refers to the enclosing enumeration.

Once you have defined an enumeration, you can create a variable of that type. However, even though enumerations define a class type, you do not instantiate an **enum** using **new**. Instead, you declare and use an enumeration variable in much the same way that you do one of the primitive types. For example, this declares **tp** as a variable of enumeration type **Transport**:

```
Transport tp;
```

Because **tp** is of type **Transport**, the only values that it can be assigned are those defined by the enumeration. For example, this assigns **tp** the value **AIRPLANE**:

```
tp = Transport.AIRPLANE;
```

Notice that the symbol **AIRPLANE** is qualified by **Transport**.

Two enumeration constants can be compared for equality by using the = = relational operator. For example, this statement compares the value in **tp** with the **TRAIN** constant:

```
if(tp == Transport.TRAIN) // ...
```

An enumeration value can also be used to control a **switch** statement. Of course, all of the **case** statements must use constants from the same **enum** as that used by the **switch** expression. For example, this **switch** is perfectly valid:

```
// Use an enum to control a switch statement.
switch(tp) {
  case CAR:
    // ...
  case TRUCK:
    // ...
```

Notice that in the **case** statements, the names of the enumeration constants are used without being qualified by their enumeration type name. That is, **TRUCK**, not **Transport.TRUCK**, is used. This is because the type of the enumeration in the **switch** expression has already implicitly specified the **enum** type of the **case** constants. There is no need to qualify the constants in the **case** statements with their **enum** type name. In fact, attempting to do so will cause a compilation error.

When an enumeration constant is displayed, such as in a **println( )** statement, its name is output. For example, given this statement:

```
System.out.println(Transport.BOAT);
```

the name **BOAT** is displayed.

The following program puts together all of the pieces and demonstrates the **Transport** enumeration:

```
// An enumeration of Transport varieties.
enum Transport {
   CAR, TRUCK, AIRPLANE, TRAIN, BOAT   ◄──────── Declare an enumeration.
}

class EnumDemo {
  public static void main(String[] args)
  {
    Transport tp;  ◄──────── Declare a Transport reference.

    tp = Transport.AIRPLANE;  ◄──────── Assign tp the constant AIRPLANE.

    // Output an enum value.
    System.out.println("Value of tp: " + tp);
    System.out.println();

    tp = Transport.TRAIN;

    // Compare two enum values.
    if(tp == Transport.TRAIN)   ◄──────────────── Compare two Transport
      System.out.println("tp contains TRAIN.\n");   objects for equality.

    // Use an enum to control a switch statement.
    switch(tp) {  ◄──────────────────────────── Use an enumeration to
      case CAR:                                    control a switch statement.
        System.out.println("A car carries people.");
        break;
      case TRUCK:
        System.out.println("A truck carries freight.");
        break;
      case AIRPLANE:
        System.out.println("An airplane flies.");
        break;
```

```
      case TRAIN:
        System.out.println("A train runs on rails.");
        break;
      case BOAT:
        System.out.println("A boat sails on water.");
        break;
    }
  }
}
```

The output from the program is shown here:

```
Value of tp: AIRPLANE

tp contains TRAIN.

A train runs on rails.
```

Before moving on, it's necessary to make one stylistic point. The constants in **Transport** use uppercase. (Thus, **CAR**, not **car**, is used.) However, the use of uppercase is not required. In other words, there is no rule that requires enumeration constants to be in uppercase. Because enumerations often replace **final** variables, which have traditionally used uppercase, some programmers believe that uppercasing enumeration constants is also appropriate. There are, of course, other viewpoints and styles. The examples in this book will use uppercase for enumeration constants, for consistency.

Java Enumerations Are Class Types

Although the preceding examples show the mechanics of creating and using an enumeration, they don't show all of its capabilities. Unlike the way enumerations are implemented in some other languages, *Java implements enumerations as class types*. Although you don't instantiate an **enum** using **new**, it otherwise acts much like other classes. The fact that **enum** defines a class enables the Java enumeration to have powers that enumerations in some other languages do not. For example, you can give it constructors, add instance variables and methods, and even implement interfaces.

The values( ) and valueOf( ) Methods

All enumerations automatically have two predefined methods: **values( )** and **valueOf( )**. Their general forms are shown here:

public static *enum-type*[ ] values( )

public static *enum-type* valueOf(String *str*)

The **values( )** method returns an array that contains a list of the enumeration constants. The **valueOf( )** method returns the enumeration constant whose value corresponds to the string passed in *str*. In both cases, *enum-type* is the type of the enumeration. For example, in the case of the **Transport** enumeration shown earlier, the return type of **Transport.valueOf("TRAIN")** is **Transport**. The value returned is **TRAIN**. The following program demonstrates the **values( )** and **valueOf( )** methods:

```java
// Use the built-in enumeration methods.

// An enumeration of Transport varieties.
enum Transport {
  CAR, TRUCK, AIRPLANE, TRAIN, BOAT
}

class EnumDemo2 {
  public static void main(String[] args)
  {
    Transport tp;

    System.out.println("Here are all Transport constants");

    // use values()
    Transport[] allTransports = Transport.values();   // Obtain an array of Transport constants.
    for(Transport t : allTransports)
      System.out.println(t);

    System.out.println();

    // use valueOf()
    tp = Transport.valueOf("AIRPLANE");   // Obtain the constant with the name AIRPLANE.
    System.out.println("tp contains " + tp);
  }
}
```

The output from the program is shown here:

```
Here are all Transport constants
CAR
TRUCK
AIRPLANE
TRAIN
BOAT

tp contains AIRPLANE
```

Notice that this program uses a for-each style **for** loop to cycle through the array of constants obtained by calling **values( )**. For the sake of illustration, the variable **allTransports** was created and assigned a reference to the enumeration array. However, this step is not necessary because the **for** could have been written as shown here, eliminating the need for the **allTransports** variable:

```
for(Transport t : Transport.values())
  System.out.println(t);
```

Now, notice how the value corresponding to the name **AIRPLANE** was obtained by calling **valueOf( )**:

```
tp = Transport.valueOf("AIRPLANE");
```

As explained, **valueOf( )** returns the enumeration value associated with the name of the constant represented as a string.

Constructors, Methods, Instance Variables, and Enumerations

It is important to understand that each enumeration constant is an object of its enumeration type. Thus, an enumeration can define constructors, add methods, and have instance variables. When you define a constructor for an **enum**, the constructor is called when each enumeration constant is created. Each enumeration constant can call any method defined by the enumeration. Each enumeration constant has its own copy of any instance variables defined by the enumeration. The following version of **Transport** illustrates the use of a constructor, an instance variable, and a method. It gives each type of transportation a typical speed.

```
// Use an enum constructor, instance variable, and method.
enum Transport {
    CAR(65), TRUCK(55), AIRPLANE(600), TRAIN(70), BOAT(22);   ←——— Notice the initialization values.

    private int speed; // typical speed of each transport   ←——— Add an instance variable.

    // Constructor
    Transport(int s) { speed = s; }   ←——— Add a constructor.

    int getSpeed() { return speed; }   ←——— Add a method.
}

class EnumDemo3 {
  public static void main(String[] args)
  {
    Transport tp;

      // Display speed of an airplane.
      System.out.println("Typical speed for an airplane is " +
                  Transport.AIRPLANE.getSpeed() +   ←——— Obtain the speed by calling getSpeed( ).
                  " miles per hour.\n");
```

```
      // Display all Transports and speeds.
      System.out.println("All Transport speeds: ");
      for(Transport t : Transport.values())
        System.out.println(t + " typical speed is " +
                            t.getSpeed() +
                            " miles per hour.");
    }
}
```

The output is shown here:

```
Typical speed for an airplane is 600 miles per hour.

All Transport speeds:
CAR typical speed is 65 miles per hour.
TRUCK typical speed is 55 miles per hour.
AIRPLANE typical speed is 600 miles per hour.
TRAIN typical speed is 70 miles per hour.
BOAT typical speed is 22 miles per hour.
```

This version of **Transport** adds three things. The first is the instance variable **speed**, which is used to hold the speed of each kind of transport. The second is the **Transport** constructor, which is passed the speed of a transport. The third is the method **getSpeed( )**, which returns the value of **speed**.

When the variable **tp** is declared in **main( )**, the constructor for **Transport** is called once for each constant that is specified. Notice how the arguments to the constructor are specified, by putting them inside parentheses, after each constant, as shown here:

```
CAR(65), TRUCK(55), AIRPLANE(600), TRAIN(70), BOAT(22);
```

These values are passed to the **s** parameter of **Transport( )**, which then assigns this value to **speed**. There is something else to notice about the list of enumeration constants: it is terminated by a semicolon. That is, the last constant, **BOAT**, is followed by a semicolon. When an enumeration contains other members, the enumeration list must end in a semicolon.

Because each enumeration constant has its own copy of **speed**, you can obtain the speed of a specified type of transport by calling **getSpeed( )**. For example, in **main( )** the speed of an airplane is obtained by the following call:

```
Transport.AIRPLANE.getSpeed()
```

The speed of each transport is obtained by cycling through the enumeration using a **for** loop. Because there is a copy of **speed** for each enumeration constant, the value associated with one constant is separate and distinct from the value associated with another constant. This is a powerful concept, which is available only when enumerations are implemented as classes, as Java does.

Although the preceding example contains only one constructor, an **enum** can offer two or more overloaded forms, just as can any other class.

Ask the Expert

Q: Since enumerations have been added to Java, should I avoid the use of final variables? In other words, have enumerations rendered final variables obsolete?

A: No. Enumerations are appropriate when you are working with lists of items that must be represented by identifiers. A **final** variable is appropriate when you have a constant value, such as an array size, that will be used in many places. Thus, each has its own use. The advantage of enumerations is that **final** variables don't have to be pressed into service for a job for which they are not ideally suited.

Two Important Restrictions

There are two restrictions that apply to enumerations. First, an enumeration can't inherit another class. Second, an **enum** cannot be a superclass. This means that an **enum** can't be extended. Otherwise, **enum** acts much like any other class type. The key is to remember that each of the enumeration constants is an object of the class in which it is defined.

Enumerations Inherit Enum

Although you can't inherit a superclass when declaring an **enum**, all enumerations automatically inherit one: **java.lang.Enum**. This class defines several methods that are available for use by all enumerations. Most often, you won't need to use these methods, but there are two that you may occasionally employ: **ordinal( )** and **compareTo( )**.

The **ordinal( )** method obtains a value that indicates an enumeration constant's position in the list of constants. This is called its *ordinal value*. The **ordinal( )** method is shown here:

final int ordinal( )

It returns the ordinal value of the invoking constant. Ordinal values begin at zero. Thus, in the **Transport** enumeration, **CAR** has an ordinal value of zero, **TRUCK** has an ordinal value of 1, **AIRPLANE** has an ordinal value of 2, and so on.

You can compare the ordinal value of two constants of the same enumeration by using the **compareTo( )** method. It has this general form:

final int compareTo(*enum-type e*)

Here, *enum-type* is the type of the enumeration and *e* is the constant being compared to the invoking constant. Remember, both the invoking constant and *e* must be of the same enumeration. If the invoking constant has an ordinal value less than *e*'s, then **compareTo( )** returns a negative value. If the two ordinal values are the same, then zero is returned. If the invoking constant has an ordinal value greater than *e*'s, then a positive value is returned.

The following program demonstrates **ordinal( )** and **compareTo( )**:

```java
// Demonstrate ordinal() and compareTo().

// An enumeration of Transport varieties.
enum Transport {
  CAR, TRUCK, AIRPLANE, TRAIN, BOAT
}

class EnumDemo4 {
  public static void main(String[] args)
  {
    Transport tp, tp2, tp3;

    // Obtain all ordinal values using ordinal().
    System.out.println("Here are all Transport constants" +
                       " and their ordinal values: ");
    for(Transport t : Transport.values())
      System.out.println(t + " " + t.ordinal());    ◄——— Obtain ordinal values.

    tp = Transport.AIRPLANE;
    tp2 = Transport.TRAIN;
    tp3 = Transport.AIRPLANE;

    System.out.println();
                                         Compare ordinal values.
    // Demonstrate compareTo()
    if(tp.compareTo(tp2) < 0)  ◄————┘
      System.out.println(tp + " comes before " + tp2);

    if(tp.compareTo(tp2) > 0)
      System.out.println(tp2 + " comes before " + tp);

    if(tp.compareTo(tp3) == 0)
      System.out.println(tp + " equals " + tp3);
  }
}
```

The output from the program is shown here:

```
Here are all Transport constants and their ordinal values:
CAR 0
TRUCK 1
AIRPLANE 2
TRAIN 3
BOAT 4

AIRPLANE comes before TRAIN
AIRPLANE equals AIRPLANE
```

A Computer-Controlled Traffic Light

TrafficLightDemo.java   Enumerations are particularly useful when your program needs a set of constants, but the actual values of the constants are arbitrary, as long as all differ. This type of situation comes up quite often when programming. One common instance involves handling the states in which some device can exist. For example, imagine that you are writing a program that controls a traffic light. Your traffic light code must automatically cycle through the light's three states: green, yellow, and red. It also must enable other code to know the current color of the light and let the color of the light be set to a known initial value. This means that the three states must be represented in some way. Although it would be possible to represent these three states by integer values (for example, the values 1, 2, and 3) or by strings (such as "red", "green", and "yellow"), an enumeration offers a much better approach. Using an enumeration results in code that is more efficient than if strings represented the states and more structured than if integers represented the states.

In this project, you will create a simulation of an automated traffic light, as just described. This project not only demonstrates an enumeration in action, it also shows another example of multithreading and synchronization.

1. Create a file called **TrafficLightDemo.java**.

2. Begin by defining an enumeration called **TrafficLightColor** that represents the three states of the light, as shown here:

```
// An enumeration of the colors of a traffic light.
enum TrafficLightColor {
  RED, GREEN, YELLOW
}
```

Whenever the color of the light is needed, its enumeration value is used.

3. Next, begin defining **TrafficLightSimulator**, as shown next. **TrafficLightSimulator** is the class that encapsulates the traffic light simulation.

```
// A computerized traffic light.
class TrafficLightSimulator implements Runnable {
  private TrafficLightColor tlc; // holds the traffic light color
  private boolean stop = false; // set to true to stop the simulation
  private boolean changed = false; // true when the light has changed

  TrafficLightSimulator(TrafficLightColor init) {
    tlc = init;
  }

  TrafficLightSimulator() {
    tlc = TrafficLightColor.RED;
  }
```

Notice that **TrafficLightSimulator** implements **Runnable**. This is necessary because a separate thread is used to run each traffic light. This thread will cycle through the colors.

(continued)

Two constructors are created. The first lets you specify the initial light color. The second defaults to red.

Now look at the instance variables. A reference to the traffic light thread is stored in **thrd**. The current traffic light color is stored in **tlc**. The **stop** variable is used to stop the simulation. It is initially set to **false**. The light will run until this variable is set to **true**. The **changed** variable is **true** when the light has changed.

4. Next, add the **run( )** method, shown here, which begins running the traffic light:

```
// Start up the light.
public void run() {
  while(!stop) {
    try {
      switch(tlc) {
        case GREEN:
          Thread.sleep(10000); // green for 10 seconds
          break;
        case YELLOW:
          Thread.sleep(2000); // yellow for 2 seconds
          break;
        case RED:
          Thread.sleep(12000); // red for 12 seconds
          break;
      }
    } catch(InterruptedException exc) {
      System.out.println(exc);
    }
    changeColor();
  }
}
```

This method cycles the light through the colors. First, it sleeps an appropriate amount of time, based on the current color. Then, it calls **changeColor( )** to change to the next color in the sequence.

5. Now, add the **changeColor( )** method, as shown here:

```
// Change color.
synchronized void changeColor() {
  switch(tlc) {
    case RED:
      tlc = TrafficLightColor.GREEN;
      break;
    case YELLOW:
      tlc = TrafficLightColor.RED;
      break;
    case GREEN:
      tlc = TrafficLightColor.YELLOW;
  }
}
```

```
    changed = true;
    notify(); // signal that the light has changed
  }
```

The **switch** statement examines the color currently stored in **tlc** and then assigns the next color in the sequence. Notice that this method is synchronized. This is necessary because it calls **notify( )** to signal that a color change has taken place. (Recall that **notify( )** can be called only from a synchronized context.)

6. The next method is **waitForChange( )**, which waits until the color of the light is changed.

```
// Wait until a light change occurs.
synchronized void waitForChange() {
  try {
    while(!changed)
      wait(); // wait for light to change
    changed = false;
  } catch(InterruptedException exc) {
    System.out.println(exc);
  }
}
```

This method simply calls **wait( )**. This call won't return until **changeColor( )** executes a call to **notify( )**. Thus, **waitForChange( )** won't return until the color has changed.

7. Finally, add the methods **getColor( )**, which returns the current light color, and **cancel( )**, which stops the traffic light thread by setting **stop** to **true**. These methods are shown here:

```
// Return current color.
synchronized TrafficLightColor getColor() {
  return tlc;
}

// Stop the traffic light.
synchronized void cancel() {
  stop = true;
}
```

8. Here is all the code assembled into a complete program that demonstrates the traffic light:

```
// Try This 12-1

// A simulation of a traffic light that uses
// an enumeration to describe the light's color.

// An enumeration of the colors of a traffic light.
enum TrafficLightColor {
  RED, GREEN, YELLOW
}
```

(continued)

```java
// A computerized traffic light.
class TrafficLightSimulator implements Runnable {
  private TrafficLightColor tlc; // holds the traffic light color
  private boolean stop = false; // set to true to stop the simulation
  private boolean changed = false; // true when the light has changed

  TrafficLightSimulator(TrafficLightColor init) {
    tlc = init;
  }

  TrafficLightSimulator() {
    tlc = TrafficLightColor.RED;
  }

  // Start up the light.
  public void run() {
    while(!stop) {
      try {
        switch(tlc) {
          case GREEN:
            Thread.sleep(10000); // green for 10 seconds
            break;
          case YELLOW:
            Thread.sleep(2000); // yellow for 2 seconds
            break;
          case RED:
            Thread.sleep(12000); // red for 12 seconds
            break;
        }
      } catch(InterruptedException exc) {
        System.out.println(exc);
      }
      changeColor();
    }
  }

  // Change color.
  synchronized void changeColor() {
    switch(tlc) {
      case RED:
        tlc = TrafficLightColor.GREEN;
        break;
      case YELLOW:
        tlc = TrafficLightColor.RED;
        break;
      case GREEN:
        tlc = TrafficLightColor.YELLOW;
    }
```

```
      changed = true;
      notify(); // signal that the light has changed
    }

    // Wait until a light change occurs.
    synchronized void waitForChange() {
      try {
        while(!changed)
          wait(); // wait for light to change
        changed = false;
      } catch(InterruptedException exc) {
        System.out.println(exc);
      }
    }
  }

    // Return current color.
    synchronized TrafficLightColor getColor() {
      return tlc;
    }

    // Stop the traffic light.
    synchronized void cancel() {
      stop = true;
    }
}

class TrafficLightDemo {
  public static void main(String[] args) {
    TrafficLightSimulator tl =
      new TrafficLightSimulator(TrafficLightColor.GREEN);
    Thread thrd = new Thread(tl);
    thrd.start();

    for(int i=0; i < 9; i++) {
      System.out.println(tl.getColor());
      tl.waitForChange();
    }

    tl.cancel();
  }
}
```

The following output is produced. As you can see, the traffic light cycles through the colors in order of green, yellow, and red:

```
GREEN
YELLOW
RED
```

(continued)

```
GREEN
YELLOW
RED
GREEN
YELLOW
RED
```

In the program, notice how the use of the enumeration simplifies and adds structure to the code that needs to know the state of the traffic light. Because the light can have only three states (red, green, or yellow), the use of an enumeration ensures that only these values are valid, thus preventing accidental misuse.

9. It is possible to improve the preceding program by taking advantage of the class capabilities of an enumeration. For example, by adding a constructor, instance variable, and method to **TrafficLightColor**, you can substantially improve the preceding programming. This improvement is left as an exercise. See Self Test, question 4.

Autoboxing

Modern versions of Java include two very helpful features: *autoboxing* and *auto-unboxing*. Autoboxing/unboxing greatly simplifies and streamlines code that must convert primitive types into objects, and vice versa. Because such situations are found frequently in Java code, the benefits of autoboxing/unboxing affect nearly all Java programmers. As you will see in Chapter 13, autoboxing/unboxing also contributes greatly to the usability of generics.

Autoboxing/unboxing is directly related to Java's type wrappers, and to the way that values are moved into and out of an instance of a wrapper. For this reason, we will begin with an overview of the type wrappers and the process of manually boxing and unboxing values.

Type Wrappers

As you know, Java uses primitive types, such as **int** or **double**, to hold the basic data types supported by the language. Primitive types, rather than objects, are used for these quantities for the sake of performance. Using objects for these basic types would add an unacceptable overhead to even the simplest of calculations. Thus, the primitive types are not part of the object hierarchy, and they do not inherit **Object**.

Despite the performance benefit offered by the primitive types, there are times when you will need an object representation. For example, you can't pass a primitive type by reference to a method. Also, many of the standard data structures implemented by Java operate on objects, which means that you can't use these data structures to store primitive types. To handle these (and other) situations, Java provides *type wrappers*, which are classes that encapsulate a primitive type within an object. The type wrapper classes were introduced briefly in Chapter 10. Here, we will look at them more closely.

The type wrappers are **Double**, **Float**, **Long**, **Integer**, **Short**, **Byte**, **Character**, and **Boolean**, which are packaged in **java.lang**. These classes offer a wide array of methods that allow you to fully integrate the primitive types into Java's object hierarchy.

Probably the most commonly used type wrappers are those that represent numeric values. These are **Byte**, **Short**, **Integer**, **Long**, **Float**, and **Double**. All of the numeric type wrappers inherit the abstract class **Number**. **Number** declares methods that return the value of an object in each of the different numeric types. These methods are shown here:

byte byteValue( )

double doubleValue( )

float floatValue( )

int intValue( )

long longValue( )

short shortValue( )

For example, **doubleValue( )** returns the value of an object as a **double**, **floatValue( )** returns the value as a **float**, and so on. These methods are implemented by each of the numeric type wrappers.

All of the numeric type wrappers define constructors that allow an object to be constructed from a given value, or a string representation of that value. For example, here are the constructors defined for **Integer** and **Double**:

Integer(int *num*)
Integer(String *str*) throws NumberFormatException

Double(double *num*)
Double(String *str*) throws NumberFormatException

If *str* does not contain a valid numeric value, then a **NumberFormatException** is thrown. However, beginning with JDK 9, the type-wrapper constructors were deprecated, and beginning with JDK 16, they have been deprecated for removal. Today, it is strongly recommended that you use one of the **valueOf( )** methods to obtain a wrapper object. The **valueOf( )** method is a static member of all of the wrapper classes and all numeric classes support forms that convert a numeric value or a string into an object. For example, here are two forms supported by **Integer**:

static Integer valueOf(int *val*)
static Integer valueOf(String *valStr*) throws NumberFormatException

Here, *val* specifies an integer value and *valStr* specifies a string that represents a properly formatted numeric value in string form. Each returns an **Integer** object that wraps the specified value. Here is an example:

```
Integer iOb = Integer.valueOf(100);
```

After this statement executes, the value 100 is represented by an **Integer** instance. Thus, **iOb** wraps the value 100 within an object.

All of the type wrappers override **toString( )**. It returns the human-readable form of the value contained within the wrapper. This allows you to output the value by passing a type wrapper object to **println( )**, for example, without having to convert it into its primitive type.

The process of encapsulating a value within an object is called *boxing*. In the early days of Java, all boxing took place manually, with the programmer explicitly constructing an instance of a wrapper with the desired value, as just shown. Therefore, in the preceding example, the value 100 is said to be *boxed* inside **iOb**.

The process of extracting a value from a type wrapper is called *unboxing*. Again, in the early days of Java, all unboxing also took place manually, with the programmer explicitly calling a method on the wrapper to obtain its value. For example, this manually unboxes the value in **iOb** into an **int**.

```
int i = iOb.intValue();
```

Here, **intValue( )** returns the value encapsulated within **iOb** as an **int**.

The following program demonstrates the preceding concepts:

```
// Demonstrate manual boxing and unboxing with a type wrapper.
class Wrap {
  public static void main(String[] args) {

    Integer iOb = new Integer.valueOf(100);  ←——— Manually box the value 100.

    int i = iOb.intValue();  ←——— Manually unbox the value in iOb.

    System.out.println(i + " " + iOb); // displays 100 100
  }
}
```

This program wraps the integer value 100 inside an **Integer** object called **iOb**. The program then obtains this value by calling **intValue( )** and stores the result in **i**. Finally, it displays the values of **i** and **iOb**, both of which are 100.

The same general procedure used by the preceding example to manually box and unbox values was required by all versions of Java prior to JDK 5 and may still be found in legacy code. The problem is that it is both tedious and error-prone because it requires the programmer to manually create the appropriate object to wrap a value and to explicitly obtain the proper primitive type when its value is needed. Fortunately, autoboxing/unboxing fundamentally improves on these essential procedures.

Autoboxing Fundamentals

Autoboxing is the process by which a primitive type is automatically encapsulated (boxed) into its equivalent type wrapper whenever an object of that type is needed. There is no need to explicitly obtain an object. Auto-unboxing is the process by which the value of a boxed object is automatically extracted (unboxed) from a type wrapper when its value is needed. There is no need to call a method such as **intValue( )** or **doubleValue( )**.

The addition of autoboxing and auto-unboxing greatly streamlines the coding of several algorithms, removing the tedium of manually boxing and unboxing values. It also helps prevent errors. With autoboxing it is not necessary to manually construct an object in order to wrap a primitive type. You need only assign that value to a type-wrapper reference. Java automatically constructs the object for you. For example, here is the modern way to declare an **Integer** object that has the value 100:

```
Integer iOb = 100; // autobox an int
```

Notice that the object is not explicitly boxed. Java handles this for you, automatically.

To unbox an object, simply assign that object reference to a primitive-type variable. For example, to unbox **iOb**, you can use this line:

```
int i = iOb; // auto-unbox
```

Java handles the details for you.

The following program demonstrates the preceding statements:

```
// Demonstrate autoboxing/unboxing.
class AutoBox {
  public static void main(String[] args) {

    Integer iOb = 100; // autobox an int                  Autobox and then auto-unbox
                                                          the value 100.
    int i = iOb; // auto-unbox

    System.out.println(i + " " + iOb); // displays 100 100
  }
}
```

Autoboxing and Methods

In addition to the simple case of assignments, autoboxing automatically occurs whenever a primitive type must be converted into an object, and auto-unboxing takes place whenever an object must be converted into a primitive type. Thus, autoboxing/unboxing might occur when an argument is passed to a method or when a value is returned by a method. For example, consider the following:

```
// Autoboxing/unboxing takes place with
// method parameters and return values.

class AutoBox2 {
  // This method has an Integer parameter.
  static void m(Integer v) {                    Receives an Integer.
    System.out.println("m() received " + v);
  }
```

```
// This method returns an int.
static int m2() {  ←──────────── Returns an int.
  return 10;
}
                                    Returns an Integer.
// This method returns an Integer.
static Integer m3() {  ←─────────┘
  return 99; // autoboxing 99 into an Integer.
}

public static void main(String[] args) {

  // Pass an int to m(). Because m() has an Integer
  // parameter, the int value passed is automatically boxed.
  m(199);

  // Here, iOb receives the int value returned by m2().
  // This value is automatically boxed so that it can be
  // assigned to iOb.
  Integer iOb = m2();
  System.out.println("Return value from m2() is " + iOb);

  // Next, m3() is called. It returns an Integer value
  // which is auto-unboxed into an int.
  int i = m3();
  System.out.println("Return value from m3() is " + i);

  // Next, Math.sqrt() is called with iOb as an argument.
  // In this case, iOb is auto-unboxed and its value promoted to
  // double, which is the type needed by sqrt().
  iOb = 100;
  System.out.println("Square root of iOb is " + Math.sqrt(iOb));
  }
}
```

This program displays the following result:

```
m() received 199
Return value from m2() is 10
Return value from m3() is 99
Square root of iOb is 10.0
```

In the program, notice that **m( )** specifies an **Integer** parameter. Inside **main( )**, **m( )** is passed the **int** value 199. Because **m( )** is expecting an **Integer**, this value is automatically boxed. Next, **m2( )** is called. It returns the **int** value 10. This **int** value is assigned to **iOb** in **main( )**. Because **iOb** is an **Integer**, the value returned by **m2( )** is autoboxed. Next, **m3( )** is called. It returns an **Integer** that is auto-unboxed into an **int**. Finally, **Math.sqrt( )** is called with **iOb** as an argument. In this case, **iOb** is auto-unboxed and its value promoted to **double**, since that is the type expected by **Math.sqrt( )**.

Autoboxing/Unboxing Occurs in Expressions

In general, autoboxing and unboxing take place whenever a conversion into an object or from an object is required. This applies to expressions. Within an expression, a numeric object is automatically unboxed. The outcome of the expression is reboxed, if necessary. For example, consider the following program:

```
// Autoboxing/unboxing occurs inside expressions.

class AutoBox3 {
  public static void main(String[] args) {
    Integer iOb, iOb2;
    int i;

    iOb = 99;
    System.out.println("Original value of iOb: " + iOb);

    // The following automatically unboxes iOb,
    // performs the increment, and then reboxes
    // the result back into iOb.
    ++iOb;
    System.out.println("After ++iOb: " + iOb);

    // Here, iOb is unboxed, its value is increased by 10,
    // and the result is boxed and stored back in iOb.
    iOb += 10;
    System.out.println("After iOb += 10: " + iOb);

    // Here, iOb is unboxed, the expression is
    // evaluated, and the result is reboxed and
    // assigned to iOb2.
    iOb2 = iOb + (iOb / 3);
    System.out.println("iOb2 after expression: " + iOb2);

    // The same expression is evaluated, but the
    // result is not reboxed.
    i = iOb + (iOb / 3);
    System.out.println("i after expression: " + i);
  }
}
```

Autoboxing/unboxing occurs in expressions.

The output is shown here:

```
Original value of iOb: 99
After ++iOb: 100
After iOb += 10: 110
iOb2 after expression: 146
i after expression: 146
```

In the program, pay special attention to this line:

```
++iOb;
```

This causes the value in **iOb** to be incremented. It works like this: **iOb** is unboxed, the value is incremented, and the result is reboxed.

Because of auto-unboxing, you can use integer numeric objects, such as an **Integer**, to control a **switch** statement. For example, consider this fragment:

```
Integer iOb = 2;

switch(iOb) {
  case 1: System.out.println("one");
    break;
  case 2: System.out.println("two");
    break;
  default: System.out.println("error");
}
```

When the **switch** expression is evaluated, **iOb** is unboxed and its **int** value is obtained.

As the examples in the program show, because of autoboxing/unboxing, using numeric objects in an expression is both intuitive and easy. With early versions of Java, such code would have involved casts and calls to methods such as **intValue( )**.

A Word of Warning

Because of autoboxing and auto-unboxing, one might be tempted to use objects such as **Integer** or **Double** exclusively, abandoning primitives altogether. For example, with autoboxing/unboxing it is possible to write code like this:

```
// A bad use of autoboxing/unboxing!
Double a, b, c;

a = 10.2;
b = 11.4;
c = 9.8;

Double avg = (a + b + c) / 3;
```

In this example, objects of type **Double** hold values, which are then averaged and the result assigned to another **Double** object. Although this code is technically correct and does, in fact, work properly, it is a very bad use of autoboxing/unboxing. It is far less efficient than the equivalent code written using the primitive type **double**. The reason is that each autobox and auto-unbox adds overhead that is not present if the primitive type is used.

In general, you should restrict your use of the type wrappers to only those cases in which an object representation of a primitive type is required. Autoboxing/unboxing was not added to Java as a "back door" way of eliminating the primitive types.

Static Import

Java supports an expanded use of the **import** keyword. By following **import** with the keyword **static**, an **import** statement can be used to import the static members of a class or interface. This is called *static import*. When using static import, it is possible to refer to static members directly by their names, without having to qualify them with the name of their class. This simplifies and shortens the syntax required to use a static member.

To understand the usefulness of static import, let's begin with an example that *does not* use it. The following program computes the solutions to a quadratic equation, which has this form:

$$ax^2 + bx + c = 0$$

The program uses two static methods from Java's built-in math class **Math**, which is part of **java.lang**. The first is **Math.pow( )**, which returns a value raised to a specified power. The second is **Math.sqrt( )**, which returns the square root of its argument.

```
// Find the solutions to a quadratic equation.
class Quadratic {
  public static void main(String[] args) {

    // a, b, and c represent the coefficients in the
    // quadratic equation: ax² + bx + c = 0
    double a, b, c, x;

    // Solve 4x² + x - 3 = 0 for x.
    a = 4;
    b = 1;
    c = -3;

    // Find first solution.
    x = (-b + Math.sqrt(Math.pow(b, 2) - 4 * a * c)) / (2 * a);
    System.out.println("First solution: " + x);

    // Find second solution.
    x = (-b - Math.sqrt(Math.pow(b, 2) - 4 * a * c)) / (2 * a);
    System.out.println("Second solution: " + x);
  }
}
```

Because **pow( )** and **sqrt( )** are static methods, they must be called through the use of their class' name, **Math**. This results in a somewhat unwieldy expression:

```
x = (-b + Math.sqrt(Math.pow(b, 2) - 4 * a * c)) / (2 * a);
```

Furthermore, having to specify the class name each time **pow( )** or **sqrt( )** (or any of Java's other math methods, such as **sin( )**, **cos( )**, and **tan( )**) are used can become tedious.

You can eliminate the tedium of specifying the class name through the use of static import, as shown in the following version of the preceding program:

```
// Use static import to bring sqrt() and pow() into view.
import static java.lang.Math.sqrt;       Use static import to bring sqrt()
import static java.lang.Math.pow;        and pow() into view.

class Quadratic {
  public static void main(String[] args) {

    // a, b, and c represent the coefficients in the
    // quadratic equation: ax² + bx + c = 0
    double a, b, c, x;

    // Solve 4x² + x - 3 = 0 for x.
    a = 4;
    b = 1;
    c = -3;

    // Find first solution.
    x = (-b + sqrt(pow(b, 2) - 4 * a * c)) / (2 * a);
    System.out.println("First solution: " + x);

    // Find second solution.
    x = (-b - sqrt(pow(b, 2) - 4 * a * c)) / (2 * a);
    System.out.println("Second solution: " + x);
  }
}
```

In this version, the names **sqrt** and **pow** are brought into view by these static import statements:

```
import static java.lang.Math.sqrt;
import static java.lang.Math.pow;
```

After these statements, it is no longer necessary to qualify **sqrt( )** or **pow( )** with its class name. Therefore, the expression can more conveniently be specified, as shown here:

```
x = (-b + sqrt(pow(b, 2) - 4 * a * c)) / (2 * a);
```

As you can see, this form is considerably shorter and easier to read.

There are two general forms of the **import static** statement. The first, which is used by the preceding example, brings into view a single name. Its general form is shown here:

import static *pkg.type-name.static-member-name*;

Here, *type-name* is the name of a class or interface that contains the desired static member. Its full package name is specified by *pkg*. The name of the member is specified by *static-member-name*.

The second form of static import imports all static members. Its general form is shown here:

import static *pkg.type-name*.*;

If you will be using many static methods or fields defined by a class, then this form lets you bring them into view without having to specify each individually. Therefore, the preceding program could have used this single **import** statement to bring both **pow( )** and **sqrt( )** (and *all other* static members of **Math**) into view:

```
import static java.lang.Math.*;
```

Of course, static import is not limited just to the **Math** class or just to methods. For example, this brings the static field **System.out** into view:

```
import static java.lang.System.out;
```

After this statement, you can output to the console without having to qualify **out** with **System**, as shown here:

```
out.println("After importing System.out, you can use out directly.");
```

Whether importing **System.out** as just shown is a good idea is subject to debate. Although it does shorten the statement, it is no longer instantly clear to anyone reading the program that the **out** being referred to is **System.out**.

As convenient as static import can be, it is important not to abuse it. Remember, one reason that Java organizes its libraries into packages is to avoid namespace collisions. When you import static members, you are bringing those members into the current namespace. Thus, you are increasing the potential for namespace conflicts and inadvertent name hiding. If you are using a static member once or twice in the program, it's best not to import it. Also, some static names, such as **System.out**, are so recognizable that you might not want to import them. Static import is designed for those situations in which you are using a static member repeatedly, such as when performing a series of mathematical computations. In essence, you should use, but not abuse, this feature.

Ask the Expert

Q: Using static import, can I import the static members of classes that I create?

A: Yes, you can use static import to import the static members of classes and interfaces you create. Doing so is especially convenient when you define several static members that are used frequently throughout a large program. For example, if a class defines a number of **static final** constants that define various limits, then using static import to bring them into view will save you a lot of tedious typing.

Annotations (Metadata)

Java provides a feature that enables you to embed supplemental information into a source file. This information, called an *annotation,* does not change the actions of a program. However, this information can be used by various tools, during both development and deployment. For example, an annotation might be processed by a source-code generator, by the compiler, or by a deployment tool. The term *metadata* is also used to refer to this feature, but the term annotation is the most descriptive, and more commonly used.

Annotation is a large and sophisticated topic, and it is far beyond the scope of this book to cover it in detail. However, an overview is given here so that you will be familiar with the concept.

NOTE

A more detailed discussion of annotations can be found in *Java: The Complete Reference, Twelfth Edition* (McGraw Hill, 2022).

An annotation is created through a mechanism based on the **interface**. Here is a simple example:

```
// A simple annotation type.
@interface MyAnno {
  String str();
  int val();
}
```

This declares an annotation called **MyAnno**. Notice the @ that precedes the keyword **interface**. This tells the compiler that an annotation type is being declared. Next, notice the two members **str( )** and **val( )**. All annotations consist solely of method declarations. However, you don't provide bodies for these methods. Instead, Java implements these methods. Moreover, the methods act much like fields.

All annotation types automatically extend the **Annotation** interface. Thus, **Annotation** is a super-interface of all annotations. It is declared within the **java.lang.annotation** package.

Originally, annotations were used to annotate only declarations. In this usage, any type of declaration can have an annotation associated with it. For example, classes, methods, fields, parameters, and **enum** constants can be annotated. Even an annotation can be annotated. In such cases, the annotation precedes the rest of the declaration. Beginning with JDK 8, you can also annotate a *type use*, such as a cast or a method return type.

When you apply an annotation, you give values to its members. For example, here is an example of **MyAnno** being applied to a method:

```
// Annotate a method.
@MyAnno(str = "Annotation Example", val = 100)
public static void myMeth() { // ...
```

This annotation is linked with the method **myMeth( )**. Look closely at the annotation syntax. The name of the annotation, preceded by an @, is followed by a parenthesized list of member initializations. To give a member a value, that member's name is assigned a value. Therefore, in the example, the string "Annotation Example" is assigned to the **str** member of **MyAnno**.

Notice that no parentheses follow **str** in this assignment. When an annotation member is given a value, only its name is used. Thus, annotation members look like fields in this context.

Annotations that don't have parameters are called *marker annotations*. These are specified without passing any arguments and without using parentheses. Their sole purpose is to mark an item with some attribute.

Java defines many built-in annotations. Most are specialized, but nine are general purpose. Four are imported from **java.lang.annotation**: **@Retention**, **@Documented**, **@Target**, and **@Inherited**. Five, **@Override**, **@Deprecated**, **@SafeVarargs**, **@FunctionalInterface**, and **@SuppressWarnings**, are included in **java.lang**. These are shown in Table 12-1.

Annotation	Description
@Retention	Specifies the retention policy that will be associated with the annotation. The retention policy determines how long an annotation is present during the compilation and deployment process.
@Documented	A marker annotation that tells a tool that an annotation is to be documented. It is designed to be used only as an annotation to an annotation declaration.
@Target	Specifies the types of items to which an annotation can be applied. It is designed to be used only as an annotation to another annotation. **@Target** takes one argument, which must be a constant or array of constants from the **ElementType** enumeration, which defines various constants, such as **CONSTRUCTOR**, **FIELD**, and **METHOD**. The argument determines the types of items to which the annotation can be applied. If **@Target** is not specified, the annotation can be used on any declaration.
@Inherited	A marker annotation that causes the annotation for a superclass to be inherited by a subclass.
@Override	A method annotated with **@Override** must override a method from a superclass. If it doesn't, a compile-time error will result. It is used to ensure that a superclass method is actually overridden, and not simply overloaded. This is a marker annotation.
@Deprecated	An annotation that indicates that an item is obsolete and not recommended for use. Beginning with JDK 9, **@Deprecated** has been enhanced to enable the Java version in which the deprecation occurred and whether the deprecated element is slated for removal to be specified.
@SafeVarargs	A marker annotation that indicates that no unsafe actions related to a varargs parameter in a method or constructor occur. Can be applied to methods and constructors, with various restrictions.
@SuppressWarnings	Specifies that one or more warnings that might be issued by the compiler are to be suppressed. The warnings to suppress are specified by name, in string form.
@FunctionalInterface	A marker annotation that is used to annotate an interface declaration. It indicates that the annotated interface is a *functional interface*, which is an interface that contains one and only one abstract method. Functional interfaces are used by lambda expressions. (See Chapter 14 for details on functional interfaces.) It is important to understand that **@FunctionalInterface** is purely informational. Any interface with exactly one abstract method is, by definition, a functional interface.

Table 12-1 The General Purpose Built-in Annotations

NOTE

Beginning with JDK 8, **java.lang.annotation** also includes the annotations
@Repeatable and **@Native**. **@Repeatable** supports repeatable annotations,
which are annotations that can be applied more than once to a single item.
@Native is used to annotate a constant field accessed by executable (i.e., native) code.
Both are special-use annotations that are beyond the scope of this book.

Here is an example that uses **@Deprecated** to mark the **MyClass** class and the **getMsg( )**
method. When you try to compile this program, warnings will report the use of these deprecated
elements.

```java
// An example that uses @Deprecated.

// Deprecate a class.
@Deprecated  ←─────────────── Mark a class as deprecated.
class MyClass {
  private String msg;

  MyClass(String m) {
    msg = m;
  }

  // Deprecate a method within a class.
  @Deprecated  ←────────────┐
  String getMsg() {
    return msg;          Mark a method as deprecated.
  }

  // ...
}

class AnnoDemo {
  public static void main(String[] args) {
    MyClass myObj = new MyClass("test");

    System.out.println(myObj.getMsg());
  }
}
```

As a point of interest, over the years several elements in Java's API library have been
deprecated, and additional deprecations may occur as Java continues to evolve. Remember,
although deprecated API elements are still available, they are not recommended for use.
Typically, an alternative to the deprecated API element is offered.

Introducing instanceof

Sometimes it is useful to know the type of an object at run time. For example, you might have one thread of execution that generates various types of objects and another thread that processes these objects. In this situation, it might be useful for the processing thread to know the type of each object when it receives it so that it can take action appropriate to the object's type. Another situation in which knowledge of an object's type at run time is important involves casting. In Java, an invalid cast can cause a run-time error. Although many invalid casts can be caught at compile time, casts involving class hierarchies can produce invalid casts that can only be detected at run time. As you saw in Chapter 7, a superclass reference can refer to a subclass object. Thus, in some cases it may not be possible to know at compile time whether or not a cast is valid when it involves a superclass reference and subclass objects.

For example, consider a superclass called Alpha that has two subclasses, called Beta and Gamma. In this situation, casting a Beta object to Alpha or casting a Gamma object to Alpha is legal since both inherit Alpha. (In other words, because both Beta and Gamma contain an Alpha, a cast to Alpha is valid.) However, casting a Beta object to Gamma (or vice versa) isn't legal because even though each inherits Alpha, they are otherwise unrelated to each other. Furthermore, casting an Alpha object to Beta or Gamma is also illegal, because the Alpha object does not contain the Beta or Gamma portion. Because an Alpha reference can refer to objects of either Alpha, Beta, or Gamma, how can you know at run time what type of object is actually being referred to before attempting to cast the object to Gamma, for example? It could be an object of type Gamma, which would be a legal cast, but it could also be an object of type Alpha or Beta. If it is an Alpha or Beta object, then a cast to Gamma would be illegal and a run-time exception would be thrown. The **instanceof** operator addresses this and other similar situations.

Before we continue, it is necessary to state that **instanceof** was significantly enhanced by JDK 17 with a powerful new feature based on pattern matching. Here, the traditional form of **instanceof** is introduced. The enhanced form is covered in Chapter 16.

The traditional form of the **instanceof** operator has this general form:

objref instanceof *type*

Here, *objref* is a reference to an object and *type* is a class or interface type. If the object referred to by *objref* is of the specified type or can be cast to the specified type, then the **instanceof** operator evaluates to **true**. Otherwise, its result is **false**. (If *objref* is **null**, the result is also **false**.) Thus, **instanceof** is the means by which your program can determine if an object is an *instance of* a specified type at run time.

The following program demonstrates the traditional form of the **instanceof** operator:

```
// Demonstrate the traditional form of the instanceof operator.
class Alpha {
  // ...
}
class Beta extends Alpha {
  // ...
}
class Gamma extends Alpha{
```

```
    // ...
  }

class InstanceOfDemo {
  public static void main(String[] args) {
    Alpha alpha = new Alpha();
    Beta beta = new Beta();
    Gamma gamma = new Gamma();

    // instanceof succeeds when the object is the same
    // type as the specified type.
    if(alpha instanceof Alpha)
      System.out.println("alpha is instance of Alpha");
    if(beta instanceof Beta)
      System.out.println("beta is instance of Beta");
    if(gamma instanceof Gamma)
      System.out.println("gamma is instance of Gamma");

    // instanceof succeeds when the object is an instance
    // of a subclass of the specified type.
    if(beta instanceof Alpha)
      System.out.println("beta is also an instance of Alpha");
    if(gamma instanceof Alpha)
      System.out.println("gamma is also an instance of Alpha");

    // This won't compile because gamma is not an instance of Beta
    // or a subclass of Beta.
//    if(gamma instanceof Beta) System.out.println("Wrong");

    // Now, make an Alpha reference refer to a Beta object.
    alpha = beta;

    // Because alpha refers to a Beta, this if will succeed and
    // alpha can be cast to Beta.
    if(alpha instanceof Beta) {
      System.out.println("alpha can be cast to Beta");
      beta = (Beta) alpha;
    }

    // This instanceof will fail because alpha refers to a Beta
    // object, which cannot be cast to Gamma. Thus, it prevents
    // a runtime error/
    if(alpha instanceof Gamma) {
      // This won't execute.
      gamma = (Gamma) alpha; // error
    }
  }
}
```

The output is shown here:

```
alpha is instance of Alpha
beta is instance of Beta
gamma is instance of Gamma
beta is also an instance of Alpha
gamma is also an instance of Alpha
alpha refers to a Beta object and can be cast to Beta
```

Although most simple programs will not need to use **instanceof** because you know the type of objects with which you are working, it can be very useful when objects of a complex class hierarchy are involved. As you will see, the pattern matching enhancements described in Chapter 16 streamline its use.

Chapter 12 Self Test

1. Enumeration constants are said to be *self-typed*. What does this mean?

2. What class do all enumerations automatically inherit?

3. Given the following enumeration, write a program that uses **values( )** to show a list of the constants and their ordinal values.

```
enum Tools {
   SCREWDRIVER, WRENCH, HAMMER, PLIERS
}
```

4. The traffic light simulation developed in Try This 12-1 can be improved with a few simple changes that take advantage of an enumeration's class features. In the version shown, the duration of each color was controlled by the **TrafficLightSimulator** class by hard-coding these values into the **run( )** method. Change this so that the duration of each color is stored by the constants in the **TrafficLightColor** enumeration. To do this, you will need to add a constructor, a private instance variable, and a method called **getDelay( )**. After making these changes, what improvements do you see? On your own, can you think of other improvements? (Hint: Try using ordinal values to switch light colors rather than relying on a **switch** statement.)

5. Define boxing and unboxing. How does autoboxing/unboxing affect these actions?

6. Change the following fragment so that it uses autoboxing.

```
Double val = Double.valueOf(123.0);
```

7. In your own words, what does static import do?

8. What does this statement do?

```
import static java.lang.Integer.parseInt;
```

9. Is static import designed for special-case situations, or is it good practice to bring all static members of all classes into view?

10. An annotation is syntactically based on a/an _____ .

11. What is a marker annotation?

12. An annotation can be applied only to methods. True or False?

13. What operator determines if an object is of a specified type?

14. Will an invalid cast that occurs at run time result in an exception?

Chapter 13

Generics

Key Skills & Concepts

- Understand the benefits of generics

- Create a generic class

- Apply bounded type parameters

- Use wildcard arguments

- Apply bounded wildcards

- Create a generic method

- Create a generic constructor

- Create a generic interface

- Utilize raw types

- Apply type inference with the diamond operator

- Understand erasure

- Avoid ambiguity errors

- Know generics restrictions

Since its original 1.0 version, many new features have been added to Java. All have enhanced and expanded the scope of the language, but one that has had an especially profound and far-reaching impact is *generics* because its effects were felt throughout the entire Java language. For example, generics added a completely new syntax element and caused changes to many of the classes and methods in the core API. It is not an overstatement to say that the inclusion of generics fundamentally reshaped the character of Java.

The topic of generics is quite large, and some of it is sufficiently advanced to be beyond the scope of this book. However, a basic understanding of generics is necessary for all Java programmers. At first glance, the generics syntax may look a bit intimidating, but don't worry. Generics are surprisingly simple to use. By the time you finish this chapter, you will have a grasp of the key concepts that underlie generics and sufficient knowledge to use generics effectively in your own programs.

Generics Fundamentals

At its core, the term *generics* means *parameterized types*. Parameterized types are important because they enable you to create classes, interfaces, and methods in which the type of data

Ask the Expert

Q: I have heard that Java's generics are similar to templates in C++. Is this the case?

A: Java generics are similar to templates in C++. What Java calls a parameterized type, C++ calls a template. However, Java generics and C++ templates are not the same, and there are some fundamental differences between the two approaches to generic types. For the most part, Java's approach is simpler to use.

A word of warning: If you have a background in C++, it is important not to jump to conclusions about how generics work in Java. The two approaches to generic code differ in subtle but fundamental ways.

upon which they operate is specified as a parameter. A class, interface, or method that operates on a type parameter is called *generic,* as in *generic class* or *generic method.*

A principal advantage of generic code is that it will automatically work with the type of data passed to its type parameter. Many algorithms are logically the same no matter what type of data they are being applied to. For example, a Quicksort is the same whether it is sorting items of type **Integer**, **String**, **Object**, or **Thread**. With generics, you can define an algorithm once, independently of any specific type of data, and then apply that algorithm to a wide variety of data types without any additional effort.

It is important to understand that Java has always given you the ability to create generalized classes, interfaces, and methods by operating through references of type **Object**. Because **Object** is the superclass of all other classes, an **Object** reference can refer to any type of object. Thus, in pre-generics code, generalized classes, interfaces, and methods used **Object** references to operate on various types of data. The problem was that they could not do so with *type safety* because casts were needed to explicitly convert from **Object** to the actual type of data being operated upon. Thus, it was possible to accidentally create type mismatches. Generics add the type safety that was lacking because they make these casts automatic and implicit. In short, generics expand your ability to reuse code and let you do so safely and reliably.

A Simple Generics Example

Before discussing any more theory, it's best to look at a simple generics example. The following program defines two classes. The first is the generic class **Gen**, and the second is **GenDemo**, which uses **Gen**.

```
// A simple generic class.
// Here, T is a type parameter that
// will be replaced by a real type
// when an object of type Gen is created.
class Gen<T> {                                          Declare a generic class. T is the
    T ob; // declare an object of type T                generic type parameter.
```

```
   // Pass the constructor a reference to
   // an object of type T.
   Gen(T o) {
     ob = o;
   }

   // Return ob.
   T getOb() {
     return ob;
   }

   // Show type of T.
   void showType() {
     System.out.println("Type of T is " +
                         ob.getClass().getName());

   }
 }

// Demonstrate the generic class.
class GenDemo {
  public static void main(String[] args) {
    // Create a Gen reference for Integers.
    Gen<Integer> iOb;

    // Create a Gen<Integer> object and assign its
    // reference to iOb. Notice the use of autoboxing
    // to encapsulate the value 88 within an Integer object.
    iOb = new Gen<Integer>(88);

    // Show the type of data used by iOb.
    iOb.showType();

    // Get the value in iOb. Notice that
    // no cast is needed.
    int v = iOb.getOb();
    System.out.println("value: " + v);

    System.out.println();

    // Create a Gen object for Strings.
    Gen<String> strOb = new Gen<String>("Generics Test");

    // Show the type of data used by strOb.
    strOb.showType();
```

Create a reference to an object of type **Gen\<Integer\>**.

Instantiate an object of type **Gen\<Integer\>**.

Create a reference and an object of type **Gen\<String\>**.

```
    // Get the value of strOb. Again, notice
    // that no cast is needed.
    String str = strOb.getOb();
    System.out.println("value: " + str);
  }
}
```

The output produced by the program is shown here:

```
Type of T is java.lang.Integer
value: 88

Type of T is java.lang.String
value: Generics Test
```

Let's examine this program carefully. First, notice how **Gen** is declared by the following line:

```
class Gen<T> {
```

Here, **T** is the name of a *type parameter*. This name is used as a placeholder for the actual type that will be passed to **Gen** when an object is created. Thus, **T** is used within **Gen** whenever the type parameter is needed. Notice that **T** is contained within **<>**. This syntax can be generalized. Whenever a type parameter is being declared, it is specified within angle brackets. Because **Gen** uses a type parameter, **Gen** is a generic class.

In the declaration of **Gen**, there is no special significance to the name **T**. Any valid identifier could have been used, but **T** is traditional. Furthermore, it is recommended that type parameter names be single-character, capital letters. Other commonly used type parameter names are **V** and **E**. One other point about type parameter names: Beginning with JDK 10, you cannot use **var** as the name of a type parameter.

Next, **T** is used to declare an object called **ob**, as shown here:

```
T ob; // declare an object of type T
```

As explained, **T** is a placeholder for the actual type that will be specified when a **Gen** object is created. Thus, **ob** will be an object of the type passed to **T**. For example, if type **String** is passed to **T**, then in that instance, **ob** will be of type **String**.

Now consider **Gen**'s constructor:

```
Gen(T o) {
  ob = o;
}
```

Notice that its parameter, **o**, is of type **T**. This means that the actual type of **o** is determined by the type passed to **T** when a **Gen** object is created. Also, because both the parameter **o** and the member variable **ob** are of type **T**, they will both be of the same actual type when a **Gen** object is created.

The type parameter **T** can also be used to specify the return type of a method, as is the case with the **getOb( )** method, shown here:

```
T getOb() {
  return ob;
}
```

Because **ob** is also of type **T**, its type is compatible with the return type specified by **getOb( )**.

The **showType( )** method displays the type of **T**. It does this by calling **getName( )** on the **Class** object returned by the call to **getClass( )** on **ob**. We haven't used this feature before, so let's examine it closely. As you should recall from Chapter 7, the **Object** class defines the method **getClass( )**. Thus, **getClass( )** is a member of all class types. It returns a **Class** object that corresponds to the class type of the object on which it is called. **Class** is a class defined within **java.lang** that encapsulates information about a class. **Class** defines several methods that can be used to obtain information about a class at run time. Among these is the **getName( )** method, which returns a string representation of the class name.

The **GenDemo** class demonstrates the generic **Gen** class. It first creates a version of **Gen** for integers, as shown here:

```
Gen<Integer> iOb;
```

Look carefully at this declaration. First, notice that the type **Integer** is specified within the angle brackets after **Gen**. In this case, **Integer** is a *type argument* that is passed to **Gen**'s type parameter, **T**. This effectively creates a version of **Gen** in which all references to **T** are translated into references to **Integer**. Thus, for this declaration, **ob** is of type **Integer**, and the return type of **getOb( )** is of type **Integer**.

Before moving on, it's necessary to state that the Java compiler does not actually create different versions of **Gen**, or of any other generic class. Although it's helpful to think in these terms, it is not what actually happens. Instead, the compiler removes all generic type information, substituting the necessary casts, to make your code *behave as if* a specific version of **Gen** was created. Thus, there is really only one version of **Gen** that actually exists in your program. The process of removing generic type information is called *erasure,* which is discussed later in this chapter.

The next line assigns to **iOb** a reference to an instance of an **Integer** version of the **Gen** class.

```
iOb = new Gen<Integer>(88);
```

Notice that when the **Gen** constructor is called, the type argument **Integer** is also specified. This is because the type of the object (in this case **iOb**) to which the reference is being assigned is of type **Gen<Integer>**. Thus, the reference returned by **new** must also be of type **Gen<Integer>**. If it isn't, a compile-time error will result. For example, the following assignment will cause a compile-time error:

```
iOb = new Gen<Double>(88.0); // Error!
```

Because **iOb** is of type **Gen<Integer>**, it can't be used to refer to an object of **Gen<Double>**. This type of checking is one of the main benefits of generics because it ensures type safety.

As the comments in the program state, the assignment

```
iOb = new Gen<Integer>(88);
```

makes use of autoboxing to encapsulate the value 88, which is an **int**, into an **Integer**. This works because **Gen<Integer>** creates a constructor that takes an **Integer** argument. Because an **Integer** is expected, Java will automatically box 88 inside one. Of course, the assignment could also have been written explicitly, like this:

```
iOb = new Gen<Integer>(Integer.valueOf(88));
```

However, there would be no benefit to using this version.

The program then displays the type of **ob** within **iOb**, which is **Integer**. Next, the program obtains the value of **ob** by use of the following line:

```
int v = iOb.getOb();
```

Because the return type of **getOb( )** is **T**, which was replaced by **Integer** when **iOb** was declared, the return type of **getOb( )** is also **Integer**, which auto-unboxes into **int** when assigned to **v** (which is an **int**). Thus, there is no need to cast the return type of **getOb( )** to **Integer**.

Next, **GenDemo** declares an object of type **Gen<String>**:

```
Gen<String> strOb = new Gen<String>("Generics Test");
```

Because the type argument is **String**, **String** is substituted for **T** inside **Gen**. This creates (conceptually) a **String** version of **Gen**, as the remaining lines in the program demonstrate.

Generics Work Only with Reference Types

When declaring an instance of a generic type, the type argument passed to the type parameter must be a reference type. You cannot use a primitive type, such as **int** or **char**. For example, with **Gen**, it is possible to pass any class type to **T**, but you cannot pass a primitive type to **T**. Therefore, the following declaration is illegal:

```
Gen<int> intOb = new Gen<int>(53); // Error, can't use primitive type
```

Of course, not being able to specify a primitive type is not a serious restriction because you can use the type wrappers (as the preceding example did) to encapsulate a primitive type. Further, Java's autoboxing and auto-unboxing mechanism makes the use of the type wrapper transparent.

Generic Types Differ Based on Their Type Arguments

A key point to understand about generic types is that a reference of one specific version of a generic type is not type-compatible with another version of the same generic type. For example, assuming the program just shown, the following line of code is in error and will not compile:

```
iOb = strOb; // Wrong!
```

Even though both **iOb** and **strOb** are of type **Gen<T>**, they are references to different types because their type arguments differ. This is part of the way that generics add type safety and prevent errors.

A Generic Class with Two Type Parameters

You can declare more than one type parameter in a generic type. To specify two or more type parameters, simply use a comma-separated list. For example, the following **TwoGen** class is a variation of the **Gen** class that has two type parameters:

```
// A simple generic class with two type
// parameters: T and V.
class TwoGen<T, V> {          ◄──────────── Use two type parameters.
  T ob1;
  V ob2;

  // Pass the constructor references to
  // objects of type T and V.
  TwoGen(T o1, V o2) {
    ob1 = o1;
    ob2 = o2;
  }

  // Show types of T and V.
  void showTypes() {
    System.out.println("Type of T is " +
                       ob1.getClass().getName());

    System.out.println("Type of V is " +
                       ob2.getClass().getName());
  }

  T getOb1() {
    return ob1;
  }

  V getOb2() {
    return ob2;
  }
}

// Demonstrate TwoGen.                    Here, Integer is passed to T,
class SimpGen {                           and String is passed to V.
  public static void main(String[] args) {

    TwoGen<Integer, String> tgObj = ◄─────────┘
      new TwoGen<Integer, String>(88, "Generics");
```

```
    // Show the types.
    tgObj.showTypes();

    // Obtain and show values.
    int v = tgObj.getOb1();
    System.out.println("value: " + v);

    String str = tgObj.getOb2();
    System.out.println("value: " + str);
  }
}
```

The output from this program is shown here:

```
Type of T is java.lang.Integer
Type of V is java.lang.String
value: 88
value: Generics
```

Notice how **TwoGen** is declared:

```
class TwoGen<T, V> {
```

It specifies two type parameters, **T** and **V**, separated by a comma. Because it has two type parameters, two type arguments must be passed to **TwoGen** when an object is created, as shown next:

```
TwoGen<Integer, String> tgObj =
  new TwoGen<Integer, String>(88, "Generics");
```

In this case, **Integer** is substituted for **T**, and **String** is substituted for **V**. Although the two type arguments differ in this example, it is possible for both types to be the same. For example, the following line of code is valid:

```
TwoGen<String, String> x = new TwoGen<String, String>("A", "B");
```

In this case, both **T** and **V** would be of type **String**. Of course, if the type arguments were always the same, then two type parameters would be unnecessary.

The General Form of a Generic Class

The generics syntax shown in the preceding examples can be generalized. Here is the syntax for declaring a generic class:

class *class-name<type-param-list>* { // ...

Here is the full syntax for declaring a reference to a generic class and creating a generic instance:

class-name<type-arg-list> var-name =
    new *class-name<type-arg-list>(cons-arg-list)*;

Bounded Types

In the preceding examples, the type parameters could be replaced by any class type. This is fine for many purposes, but sometimes it is useful to limit the types that can be passed to a type parameter. For example, assume that you want to create a generic class that stores a numeric value and is capable of performing various mathematical functions, such as computing the reciprocal or obtaining the fractional component. Furthermore, you want to use the class to compute these quantities for any type of number, including integers, **float**s, and **double**s. Thus, you want to specify the type of the numbers generically, using a type parameter. To create such a class, you might try something like this:

```
// NumericFns attempts (unsuccessfully) to create
// a generic class that can compute various
// numeric functions, such as the reciprocal or the
// fractional component, given any type of number.
class NumericFns<T> {
  T num;

  // Pass the constructor a reference to
  // a numeric object.
  NumericFns(T n) {
    num = n;
  }

  // Return the reciprocal.
  double reciprocal() {
    return 1 / num.doubleValue(); // Error!
  }

  // Return the fractional component.
  double fraction() {
    return num.doubleValue() - num.intValue(); // Error!
  }

  // ...
}
```

Unfortunately, **NumericFns** will not compile as written because both methods will generate compile-time errors. First, examine the **reciprocal( )** method, which attempts to return the reciprocal of **num**. To do this, it must divide 1 by the value of **num**. The value of **num** is obtained by calling **doubleValue( )**, which obtains the **double** version of the numeric object stored in **num**. Because all numeric classes, such as **Integer** and **Double**, are subclasses of **Number**, and **Number** defines the **doubleValue( )** method, this method is available to all numeric wrapper classes. The trouble is that the compiler has no way to know that you are intending to create **NumericFns** objects using only numeric types. Thus, when you try to compile **NumericFns**, an error is reported that indicates that the **doubleValue( )** method is unknown. The same type of error occurs twice in **fraction( )**, which needs to call both

doubleValue( ) and **intValue( )**. Both calls result in error messages stating that these methods are unknown. To solve this problem, you need some way to tell the compiler that you intend to pass only numeric types to **T**. Furthermore, you need some way to *ensure* that *only* numeric types are actually passed.

To handle such situations, Java provides *bounded types*. When specifying a type parameter, you can create an upper bound that declares the superclass from which all type arguments must be derived. This is accomplished through the use of an **extends** clause when specifying the type parameter, as shown here:

<*T* extends *superclass*>

This specifies that *T* can be replaced only by *superclass,* or subclasses of *superclass.* Thus, *superclass* defines an inclusive, upper limit.

You can use an upper bound to fix the **NumericFns** class shown earlier by specifying **Number** as an upper bound, as shown here:

```
// In this version of NumericFns, the type argument
// for T must be either Number, or a class derived
// from Number.
class NumericFns<T extends Number> {
  T num;

  // Pass the constructor a reference to
  // a numeric object.
  NumericFns(T n) {
    num = n;
  }

  // Return the reciprocal.
  double reciprocal() {
    return 1 / num.doubleValue();
  }

  // Return the fractional component.
  double fraction() {
    return num.doubleValue() - num.intValue();
  }

  // ...
}

// Demonstrate NumericFns.
class BoundsDemo {
  public static void main(String[] args) {

    NumericFns<Integer> iOb =
                new NumericFns<Integer>(5);
```

In this case, the type argument must be either **Number** or a subclass of **Number**.

Integer is OK because it is a subclass of **Number**.

```
        System.out.println("Reciprocal of iOb is " +
                           iOb.reciprocal());
        System.out.println("Fractional component of iOb is " +
                           iOb.fraction());

        System.out.println();

        NumericFns<Double> dOb =                          Double is also OK.
                    new NumericFns<Double>(5.25);

        System.out.println("Reciprocal of dOb is " +
                           dOb.reciprocal());
        System.out.println("Fractional component of dOb is " +
                           dOb.fraction());


        // This won't compile because String is not a
        // subclass of Number.
  //  NumericFns<String> strOb = new NumericFns<String>("Error");
    }
}
```

String is illegal because it is not a subclass of **Number**.

The output is shown here:

```
Reciprocal of iOb is 0.2
Fractional component of iOb is 0.0

Reciprocal of dOb is 0.19047619047619047
Fractional component of dOb is 0.25
```

Notice how **NumericFns** is now declared by this line:

```
class NumericFns<T extends Number> {
```

Because the type **T** is now bounded by **Number**, the Java compiler knows that all objects of type **T** can call **doubleValue( )** because it is a method declared by **Number**. This is, by itself, a major advantage. However, as an added bonus, the bounding of **T** also prevents nonnumeric **NumericFns** objects from being created. For example, if you remove the comments from the line at the end of the program, and then try re-compiling, you will receive compile-time errors because **String** is not a subclass of **Number**.

Bounded types are especially useful when you need to ensure that one type parameter is compatible with another. For example, consider the following class called **Pair**, which stores two objects that must be compatible with each other:

```
class Pair<T, V extends T> {              Here, V must be either the same
  T first;                                type as T, or a subclass of T.
  V second;
```

```
Pair(T a, V b) {
    first = a;
    second = b;
}

// ...
}
```

Notice that **Pair** uses two type parameters, **T** and **V**, and that **V** extends **T**. This means that **V** will either be the same as **T** or a subclass of **T**. This ensures that the two arguments to **Pair**'s constructor will be objects of the same type or of related types. For example, the following constructions are valid:

```
// This is OK because both T and V are Integer.
Pair<Integer, Integer> x = new Pair<Integer, Integer>(1, 2);

// This is OK because Integer is a subclass of Number.
Pair<Number, Integer> y = new Pair<Number, Integer>(10.4, 12);
```

However, the following is invalid:

```
// This causes an error because String is not
// a subclass of Number
Pair<Number, String> z = new Pair<Number, String>(10.4, "12");
```

In this case, **String** is not a subclass of **Number**, which violates the bound specified by **Pair**.

Using Wildcard Arguments

As useful as type safety is, sometimes it can get in the way of perfectly acceptable constructs. For example, given the **NumericFns** class shown at the end of the preceding section, assume that you want to add a method called **absEqual( )** that returns true if two **NumericFns** objects contain numbers whose absolute values are the same. Furthermore, you want this method to be able to work properly no matter what type of number each object holds. For example, if one object contains the **Double** value 1.25 and the other object contains the **Float** value −1.25, then **absEqual( )** would return true. One way to implement **absEqual( )** is to pass it a **NumericFns** argument, and then compare the absolute value of that argument against the absolute value of the invoking object, returning true only if the values are the same. For example, you want to be able to call **absEqual( )**, as shown here:

```
NumericFns<Double> dOb = new NumericFns<Double>(1.25);
NumericFns<Float> fOb = new NumericFns<Float>(-1.25);

if(dOb.absEqual(fOb))
    System.out.println("Absolute values are the same.");
else
    System.out.println("Absolute values differ.");
```

At first, creating **absEqual( )** seems like an easy task. Unfortunately, trouble starts as soon as you try to declare a parameter of type **NumericFns**. What type do you specify for **NumericFns'** type parameter? At first, you might think of a solution like this, in which **T** is used as the type parameter:

```
// This won't work!
// Determine if the absolute values of two objects are the same.
boolean absEqual(NumericFns<T> ob) {
  if(Math.abs(num.doubleValue()) ==
      Math.abs(ob.num.doubleValue()) return true;

  return false;
}
```

Here, the standard method **Math.abs( )** is used to obtain the absolute value of each number, and then the values are compared. The trouble with this attempt is that it will work only with other **NumericFns** objects whose type is the same as the invoking object. For example, if the invoking object is of type **NumericFns<Integer>**, then the parameter **ob** must also be of type **NumericFns<Integer>**. It can't be used to compare an object of type **NumericFns<Double>**, for example. Therefore, this approach does not yield a general (i.e., generic) solution.

To create a generic **absEqual( )** method, you must use another feature of Java generics: the *wildcard argument*. The wildcard argument is specified by the **?**, and it represents an unknown type. Using a wildcard, here is one way to write the **absEqual( )** method:

```
// Determine if the absolute values of two
// objects are the same.
boolean absEqual(NumericFns<?> ob) {          Notice the wildcard.
  if(Math.abs(num.doubleValue()) ==
      Math.abs(ob.num.doubleValue())) return true;

  return false;
}
```

Here, **NumericFns<?>** matches any type of **NumericFns** object, allowing any two **NumericFns** objects to have their absolute values compared. The following program demonstrates this:

```
// Use a wildcard.
class NumericFns<T extends Number> {
  T num;

  // Pass the constructor a reference to
  // a numeric object.
  NumericFns(T n) {
    num = n;
  }

  // Return the reciprocal.
  double reciprocal() {
```

```
    return 1 / num.doubleValue();
  }

  // Return the fractional component.
  double fraction() {
    return num.doubleValue() - num.intValue();
  }

  // Determine if the absolute values of two
  // objects are the same.
  boolean absEqual(NumericFns<?> ob) {
    if(Math.abs(num.doubleValue()) ==
        Math.abs(ob.num.doubleValue())) return true;

    return false;
  }

  // ...
}

// Demonstrate a wildcard.
class WildcardDemo {
  public static void main(String[] args) {

    NumericFns<Integer> iOb =
                     new NumericFns<Integer>(6);

    NumericFns<Double> dOb =
                     new NumericFns<Double>(-6.0);

    NumericFns<Long> lOb =
                     new NumericFns<Long>(5L);

    System.out.println("Testing iOb and dOb.");
    if(iOb.absEqual(dOb))
      System.out.println("Absolute values are equal.");
    else
      System.out.println("Absolute values differ.");

    System.out.println();

    System.out.println("Testing iOb and lOb.");
    if(iOb.absEqual(lOb))
      System.out.println("Absolute values are equal.");
    else
      System.out.println("Absolute values differ.");

  }
}
```

In this call, the wildcard type matches **Double**.

In this call, the wildcard matches **Long**.

The output is shown here:

```
Testing iOb and dOb.
Absolute values are equal.

Testing iOb and lOb.
Absolute values differ.
```

In the program, notice these two calls to **absEqual( )**:

```
if(iOb.absEqual(dOb))

if(iOb.absEqual(lOb))
```

In the first call, **iOb** is an object of type **NumericFns<Integer>** and **dOb** is an object of type **NumericFns<Double>**. However, through the use of a wildcard, it is possible for **iOb** to pass **dOb** in the call to **absEqual( )**. The same applies to the second call, in which an object of type **NumericFns<Long>** is passed.

One last point: It is important to understand that the wildcard does not affect what type of **NumericFns** objects can be created. This is governed by the **extends** clause in the **NumericFns** declaration. The wildcard simply matches any *valid* **NumericFns** object.

Bounded Wildcards

Wildcard arguments can be bounded in much the same way that a type parameter can be bounded. A bounded wildcard is especially important when you are creating a method that is designed to operate only on objects that are subclasses of a specific superclass. To understand why, let's work through a simple example. Consider the following set of classes:

```
class A {
  // ...
}

class B extends A {
  // ...
}

class C extends A {
  // ...
}

// Note that D does NOT extend A.
class D {
  // ...
}
```

Here, class **A** is extended by classes **B** and **C**, but not by **D**.

Next, consider the following very simple generic class:

```
// A simple generic class.
class Gen<T> {
  T ob;

  Gen(T o) {
    ob = o;
  }
}
```

Gen takes one type parameter, which specifies the type of object stored in **ob**. Because **T** is unbounded, the type of **T** is unrestricted. That is, **T** can be of any class type.

Now, suppose that you want to create a method that takes as an argument any type of **Gen** object so long as its type parameter is **A** or a subclass of **A**. In other words, you want to create a method that operates only on objects of **Gen<***type***>**, where *type* is either **A** or a subclass of **A**. To accomplish this, you must use a bounded wildcard. For example, here is a method called **test( )** that accepts as an argument only **Gen** objects whose type parameter is **A** or a subclass of **A**:

```
// Here, the ? will match A or any class type
// that extends A.
static void test(Gen<? extends A> o) {
  // ...
}
```

The following class demonstrates the types of **Gen** objects that can be passed to **test( )**.

```
class UseBoundedWildcard {
  // Here, the ? will match A or any class type
  // that extends A
  static void test(Gen<? extends A> o) {  ◄──────────  Use a bounded wildcard.
    // ...
  }

  public static void main(String[] args) {
    A a = new A();
    B b = new B();
    C c = new C();
    D d = new D();

    Gen<A> w = new Gen<A>(a);
    Gen<B> w2 = new Gen<B>(b);
    Gen<C> w3 = new Gen<C>(c);
    Gen<D> w4 = new Gen<D>(d);
```

```
            // These calls to test() are OK.
            test(w);
            test(w2);                      These are legal because w, w2, and w3 are subclasses of A.
            test(w3);

            // Can't call test() with w4 because
            // it is not an object of a class that
            // inherits A.
//          test(w4); // Error!                This is illegal because w4 is not a subclass of A.
        }
    }
```

In **main( )**, objects of type **A**, **B**, **C**, and **D** are created. These are then used to create four **Gen** objects, one for each type. Finally, four calls to **test( )** are made, with the last call commented out. The first three calls are valid because **w**, **w2**, and **w3** are **Gen** objects whose type is either **A** or a subclass of **A**. However, the last call to **test( )** is illegal because **w4** is an object of type **D**, which is not derived from **A**. Thus, the bounded wildcard in **test( )** will not accept **w4** as an argument.

In general, to establish an upper bound for a wildcard, use the following type of wildcard expression:

<? extends *superclass*>

Ask the Expert

Q: Can I cast one instance of a generic class into another?

A: Yes, you can cast one instance of a generic class into another, but only if the two are otherwise compatible and their type arguments are the same. For example, assume a generic class called **Gen** that is declared like this:

```
class Gen<T> { // ...
```

Next, assume that **x** is declared as shown here:

```
Gen<Integer> x = new Gen<Integer>();
```

Then, this cast is legal

```
(Gen<Integer>) x // legal
```

because **x** is an instance of **Gen<Integer>**. But, this cast

```
(Gen<Long>) x // illegal
```

is not legal because **x** is not an instance of **Gen<Long>**.

where *superclass* is the name of the class that serves as the upper bound. Remember, this is an inclusive clause because the class forming the upper bound (specified by *superclass*) is also within bounds.

You can also specify a lower bound for a wildcard by adding a **super** clause to a wildcard declaration. Here is its general form:

<? super *subclass*>

In this case, only classes that are superclasses of *subclass* are acceptable arguments. This is an inclusive clause.

Generic Methods

As the preceding examples have shown, methods inside a generic class can make use of a class' type parameter and are, therefore, automatically generic relative to the type parameter. However, it is possible to declare a generic method that uses one or more type parameters of its own. Furthermore, it is possible to create a generic method that is enclosed within a nongeneric class.

The following program declares a nongeneric class called **GenericMethodDemo** and a static generic method within that class called **arraysEqual( )**. This method determines if two arrays contain the same elements, in the same order. It can be used to compare any two arrays as long as the arrays are of the same or compatible types and the array elements are, themselves, comparable.

```
// Demonstrate a simple generic method.
class GenericMethodDemo {

  // Determine if the contents of two arrays are the same.
  static <T extends Comparable<T>, V extends T> boolean
    arraysEqual(T[] x, V[] y) {          A generic method.
    // If array lengths differ, then the arrays differ.
    if(x.length != y.length) return false;

    for(int i=0; i < x.length; i++)
      if(!x[i].equals(y[i])) return false; // arrays differ

    return true; // contents of arrays are equivalent
  }

  public static void main(String[] args) {

    Integer[] nums = { 1, 2, 3, 4, 5 };        The type arguments for T and V
    Integer[] nums2 = { 1, 2, 3, 4, 5 };       are implicitly determined when
    Integer[] nums3 = { 1, 2, 7, 4, 5 };       the method is called.
    Integer[] nums4 = { 1, 2, 7, 4, 5, 6 };

    if(arraysEqual(nums, nums))
      System.out.println("nums equals nums");

    if(arraysEqual(nums, nums2))
      System.out.println("nums equals nums2");
```

```
      if(arraysEqual(nums, nums3))
        System.out.println("nums equals nums3");

      if(arraysEqual(nums, nums4))
        System.out.println("nums equals nums4");

      // Create an array of Doubles
      Double[] dvals = { 1.1, 2.2, 3.3, 4.4, 5.5 };

      // This won't compile because nums and dvals
      // are not of the same type.
//      if(arraysEqual(nums, dvals))
//        System.out.println("nums equals dvals");
    }
}
```

The output from the program is shown here:

```
nums equals nums
nums equals nums2
```

Let's examine **arraysEqual( )** closely. First, notice how it is declared by this line:

```
static <T extends Comparable<T>, V extends T> boolean arraysEqual(T[] x, V[] y) {
```

The type parameters are declared *before* the return type of the method. Also note that **T** extends **Comparable<T>**. **Comparable** is an interface declared in **java.lang**. A class that implements **Comparable** defines objects that can be ordered. Thus, requiring an upper bound of **Comparable** ensures that **arraysEqual( )** can be used only with objects that are capable of being compared. **Comparable** is generic, and its type parameter specifies the type of objects that it compares. (Shortly, you will see how to create a generic interface.) Next, notice that the type **V** is upper-bounded by **T**. Thus, **V** must be either the same as type **T** or a subclass of **T**. This relationship enforces that **arraysEqual( )** can be called only with arguments that are comparable with each other. Also notice that **arraysEqual( )** is static, enabling it to be called independently of any object. Understand, though, that generic methods can be either static or nonstatic. There is no restriction in this regard.

Now, notice how **arraysEqual( )** is called within **main( )** by use of the normal call syntax, without the need to specify type arguments. This is because the types of the arguments are automatically discerned, and the types of **T** and **V** are adjusted accordingly. For example, in the first call:

```
if(arraysEqual(nums, nums))
```

the element type of the first argument is **Integer**, which causes **Integer** to be substituted for **T**. The element type of the second argument is also **Integer**, which makes **Integer** a substitute for **V**, too. Thus, the call to **arraysEqual( )** is legal, and the two arrays can be compared.

Now, notice the commented-out code, shown here:

```
//    if(arraysEqual(nums, dvals))
//      System.out.println("nums equals dvals");
```

If you remove the comments and then try to compile the program, you will receive an error. The reason is that the type parameter **V** is bounded by **T** in the **extends** clause in **V**'s declaration. This means that **V** must be either type **T** or a subclass of **T**. In this case, the first argument is of type **Integer**, making **T** into **Integer**, but the second argument is of type **Double**, which is not a subclass of **Integer**. This makes the call to **arraysEqual( )** illegal, and results in a compile-time type-mismatch error.

The syntax used to create **arraysEqual( )** can be generalized. Here is the syntax for a generic method:

<type-param-list> ret-type meth-name(param-list) { // ...

In all cases, *type-param-list* is a comma-separated list of type parameters. Notice that for a generic method, the type parameter list precedes the return type.

Generic Constructors

A constructor can be generic, even if its class is not. For example, in the following program, the class **Summation** is not generic, but its constructor is.

```
// Use a generic constructor.
class Summation {
  private int sum;

  <T extends Number> Summation(T arg) {  ◀──────── A generic constructor
    sum = 0;

    for(int i=0; i <= arg.intValue(); i++)
      sum += i;
  }

  int getSum() {
    return sum;
  }
}

class GenConsDemo {
  public static void main(String[] args) {
    Summation ob = new Summation(4.0);

    System.out.println("Summation of 4.0 is " +
                        ob.getSum());
  }
}
```

The **Summation** class computes and encapsulates the summation of the numeric value passed to its constructor. Recall that the summation of *N* is the sum of all the whole numbers between

0 and *N*. Because **Summation( )** specifies a type parameter that is bounded by **Number**, a **Summation** object can be constructed using any numeric type, including **Integer**, **Float**, or **Double**. No matter what numeric type is used, its value is converted to **Integer** by calling **intValue( )**, and the summation is computed. Therefore, it is not necessary for the class **Summation** to be generic; only a generic constructor is needed.

Generic Interfaces

As you saw in the **GenericMethodDemo** program presented earlier, an interface can be generic. In that example, the standard interface **Comparable<T>** was used to ensure that elements of two arrays could be compared. Of course, you can also define your own generic interface. Generic interfaces are specified just like generic classes. Here is an example. It creates an interface called **Containment**, which can be implemented by classes that store one or more values. It declares a method called **contains( )** that determines if a specified value is contained by the invoking object.

```
// A generic interface example.

// A generic containment interface.
// This interface implies that an implementing
// class contains one or more values.
interface Containment<T> {                          ◄——————————————— A generic interface
  // The contains() method tests if a
  // specific item is contained within
  // an object that implements Containment.
  boolean contains(T o);
}

// Implement Containment using an array to
// hold the values.
class MyClass<T> implements Containment<T> {        ◄——————— Any class that implements
  T[] arrayRef;                                             a generic interface must
                                                            itself be generic.
  MyClass(T[] o) {
    arrayRef = o;
  }

  // Implement contains()
  public boolean contains(T o) {
    for(T x : arrayRef)
      if(x.equals(o)) return true;
    return false;
  }
}

class GenIFDemo {
  public static void main(String[] args) {
    Integer[] x = { 1, 2, 3 };

    MyClass<Integer> ob = new MyClass<Integer>(x);
```

```
    if(ob.contains(2))
      System.out.println("2 is in ob");
    else
      System.out.println("2 is NOT in ob");

    if(ob.contains(5))
      System.out.println("5 is in ob");
    else
      System.out.println("5 is NOT in ob");

    // The following is illegal because ob
    // is an Integer Containment and 9.25 is
    // a Double value.
//    if(ob.contains(9.25)) // Illegal!
//      System.out.println("9.25 is in ob");
  }
}
```

The output is shown here:

```
2 is in ob
5 is NOT in ob
```

Although most aspects of this program should be easy to understand, a couple of key points need to be made. First, notice that **Containment** is declared like this:

```
interface Containment<T> {
```

In general, a generic interface is declared in the same way as a generic class. In this case, the type parameter **T** specifies the type of objects that are contained.

Next, **Containment** is implemented by **MyClass**. Notice the declaration of **MyClass**, shown here:

```
class MyClass<T> implements Containment<T> {
```

In general, if a class implements a generic interface, then that class must also be generic, at least to the extent that it takes a type parameter that is passed to the interface. For example, the following attempt to declare **MyClass** is in error:

```
class MyClass implements Containment<T> { // Wrong!
```

This declaration is wrong because **MyClass** does not declare a type parameter, which means that there is no way to pass one to **Containment**. In this case, the identifier **T** is simply unknown and the compiler reports an error. Of course, if a class implements a *specific type* of generic interface, such as shown here:

```
class MyClass implements Containment<Double> { // OK
```

then the implementing class does not need to be generic.

As you might expect, the type parameter(s) specified by a generic interface can be bounded. This lets you limit the type of data for which the interface can be implemented. For example, if you wanted to limit **Containment** to numeric types, then you could declare it like this:

```
interface Containment<T extends Number> {
```

Now, any implementing class must pass to **Containment** a type argument also having the same bound. For example, now **MyClass** must be declared as shown here:

```
class MyClass<T extends Number> implements Containment<T> {
```

Pay special attention to the way the type parameter **T** is declared by **MyClass** and then passed to **Containment**. Because **Containment** now requires a type that extends **Number**, the implementing class (**MyClass** in this case) must specify the same bound. Furthermore, once this bound has been established, there is no need to specify it again in the **implements** clause. In fact, it would be wrong to do so. For example, this declaration is incorrect and won't compile:

```
// This is wrong!
class MyClass<T extends Number>
  implements Containment<T extends Number> { // Wrong!
```

Once the type parameter has been established, it is simply passed to the interface without further modification.

Here is the generalized syntax for a generic interface:

interface *interface-name<type-param-list>* { // ...

Here, *type-param-list* is a comma-separated list of type parameters. When a generic interface is implemented, you must specify the type arguments, as shown here:

class *class-name<type-param-list>*
     implements *interface-name<type-param-list>* {

Try This 13-1   Create a Generic Queue

```
IGenQ.java
QueueFullException.java
QueueEmptyException.java
GenQueue.java
GenQDemo.java
```

One of the most powerful advantages that generics bring to programming is the ability to construct reliable, reusable code. As mentioned at the start of this chapter, many algorithms are the same no matter what type of data they are used on. For example, a queue works the same way whether that queue is for integers, strings, or **File** objects. Instead of creating a separate queue class for each type of object, you can craft a single, generic solution that can be used with any type of object. Thus, the development cycle of design, code, test, and debug occurs only once when you create a generic solution—not repeatedly, each time a queue is needed for a new data type.

In this project, you will adapt the queue example that has been evolving since Try This 5-2, making it generic. This project represents the final evolution of the queue. It includes a generic interface that defines the queue operations, two exception classes, and one queue implementation: a fixed-size queue. Of course, you can experiment with other types of generic queues, such as a generic dynamic queue or a generic circular queue. Just follow the lead of the example shown here.

Like the previous version of the queue shown in Try This 9-1, this project organizes the queue code into a set of separate files: one for the interface, one for each queue exception, one for the fixed-queue implementation, and one for the program that demonstrates it. This organization reflects the way that this project would normally be organized in the real world.

1. The first step in creating a generic queue is to create a generic interface that describes the queue's two operations: put and get. The generic version of the queue interface is called **IGenQ** and it is shown here. Put this interface into a file called **IGenQ.java**.

```java
// A generic queue interface.
public interface IGenQ<T> {
  // Put an item into the queue.
  void put(T ch) throws QueueFullException;

  // Get an item from the queue.
  T get() throws QueueEmptyException;
}
```

   Notice that the type of data stored by the queue is specified by the generic type parameter **T**.

2. Next, create the files **QueueFullException.java** and **QueueEmptyException.java**. Put in each file its corresponding class, shown here:

```java
// An exception for queue-full errors.
public class QueueFullException extends Exception {
  int size;

  QueueFullException(int s) { size = s; }

  public String toString() {
   return "\nQueue is full. Maximum size is " +
         size;
  }
}

// An exception for queue-empty errors.
public class QueueEmptyException extends Exception {

  public String toString() {
    return "\nQueue is empty.";
  }
}
```

   These classes encapsulate the two queue errors: full or empty. They are not generic classes because they are the same no matter what type of data is stored in a queue. Thus, these two files will be the same as those you used with Try This 9-1.

(continued)

3. Now, create a file called **GenQueue.java**. Into that file, put the following code, which implements a fixed-size queue:

```java
// A generic, fixed-size queue class.
class GenQueue<T> implements IGenQ<T> {
  private T[] q; // this array holds the queue
  private int putloc, getloc; // the put and get indices

  // Construct an empty queue with the given array.
  public GenQueue(T[] aRef) {
    q = aRef;
    putloc = getloc = 0;
  }

  // Put an item into the queue.
  public void put(T obj)
    throws QueueFullException {

    if(putloc==q.length)
      throw new QueueFullException(q.length);

    q[putloc++] = obj;
  }

  // Get a character from the queue.
  public T get()
    throws QueueEmptyException {

    if(getloc == putloc)
      throw new QueueEmptyException();

    return q[getloc++];
  }
}
```

GenQueue is a generic class with type parameter **T**, which specifies the type of data stored in the queue. Notice that **T** is also passed to the **IGenQ** interface.

Notice that the **GenQueue** constructor is passed a reference to an array that will be used to hold the queue. Thus, to construct a **GenQueue**, you will first create an array whose type is compatible with the objects that you will be storing in the queue and whose size is long enough to store the number of objects that will be placed in the queue.

For example, the following sequence shows how to create a queue that holds strings:

```java
String[] strArray = new String[10];
GenQueue<String> strQ = new GenQueue<String>(strArray);
```

4. Create a file called **GenQDemo.java** and put the following code into it. This program demonstrates the generic queue.

```
/*
    Try This 13-1

    Demonstrate a generic queue class.
*/
class GenQDemo {
  public static void main(String[] args) {
    // Create an integer queue.
    Integer[] iStore = new Integer[10];
    GenQueue<Integer> q = new GenQueue<Integer>(iStore);

    Integer iVal;

    System.out.println("Demonstrate a queue of Integers.");
    try {
      for(int i=0; i < 5; i++) {
        System.out.println("Adding " + i + " to q.");
        q.put(i); // add integer value to q
      }
    }
    catch (QueueFullException exc) {
      System.out.println(exc);
    }
    System.out.println();

    try {
      for(int i=0; i < 5; i++) {
        System.out.print("Getting next Integer from q: ");
        iVal = q.get();
        System.out.println(iVal);
      }
    }
    catch (QueueEmptyException exc) {
      System.out.println(exc);
    }

    System.out.println();

    // Create a Double queue.
    Double[] dStore = new Double[10];
    GenQueue<Double> q2 = new GenQueue<Double>(dStore);
```

(continued)

```
      Double dVal;

      System.out.println("Demonstrate a queue of Doubles.");
      try {
        for(int i=0; i < 5; i++) {
          System.out.println("Adding " + (double)i/2 +
                             " to q2.");
          q2.put((double)i/2); // add double value to q2
        }
      }
      catch (QueueFullException exc) {
        System.out.println(exc);
      }
      System.out.println();

      try {
        for(int i=0; i < 5; i++) {
          System.out.print("Getting next Double from q2: ");
          dVal = q2.get();
          System.out.println(dVal);
        }
      }
      catch (QueueEmptyException exc) {
        System.out.println(exc);
      }
    }
  }
```

5. Compile the program and run it. You will see the output shown here:

```
Demonstrate a queue of Integers.
Adding 0 to q.
Adding 1 to q.
Adding 2 to q.
Adding 3 to q.
Adding 4 to q.

Getting next Integer from q: 0
Getting next Integer from q: 1
Getting next Integer from q: 2
Getting next Integer from q: 3
Getting next Integer from q: 4

Demonstrate a queue of Doubles.
Adding 0.0 to q2.
Adding 0.5 to q2.
Adding 1.0 to q2.
Adding 1.5 to q2.
Adding 2.0 to q2.
```

```
Getting next Double from q2: 0.0
Getting next Double from q2: 0.5
Getting next Double from q2: 1.0
Getting next Double from q2: 1.5
Getting next Double from q2: 2.0
```

6. On your own, try converting the **CircularQueue** and **DynQueue** classes from Try This 8-1 into generic classes.

Raw Types and Legacy Code

Because support for generics did not exist prior to JDK 5, it was necessary for Java to provide some transition path from old, pre-generics code. Simply put, pre-generics legacy code had to remain both functional and compatible with generics. This meant that pre-generics code must be able to work with generics, and generic code must be able to work with pre-generics code.

To handle the transition to generics, Java allows a generic class to be used without any type arguments. This creates a *raw type* for the class. This raw type is compatible with legacy code, which has no knowledge of generics. The main drawback to using the raw type is that the type safety of generics is lost.

Here is an example that shows a raw type in action:

```
// Demonstrate a raw type.
class Gen<T> {
  T ob; // declare an object of type T

  // Pass the constructor a reference to
  // an object of type T.
  Gen(T o) {
    ob = o;
  }

  // Return ob.
  T getOb() {
    return ob;
  }
}

// Demonstrate raw type.
class RawDemo {
  public static void main(String[] args) {

    // Create a Gen object for Integers.
    Gen<Integer> iOb = new Gen<Integer>(88);
```

```
// Create a Gen object for Strings.
Gen<String> strOb = new Gen<String>("Generics Test");

// Create a raw-type Gen object and give it
// a Double value.
Gen raw = new Gen(98.6);
```
When no type argument is
supplied, a raw type is created.

```
// Cast here is necessary because type is unknown.
double d = (Double) raw.getOb();
System.out.println("value: " + d);

// The use of a raw type can lead to run-time.
// exceptions. Here are some examples.

// The following cast causes a run-time error!
//    int i = (Integer) raw.getOb(); // run-time error

// This assignment overrides type safety.
strOb = raw; // OK, but potentially wrong
//    String str = strOb.getOb(); // run-time error
```
Raw types override
type safety.

```
// This assignment also overrides type safety.
raw = iOb; // OK, but potentially wrong
//    d = (Double) raw.getOb(); // run-time error
  }
}
```

This program contains several interesting things. First, a raw type of the generic **Gen** class is created by the following declaration:

```
Gen raw = new Gen(98.6);
```

Notice that no type arguments are specified. In essence, this creates a **Gen** object whose type **T** is replaced by **Object**.

A raw type is not type safe. Thus, a variable of a raw type can be assigned a reference to any type of **Gen** object. The reverse is also allowed, in which a variable of a specific **Gen** type can be assigned a reference to a raw **Gen** object. However, both operations are potentially unsafe because the type checking mechanism of generics is circumvented.

This lack of type safety is illustrated by the commented-out lines at the end of the program. Let's examine each case. First, consider the following situation:

```
//    int i = (Integer) raw.getOb(); // run-time error
```

In this statement, the value of **ob** inside **raw** is obtained, and this value is cast to **Integer**. The trouble is that **raw** contains a **Double** value, not an integer value. However, this cannot be detected at compile time because the type of **raw** is unknown. Thus, this statement fails at run time.

The next sequence assigns to **strOb** (a reference of type **Gen<String>**) a reference to a raw **Gen** object:

```
   strOb = raw; // OK, but potentially wrong
//     String str = strOb.getOb(); // run-time error
```

The assignment itself is syntactically correct, but questionable. Because **strOb** is of type **Gen<String>**, it is assumed to contain a **String**. However, after the assignment, the object referred to by **strOb** contains a **Double**. Thus, at run time, when an attempt is made to assign the contents of **strOb** to **str**, a run-time error results because **strOb** now contains a **Double**. Thus, the assignment of a raw reference to a generic reference bypasses the type-safety mechanism.

The following sequence inverts the preceding case:

```
   raw = iOb; // OK, but potentially wrong
//     d = (Double) raw.getOb(); // run-time error
```

Here, a generic reference is assigned to a raw reference variable. Although this is syntactically correct, it can lead to problems, as illustrated by the second line. In this case, **raw** now refers to an object that contains an **Integer** object, but the cast assumes that it contains a **Double**. This error cannot be prevented at compile time. Rather, it causes a run-time error.

Because of the potential for danger inherent in raw types, **javac** displays *unchecked warnings* when a raw type is used in a way that might jeopardize type safety. In the preceding program, these lines generate unchecked warnings:

```
Gen raw = new Gen(98.6);

strOb = raw; // OK, but potentially wrong
```

In the first line, it is the use of **Gen** without a type argument that causes the warning. In the second line, it is the assignment of a raw reference to a generic variable that generates the warning.

At first, you might think that this line should also generate an unchecked warning, but it does not:

```
raw = iOb; // OK, but potentially wrong
```

No compiler warning is issued because the assignment does not cause any *further* loss of type safety than had already occurred when **raw** was created.

One final point: You should limit the use of raw types to those cases in which you must mix legacy code with modern, generic code. Raw types are simply a transitional feature and not something that should be used for new code.

Type Inference with the Diamond Operator

Beginning with JDK 7, it is possible to shorten the syntax used to create an instance of a generic type. To begin, think back to the **TwoGen** class shown earlier in this chapter. A portion is shown here for convenience. Notice that it uses two generic types.

```
class TwoGen<T, V> {
  T ob1;
  V ob2;

  // Pass the constructor a reference to
  // an object of type T.
  TwoGen(T o1, V o2) {
    ob1 = o1;
    ob2 = o2;
  }
  // ...
}
```

For versions of Java prior to JDK 7, to create an instance of **TwoGen**, you must use a statement similar to the following:

```
TwoGen<Integer, String> tgOb =
  new TwoGen<Integer, String>(42, "testing");
```

Here, the type arguments (which are **Integer** and **String**) are specified twice: first, when **tgOb** is declared, and second, when a **TwoGen** instance is created via **new**. While there is nothing wrong, per se, with this form, it is a bit more verbose than it needs to be. Since, in the **new** clause, the type of the type arguments can be readily inferred, there is really no reason that they need to be specified a second time. To address this situation, JDK 7 added a syntactic element that lets you avoid the second specification.

Today, the preceding declaration can be rewritten as shown here:

```
TwoGen<Integer, String> tgOb = new TwoGen<>(42, "testing");
```

Notice that the instance creation portion simply uses < >, which is an empty type argument list. This is referred to as the *diamond* operator. It tells the compiler to infer the type arguments needed by the constructor in the **new** expression. The principal advantage of this type-inference syntax is that it shortens what are sometimes quite long declaration statements. This is especially helpful for generic types that specify bounds.

The preceding example can be generalized. When type inference is used, the declaration syntax for a generic reference and instance creation has this general form:

class-name<type-arg-list> var-name = new *class-name< >(cons-arg-list)*;

Here, the type argument list of the **new** clause is empty.

Although mostly for use in declaration statements, type inference can also be applied to parameter passing. For example, if the following method is added to **TwoGen**:

```
boolean isSame(TwoGen<T, V> o) {
  if(ob1 == o.ob1 && ob2 == o.ob2) return true;
  else return false;
}
```

then the following call is legal:

```
if(tgOb.isSame(new TwoGen<>(42, "testing"))) System.out.println("Same");
```

In this case, the type arguments for the arguments passed to **isSame( )** can be inferred from the parameters' types. They don't need to be specified again.

Although the diamond operator offers convenience, in general, the remaining examples of generics in this book will continue to use the full syntax when declaring instances of generic classes. There are two reasons for this. First, and most importantly, using the full-length syntax makes it very clear precisely what is being created, which is helpful when example code is shown. Second, the code will work in environments that are using an older compiler. Of course, in your own code, the use of the type inference syntax will streamline your declarations.

Local Variable Type Inference and Generics

As just explained, type inference is already supported for generics through the use of the diamond operator. However, you can also use the local variable type inference feature added by JDK 10 with a generic class. For example, again assuming the **TwoGen** class, this declaration:

```
TwoGen<Integer, String> tgObj =
  new TwoGen<Integer, String>(42, "testing");
```

can be rewritten like this using local variable type inference:

```
var tgOb = new TwoGen<Integer, String>(42, "testing");
```

In this case, the type of **tgOb** is inferred to be **TwoGen<Integer, String>** because that is the type of its initializer. Also notice that the use of **var** results in a shorter declaration than would be the case otherwise. In general, generic type names can often be quite long and (in some cases) complicated. The use of **var** is another way to substantially shorten such declarations. For the same reasons as just explained for the diamond operator, the remaining examples in this book will continue to use the full generic syntax, but in your own code the use of local variable type inference can be quite helpful.

Erasure

Usually, it is not necessary for the programmer to know the details about how the Java compiler transforms your source code into object code. However, in the case of generics, some general understanding of the process is important because it explains why the generic features work as they do—and why their behavior is sometimes a bit surprising. For this reason, a brief discussion of how generics are implemented in Java is in order.

An important constraint that governed the way generics were added to Java was the need for compatibility with previous versions of Java. Simply put: generic code had to be compatible with preexisting, nongeneric code. Thus, any changes to the syntax of the Java language, or to the JVM, had to avoid breaking older code. The way Java implements generics while satisfying this constraint is through the use of *erasure.*

In general, here is how erasure works. When your Java code is compiled, all generic type information is removed (erased). This means replacing type parameters with their bound type, which is **Object** if no explicit bound is specified, and then applying the appropriate casts (as determined by the type arguments) to maintain type compatibility with the types specified by the type arguments. The compiler also enforces this type compatibility. This approach to generics means that no type parameters exist at run time. They are simply a source-code mechanism.

Ambiguity Errors

The inclusion of generics gives rise to a new type of error that you must guard against: *ambiguity.* Ambiguity errors occur when erasure causes two seemingly distinct generic declarations to resolve to the same erased type, causing a conflict. Here is an example that involves method overloading:

```
// Ambiguity caused by erasure on
// overloaded methods.
class MyGenClass<T, V> {
  T ob1;
  V ob2;

  // ...

  // These two overloaded methods are ambiguous
  // and will not compile.
  void set(T o) {
    ob1 = o;
  }

  void set(V o) {
    ob2 = o;
  }
}
```

These two methods are inherently ambiguous.

Notice that **MyGenClass** declares two generic types: **T** and **V**. Inside **MyGenClass**, an attempt is made to overload **set( )** based on parameters of type **T** and **V**. This looks reasonable because **T** and **V** appear to be different types. However, there are two ambiguity problems here.

First, as **MyGenClass** is written there is no requirement that **T** and **V** actually be different types. For example, it is perfectly correct (in principle) to construct a **MyGenClass** object as shown here:

```
MyGenClass<String, String> obj = new MyGenClass<String, String>()
```

In this case, both **T** and **V** will be replaced by **String**. This makes both versions of **set( )** identical, which is, of course, an error.

Second, and more fundamental, is that the type erasure of **set( )** effectively reduces both versions to the following:

```
void set(Object o) { // ...
```

Thus, the overloading of **set( )** as attempted in **MyGenClass** is inherently ambiguous. The solution in this case is to use two separate method names rather than trying to overload **set( )**.

Some Generic Restrictions

There are a few restrictions that you need to keep in mind when using generics. They involve creating objects of a type parameter, static members, exceptions, and arrays. Each is examined here.

Type Parameters Can't Be Instantiated

It is not possible to create an instance of a type parameter. For example, consider this class:

```
// Can't create an instance of T.
class Gen<T> {
  T ob;
  Gen() {
    ob = new T(); // Illegal!!!
  }
}
```

Here, it is illegal to attempt to create an instance of **T**. The reason should be easy to understand: the compiler has no way to know what type of object to create. **T** is simply a placeholder.

Restrictions on Static Members

No **static** member can use a type parameter declared by the enclosing class. For example, both of the **static** members of this class are illegal:

```
class Wrong<T> {
  // Wrong, no static variables of type T.
  static T ob;

  // Wrong, no static method can use T.
  static T getOb() {
    return ob;
  }
}
```

Although you can't declare **static** members that use a type parameter declared by the enclosing class, you *can* declare **static** generic methods, which define their own type parameters, as was done earlier in this chapter.

Generic Array Restrictions

There are two important generics restrictions that apply to arrays. First, you cannot instantiate an array whose element type is a type parameter. Second, you cannot create an array of type-specific generic references. The following short program shows both situations:

```
// Generics and arrays.
class Gen<T extends Number> {
  T ob;

  T[] vals; // OK

  Gen(T o, T[] nums) {
    ob = o;

    // This statement is illegal.
//  vals = new T[10]; // can't create an array of T

    // But, this statement is OK.
    vals = nums; // OK to assign reference to existent array
  }
}

class GenArrays {
  public static void main(String[] args) {
    Integer[] n = { 1, 2, 3, 4, 5 };

    Gen<Integer> iOb = new Gen<Integer>(50, n);

    // Can't create an array of type-specific generic references.
    // Gen<Integer>[] gens = new Gen<Integer>[10]; // Wrong!

    // This is OK.
    Gen<?>[] gens = new Gen<?>[10]; // OK
  }
}
```

As the program shows, it's valid to declare a reference to an array of type **T**, as this line does:

```
T[] vals; // OK
```

But, you cannot instantiate an array of **T**, as this commented-out line attempts:

```
// vals = new T[10]; // can't create an array of T
```

The reason you can't create an array of **T** is that there is no way for the compiler to know what type of array to actually create. However, you can pass a reference to a type-compatible array to **Gen( )** when an object is created and assign that reference to **vals**, as the program does in this line:

```
vals = nums; // OK to assign reference to existent array
```

This works because the array passed to **Gen()** has a known type, which will be the same type as **T** at the time of object creation. Inside **main( )**, notice that you can't declare an array of references to a specific generic type. That is, this line

```
// Gen<Integer>[] gens = new Gen<Integer>[10]; // Wrong!
```

won't compile.

Generic Exception Restriction

A generic class cannot extend **Throwable**. This means that you cannot create generic exception classes.

Continuing Your Study of Generics

As mentioned at the start, this chapter gives you sufficient knowledge to use generics effectively in your own programs. However, there are many side issues and special cases that are not covered here. Readers especially interested in generics will want to learn about how generics affect class hierarchies, run-time type comparisons, and overriding, for example. Discussions of these and other topics are found in *Java: The Complete Reference, Twelfth Edition* (McGraw Hill, 2022).

Chapter 13 Self Test

1. Generics are important to Java because they enable the creation of code that is

    A. Type-safe

    B. Reusable

    C. Reliable

    D. All of the above

2. Can a primitive type be used as a type argument?

3. Show how to declare a class called **FlightSched** that takes two generic parameters.

4. Beginning with your answer to question 3, change **FlightSched**'s second type parameter so that it must extend **Thread**.

5. Now, change **FlightSched** so that its second type parameter must be a subclass of its first type parameter.

6. As it relates to generics, what is the **?** and what does it do?

7. Can the wildcard argument be bounded?

8. A generic method called **MyGen( )** has one type parameter. Furthermore, **MyGen( )** has one parameter whose type is that of the type parameter. It also returns an object of that type parameter. Show how to declare **MyGen( )**.

9. Given this generic interface

```
interface IGenIF<T, V extends T> { // ...
```

show the declaration of a class called **MyClass** that implements **IGenIF**.

10. Given a generic class called **Counter<T>**, show how to create an object of its raw type.

11. Do type parameters exist at run time?

12. Convert your solution to question 10 of the Self Test for Chapter 9 so that it is generic. In the process, create a stack interface called **IGenStack** that generically defines the operations **push( )** and **pop( )**.

13. What is **< >**?

14. How can the following be simplified?

```
MyClass<Double,String> obj = new MyClass<Double,String>(1.1,"Hi");
```

Chapter 14

Lambda Expressions and Method References

Key Skills & Concepts

- Know the general form of a lambda expression

- Understand the definition of a functional interface

- Use expression lambdas

- Use block lambdas

- Use generic functional interfaces

- Understand variable capture in a lambda expression

- Throw an exception from a lambda expression

- Understand the method reference

- Understand the constructor reference

- Know about the predefined functional interfaces in **java.util.function**

Beginning with JDK 8, a feature was added to Java that profoundly enhanced the expressive power of the language. This feature is the *lambda expression*. Not only did lambda expressions add new syntax elements to the language, they also streamlined the way that certain common constructs are implemented. In much the same way that the addition of generics reshaped Java years ago, lambda expressions continue to reshape Java today. They truly are that important.

The addition of lambda expressions also provided the catalyst for other Java features. You have already seen one of them—the default method—which was described in Chapter 8. It lets you define default behavior for an **interface** method. Another example is the method reference, described later in this chapter, which lets you refer to a method without executing it. Furthermore, the inclusion of lambda expressions resulted in new capabilities being incorporated into the API library.

Beyond the benefits that lambda expressions bring to the language, there is another reason why they constitute such an important part of Java. Over the past few years, lambda expressions have become a major focus of computer language design. For example, they have been added to languages such as C# and C++. Their inclusion in Java helps it remain the vibrant, innovative language that programmers have come to expect. This chapter presents an introduction to this important feature.

Introducing Lambda Expressions

Key to understanding the lambda expression are two constructs. The first is the lambda expression, itself. The second is the functional interface. Let's begin with a simple definition of each.

A *lambda expression* is, essentially, an anonymous (that is, unnamed) method. However, this method is not executed on its own. Instead, it is used to implement a method defined by a functional interface. Thus, a lambda expression results in a form of anonymous class. Lambda expressions are also commonly referred to as *closures*.

A *functional interface* is an interface that contains one and only one abstract method. Normally, this method specifies the intended purpose of the interface. Thus, a functional interface typically represents a single action. For example, the standard interface **Runnable** is a functional interface because it defines only one method: **run( )**. Therefore, **run( )** defines the action of **Runnable**. Furthermore, a functional interface defines the *target type* of a lambda expression. Here is a key point: a lambda expression can be used only in a context in which a target type is specified. One other thing: a functional interface is sometimes referred to as a *SAM type,* where SAM stands for *Single Abstract Method.*

Let's now look more closely at both lambda expressions and functional interfaces.

NOTE

A functional interface may specify any public method defined by **Object**, such as **equals( )**, without affecting its "functional interface" status. The public **Object** methods are considered implicit members of a functional interface because they are automatically implemented by an instance of a functional interface.

Lambda Expression Fundamentals

The lambda expression relies on a syntax element and operator that differ from what you have seen in the preceding chapters. The operator, sometimes referred to as the *lambda operator* or the *arrow operator*, is –>. It divides a lambda expression into two parts. The left side specifies any parameters required by the lambda expression. On the right side is the *lambda body,* which specifies the actions of the lambda expression. Java defines two types of lambda bodies. One type consists of a single expression, and the other type consists of a block of code. We will begin with lambdas that define a single expression.

At this point, it will be helpful to look at a few examples of lambda expressions before continuing. Let's begin with what is probably the simplest type of lambda expression you can write. It evaluates to a constant value and is shown here:

```
() -> 98.6
```

This lambda expression takes no parameters, thus the parameter list is empty. It returns the constant value 98.6. The return type is inferred to be **double**. Therefore, it is similar to the following method:

```
double myMeth() { return 98.6; }
```

Of course, the method defined by a lambda expression does not have a name.

A slightly more interesting lambda expression is shown here:

```
() -> Math.random() * 100
```

This lambda expression obtains a pseudo-random value from **Math.random( )**, multiplies it by 100, and returns the result. It, too, does not require a parameter.

When a lambda expression requires a parameter, it is specified in the parameter list on the left side of the lambda operator. Here is a simple example:

```
(n) -> 1.0 / n
```

This lambda expression returns the reciprocal of the value of parameter **n**. Thus, if **n** is 4.0, the reciprocal is 0.25. Although it is possible to explicitly specify the type of a parameter, such as **n** in this case, often you won't need to because, in many cases, its type can be inferred. Like a named method, a lambda expression can specify as many parameters as needed.

Any valid type can be used as the return type of a lambda expression. For example, this lambda expression returns **true** if the value of parameter **n** is even and **false** otherwise.

```
(n) -> (n % 2)==0
```

Thus, the return type of this lambda expression is **boolean**.

One other point before moving on. When a lambda expression has only one parameter, it is not necessary to surround the parameter name with parentheses when it is specified on the left side of the lambda operator. For example, this is also a valid way to write the lambda expression just shown:

```
n -> (n % 2)==0
```

For consistency, this book will surround all lambda expression parameter lists with parentheses, even those containing only one parameter. Of course, you are free to adopt a different style.

Functional Interfaces

As stated, a functional interface is an interface that specifies only one abstract method. Before continuing, recall from Chapter 8 that not all interface methods are abstract. Beginning with JDK 8, it is possible for an interface to have one or more default methods. Default methods are *not* abstract. Neither are **static** or **private** interface methods. Thus, an interface method is abstract only if it is does not specify an implementation. This means that a functional interface can include default, **static**, or **private** methods, but in all cases it must have one and only one abstract method. Because non-default, non-**static**, non-**private** interface methods are implicitly abstract, there is no need to use the **abstract** modifier (although you can specify it, if you like).

Here is an example of a functional interface:

```
interface MyValue {
  double getValue();
}
```

In this case, the method **getValue( )** is implicitly abstract, and it is the only method defined by **MyValue**. Thus, **MyValue** is a functional interface, and its function is defined by **getValue( )**.

As mentioned earlier, a lambda expression is not executed on its own. Rather, it forms the implementation of the abstract method defined by the functional interface that specifies its target type. As a result, a lambda expression can be specified only in a context in which a target type is defined. One of these contexts is created when a lambda expression is assigned to a functional interface reference. Other target type contexts include variable initialization, **return** statements, and method arguments, to name a few.

Let's work through a simple example. First, a reference to the functional interface **MyValue** is declared:

```
// Create a reference to a MyValue instance.
MyValue myVal;
```

Next, a lambda expression is assigned to that interface reference:

```
// Use a lambda in an assignment context.
myVal = () -> 98.6;
```

This lambda expression is compatible with **getValue( )** because, like **getValue( )**, it has no parameters and returns a **double** result. In general, the type of the abstract method defined by the functional interface and the type of the lambda expression must be compatible. If they aren't, a compile-time error will result.

As you can probably guess, the two steps just shown can be combined into a single statement, if desired:

```
MyValue myVal = () -> 98.6;
```

Here, **myVal** is initialized with the lambda expression.

When a lambda expression occurs in a target type context, an instance of a class is automatically created that implements the functional interface, with the lambda expression defining the behavior of the abstract method declared by the functional interface. When that method is called through the target, the lambda expression is executed. Thus, a lambda expression gives us a way to transform a code segment into an object.

In the preceding example, the lambda expression becomes the implementation for the **getValue( )** method. As a result, the following displays the value 98.6:

```
// Call getValue(), which is implemented by the previously assigned
// lambda expression.
System.out.println("A constant value: " + myVal.getValue());
```

Because the lambda expression assigned to **myVal** returns the value 98.6, that is the value obtained when **getValue( )** is called.

If the lambda expression takes one or more parameters, then the abstract method in the functional interface must also take the same number of parameters. For example, here is a functional interface called **MyParamValue**, which lets you pass a value to **getValue( )**:

```
interface MyParamValue {
  double getValue(double v);
}
```

You can use this interface to implement the reciprocal lambda shown in the previous section. For example:

```
MyParamValue myPval = (n) -> 1.0 / n;
```

You can then use **myPval** like this:

```
System.out.println("Reciprocal of 4 is " + myPval.getValue(4.0));
```

Here, **getValue( )** is implemented by the lambda expression referred to by **myPval**, which returns the reciprocal of the argument. In this case, 4.0 is passed to **getValue( )**, which returns 0.25.

There is something else of interest in the preceding example. Notice that the type of **n** is not specified. Rather, its type is inferred from the context. In this case, its type is inferred from the parameter type of **getValue( )** as defined by the **MyParamValue** interface, which is **double**. It is also possible to explicitly specify the type of a parameter in a lambda expression. For example, this is also a valid way to write the preceding:

```
(double n) -> 1.0 / n;
```

Here, **n** is explicitly specified as **double**. Usually it is not necessary to explicitly specify the type.

NOTE

As a point of interest, beginning with JDK 11, you can also explicitly indicate type inference for a lambda expression parameter by use of **var**. For example, you could write:

```
(var n) -> 1.0 / n;
```

Of course, the use of **var** here is redundant. Its use would, however, allow an annotation to be added.

Before moving on, it is important to emphasize a key point: For a lambda expression to be used in a target type context, the type of the abstract method and the type of the lambda expression must be compatible. For example, if the abstract method specifies two **int** parameters, then the lambda must specify two parameters whose type either is explicitly **int** or can be implicitly inferred as **int** by the context. In general, the type and number of the lambda expression's parameters must be compatible with the method's parameters and its return type.

Lambda Expressions in Action

With the preceding discussion in mind, let's look at some simple examples that put the basic lambda expression concepts into action. The first example assembles the pieces shown in the foregoing section into a complete program that you can run and experiment with.

```
// Demonstrate two simple lambda expressions.

// A functional interface.
```

```
interface MyValue {
  double getValue();
}

// Another functional interface.
interface MyParamValue {
  double getValue(double v);
}

class LambdaDemo {
  public static void main(String[] args)
  {
    MyValue myVal;  // declare an interface reference

    // Here, the lambda expression is simply a constant expression.
    // When it is assigned to myVal, a class instance is
    // constructed in which the lambda expression implements
    // the getValue() method in MyValue.
    myVal = () -> 98.6;

    // Call getValue(), which is provided by the previously assigned
    // lambda expression.
    System.out.println("A constant value: " + myVal.getValue());

    // Now, create a parameterized lambda expression and assign it to
    // a MyParamValue reference. This lambda expression returns
    // the reciprocal of its argument.
    MyParamValue myPval = (n) -> 1.0 / n;

    // Call getValue(v) through the myPval reference.
    System.out.println("Reciprocal of 4 is " + myPval.getValue(4.0));
    System.out.println("Reciprocal of 8 is " + myPval.getValue(8.0));

    // A lambda expression must be compatible with the method
    // defined by the functional interface. Therefore, these won't work:
//  myVal = () -> "three"; // Error! String not compatible with double!
//  myPval = () -> Math.random(); // Error! Parameter required!
  }
}
```

Functional interfaces → (points to `interface MyValue {` and `interface MyParamValue {`)

A simple lambda expression → (points to `myVal = () -> 98.6;`)

A lambda expression that has a parameter → (points to `MyParamValue myPval = (n) -> 1.0 / n;`)

Sample output from the program is shown here:

```
A constant value: 98.6
Reciprocal of 4 is 0.25
Reciprocal of 8 is 0.125
```

As mentioned, the lambda expression must be compatible with the abstract method that it is intended to implement. For this reason, the commented-out lines at the end of the preceding program are illegal. The first, because a value of type **String** is not compatible with **double**, which is the return type required by **getValue( )**. The second, because **getValue(int)** in **MyParamValue** requires a parameter, and one is not provided.

A key aspect of a functional interface is that it can be used with any lambda expression that is compatible with it. For example, consider the following program. It defines a functional interface called **NumericTest** that declares the abstract method **test( )**. This method has two **int** parameters and returns a **boolean** result. Its purpose is to determine if the two arguments passed to **test( )** satisfy some condition. It returns the result of the test. In **main( )**, three different tests are created through the use of lambda expressions. One tests if the first argument can be evenly divided by the second; the second determines if the first argument is less than the second; and the third returns **true** if the absolute values of the arguments are equal. Notice that the lambda expressions that implement these tests have two parameters and return a **boolean** result. This is, of course, necessary since **test( )** has two parameters and returns a **boolean** result.

```java
// Use the same functional interface with three different lambda expressions.

// A functional interface that takes two int parameters and returns
// a boolean result.
interface NumericTest {
  boolean test(int n, int m);
}

class LambdaDemo2 {
  public static void main(String[] args)
  {
    // This lambda expression determines if one number is
    // a factor of another.
    NumericTest isFactor = (n, d) -> (n % d) == 0;

    if(isFactor.test(10, 2))
      System.out.println("2 is a factor of 10");
    if(!isFactor.test(10, 3))
      System.out.println("3 is not a factor of 10");
    System.out.println();

    // This lambda expression returns true if the first
    // argument is less than the second.
    NumericTest lessThan = (n, m) -> (n < m);

    if(lessThan.test(2, 10))
      System.out.println("2 is less than 10");
    if(!lessThan.test(10, 2))
      System.out.println("10 is not less than 2");
    System.out.println();

    // This lambda expression returns true if the absolute
    // values of the arguments are equal.
    NumericTest absEqual = (n, m) -> (n < 0 ? -n : n) == (m < 0 ? -m : m);

    if(absEqual.test(4, -4))
      System.out.println("Absolute values of 4 and -4 are equal.");
```

Use the same functional interface with three different lambda expressions.

```
    if(!lessThan.test(4, -5))
      System.out.println("Absolute values of 4 and -5 are not equal.");
    System.out.println();
  }
}
```

The output is shown here:

```
2 is a factor of 10
3 is not a factor of 10

2 is less than 10
10 is not less than 2

Absolute values of 4 and -4 are equal.
Absolute values of 4 and -5 are not equal.
```

As the program illustrates, because all three lambda expressions are compatible with **test( )**, all can be executed through a **NumericTest** reference. In fact, there is no need to use three separate **NumericTest** reference variables because the same one could have been used for all three tests. For example, you could create the variable **myTest** and then use it to refer to each test, in turn, as shown here:

```
NumericTest myTest;
```

```
myTest = (n, d) -> (n % d) == 0;
if(myTest.test(10, 2))
  System.out.println("2 is a factor of 10");
// ...
myTest = (n, m) -> (n < m);
if(myTest.test(2, 10))
  System.out.println("2 is less than 10");
//...
myTest = (n, m) -> (n < 0 ? -n : n) == (m < 0 ? -m : m);
if(myTest.test(4, -4))
  System.out.println("Absolute values of 4 and -4 are equal.");
// ...
```

Of course, using different reference variables called **isFactor**, **lessThan**, and **absEqual**, as the original program does, makes it very clear to which lambda expression each variable refers.

There is one other point of interest in the preceding program. Notice how the two parameters are specified for the lambda expressions. For example, here is the one that determines if one number is a factor of another:

```
(n, d) -> (n % d) == 0
```

Notice that **n** and **d** are separated by commas. In general, whenever more than one parameter is required, the parameters are specified, separated by commas, in a parenthesized list on the left side of the lambda operator.

Although the preceding examples used primitive values as the parameter types and return type of the abstract method defined by a functional interface, there is no restriction in this regard. For example, the following program declares a functional interface called **StringTest**. It has a method called **test( )** that takes two **String** parameters and returns a **boolean** result. Thus, it can be used to test some condition related to strings. Here, a lambda expression is created that determines if one string is contained within another:

```
// A functional interface that tests two strings.
interface StringTest {
  boolean test(String aStr, String bStr);
}

class LambdaDemo3 {
  public static void main(String[] args)
  {
    // This lambda expression determines if one string is
    // part of another.
    StringTest isIn = (a, b) -> a.indexOf(b) != -1;

    String str = "This is a test";

    System.out.println("Testing string: " + str);

    if(isIn.test(str, "is a"))
      System.out.println("'is a' found.");
    else
      System.out.println("'is a' not found.");

    if(isIn.test(str, "xyz"))
      System.out.println("'xyz' Found");
    else
      System.out.println("'xyz' not found");
  }
}
```

The output is shown here:

```
Testing string: This is a test
'is a' found.
'xyz' not found
```

Notice that the lambda expression uses the **indexOf( )** method defined by the **String** class to determine if one string is part of another. This works because the parameters **a** and **b** are determined by type inference to be of type **String**. Thus, it is permissible to call a **String** method on **a**.

Ask the Expert

Q: Earlier you mentioned that I can explicitly declare the type of a parameter in a lambda expression if needed. In cases in which a lambda expression requires two or more parameters, must I specify the types of all parameters, or can I let one or more use type inference?

A: In cases in which you need to explicitly declare the type of a parameter, then all of the parameters in the list must have declared types. For example, this is legal:

```
(int n, int d) -> (n % d) == 0
```

But this is not legal:

```
(int n, d) -> (n % d) == 0
```

Nor is this legal:

```
(n, int d) -> (n % d) == 0
```

Block Lambda Expressions

The body of the lambdas shown in the preceding examples consist of a single expression. These types of lambda bodies are referred to as *expression bodies,* and lambdas that have expression bodies are sometimes called *expression lambdas.* In an expression body, the code on the right side of the lambda operator must consist of a single expression, which becomes the lambda's value. Although expression lambdas are quite useful, sometimes the situation will require more than a single expression. To handle such cases, Java supports a second type of lambda expression in which the code on the right side of the lambda operator consists of a block of code that can contain more than one statement. This type of lambda body is called a *block body.* Lambdas that have block bodies are sometimes referred to as *block lambdas.*

A block lambda expands the types of operations that can be handled within a lambda expression because it allows the body of the lambda to contain multiple statements. For example, in a block lambda you can declare variables, use loops, specify **if** and **switch** statements, create nested blocks, and so on. A block lambda is easy to create. Simply enclose the body within braces as you would any other block of statements.

Aside from allowing multiple statements, block lambdas are used much like the expression lambdas just discussed. One key difference, however, is that you must explicitly use a **return** statement to return a value. This is necessary because a block lambda body does not represent a single expression.

Here is an example that uses a block lambda to find the smallest positive factor of an **int** value. It uses an interface called **NumericFunc** that has a method called **func( )**, which takes one **int** argument and returns an **int** result. Thus, **NumericFunc** supports a numeric function on values of type **int**.

```
// A block lambda that finds the smallest positive factor
// of an int value.

interface NumericFunc {
  int func(int n);
}

class BlockLambdaDemo {
  public static void main(String[] args)
  {

    // This block lambda returns the smallest positive factor of a value.
    NumericFunc smallestF = (n) ->  {
      int result = 1;

      // Get absolute value of n.
      n =  n < 0 ? -n : n;

      for(int i=2; i <= n/i; i++)
        if((n % i) == 0) {
          result = i;
          break;
        }

      return result;
    };

    System.out.println("Smallest factor of 12 is " + smallestF.func(12));
    System.out.println("Smallest factor of 11 is " + smallestF.func(11));
  }
}
```

——— A block lambda expression

The output is shown here:

```
Smallest factor of 12 is 2
Smallest factor of 11 is 1
```

In the program, notice that the block lambda declares a variable called **result**, uses a **for** loop, and has a **return** statement. These are legal inside a block lambda body. In essence, the block body of a lambda is similar to a method body. One other point. When a **return** statement occurs within a lambda expression, it simply causes a return from the lambda. It does not cause an enclosing method to return.

Generic Functional Interfaces

A lambda expression, itself, cannot specify type parameters. Thus, a lambda expression cannot be generic. (Of course, because of type inference, all lambda expressions exhibit some "generic-like" qualities.) However, the functional interface associated with a lambda expression can be generic.

In this case, the target type of the lambda expression is determined, in part, by the type argument or arguments specified when a functional interface reference is declared.

To understand the value of generic functional interfaces, consider this. Earlier in this chapter, two different functional interfaces were created, one called **NumericTest** and the other called **StringTest**. They were used to determine if two values satisfied some condition. To do this, both defined a method called **test( )** that took two parameters and returned a **boolean** result. In the case of **NumericTest**, the values being tested were integers. For **StringTest**, the values were of type **String**. Thus, the only difference between the two methods was the type of data they operated on. Such a situation is perfect for generics. Instead of having two functional interfaces whose methods differ only in their data types, it is possible to declare one generic interface that can be used to handle both circumstances. The following program shows this approach:

```
// Use a generic functional interface.

// A generic functional interface with two parameters
// that returns a boolean result.
interface SomeTest<T> {  ◀──────────────── A generic functional interface
  boolean test(T n, T m);
}

class GenericFunctionalInterfaceDemo {
  public static void main(String[] args)
  {
    // This lambda expression determines if one integer is
    // a factor of another.
    SomeTest<Integer> isFactor = (n, d) -> (n % d) == 0;

    if(isFactor.test(10, 2))
      System.out.println("2 is a factor of 10");
    System.out.println();

    // The next lambda expression determines if one Double is
    // a factor of another.
    SomeTest<Double> isFactorD = (n, d) -> (n % d) == 0;

    if(isFactorD.test(212.0, 4.0))
      System.out.println("4.0 is a factor of 212.0");
    System.out.println();

    // This lambda expression determines if one string is
    // part of another.
    SomeTest<String> isIn = (a, b) -> a.indexOf(b) != -1;

    String str = "Generic Functional Interface";

    System.out.println("Testing string: " + str);
```

```
    if(isIn.test(str, "face"))
      System.out.println("'face' is found.");
    else
      System.out.println("'face' not found.");
  }
}
```

The output is shown here:

```
2 is a factor of 10

4.0 is a factor of 212.0

Testing string: Generic Functional Interface
'face' is found.
```

In the program, the generic functional interface **SomeTest** is declared as shown here:

```
interface SomeTest<T> {
  boolean test(T n, T m);
}
```

Here, **T** specifies the type of both parameters for **test( )**. This means that it is compatible with any lambda expression that takes two parameters of the same type and returns a **boolean** result.

The **SomeTest** interface is used to provide a reference to three different types of lambdas. The first uses type **Integer**, the second uses type **Double**, and the third uses type **String**. Thus, the same functional interface can be used to refer to the **isFactor**, **isFactorD**, and **isIn** lambdas. Only the type argument passed to **SomeTest** differs.

As a point of interest, the **NumericFunc** interface shown in the previous section can also be rewritten as a generic interface. This is an exercise in "Chapter 14 Self Test," at the end of this chapter.

Try This 14-1   Pass a Lambda Expression as an Argument

LambdaArgumentDemo.java

A lambda expression can be used in any context that provides a target type. The target contexts used by the preceding examples are assignment and initialization. Another one is when a lambda expression is passed as an argument. In fact, passing a lambda expression as an argument is a common use of lambdas. Moreover, it is a very powerful use because it gives you a way to pass executable code as an argument to a method. This greatly enhances the expressive power of Java.

To illustrate the process, this project creates three string functions that perform the following operations: reverse a string, reverse the case of letters within a string, and replace spaces with hyphens. These functions are implemented as lambda expressions of the functional interface **StringFunc**. They are then passed as the first argument to a method called **changeStr( )**. This method applies the string function to the string passed as the second argument to

changeStr( ) and returns the result. Thus, **changeStr( )** can be used to apply a variety of different string functions.

1. Create a file called **LambdaArgumentDemo.java**.

2. To the file, add the functional interface **StringFunc**, as shown here:

```
interface StringFunc {
   String func(String str);
}
```

This interface defines the method **func( )**, which takes a **String** argument and returns a **String**. Thus, **func( )** can act on a string and return the result.

3. Begin the **LambdaArgumentDemo** class, as shown here, by defining the **changeStr( )** method:

```
class LambdaArgumentDemo {

   // This method has a functional interface as the type of its
   // first parameter. Thus, it can be passed a reference to any
   // instance of that interface, including an instance created
   // by a lambda expression. The second parameter specifies the
   // string to operate on.
   static String changeStr(StringFunc sf, String s) {
      return sf.func(s);
   }
```

As the comment indicates, **changeStr( )** has two parameters. The type of the first is **StringFunc**. This means it can be passed a reference to any **StringFunc** instance. Thus, it can be passed a reference to an instance created by a lambda expression that is compatible with **StringFunc**. The string to be acted on is passed to **s**. The resulting string is returned.

4. Begin the **main( )** method, as shown here:

```
public static void main(String[] args)
{
   String inStr = "Lambda Expressions Expand Java";
   String outStr;

   System.out.println("Here is input string: " + inStr);
```

Here, **inStr** refers to the string that will be acted on, and **outStr** will receive the modified string.

5. Define a lambda expression that reverses the characters in a string and assign it to a **StringFunc** reference. Notice that this is another example of a block lambda.

```
// Define a lambda expression that reverses the contents
// of a string and assign it to a StringFunc reference variable.
StringFunc reverse = (str) -> {
   String result = "";
```

(continued)

```
   for(int i = str.length()-1; i >= 0; i--)
     result += str.charAt(i);

   return result;
};
```

6. Call **changeStr( )**, passing in the **reverse** lambda and **inStr**. Assign the result to **outStr**, and display the result.

```
// Pass reverse to the first argument to changeStr().
// Pass the input string as the second argument.
outStr = changeStr(reverse, inStr);
System.out.println("The string reversed: " + outStr);
```

Because the first parameter to **changeStr( )** is of type **StringFunc**, the **reverse** lambda can be passed to it. Recall that a lambda expression causes an instance of its target type to be created, which in this case is **StringFunc**. Thus, a lambda expression gives you a way to effectively pass a code sequence to a method.

7. Finish the program by adding lambdas that replace spaces with hyphens and invert the case of the letters, as shown next. Notice that both of these lambdas are embedded in the call to **changeStr( )**, itself, rather than using a separate **StringFunc** variable.

```
// This lambda expression replaces spaces with hyphens.
// It is embedded directly in the call to changeStr().
outStr = changeStr((str) -> str.replace(' ', '-'), inStr);
System.out.println("The string with spaces replaced: " + outStr);

// This block lambda inverts the case of the characters in the
// string. It is also embedded directly in the call to changeStr().
outStr = changeStr((str) -> {
                      String result = "";
                      char ch;

                      for(int i = 0; i < str.length(); i++ ) {
                        ch = str.charAt(i);
                        if(Character.isUpperCase(ch))
                          result += Character.toLowerCase(ch);
                        else
                          result += Character.toUpperCase(ch);
                      }
                      return result;
                   }, inStr);

   System.out.println("The string in reversed case: " + outStr);
  }
}
```

As you can see by looking at this code, embedding the lambda that replaces spaces with hyphens in the call to **changeStr( )** is both convenient and easy to understand. This is because it is a short, expression lambda that simply calls **replace( )** to replace spaces with hyphens. The **replace( )** method is another method defined by the **String** class. The version used here takes as arguments the character to be replaced and its replacement. It returns a modified string.

For the sake of illustration, the lambda that inverts the case of the letters in a string is also embedded in the call to **changeStr( )**. However, in this case, rather unwieldy code is produced that is somewhat hard to follow. Usually, it is better to assign such a lambda to a separate reference variable (as was done for the string-reversing lambda), and then pass that variable to the method. Of course, it is technically correct to pass a block lambda as an argument, as the example shows.

One other point: notice that the invert-case lambda uses the **static** methods **isUpperCase( )**, **toUpperCase( )**, and **toLowerCase( )** defined by **Character**. Recall that **Character** is a wrapper class for **char**. The **isUpperCase( )** method returns **true** if its argument is an uppercase letter and **false** otherwise. The **toUpperCase( )** and **toLowerCase( )** perform the indicated action and return the result. In addition to these methods, **Character** defines several others that manipulate or test characters. You will want to explore them on your own.

8. Here is all the code assembled into a complete program.

```java
// Use a lambda expression as an argument to a method.

interface StringFunc {
  String func(String str);
}

class LambdaArgumentDemo {

  // This method has a functional interface as the type of its
  // first parameter. Thus, it can be passed a reference to any
  // instance of that interface, including an instance created
  // by a lambda expression. The second parameter specifies the
  // string to operate on.
  static String changeStr(StringFunc sf, String s) {
    return sf.func(s);
  }

  public static void main(String[] args)
  {
    String inStr = "Lambda Expressions Expand Java";
    String outStr;

    System.out.println("Here is input string: " + inStr);

    // Define a lambda expression that reverses the contents
    // of a string and assign it to a StringFunc reference variable.
```

(continued)

```
        StringFunc reverse = (str) ->  {
          String result = "";

          for(int i = str.length()-1; i >= 0; i--)
            result += str.charAt(i);

          return result;
        };

      // Pass reverse to the first argument to changeStr().
      // Pass the input string as the second argument.
      outStr = changeStr(reverse, inStr);
      System.out.println("The string reversed: " + outStr);


      // This lambda expression replaces spaces with hyphens.
      // It is embedded directly in the call to changeStr().
      outStr = changeStr((str) -> str.replace(' ', '-'), inStr);
      System.out.println("The string with spaces replaced: " + outStr);

      // This block lambda inverts the case of the characters in the
      // string. It is also embedded directly in the call to changeStr().
      outStr = changeStr((str) -> {
                           String result = "";
                           char ch;

                           for(int i = 0; i < str.length(); i++ ) {
                             ch = str.charAt(i);
                             if(Character.isUpperCase(ch))
                               result += Character.toLowerCase(ch);
                             else
                               result += Character.toUpperCase(ch);
                           }
                           return result;
                         }, inStr);

      System.out.println("The string in reversed case: " + outStr);
    }
  }
```

The following output is produced:

```
Here is input string: Lambda Expressions Expand Java
The string reversed: avaJ dnapxE snoisserpxE adbmaL
The string with spaces replaced: Lambda-Expressions-Expand-Java
The string in reversed case: lAMBDA eXPRESSIONS eXPAND jAVA
```

Ask the Expert

Q: In addition to variable initialization, assignment, and argument passing, what other places constitute a target type context for a lambda expression?

A: Casts, the ? operator, array initializers, **return** statements, and lambda expressions, themselves, can also serve as target type contexts.

Lambda Expressions and Variable Capture

Variables defined by the enclosing scope of a lambda expression are accessible within the lambda expression. For example, a lambda expression can use an instance variable or **static** variable defined by its enclosing class. A lambda expression also has access to **this** (both explicitly and implicitly), which refers to the invoking instance of the lambda expression's enclosing class. Thus, a lambda expression can obtain or set the value of an instance variable or **static** variable and call a method defined by its enclosing class.

However, when a lambda expression uses a local variable from its enclosing scope, a special situation is created that is referred to as a *variable capture*. In this case, a lambda expression may only use local variables that are *effectively final*. An effectively final variable is one whose value does not change after it is first assigned. There is no need to explicitly declare such a variable as **final**, although doing so would not be an error. (The **this** parameter of an enclosing scope is automatically effectively final, and lambda expressions do not have a **this** of their own.)

It is important to understand that a local variable of the enclosing scope cannot be modified by the lambda expression. Doing so would remove its effectively final status, thus rendering it illegal for capture.

The following program illustrates the difference between effectively final and mutable local variables:

```
// An example of capturing a local variable from the enclosing scope.

interface MyFunc {
  int func(int n);
}

class VarCapture {
  public static void main(String[] args)
  {
    // A local variable that can be captured.
    int num = 10;

    MyFunc myLambda = (n) -> {
      // This use of num is OK. It does not modify num.
      int v = num + n;
```

```
        // However, the following is illegal because it attempts
        // to modify the value of num.
//      num++;

        return v;
    };

    // Use the lambda. This will display 18.
    System.out.println(myLambda.func(8));

    // The following line would also cause an error, because
    // it would remove the effectively final status from num.
//   num = 9;
    }
}
```

As the comments indicate, **num** is effectively final and can, therefore, be used inside **myLambda**. This is why the **println( )** statement outputs the number 18. When **func( )** is called with the argument 8, the value of **v** inside the lambda is set by adding **num** (which is 10) to the value passed to **n** (which is 8). Thus, **func( )** returns 18. This works because **num** is not modified after it is initialized. However, if **num** were to be modified, either inside the lambda or outside of it, **num** would lose its effectively **final** status. This would cause an error, and the program would not compile.

It is important to emphasize that a lambda expression can use and modify an instance variable from its invoking class. It just can't use a local variable of its enclosing scope unless that variable is effectively final.

Throw an Exception from Within a Lambda Expression

A lambda expression can throw an exception. If it throws a checked exception, however, then that exception must be compatible with the exception(s) listed in the **throws** clause of the abstract method in the functional interface. For example, if a lambda expression throws an **IOException**, then the abstract method in the functional interface must list **IOException** in a **throws** clause. This situation is demonstrated by the following program:

```
import java.io.*;

interface MyIOAction {
  boolean ioAction(Reader rdr) throws IOException;
}

class LambdaExceptionDemo {

  public static void main(String[] args)
  {
```

```
    // This block lambda could throw an IOException.
    // Thus, IOException must be specified in a throws
    // clause of ioAction() in MyIOAction.
    MyIOAction myIO = (rdr) -> {                      ─────── This lambda might
      int ch = rdr.read(); // could throw IOException          throw an exception.
      // ...
      return true;
    };
  }
}
```

Because a call to **read( )** could result in an **IOException**, the **ioAction( )** method of the functional interface **MyIOAction** must include **IOException** in a **throws** clause. Without it, the program will not compile because the lambda expression will no longer be compatible with **ioAction( )**. To prove this, simply remove the **throws** clause and try compiling the program. As you will see, an error will result.

Ask the Expert

Q: Can a lambda expression use a parameter that is an array?

A: Yes. However, when the type of the parameter is inferred, the parameter to the lambda expression is *not* specified using the normal array syntax. Rather, the parameter is specified as a simple name, such as **n**, not as **n[ ]**. Remember, the type of a lambda expression parameter will be inferred from the target context. Thus, if the target context requires an array, then the parameter's type will automatically be inferred as an array. To better understand this, let's work through a short example.

Here is a generic functional interface called **MyTransform**, which can be used to apply some transform to the elements of an array:

```
// A functional interface.
interface MyTransform<T> {
  void transform(T[] a);
}
```

Notice that the parameter to the **transform( )** method is an array of type **T**. Now, consider the following lambda expression that uses **MyTransform** to convert the elements of an array of **Double** values into their square roots:

```
MyTransform<Double> sqrts = (v) -> {
  for(int i=0; i < v.length; i++) v[i] = Math.sqrt(v[i]);
};
```

Here, the type of **a** in **transform( )** is **Double[ ]**, because **Double** is specified as the type parameter for **MyTransform** when **sqrts** is declared. Therefore, the type of **v** in the lambda expression is inferred as **Double[ ]**. It is not necessary (or legal) to specify it as **v[ ]**.

One last point: It is legal to declare the lambda parameter as **Double[ ] v**, because doing so explicitly declares the type of the parameter. However, doing so gains nothing in this case.

Method References

There is an important feature related to lambda expressions called the *method reference*. A method reference provides a way to refer to a method without executing it. It relates to lambda expressions because it, too, requires a target type context that consists of a compatible functional interface. When evaluated, a method reference also creates an instance of a functional interface. There are different types of method references. We will begin with method references to **static** methods.

Method References to static Methods

A method reference to a **static** method is created by specifying the method name preceded by its class name, using this general syntax:

ClassName::*methodName*

Notice that the class name is separated from the method name by a double colon. The **::** is a separator that was added to Java by JDK 8 expressly for this purpose. This method reference can be used anywhere in which it is compatible with its target type.

The following program demonstrates the **static** method reference. It does so by first declaring a functional interface called **IntPredicate** that has a method called **test( )**. This method has an **int** parameter and returns a **boolean** result. Thus, it can be used to test an integer value against some condition. The program then creates a class called **MyIntPredicates**, which defines three **static** methods, with each one checking if a value satisfies some condition. The methods are called **isPrime( )**, **isEven( )**, and **isPositive( )**, and each method performs the test indicated by its name. Inside **MethodRefDemo**, a method called **numTest( )** is created that has as its first parameter, a reference to **IntPredicate**. Its second parameter specifies the integer being tested. Inside **main( )**, three different tests are performed by calling **numTest( )**, passing in a method reference to the test to perform.

```
// Demonstrate a method reference for a static method.

// A functional interface for numeric predicates that operate
// on integer values.
interface IntPredicate {
  boolean test(int n);
}

// This class defines three static methods that check an integer
// against some condition.
class MyIntPredicates {
  // A static method that returns true if a number is prime.
  static boolean isPrime(int n) {

    if(n < 2) return false;
```

```
    for(int i=2; i <= n/i; i++) {
      if((n % i) == 0)
        return false;
    }
    return true;
  }

  // A static method that returns true if a number is even.
  static boolean isEven(int n) {
    return (n % 2) == 0;
  }

  // A static method that returns true if a number is positive.
  static boolean isPositive(int n) {
    return n > 0;
  }
}

class MethodRefDemo {

  // This method has a functional interface as the type of its
  // first parameter. Thus, it can be passed a reference to any
  // instance of that interface, including one created by a
  // method reference.
  static boolean numTest(IntPredicate p, int v) {
    return p.test(v);
  }

  public static void main(String[] args)
  {
    boolean result;

    // Here, a method reference to isPrime is passed to numTest().
    result = numTest(MyIntPredicates::isPrime, 17);
    if(result) System.out.println("17 is prime.");

    // Next, a method reference to isEven is used.
    result = numTest(MyIntPredicates::isEven, 12);
    if(result) System.out.println("12 is even.");

    // Now, a method reference to isPositive is passed.
    result = numTest(MyIntPredicates::isPositive, 11);
    if(result) System.out.println("11 is positive.");
  }
}
```

Use method references to a **static** method.

The output is shown here:

```
17 is prime.
12 is even.
11 is positive.
```

In the program, pay special attention to this line:

```
result = numTest(MyIntPredicates::isPrime, 17);
```

Here, a reference to the **static** method **isPrime( )** is passed as the first argument to **numTest( )**. This works because **isPrime** is compatible with the **IntPredicate** functional interface. Thus, the expression **MyIntPredicates::isPrime** evaluates to a reference to an object in which **isPrime( )** provides the implementation of **test( )** in **IntPredicate**. The other two calls to **numTest( )** work in the same way.

Method References to Instance Methods

A reference to an instance method on a specific object is created by this basic syntax:

objRef::*methodName*

As you can see, the syntax is similar to that used for a **static** method, except that an object reference is used instead of a class name. Thus, the method referred to by the method reference operates relative to *objRef*. The following program illustrates this point. It uses the same **IntPredicate** interface and **test( )** method as the previous program. However, it creates a class called **MyIntNum**, which stores an **int** value and defines the method **isFactor( )**, which determines if the value passed is a factor of the value stored by the **MyIntNum** instance. The **main( )** method then creates two **MyIntNum** instances. It then calls **numTest( )**, passing in a method reference to the **isFactor( )** method and the value to be checked. In each case, the method reference operates relative to the specific object.

```
// Use a method reference to an instance method.

// A functional interface for numeric predicates that operate
// on integer values.
interface IntPredicate {
  boolean test(int n);
}

// This class stores an int value and defines the instance
// method isFactor(), which returns true if its argument
// is a factor of the stored value.
class MyIntNum {
  private int v;

  MyIntNum(int x) { v = x; }
```

```
  int getNum() { return v; }

  // Return true if n is a factor of v.
  boolean isFactor(int n) {
    return (v % n) == 0;
  }
}

class MethodRefDemo2 {

  public static void main(String[] args)
  {
    boolean result;

    MyIntNum myNum = new MyIntNum(12);
    MyIntNum myNum2 = new MyIntNum(16);

    // Here, a method reference to isFactor on myNum is created.
    IntPredicate ip = myNum::isFactor; ◄──────────────── A method reference
                                                          to an instance method
    // Now, it is used to call isFactor() via test().
    result = ip.test(3);
    if(result) System.out.println("3 is a factor of " + myNum.getNum());


    // This time, a method reference to isFactor on myNum2 is created.
    // and used to call isFactor() via test().
    ip = myNum2::isFactor; ◄──────
    result = ip.test(3);
    if(!result) System.out.println("3 is not a factor of " + myNum2.getNum());
  }
}
```

This program produces the following output:

```
3 is a factor of 12
3 is not a factor of 16
```

In the program, pay special attention to the line

```
IntPredicate ip = myNum::isFactor;
```

Here, the method reference assigned to **ip** refers to an instance method **isFactor( )** on **myNum**. Thus, when **test( )** is called through that reference, as shown here:

```
result = ip.test(3);
```

the method will call **isFactor( )** on **myNum**, which is the object specified when the method reference was created. The same situation occurs with the method reference **myNum2::isFactor**, except that **isFactor( )** will be called on **myNum2**. This is confirmed by the output.

It is also possible to handle a situation in which you want to specify an instance method that can be used with any object of a given class—not just a specified object. In this case, you will create a method reference as shown here:

ClassName::instanceMethodName

Here, the name of the class is used instead of a specific object, even though an instance method is specified. With this form, the first parameter of the functional interface method matches the invoking object and the second parameter matches the parameter (if any) specified by the instance method. Here is an example. It reworks the previous example. First, it replaces **IntPredicate** with the interface **MyIntNumPredicate**. In this case, the first parameter to **test( )** is of type **MyIntNum**. It will be used to receive the object being operated upon. This allows the program to create a method reference to the instance method **isFactor( )** that can be used with any **MyIntNum** object.

```
// Use an instance method reference to refer to any instance.

// A functional interface for numeric predicates that operate
// on an object of type MyIntNum and an integer value.
interface MyIntNumPredicate {
  boolean test(MyIntNum mv, int n);
}

// This class stores an int value and defines the instance
// method isFactor(), which returns true if its argument
// is a factor of the stored value.
class MyIntNum {
  private int v;

  MyIntNum(int x) { v = x; }

  int getNum() { return v; }

  // Return true if n is a factor of v.
  boolean isFactor(int n) {
    return (v % n) == 0;
  }
}

class MethodRefDemo3 {
  public static void main(String[] args)
  {
    boolean result;

    MyIntNum myNum = new MyIntNum(12);
    MyIntNum myNum2 = new MyIntNum(16);
```

```
      // This makes inp refer to the instance method isFactor().
      MyIntNumPredicate inp = MyIntNum::isFactor;  ◄────── A method reference to any
                                                            object of type MyIntNum
      // The following calls isFactor() on myNum.
      result = inp.test(myNum, 3);
      if(result)
        System.out.println("3 is a factor of " + myNum.getNum());

      // The following calls isFactor() on myNum2.
      result = inp.test(myNum2, 3);
      if(!result)
        System.out.println("3 is a not a factor of " + myNum2.getNum());
  }
}
```

Ask the Expert

Q: How do I specify a method reference to a generic method?

A: Often, because of type inference, you won't need to explicitly specify a type argument to a generic method when obtaining its method reference, but Java does include a syntax to handle those cases in which you do. For example, assuming the following:

```
interface SomeTest<T> {
  boolean test(T n, T m);
}

class MyClass {
  static <T> boolean myGenMeth(T x, T y) {
    boolean result = false;
    // ...
    return result;
  }
}
```

the following statement is valid:

```
SomeTest<Integer> mRef = MyClass::<Integer>myGenMeth;
```

Here, the type argument for the generic method **myGenMeth** is explicitly specified. Notice that the type argument occurs after the **::**. This syntax can be generalized: When a generic method is specified as a method reference, its type argument comes after the **::** and before the method name. In cases in which a generic class is specified, the type argument follows the class name and precedes the **::**.

The output is shown here:

```
3 is a factor of 12
3 is a not a factor of 16
```

In the program, pay special attention to this line:

```
MyIntNumPredicate inp = MyIntNum::isFactor;
```

It creates a method reference to the instance method **isFactor( )** that will work with any object of type **MyIntNum**. For example, when **test( )** is called through the **inp**, as shown here:

```
result = inp.test(myNum, 3);
```

it results in a call to **myNum.isFactor(3)**. In other words, **myNum** becomes the object on which **isFactor(3)** is called.

NOTE

A method reference can use the keyword **super** to refer to a superclass version of a method. The general forms of the syntax are **super**::*methodName* and *typeName*.**super**::*methodName*. In the second form, *typeName* must refer to the enclosing class or a superinterface.

Constructor References

Similar to the way that you can create references to methods, you can also create references to constructors. Here is the general form of the syntax that you will use:

classname::new

This reference can be assigned to any functional interface reference that defines a method compatible with the constructor. Here is a simple example:

```
// Demonstrate a Constructor reference.

// MyFunc is a functional interface whose method returns
// a MyClass reference.
interface MyFunc {
   MyClass func(String s);
}

class MyClass {
  private String str;

  // This constructor takes an argument.
  MyClass(String s) { str = s; }

  // This is the default constructor.
  MyClass() { str = ""; }
```

```
    // ...

    String getStr() { return str; }
}

class ConstructorRefDemo {
  public static void main(String[] args)
  {
    // Create a reference to the MyClass constructor.
    // Because func() in MyFunc takes an argument, new
    // refers to the parameterized constructor in MyClass,
    // not the default constructor.
    MyFunc myClassCons = MyClass::new;  ◄───────────────── A constructor reference

    // Create an instance of MyClass via that constructor reference.
    MyClass mc = myClassCons.func("Testing");

    // Use the instance of MyClass just created.
    System.out.println("str in mc is " + mc.getStr() );
  }
}
```

The output is shown here:

```
str in mc is Testing
```

In the program, notice that the **func( )** method of **MyFunc** returns a reference of type **MyClass** and has a **String** parameter. Next, notice that **MyClass** defines two constructors. The first specifies a parameter of type **String**. The second is the default, parameterless constructor. Now, examine the following line:

```
MyFunc myClassCons = MyClass::new;
```

Here, the expression **MyClass::new** creates a constructor reference to a **MyClass** constructor. In this case, because **MyFunc**'s **func( )** method takes a **String** parameter, the constructor being referred to is **MyClass(String s)** because it is the one that matches. Also notice that the reference to this constructor is assigned to a **MyFunc** reference called **myClassCons**. After this statement executes, **myClassCons** can be used to create an instance of **MyClass**, as this line shows:

```
MyClass mc = myClassCons.func("Testing");
```

In essence, **myClassCons** has become another way to call **MyClass(String s)**.

If you wanted **MyClass::new** to use **MyClass**'s default constructor, then you would need to use a functional interface that defines a method that has no parameter. For example, if you define **MyFunc2**, as shown here:

```
interface MyFunc2 {
  MyClass func();
}
```

then the following line will assign to **MyClassCons** a reference to **MyClass**'s default (i.e., parameterless) constructor:

```
MyFunc2 myClassCons = MyClass::new;
```

In general, the constructor that will be used when **::new** is specified is the one whose parameters match those specified by the functional interface.

Ask the Expert

Q: Can I declare a constructor reference that creates an array?

A: Yes. To create a constructor reference for an array, use this construct:

type[]::new

Here, *type* specifies the type of object being created. For example, assuming the form of **MyClass** shown in the preceding example and given the **MyClassArrayCreator** interface shown here:

```
interface MyClassArrayCreator {
    MyClass[] func(int n);
}
```

the following creates an array of **MyClass** objects and gives each element an initial value:

```
MyClassArrayCreator mcArrayCons = MyClass[]::new;
MyClass[] a = mcArrayCons.func(3);
for(int i=0; i < 3; i++)
  a[i] = new MyClass(i+"");
```

Here, the call to **func(3)** causes a three-element array to be created. This example can be generalized. Any functional interface that will be used to create an array must contain a method that takes a single **int** parameter and returns a reference to the array of the specified size.

As a point of interest, you can create a generic functional interface that can be used with other types of classes, as shown here:

```
interface MyArrayCreator<T> {
    T[] func(int n);
}
```

For example, you could create an array of five **Thread** objects like this:

```
MyArrayCreator<Thread> mcArrayCons = Thread[]::new;
Thread[] thrds = mcArrayCons.func(5);
```

One last point: In the case of creating a constructor reference for a generic class, you can specify the type parameter in the normal way, after the class name. For example, if **MyGenClass** is declared like this:

```
MyGenClass<T> { // ...
```

then the following creates a constructor reference with a type argument of **Integer**:

```
MyGenClass<Integer>::new;
```

Because of type inference, you won't always need to specify the type argument, but you can when necessary.

Predefined Functional Interfaces

Up to this point, the examples in this chapter have defined their own functional interfaces so that the fundamental concepts behind lambda expressions and functional interfaces could be clearly illustrated. In many cases, however, you won't need to define your own functional interface because the package **java.util.function** provides several predefined ones. Here is a sampling:

Interface	Purpose
UnaryOperator<T>	Apply a unary operation to an object of type **T** and return the result, which is also of type **T**. Its method is called **apply( )**.
BinaryOperator<T>	Apply an operation to two objects of type **T** and return the result, which is also of type **T**. Its method is called **apply( )**.
Consumer<T>	Apply an operation on an object of type **T**. Its method is called **accept( )**.
Supplier<T>	Return an object of type **T**. Its method is called **get( )**.
Function<T, R>	Apply an operation to an object of type **T** and return the result as an object of type **R**. Its method is called **apply( )**.
Predicate<T>	Determine if an object of type **T** fulfills some constraint. Returns a **boolean** value that indicates the outcome. Its method is called **test( )**.

The following program shows the **Predicate** interface in action. It uses **Predicate** as the functional interface for a lambda expression the determines if a number is even. **Predicate**'s abstract method is called **test( )**, and it is shown here:

```
boolean test(T val)
```

It must return **true** if *val* satisfies some constraint or condition. As it is used here, it will return **true** if *val* is even.

```
// Use the Predicate built-in functional interface.

// Import the Predicate interface.
import java.util.function.Predicate;

class UsePredicateInterface {
  public static void main(String[] args)
  {

    // This lambda uses Predicate<Integer> to determine
    // if a number is even.
    Predicate<Integer> isEven = (n) -> (n %2) == 0;      Use the built-in
                                                         Predicate interface.

    if(isEven.test(4)) System.out.println("4 is even");

    if(!isEven.test(5)) System.out.println("5 is odd");
  }
}
```

Ask the Expert

Q: At the start of this chapter, you mentioned that the inclusion of lambda expressions resulted in new capabilities being incorporated into the API library. Can you give me an example?

A: One example is the stream package **java.util.stream**. This package defines several stream interfaces, the most general of which is **Stream**. As it relates to **java.util.stream**, a *stream* is a conduit for data. Thus, a stream represents a sequence of objects. Furthermore, a stream supports many types of operations that let you create a *pipeline* that performs a series of actions on the data. Often, these actions are represented by lambda expressions. For example, using the stream API, you can construct sequences of actions that resemble, in concept, the type of database queries for which you might use SQL. Furthermore, in many cases, such actions can be performed in parallel, thus providing a high level of efficiency, especially when large data sets are involved. Put simply, the stream API provides a powerful means of handling data in an efficient, yet easy to use way. One last point: although the streams supported by the new stream API have some similarities with the I/O streams described in Chapter 10, they are not the same.

The program produces the following output:

```
4 is even
5 is odd
```

Chapter 14 Self Test

1. What is the lambda operator?

2. What is a functional interface?

3. How do functional interfaces and lambda expressions relate?

4. What are the two general types of lambda expressions?

5. Show a lambda expression that returns **true** if a number is between 10 and 20, inclusive.

6. Create a functional interface that can support the lambda expression you created in question 5. Call the interface **MyTest** and its abstract method **testing( )**.

7. Create a block lambda that computes the factorial of an integer value. Demonstrate its use. Use **NumericFunc**, shown in this chapter, for the functional interface.

8. Create a generic functional interface called **MyFunc<T>**. Call its abstract method **func( )**. Have **func( )** return a reference of type **T**. Have it take a parameter of type **T**. (Thus, **MyFunc** will be a generic version of **NumericFunc** shown in the chapter.) Demonstrate its use by rewriting your answer to question 7 so it uses **MyFunc<T>** rather than **NumericFunc**.

9. Using the program shown in Try This 14-1, create a lambda expression that removes all spaces from a string and returns the result. Demonstrate this method by passing it to **changeStr( )**.

10. Can a lambda expression use a local variable? If so, what constraint must be met?

11. If a lambda expression throws a checked exception, the abstract method in the functional interface must have a **throws** clause that includes that exception. True or False?

12. What is a method reference?

13. When evaluated, a method reference creates an instance of the _____ _____ supplied by its target context.

14. Given a class called **MyClass** that contains a **static** method called **myStaticMethod( )**, show how to specify a method reference to **myStaticMethod( )**.

15. Given a class called **MyClass** that contains an instance method called **myInstMethod( )** and assuming an object of **MyClass** called **mcObj**, show how to create a method reference to **myInstMethod( )** on **mcObj**.

16. To the **MethodRefDemo2** program, add a new method to **MyIntNum** called **hasCommonFactor( )**. Have it return **true** if its **int** argument and the value stored in the invoking **MyIntNum** object have at least one factor in common. For example, 9 and 12 have a common factor, which is 3, but 9 and 16 have no common factor. Demonstrate **hasCommonFactor( )** via a method reference.

17. How is a constructor reference specified?

18. Java defines several predefined functional interfaces in what package?

Chapter 15

Modules

Key Skills & Concepts

- Know the definition of a module

- Know Java's module-related keywords

- Declare a module by use of the **module** keyword

- Use **requires** and **exports**

- Understand the purpose of **module-info.java**

- Use **javac** and **java** to compile and run module-based programs

- Understand the purpose of **java.base**

- Understand how pre-module legacy code is supported

- Export a package to a specific module

- Use implied readability

- Use services in a module

Beginning with JDK 9, an important feature called *modules* was added to Java. Modules give you a way to describe the relationships and dependencies of the code that comprises an application. Modules also let you control which parts of a module are accessible to other modules and which are not. Through the use of modules you can create more reliable, scalable programs.

As a general rule, modules are most helpful to large applications because they help reduce the management complexity often associated with a large software system. However, small programs also benefit from modules because the Java API library has now been organized into modules. Thus, it is now possible to specify which parts of the API are required by your program and which are not. This makes it possible to deploy programs with a smaller run-time footprint, which is especially important when creating code for small devices, such as those intended to be part of the Internet of Things (IoT).

Support for modules is provided both by language elements, including several keywords, and by enhancements to **javac**, **java**, and other JDK tools. Furthermore, new tools and file formats were introduced. As a result, the JDK and the run-time system were substantially upgraded to support modules. In short, modules constitute a major addition to, and evolution of, the Java language. This chapter introduces the key aspects of this important, recently added capability.

Module Basics

In its most fundamental sense, a *module* is a grouping of packages and resources that can be collectively referred to by the module's name. A *module declaration* specifies the name of a module and defines the relationship a module and its packages have to other modules. Module declarations are program statements in a Java source file and are supported by several module-related keywords added to Java by JDK 9. They are shown here:

exports	module	open	opens
provides	requires	to	transitive
uses	with		

It is important to understand that these keywords are recognized *as keywords* only in the context of a module declaration. Otherwise, they are interpreted as identifiers in other situations. Thus, the keyword **module** could, for example, also be used as a parameter name, but such a use is certainly not now recommended.

A module declaration is contained in a file called **module-info.java**. Thus, a module is defined in a Java source file. This file is then compiled by **javac** into a class file and is known as a *module descriptor*. The **module-info.java** file must contain only a module definition. It is not a general-purpose file.

A module declaration begins with the keyword **module**. Here is its general form:

```
module moduleName {
    // module definition
}
```

The name of the module is specified by *moduleName*, which must be a valid Java identifier or a sequence of identifiers separated by periods. The module definition is specified within the braces.

Ask the Expert

Q: Why are the module-related keywords, such as module and requires, recognized as keywords only in the context of a module declaration?

A: Restricting their use as keywords to a module declaration prevents problems with preexisting code that uses one or more of them as identifiers. For example, consider a situation in which a pre-JDK 9 program uses **requires** as the name of a variable. When that program is ported to a modern version of Java, if **requires** were recognized as a keyword outside a module declaration, then any other place in which it is used would result in a compilation error. By recognizing **requires** as a keyword only within a module declaration, any other uses of **requires** in the program are unaffected and remain valid. Of course, the same goes for the other module-related keywords.

Although a module definition may be empty (which results in a declaration that simply names the module), typically it specifies one or more clauses that define the characteristics of the module.

A Simple Module Example

At the foundation of a module's capabilities are two key features. The first is a module's ability to specify that it requires another module. In other words, one module can specify that it *depends* on another. A dependence relationship is specified by use of a **requires** statement. By default, the presence of the required module is checked at both compile time and run time. The second key feature is a module's ability to control which, if any, of its packages are accessible by another module. This is accomplished by use of the **exports** keyword. The public and protected types within a package are accessible to other modules only if they are explicitly exported. Here we will develop an example that introduces both of these features.

The following example creates a modular application that demonstrates some simple mathematical functions. Although this application is purposely very small, it illustrates the core concepts and procedures required to create, compile, and run module-based code. Furthermore, the general approach shown here also applies to larger, real-world applications. It is strongly recommended that you work through the example on your computer, carefully following each step.

NOTE

This chapter shows the process of creating, compiling, and running module-based code by use of the command-line tools. This approach has two advantages. First, it works for all Java programmers, because no IDE is required. Second, it very clearly shows the fundamentals of the module system, including how it utilizes directories. To follow along, you will need to manually create a number of directories and ensure that each file is placed in its proper directory. As you might expect, when creating real-world, module-based applications you will likely find a module-aware IDE easier to use because, typically, it will automate much of the process. However, learning the fundamentals of modules using the command-line tools ensures that you have a solid understanding of the topic.

The application defines two modules. The first module is called **appstart**. It contains a package called **appstart.mymodappdemo** that defines the application's entry point in a class called **MyModAppDemo**. Thus, **MyModAppDemo** contains the application's **main( )** method. The second module is called **appfuncs**. It contains a package called **appfuncs.simplefuncs** that includes the class **SimpleMathFuncs**. This class defines three static methods that implement some simple mathematical functions. The entire application will be contained in a directory tree that begins at **mymodapp**.

Before continuing, a few words about module names are appropriate. First, in the examples that follow, the name of a module (such as **appfuncs**) is the prefix of the name of a package (such as **appfuncs.simplefuncs**) that it contains. This is *not* required, but is used here as a way of clearly indicating to what module a package belongs. In general, when learning about and experimenting with modules, short, simple names, such as those used in this chapter, are helpful, and you can use any sort of convenient names that you like. However, when creating modules suitable for distribution, you must be careful with the names you choose because you will want those names to be unique. At the time of this writing, the suggested way to achieve this is to use the reverse domain name method. In this method, the reverse domain name of the domain that "owns" the project is used as a prefix for the module. For example, a project associated with **herbschildt.com** would use **com.herbschildt** as the module prefix. (The same goes for package names.) Because naming conventions may evolve over time, you will want to check the Java documentation for current recommendations.

Let's now begin. Start by creating the necessary source code directories by following these steps:

1. Create a directory called **mymodapp**. This is the top-level directory for the entire application.

2. Under **mymodapp**, create a subdirectory called **appsrc**. This is the top-level directory for the application's source code.

3. Under **appsrc**, create the subdirectory **appstart**. Under this directory, create a subdirectory also called **appstart**. Under this directory, create the directory **mymodappdemo**. Thus, beginning with **appsrc**, you will have created this tree:

   ```
   appsrc\appstart\appstart\mymodappdemo
   ```

4. Also under **appsrc**, create the subdirectory **appfuncs**. Under this directory, create a subdirectory also called **appfuncs**. Under this directory, create the directory called **simplefuncs**. Thus, beginning with **appsrc**, you will have created this tree:

   ```
   appsrc\appfuncs\appfuncs\simplefuncs
   ```

Your directory tree should look like that shown here.

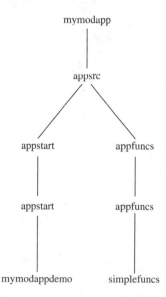

After you have set up these directories, you can create the application's source files.

This example will use four source files. Two are the source files that define the application. The first is **SimpleMathFuncs.java**, shown here. Notice that **SimpleMathFuncs** is packaged in **appfuncs.simplefuncs**.

```java
// Some simple math functions.

package appfuncs.simplefuncs;          Notice the package
                                       declaration.

public class SimpleMathFuncs {

  // Determine if a is a factor of b.
  public static boolean isFactor(int a, int b) {
    if((b%a) == 0) return true;
    return false;
  }

  // Return the smallest positive factor that a and b have in common.
  public static int lcf(int a, int b) {
    // Factor using positive values.
    a = Math.abs(a);
    b = Math.abs(b);

    int min = a < b ? a : b;

    for(int i = 2; i <= min/2; i++) {
      if(isFactor(i, a) && isFactor(i, b))
        return i;
    }

    return 1;
  }

  // Return the largest positive factor that a and b have in common.
  public static int gcf(int a, int b) {
    // Factor using positive values.
    a = Math.abs(a);
    b = Math.abs(b);

    int min = a < b ? a : b;

    for(int i = min/2; i >= 2; i--) {
      if(isFactor(i, a) && isFactor(i, b))
        return i;
    }

    return 1;
  }
}
```

SimpleMathFuncs defines three simple **static** math functions. The first, **isFactor( )**, returns true if **a** is a factor of **b**. The **lcf( )** method returns the smallest factor common to both **a** and **b**. In other words, it returns the least common factor of **a** and **b**. The **gcf( )** method returns the greatest common factor of **a** and **b**. In both cases, 1 is returned if no common factors are found. This file must be put in the following directory:

```
appsrc\appfuncs\appfuncs\simplefuncs
```

This is the **appfuncs.simplefuncs** package directory.

The second source file is **MyModAppDemo.java**, shown next. It uses the methods in **SimpleMathFuncs**. Notice that it is packaged in **appstart.mymodappdemo**. Also note that it imports the **SimpleMathFuncs** class because it depends on **SimpleMathFuncs** for its operation.

```
// Demonstrate a simple module-based application.
package appstart.mymodappdemo;              ←——————————— Notice the package
                                                         declaration and the
import appfuncs.simplefuncs.SimpleMathFuncs; ←————— import statement.

public class MyModAppDemo {
  public static void main(String[] args) {

    if(SimpleMathFuncs.isFactor(2, 10))
      System.out.println("2 is a factor of 10");

    System.out.println("Smallest factor common to both 35 and 105 is " +
                    SimpleMathFuncs.lcf(35, 105));

    System.out.println("Largest factor common to both 35 and 105 is " +
                    SimpleMathFuncs.gcf(35, 105));

  }
}
```

This file must be put in the following directory:

```
appsrc\appstart\appstart\mymodappdemo
```

This is the directory for the **appstart.mymodappdemo** package.

Next, you will need to add **module-info.java** files for each module. These files contain the module definitions. First, add this one, which defines the **appfuncs** module:

```
// Module definition for the functions module.
module appfuncs {                 ←——————————————— Define a module for appfuncs.
  // Exports the package appfuncs.simplefuncs.
  exports appfuncs.simplefuncs;
}
```

Notice that **appfuncs** exports the package **appfuncs.simplefuncs**, which makes it accessible to other modules. This file must be put into this directory:

```
appsrc\appfuncs
```

Thus, it goes in the **appfuncs** module directory, which is above the package directories.

Finally, add the **module-info.java** file for the **appstart** module. It is shown here. Notice that **appstart** requires the module **appfuncs**.

```
// Module definition for the main application module.
module appstart {   ◄──────────────────────────────── Define a module for appstart.
  // Requires the module appfuncs.
  requires appfuncs;
}
```

This file must be put into its module directory:

```
appsrc\appstart
```

Before examining the **requires**, **exports**, and **module** statements more closely, let's first compile and run this example. Be sure that you have correctly created the directories and entered each file into its proper directory, as just explained.

Compile and Run the First Module Example

Beginning with JDK 9, **javac** has been updated to support modules. Thus, like all other Java programs, module-based programs are compiled using **javac**. The process is easy, with the primary difference being that you will usually explicitly specify a *module path*. A module path tells the compiler where the compiled files will be located. When following along with this example, be sure that you execute the **javac** commands from the **mymodapp** directory in order for the paths to be correct. Recall that **mymodapp** is the top-level directory for the entire module application.

To begin, compile the **SimpleMathFuncs.java** file, using this command:

```
javac -d appmodules\appfuncs
  appsrc\appfuncs\appfuncs\simplefuncs\SimpleMathFuncs.java
```

Remember, this command *must be* executed from the **mymodapp** directory. Notice the use of the **-d** option. This tells **javac** where to put the output **.class** file. For the examples in this chapter, the top of the directory tree for compiled code is **appmodules**. This command will automatically create the output package directories for **appfuncs.simplefuncs** under **appmodules\appfuncs** as needed.

Next, here is the **javac** command that compiles the **module-info.java** file for the **appfuncs** module:

```
javac -d appmodules\appfuncs appsrc\appfuncs\module-info.java
```

This puts the **module-info.class** file into the **appmodules\appfuncs** directory.

Although the preceding two-step process works, it was shown primarily for the sake of discussion. It is usually easier to compile a module's **module-info.java** file and its source files in one command line. Here, the preceding two **javac** commands are combined into one:

```
javac -d appmodules\appfuncs appsrc\appfuncs\module-info.java
   appsrc\appfuncs\appfuncs\simplefuncs\SimpleMathFuncs.java
```

In this case, each compiled file is put in its proper module or package directory.

Now, compile the **module-info.java** and **MyModAppDemo.java** files for the **appstart** module, using this command:

```
javac --module-path appmodules -d appmodules\appstart
   appsrc\appstart\module-info.java
   appsrc\appstart\appstart\mymodappdemo\MyModAppDemo.java
```

Notice the **--module-path** option. It specifies the module path, which is the path on which the compiler will look for the user-defined modules required by the **module-info.java** file. In this case, it will look for the **appfuncs** module because it is needed by the **appstart** module. Also, notice that it specifies the output directory as **appmodules\appstart**. This means that the **module-info.class** file will be in the **appmodules\appstart** module directory and **MyModAppDemo.class** will be in the **appmodules\appstart\appstart\mymodappdemo** package directory.

Once you have completed the compilation, you can run the application with this **java** command:

```
java --module-path appmodules -m appstart/appstart.mymodappdemo.MyModAppDemo
```

Here, the **--module-path** option specifies the path to the application's modules. As mentioned, **appmodules** is the directory at the top of the compiled modules tree. The **-m** option specifies the class that contains the entry point of the application and, in this case, the name of the class that contains the **main( )** method. When you run the program, you will see the following output:

```
2 is a factor of 10
Smallest factor common to both 35 and 105 is 5
Largest factor common to both 35 and 105 is 7
```

A Closer Look at requires and exports

The preceding module-based example relies on the two foundational features of the module system: the ability to specify a dependence and the ability to satisfy that dependence. These capabilities are specified through the use of the **requires** and **exports** statements within a **module** declaration. Each merits a closer examination at this time.

Here is the form of the **requires** statement used in the example:

requires *moduleName*;

Here, *moduleName* specifies the name of a module that is required by the module in which the **requires** statement occurs. This means that the required module must be present in order for the current module to compile. In the language of modules, the current module is said to *read* the module specified in the **requires** statement. In general, the **requires** statement gives you a way to ensure that your program has access to the modules that it needs.

Here is the general form of the **exports** statement used in the example:

exports *packageName*;

Here, *packageName* specifies the name of the package that is exported by the module in which this statement occurs. When a module exports a package, it makes all of the public and protected types in the package accessible to other modules. Furthermore, the public and protected members of those types are also accessible. However, if a package within a module is not exported, then it is private to that module, including all of its public types. For example, even though a class is declared as **public** within a package, if that package is not explicitly exported by an **exports** statement, then that class is not accessible to other modules. It is important to understand that the public and protected types of a package, whether exported or not, are always accessible within that package's module. The **exports** statement simply makes them accessible to outside modules. Thus, any nonexported package is only for the internal use of its module.

The key to understanding **requires** and **exports** is that they work together. If one module depends on another, then it must specify that dependence with **requires**. The module on which another depends must explicitly export (i.e., make accessible) the packages that the dependent module needs. If either side of this dependence relationship is missing, the dependent module will not compile. As it relates to the foregoing example, **MyModAppDemo** uses the functions in **SimpleMathFuncs**. As a result, the **appstart** module declaration contains a **requires** statement that names the **appfuncs** module. The **appfuncs** module declaration exports the **appfuncs .simplefuncs** package, thus making the public types in the **SimpleMathFuncs** class available. Since both sides of the dependence relationship have been fulfilled, the application can compile and run. If either is missing, the compilation will fail. (You will see the results of a missing **exports** statement when you answer exercise 10 in the self-test at the end of this chapter.)

It is important to emphasize that **requires** and **exports** statements must occur only within a **module** statement. Furthermore, a **module** statement must occur by itself in a file called **module-info.java**.

java.base and the Platform Modules

As mentioned at the start of this chapter, beginning with JDK 9 the Java API packages have been incorporated into modules. In fact, the modularization of the API is one of the primary benefits realized by the addition of the modules. Because of their special role, the API modules are referred to as *platform modules,* and their names all begin with the prefix **java**. Here are some examples: **java.base**, **java.desktop**, and **java.xml**. By modularizing the API, it becomes possible to deploy an application with only the packages that it requires, rather than the entire Java Runtime Environment (JRE). Because of the size of the full JRE, this is a very important improvement.

The fact that all of the Java API library packages are now in modules gives rise to the following question: How can the **main( )** method in **MyModAppDemo** in the preceding example use **System.out.println( )** without specifying a **requires** statement for the module that contains the **System** class? Obviously, the program will not compile and run unless **System** is present. The same question also applies to the use of the **Math** class in **SimpleMathFuncs**. The answer to this question is found in **java.base**.

Of the platform modules, the most important is **java.base**. It includes and exports those packages fundamental to Java, such as **java.lang**, **java.io**, and **java.util**, among many others. Because of its importance, **java.base** is *automatically accessible* to all modules. Furthermore, all other modules automatically require **java.base**. There is no need to include a **requires java.base** statement in a module declaration. (As a point of interest, it is not wrong to explicitly specify **java.base**; it's just not necessary.) Thus, in much the same way that **java.lang** is automatically available to all programs without the use of an **import** statement, the **java.base** module is automatically accessible to all module-based programs without explicitly requesting it.

Because **java.base** contains the **java.lang** package, and **java.lang** contains the **System** class, **MyModAppDemo** in the preceding example can automatically use **System.out.println( )** without an explicit **requires** statement. The same applies to the use of the **Math** class in **SimpleMathFuncs**, because the **Math** class is also in **java.lang**. As you will see when you begin to create your own module-based applications, many of the API classes you will commonly need are in the packages included in **java.base**. Thus, the automatic inclusion of **java.base** simplifies the creation of module-based code because Java's core packages are automatically accessible.

One last point: Beginning with JDK 9, the documentation for the Java API now tells you the name of the module in which a package is contained. If the module is **java.base**, then you can use the contents of that package directly. Otherwise, your module declaration must include a **requires** clause for the desired module.

Ask the Expert

Q: I recall that JDK 8 had the ability to use a feature called *compact profiles*. Are compact profiles a part of modules?

A: Compact profiles are a feature that, in some situations, let you specify a subset of the API library. They are not part of the module system. Moreover, the module system introduced by JDK 9 fully supersedes them.

Legacy Code and the Unnamed Module

Another question may have occurred to you when working through the first example module program. Because Java now supports modules, and the API packages are also contained in modules, why do all of the other programs in the preceding chapters compile and run without error even though they do not use modules? More generally, since there is now over 20 years of Java code in existence and (at the time of this writing) much of that code does not use modules, how is it possible to compile, run, and maintain that legacy code with a modern compiler? Given Java's original philosophy of "write once, run everywhere," this is a very important question because backward capability must be maintained. As you will see, Java answers this question by providing an elegant, nearly transparent means of ensuring backward compatibility with preexisting code.

Support for legacy code is provided by two key features. The first is the *unnamed module*. When you use code that is not part of a named module, it automatically becomes part of the unnamed module. The unnamed module has two important attributes. First, all of the packages in the unnamed module are automatically exported. Second, the unnamed module can access any and all other modules. Thus, when a program does not use modules, all API modules in the Java platform are automatically accessible through the unnamed module.

The second key feature that supports legacy code is the automatic use of the class path, rather than the module path. When you compile a program that does not use modules, the class path mechanism is employed, just as it has been since Java's original release. As a result, the program is compiled and run in the same way it was prior to the advent of modules.

Because of the unnamed module and the automatic use of the class path, there was no need to declare any modules for the sample programs shown elsewhere in this book. They run properly whether you compile them with a modern compiler or an earlier one, such as JDK 8. Thus, even though modules are a feature that has a significant impact on Java, compatibility with legacy code is maintained. This approach also provides a smooth, nonintrusive, nondisruptive transition path to modules. Thus, it enables you to move a legacy application to modules at your own pace. Furthermore, it allows you to avoid the use of modules when they are not needed.

Before moving on, an important point needs to be made. For the types of example programs used elsewhere in this book, and for example programs in general, there is no benefit in using modules. Modularizing them would simply add clutter and complicate them for no reason or benefit. Furthermore, for many simple programs that you will write when learning the essentials of Java, there is no need to contain them in modules. For the reasons stated at the start of this chapter, modules are often of the greatest benefit when creating commercial programs. Therefore, no examples outside this chapter will use modules. This also allows the examples to be compiled and run in a pre-JDK 9 environment, which is important to readers using an older version of Java. Thus, except for the examples in this chapter, the examples in this book work for both pre-module and post-module JDKs.

Exporting to a Specific Module

The basic form of the **exports** statement makes a package accessible to any and all other modules. This is often exactly what you want. However, in some specialized development situations, it can be desirable to make a package accessible to only a *specific set* of modules, not *all* other modules. For example, a library developer might want to export a support package to certain other modules within the library, but not make it available for general use. Adding a **to** clause to the **exports** statement provides a means by which this can be accomplished.

In an **exports** statement, the **to** clause specifies a list of one or more modules that have access to the exported package. Furthermore, only those modules named in the **to** clause will have access. In the language of modules, the **to** clause creates what is known as a *qualified export*.

The form of **exports** that includes **to** is shown here:

exports *packageName* to *moduleNames*;

Here, *moduleNames* is a comma-separated list of modules to which the exporting module grants access.

You can try the **to** clause by changing the **module-info.java** file for the **appfuncs** module, as shown here:

```
// Module definition that uses a to clause.
module appfuncs {
  // Exports the package appfuncs.simplefuncs to appstart.
  exports appfuncs.simplefuncs to appstart;  ◄────── A qualified export.
}
```

Now, **simplefuncs** is exported only to **appstart** and to no other modules. After making this change, you can recompile the application by using this **javac** command:

```
javac -d appmodules --module-source-path appsrc
   appsrc\appstart\appstart\mymodappdemo\MyModAppDemo.java
```

After compiling, you can run the application as shown earlier.

This example also uses another module-related feature. Look closely at the preceding **javac** command. First, notice that it specifies the --**module-source-path** option. The module source path specifies the top of the module source tree. The --**module-source-path** option automatically compiles the files in the tree under the specified directory, which is **appsrc** in this example. The --**module-source-path** option must be used with the -**d** option to ensure that the compiled modules are stored in their proper directories under **appmodules**. This form of **javac** is called *multimodule mode* because it enables more than one module to be compiled at a time. The multimodule compilation mode is especially helpful here because the **to** clause refers to a specific module, and the requiring module must have access to the exported package. Thus, in this case, both **appstart** and **appfuncs** are needed to avoid warnings and/or errors during compilation. Multimodule mode avoids this problem because both modules are being compiled at the same time.

The multimodule mode of **javac** has another advantage. It automatically finds and compiles all source files for the application, creating the necessary output directories. Because of the advantages that multimodule compilation mode offers, it will be used for the subsequent examples.

NOTE

As a general rule, qualified export is a special case feature. Most often, your modules will either provide unqualified export of a package or not export the package at all, keeping it inaccessible. As such, qualified export is discussed here primarily for the sake of completeness. Also, qualified export by itself does not prevent the exported package from being misused by malicious code in a module that masquerades as the targeted module. The security techniques required to prevent this from happening are beyond the scope of this book. Consult the Oracle documentation for details on security in this regard and Java security details in general.

Using requires transitive

Consider a situation in which there are three modules, A, B, and C, that have the following dependences:

- A requires B.
- B requires C.

Given this situation, it is clear that since A depends on B and B depends on C, A has an indirect dependence on C. As long as A does not directly use any of the contents of C, then you can simply have A require B in its module-info file, and have B export the packages required by A in its module-info file, as shown here:

```
// A's module-info file:
module A {
  requires B;
}
```

```
// B's module-info file.
module B {
  exports somepack;
  requires C;
}
```

Here, *somepack* is a placeholder for the package exported by B and used by A. Although this works as long as A does not need to use anything defined in C, a problem occurs if A *does* want to access a type in C. In this case, there are two solutions.

The first solution is to simply add a **requires C** statement to A's file, as shown here:

```
// A's module-info file updated to explicitly require C:
module A {
  requires B;
  requires C; // also require C
}
```

This solution certainly works, but if B will be used by many modules, you must add **requires C** to all module definitions that require B. This is not only tedious; it is also error prone. Fortunately, there is a better solution. You can create an *implied dependence* on C. Implied dependence is also referred to as *implied readability*.

To create an implied dependence, add the **transitive** keyword after **requires** in the clause that requires the module upon which an implied readability is needed. In the case of this example, you would change B's module-info file as shown here:

```
// B's module-info file.
module B {
  exports somepack;
  requires transitive C;
}
```

Here, C is now required as transitive. After making this change, any module that depends on B will also automatically depend on C. Thus, A would automatically have access to C.

As a point of interest, because of a special exception in the Java syntax, in a **requires** statement, if **transitive** is immediately followed by a separator (such as a semicolon), it is interpreted as an identifier (for example, as a module name) rather than a keyword.

Try This 15-1   Experiment with requires transitive

```
MyModAppDemo.java
SimpleFuncs.java
SupportFuncs.java
module-info.java
```

You can experiment with **requires transitive** by reworking the preceding modular application example. Here, you will remove the **isFactor( )** method from the **SimpleMathFuncs** class in the **appfuncs.simplefuncs** package and put it into a new class, module, and package. The new class will be called **SupportFuncs**, the module will be called **appsupport**, and the package will be called **appsupport.supportfuncs**. The **appfuncs** module will then add a dependence on the **appsupport** module by use of **requires transitive**. This will enable both the **appfuncs** and **appstart** modules to access it without **appstart** having to provide its own **requires** statement. This works because **appstart** receives access to it through an **appfuncs** **requires transitive** statement.

1. To begin, create the source directories that support the new **appsupport** module. To do so, create **appsupport** under the **appsrc** directory. This is the module directory for the support functions. Under **appsupport**, create the package directory by adding the **appsupport** subdirectory followed by the **supportfuncs** subdirectory. Thus, the directory tree for **appsupport** should now look like this:

   ```
   appsrc\appsupport\appsupport\supportfuncs
   ```

   (continued)

2. Add the following **module-info.java** file to the module source directory for **appsupport**, which is **appsrc\appsupport**:

```
// Module definition for appsupport.
module appsupport {
  exports appsupport.supportfuncs;
}
```

3. In the **appsupport.supportfuncs** package directory, add the following file called **SupportFuncs.java**:

```
// Support functions.

package appsupport.supportfuncs;

public class SupportFuncs {

  // Determine if a is a factor of b.
  public static boolean isFactor(int a, int b) {
    if((b%a) == 0) return true;
    return false;
  }
}
```

As you can see, the **isFactor( )** method is now in **SupportFuncs** rather than in **SimpleMathFuncs**.

4. Remove **isFactor( )** from **SimpleMathFuncs**. Thus, **SimpleMathFuncs.java** will now look like this:

```
// Some simple math functions, with isFactor() removed.

package appfuncs.simplefuncs;
import appsupport.supportfuncs.SupportFuncs;

public class SimpleMathFuncs {

  // Return the smallest positive factor that a and b have in common.
  public static int lcf(int a, int b) {
    // Factor using positive values.
    a = Math.abs(a);
    b = Math.abs(b);

    int min = a < b ? a : b;

    for(int i = 2; i <= min/2; i++) {
      if(SupportFuncs.isFactor(i, a) && SupportFuncs.isFactor(i, b))
        return i;
    }

    return 1;
  }
```

```
      // Return the largest positive factor that a and b have in common.
      public static int gcf(int a, int b) {
        // Factor using positive values.
        a = Math.abs(a);
        b = Math.abs(b);

        int min = a < b ? a : b;

        for(int i = min/2; i >= 2; i--) {
          if(SupportFuncs.isFactor(i, a) && SupportFuncs.isFactor(i, b))
            return i;
        }

        return 1;
      }
    }
```

Notice that now the **SupportFuncs** class is imported, and calls to **isFactor( )** are referred to through the class name **SupportFuncs**.

5. Change the **module-info.java** file for **appfuncs** so that in its **requires** statement, **appsupport** is specified as **transitive**, as shown here:

```
    // Module definition for appfuncs.
    module appfuncs {
      // Exports the package appfuncs.simplefuncs.
      exports appfuncs.simplefuncs;

      // Requires appsupport and makes it transitive.
      requires transitive appsupport;
    }
```

6. Because **appfuncs** requires **appsupport** as **transitive**, there is no need for the **module-info .java** file for **appstart** to also require it. Its dependence on **appsupport** is implied. Thus, no changes to the **module-info.java** file for **appstart** are required.

7. Update **MyModAppDemo.java** to reflect these changes. Specifically, it must now import the **SupportFuncs** class and specify it when invoking **isFactor( )**, as shown here:

```
    // Updated to use SupportFuncs.
    package appstart.mymodappdemo;

    import appfuncs.simplefuncs.SimpleMathFuncs;
    import appsupport.supportfuncs.SupportFuncs;

    public class MyModAppDemo {
      public static void main(String[] args) {

        // Now, isFactor() is referred to via SupportFuncs,
        // not SimpleMathFuncs.
        if(SupportFuncs.isFactor(2, 10))
          System.out.println("2 is a factor of 10");
```

(continued)

```
System.out.println("Smallest factor common to both 35 and 105 is " +
                   SimpleMathFuncs.lcf(35, 105));

System.out.println("Largest factor common to both 35 and 105 is " +
                   SimpleMathFuncs.gcf(35, 105));

  }
}
```

8. Recompile the entire program using this multimodule compilation command:

```
javac -d appmodules --module-source-path appsrc
    appsrc\appstart\appstart\mymodappdemo\MyModAppDemo.java
```

As explained earlier, the multimodule compilation will automatically create the parallel module subdirectories under the **appmodules** directory.

9. Run the application as before, using this command:

```
java --module-path appmodules -m appstart/appstart.mymodappdemo.MyModAppDemo
```

It will produce the same output as before.

10. As an experiment, remove the **transitive** specifier from the **module-info.java** file for **appfuncs** and then try recompiling. As you will see, an error will result because **appsupport** is no longer accessible by **appstart**.

11. Here is another experiment. In the module-info file for **appsupport**, try exporting the **appsupport.supportfuncs** package to only **appfuncs** by use of a qualified export, as shown here:

```
exports appsupport.supportfuncs to appfuncs;
```

Next, try recompiling the program. As you can see, the program will not compile because now the support function **isFactor( )** is not available to the **MyModAppDemo**, which is in the **appstart** module. As explained previously, a qualified export restricts access to a package to only those modules specified by the **to** clause.

Use Services

In programming, it is often useful to separate *what* must be done from *how* it is done. As you learned in Chapter 8, one way this is accomplished in Java is through the use of interfaces. The interface specifies the *what,* and the implementing class specifies the *how*. This concept can be expanded so that the implementing class is provided by code that is outside your program, through the use of a *plug-in*. Using such an approach, the capabilities of an application can be enhanced, upgraded, or altered by simply changing the plug-in. The core of the application itself remains unchanged. One way that Java supports a pluggable application architecture is through the use of *services* and *service providers*. Because of their importance, especially in large, commercial applications, Java's module system provides support for them.

Before we begin, it is necessary to state that applications that use services and service providers are typically fairly sophisticated. Therefore, you may find that you do not often need the service-based module features. However, because support for services constitutes a rather significant part of the module system, it is important that you have a general understanding of how these features work. Also, a simple example is presented that illustrates the core techniques needed to use them.

Service and Service Provider Basics

In Java, a *service* is a program unit whose functionality is defined by an interface or an abstract class. Thus, a service specifies in a general way some form of program activity. A concrete implementation of a service is supplied by a *service provider*. In other words, a service defines the form of some action, and the service provider supplies that action.

As mentioned, services are often used to support a pluggable architecture. For example, a service might be used to support the translation of one language into another. In this case, the service supports translation in general. The service provider supplies a specific translation, such as German to English or French to Chinese. Because all service providers implement the same interface, different translators can be used to translate different languages without having to change the core of the application. You can simply change the service provider.

Service providers are supported by the **ServiceLoader** class. **ServiceLoader** is a generic class packaged in **java.util**. It is declared like this:

class ServiceLoader<S>

Here, **S** specifies the service type. Service providers are loaded by the **load( )** method. It has several forms; the one we will use is shown here:

public static <S> ServiceLoader<S> load(Class <S> *serviceType*)

Here, *serviceType* specifies the **Class** object for the desired service type. Recall from Chapter 13 that **Class** is a class that encapsulates information about a class. There are a variety of ways to obtain a **Class** instance. The way we will use here is called a *class literal*. A class literal has this general form:

className.class

Here, *className* specifies the name of the class.

When **load( )** is called, it returns a **ServiceLoader** instance for the application. This object supports iteration and can be cycled through by use of a for-each **for** loop. Therefore, to find a specific provider, simply search for it using a loop.

The Service-Based Keywords

Modules support services through the use of the keywords **provides**, **uses**, and **with**. Essentially, a module specifies that it provides a service with a **provides** statement. A module indicates that it requires a service with a **uses** statement. The specific type of service provider is declared by **with**. When used together, they enable you to specify a module that provides a service, a module that needs that service, and the specific implementation of that service. Furthermore, the module system ensures that the service and service providers are available and will be found.

Here is the general form of **provides**:

provides *serviceType* with *implementationTypes*;

Here, *serviceType* specifies the type of the service, which is often an interface, although abstract classes are also used. A comma-separated list of the implementation types is specified by *implementationTypes*. Therefore, to provide a service, the module indicates both the name of the service and its implementation.

Here is the general form of the **uses** statement:

uses *serviceType*;

Here, *serviceType* specifies the type of the service required.

A Module-Based Service Example

To demonstrate the use of services, we will add a service to the modular application example that we have been using. For simplicity, we will begin with the first version of the application shown at the start of this chapter. To it we will add two new modules. The first is called **userfuncs**. It will define interfaces that support functions that perform binary operations in which each argument is an **int** and the result is an **int**. The second module is called **userfuncsimp**, and it contains concrete implementations of the interfaces.

Begin by creating the necessary source directories.

1. Under the **appsrc** directory, add directories called **userfuncs** and **userfuncsimp**.

2. Under **userfuncs**, add the subdirectory also called **userfuncs**. Under that directory, add the subdirectory **binaryfuncs**. Thus, beginning with **appsrc**, you will have created this tree:

   ```
   appsrc\userfuncs\userfuncs\binaryfuncs
   ```

3. Under **userfuncsimp**, add the subdirectory also called **userfuncsimp**. Under that directory, add the subdirectory **binaryfuncsimp**. Thus, beginning with **appsrc**, you will have created this tree:

   ```
   appsrc\userfuncsimp\userfuncsimp\binaryfuncsimp
   ```

This example expands the original version of the application by providing support for functions beyond those built into the application. Recall that the **SimpleMathFuncs** class supplies three built-in functions: **isFactor( )**, **lcf( )**, and **gcf( )**. Although it would be possible to add more functions to this class, doing so requires modifying and recompiling the application. By implementing services, it becomes possible to "plug in" new functions at run time without modifying the application, and that is what this example will do. In this case, the service supplies functions that take two **int** arguments and return an **int** result. Of course, other types of functions can be supported if additional interfaces are provided, but support for binary integer functions is sufficient for our purposes and keeps the source code size of the example manageable.

The Service Interfaces

Two service-related interfaces are needed. One specifies the form of an action, and the other specifies the form of the provider of that action. Both go in the **binaryfuncs** directory, and both are in the **userfuncs.binaryfuncs** package. The first, called **BinaryFunc**, declares the form of a binary function. It is shown here:

```
// This interface defines a function that takes two int
// arguments and returns an int result. Thus, it can
// describe any binary operation on two ints that
// returns an int.

package userfuncs.binaryfuncs;

public interface BinaryFunc {
  // Obtain the name of the function.
  public String getName();

  // This is the function to perform. It will be
  // provided by specific implementations.
  public int func(int a, int b);
}
```

BinaryFunc declares the form of an object that can implement a binary integer function. This is specified by the **func( )** method. The name of the function is obtainable from **getName( )**. The name will be used to determine what type of function is implemented. This interface is implemented by a class that supplies a binary function.

The second interface declares the form of the service provider. It is called **BinFuncProvider** and is shown here:

```
// This interface defines the form of a service provider that
// obtains BinaryFunc instances.
package userfuncs.binaryfuncs;

import userfuncs.binaryfuncs.BinaryFunc;

public interface BinFuncProvider {

  // Obtain a BinaryFunc.
  public BinaryFunc get();
}
```

BinFuncProvider declares only one method, **get( )**, which is used to obtain an instance of **BinaryFunc**. This interface must be implemented by a class that wants to provide instances of **BinaryFunc**.

The Implementation Classes

In this example, two concrete implementations of **BinaryFunc** are supported. The first is **AbsPlus**, which returns the sum of the absolute values of its arguments. The second is **AbsMinus**, which returns the result of subtracting the absolute value of the second argument from the absolute value of the first argument. These are provided by the classes **AbsPlusProvider** and **AbsMinusProvider**. The source code for these classes must be stored in the **binaryfuncsimp** directory, and they are all part of the **userfuncsimp.binaryfuncsimp** package.

The code for **AbsPlus** is shown here:

```
// AbsPlus provides a concrete implementation of
// BinaryFunc. It returns the result of abs(a) + abs(b).
package userfuncsimp.binaryfuncsimp;

import userfuncs.binaryfuncs.BinaryFunc;

public class AbsPlus implements BinaryFunc {

  // Return name of this function.
  public String getName() {
    return "absPlus";
  }
```
Implement **func( )** for absolute-value addition.
```
  // Implement the AbsPlus function.
  public int func(int a, int b) { return Math.abs(a) + Math.abs(b); }
}
```

AbsPlus implements **func( )** such that it returns the result of adding the absolute values of **a** and **b**. Notice that **getName( )** returns the "absPlus" string. It identifies this function.

The **AbsMinus** class is shown next:

```
// AbsMinus provides a concrete implementation of
// BinaryFunc. It returns the result of abs(a) - abs(b).

package userfuncsimp.binaryfuncsimp;

import userfuncs.binaryfuncs.BinaryFunc;

public class AbsMinus implements BinaryFunc {

  // Return name of this function.
  public String getName() {
    return "absMinus";
  }
```
Implement **func( )** for absolute-value subtraction.
```
  // Implement the AbsMinus function.
  public int func(int a, int b) { return Math.abs(a) - Math.abs(b); }
}
```

Here, **func( )** is implemented to return the difference between the absolute values of **a** and **b**, and the string "absMinus" is returned by **getName( )**.

To obtain an instance of **AbsPlus**, the **AbsPlusProvider** is used. It implements **BinFuncProvider** and is shown here:

```
// This is a provider for the AbsPlus function.

package userfuncsimp.binaryfuncsimp;

import userfuncs.binaryfuncs.*;

public class AbsPlusProvider implements BinFuncProvider {

  // Provide an AbsPlus object.
  public BinaryFunc get() { return new AbsPlus(); }
}
```
Returns an **AbsPlus** object.

The **get( )** method simply returns a new **AbsPlus( )** object. Although this provider is very simple, it is important to point out that some service providers will be much more complex.

The provider for **AbsMinus** is called **AbsMinusProvider** and is shown next:

```
// This is a provider for the AbsMinus function.

package userfuncsimp.binaryfuncsimp;

import userfuncs.binaryfuncs.*;

public class AbsMinusProvider implements BinFuncProvider {

  // Provide an AbsMinus object.
  public BinaryFunc get() { return new AbsMinus(); }
}
```
Returns an **AbsMinus** object.

Its **get( )** method returns an object of **AbsMinus**.

The Module Definition Files

Next, two module definition files are needed. The first is for the **userfuncs** module. It is shown here:

```
module userfuncs {
  exports userfuncs.binaryfuncs;
}
```

This code must be contained in a **module-info.java** file that is in the **userfuncs** module directory. Notice that it exports the **userfuncs.binaryfuncs** package. This is the package that defines the **BinaryFunc** and **BinFuncProvider** interfaces.

The second **module-info.java** file is shown next. It defines the module that contains the implementations. It goes in the **userfuncsimp** module directory.

```
module userfuncsimp {
  requires userfuncs;

  provides userfuncs.binaryfuncs.BinFuncProvider with
    userfuncsimp.binaryfuncsimp.AbsPlusProvider,
    userfuncsimp.binaryfuncsimp.AbsMinusProvider;
}
```

This module requires **userfuncs** because that is where **BinaryFunc** and **BinFuncProvider** are contained, and those interfaces are needed by the implementations. The module provides **BinFuncProvider** implementations with the classes **AbsPlusProvider** and **AbsMinusProvider**.

Demonstrate the Service Providers in MyModAppDemo

To demonstrate the use of the services, the **main( )** method of **MyModAppDemo** is expanded to use **AbsPlus** and **AbsMinus**. It does so by loading them at run time by use of **ServiceLoader** **.load( )**. Here is the updated code:

```
// A module-based application that demonstrates services
// and service providers.

package appstart.mymodappdemo;

import java.util.ServiceLoader;

import appfuncs.simplefuncs.SimpleMathFuncs;
import userfuncs.binaryfuncs.*;

public class MyModAppDemo {
  public static void main(String[] args) {

    // First, use built-in functions as before.
    if(SimpleMathFuncs.isFactor(2, 10))
      System.out.println("2 is a factor of 10");

    System.out.println("Smallest factor common to both 35 and 105 is " +
                    SimpleMathFuncs.lcf(35, 105));

    System.out.println("Largest factor common to both 35 and 105 is " +
                    SimpleMathFuncs.gcf(35, 105));


    // Now, use service-based, user-defined operations.

    // Get a service loader for binary functions.
    ServiceLoader<BinFuncProvider> ldr =          ─────────Load services.
```

```
        ServiceLoader.load(BinFuncProvider.class);

    BinaryFunc binOp = null;

    // Find the provider for absPlus and obtain the function.
    for(BinFuncProvider bfp : ldr) {
      if(bfp.get().getName().equals("absPlus")) {  ◄────── Find provider for
        binOp = bfp.get();                                  absolute-value addition.
        break;
      }
    }

    if(binOp != null)
      System.out.println("Result of absPlus function: " +
                            binOp.func(12, -4));
    else
      System.out.println("absPlus function not found");

    binOp = null;

    // Now, find the provider for absMinus and obtain the function.
    for(BinFuncProvider bfp : ldr) {
      if(bfp.get().getName().equals("absMinus")) {  ◄────── Find provider
        binOp = bfp.get();                                   for absolute-value
        break;                                               subtraction.
      }
    }

    if(binOp != null)
      System.out.println("Result of absMinus function: " +
                            binOp.func(12, -4));
    else
      System.out.println("absMinus function not found");

  }
}
```

Let's take a close look at how a service is loaded and executed by the preceding code. First, a service loader for services of type **BinFuncProvider** is created with this statement:

```
ServiceLoader<BinFuncProvider> ldr =
  ServiceLoader.load(BinFuncProvider.class);
```

Notice that the type parameter to **ServiceLoader** is **BinFuncProvider**. This is also the type used in the call to **load( )**. This means that providers that implement this interface will be found. Thus, after this statement executes, **BinFuncProvider** classes in the module will be available through **ldr**. In this case, both **AbsPlusProvider** and **AbsMinusProvider** will be available.

Next, a reference of type **BinaryFunc** called **binOp** is declared and initialized to **null**. It will be used to refer to an implementation that supplies a specific type of binary function. Next, the following loop searches **ldr** for one that has the "absPlus" name.

```
// Find the provider for absPlus and obtain the function.
for(BinFuncProvider bfp : ldr) {
  if(bfp.get().getName().equals("absPlus")) {
    binOp = bfp.get();
    break;
  }
}
```

Here, a for-each loop iterates through **ldr**. Inside the loop, the name of the function supplied by the provider is checked. If it matches "absPlus", that function is assigned to **binOp** by calling the provider's **get( )** method.

Finally, if the function is found, as it will be in this example, it is executed by this statement:

```
if(binOp != null)
  System.out.println("Result of absPlus function: " +
                        binOp.func(12, -4));
```

In this case, because **binOp** refers to an instance of **AbsPlus**, the call to **func( )** performs an absolute value addition. A similar sequence is used to find and execute **AbsMinus**.

Because **MyModAppDemo** now uses **BinFuncProvider**, its module definition file must include a **uses** statement that specifies this fact. Recall that **MyModAppDemo** is in the **appstart** module. Therefore, you must change the **module-info.java** file for **appstart** as shown here:

```
// Module definition for the main application module.
// It now uses BinFuncProvider.
module appstart {
  // Requires the modules appfuncs and userfuncs.
  requires appfuncs;
  requires userfuncs;

  // appstart now uses BinFuncProvider.
  uses userfuncs.binaryfuncs.BinFuncProvider;
}
```

Compile and Run the Module-Based Service Example

Once you have performed all of the preceding steps, you can compile and run the example by executing the following commands:

```
javac  -d appmodules --module-source-path appsrc
    appsrc\userfuncsimp\module-info.java
    appsrc\appstart\appstart\mymodappdemo\MyModAppDemo.java

java --module-path appmodules -m appstart/appstart.mymodappdemo.MyModAppDemo
```

Here is the output:

```
2 is a factor of 10
Smallest factor common to both 35 and 105 is 5
Largest factor common to both 35 and 105 is 7
Result of absPlus function: 16
Result of absMinus function: 8
```

As the output shows, the binary functions were located and executed. It is important to emphasize that if either the **provides** statement in the **userfuncsimp** module or the **uses** statement in the **appstart** module were missing, the application would fail.

Additional Module Features

Before concluding our discussion of modules, there are three more features that require a brief introduction. These are the **open** module, the **opens** statement, and the use of **requires static**. Each of these features is designed to handle a specialized situation, and each constitutes a fairly advanced aspect of the module system. That said, it is important for you to have a general understanding of their purpose. As you gain more experience with Java, you may encounter situations for which they provide elegant solutions.

Open Modules

As you learned earlier in this chapter, by default, the types in a module's packages are accessible only if they are explicitly exported via an **exports** statement. While this is usually what you will want, there can be circumstances in which it is useful to enable run-time access to all packages in the module, whether a package is exported or not. To allow this, you can create an *open module*. An open module is declared by preceding the **module** keyword with the **open** modifier, as shown here:

```
open module moduleName {
  // module definition
}
```

In an open module, types in all packages are accessible at run time. Understand, however, that only those packages that are explicitly exported are available at compile time. Thus, the **open** modifier affects only run-time accessibility.

The primary reason for an open module is to enable the packages in the module to be accessed through reflection. *Reflection* is the feature that lets a program analyze code at run time. Although the topic of and techniques required to use reflection are beyond the scope of this book, it can be quite important to certain types of programs that require run-time access to a third-party library.

NOTE

Information about reflection can be found in *Java: The Complete Reference, Twelfth Edition* (McGraw Hill, 2022).

The opens Statement

It is possible for a module to open a specific package for run-time access by other modules and for reflective access rather than opening an entire module. To do so, use the **opens** statement, shown here:

opens *packageName*;

Here, *packageName* specifies the package to open. It is also possible to include a **to** clause, which names those modules for which the package is opened.

It is important to understand that **opens** does not grant compile-time access. It is used only to open a package for run-time and reflective access. One other point: an **opens** statement cannot be used in an open module. Remember, all packages in an open module are already open.

requires static

As you know, **requires** specifies a dependence that, by default, is enforced both during compilation and at run time. However, it is possible to relax this requirement in such a way that a module is not required at run time. This is accomplished by use of the **static** modifier in a **requires** statement. For example, this specifies that **mymod** is required for compilation, but not at run time:

```
requires static mymod;
```

In this case, the addition of **static** makes **mymod** optional at run time. This can be helpful in a situation in which a program can utilize functionality if it is present, but not require it.

Continuing Your Study of Modules

The preceding discussions have introduced and demonstrated the core elements of Java's module system. They are the features that are directly supported by keywords in the Java language. Thus, they are the features about which every Java programmer should have at least a basic understanding. As you might guess, the module system provides additional features that you will want to learn about as you advance in your study of Java. A good place to begin is with **javac** and **java**. Both have more options related to modules.

Here are some other areas that you will want to explore. Beginning with JDK 9, the JDK includes the **jlink** tool that assembles a modular application into a run-time image that has only those modules related to the application. This saves both space and download time. A modular application can be packaged into a JAR file. (JAR stands for *Java ARchive*. It is a file format typically used for application deployment.) As a result, the **jar** tool now has options that support modules. For example, it can now recognize a module path. A JAR file that contains a **module-info.class** file is called a *modular JAR file*. For specialized advanced work with modules, you will want to learn about layers of modules, automatic modules, and the technique by which modules can be added during compilation or execution.

In conclusion, modules are expected to play an important role in Java programming. Although their use is not required at this time, they offer important benefits for commercial applications that no Java programmer can afford to ignore. It is likely that module-based development will be in nearly every Java programmer's future.

Ask the Expert

Q: I have heard the term *module graph* used in discussions of modules. What does it mean?

A: During compilation, the compiler resolves the dependence relationships between modules by creating a module graph that represents the dependences. The process ensures that *all dependences* are resolved, including those that occur indirectly. For example, if module A requires module B and B requires module C, then the module graph will contain module C even if A does not use it directly.

Module graphs can be depicted visually in a drawing to illustrate the relationship between modules, and you will likely encounter one as you continue on in Java. Here is a simple example. It is the graph for the first module example in this chapter. (Because **java .base** is automatically included, it is not shown in the diagram.)

appstart

appfuncs

In Java, the arrows point from the dependent module to the required module. Thus, a drawing of a module graph depicts what modules have access to what other modules. Frankly, only the smallest applications can have their module graphs visually represented because of the complexity typically involved in many commercial applications.

Chapter 15 Self Test

1. In general terms, modules give you a way to specify when one unit of code depends on another. True or False?

2. A module is declared using what keyword?

3. The keywords that support modules are context sensitive. Explain what this means.

4. What is **module-info.java** and why is it important?

5. To declare that one module depends on another module, what keyword do you use?

6. To make the public members of a package accessible outside the module in which it is contained, it must be specified in an _____ statement.

7. When compiling or running a module-based application, why is the module path important?

8. What does **requires transitive** do?

9. Does an **exports** statement export another module, or does it export a package?

10. In the first module example, if you remove

```
exports appfuncs.simplefuncs;
```

from the **appfuncs** module-info file and then attempt to compile the program, what error do you see?

11. Module-based services are supported by what keywords?

12. A service specifies the general form of a unit of program functionality using either an interface or abstract class. True or False?

13. A service provider _____ a service.

14. To load a service, what class do you use?

15. Can a module dependency be made optional at run time? If so, how?

16. Briefly describe what **open** and **opens** do.

Chapter 16

Switch Expressions, Records, and Other Recently Added Features

Key Skills & Concepts

- Know the features of the expanded **switch** statement
- Use a list of **case** constants with **switch**
- Understand the **switch** expression
- Use an arrow **case** with the **switch**
- Know the fundamentals of **record**
- Use **record** canonical constructors
- Use **record** non-canonical constructors
- Understand how patterns are used with **instanceof**
- Know the fundamentals of sealed classes and interfaces
- Gain insight into Java's future directions

One of the key ingredients of Java's long-term success has been its ability to adapt to the fast-paced evolution of the modern computing environment. Over the years, Java has incorporated many new features, each responding to changes in hardware, software, or usage patterns and to innovations in computer language design. This ongoing process has enabled Java to remain one of the world's most important and popular computer languages. As explained earlier, this book has been updated for JDK 17, which is a long-term support (LTS) version of Java. JDK 17 incorporates a number of new language features that have been added to Java since the previous LTS version, which was JDK 11. A few of the smaller additions, such as text blocks, have been described in the preceding chapters. Here, the major additions are examined. They are

- Enhancements to **switch**
- Records
- Patterns in **instanceof**
- Sealed classes and interfaces

Here is a brief description of each. The **switch** has been enhanced in a number of ways, the most impacting of which is the *switch expression*. A **switch** expression enables a **switch** to produce a value. Supported by the new keyword **record**, records enable you to create a new kind of class that is specifically designed to hold a group of values. A second form of

instanceof has been added that uses a type pattern. With this form, you can specify a variable that receives an instance of the type being tested if **instanceof** succeeds. It is now possible to specify a *sealed* class or interface. A sealed class can be inherited by only explicitly specified subclasses. A scaled interface can be implemented by only explicitly specified classes, or extended by only explicitly specified interfaces. Thus, sealing a class or interface gives you detailed control over its inheritance and implementation.

Enhancements to switch

The **switch** statement has been part of Java since the start. It is a crucial element of Java's program control statements and provides for a multiway branch. Moreover, **switch** is so fundamental to programming that it is found in one form or another in other popular programming languages. The traditional form of **switch** was described in Chapter 3. This is the form of **switch** that has always been part of Java. Beginning with JDK 14, **switch** has been substantially enhanced with the addition of four new features, shown here:

- The **switch** expression
- The **yield** statement
- The **case** with an arrow
- Support for a list of **case** constants

The **switch** expression is, essentially, a **switch** that produces a value. Thus, a **switch** expression can be used on the right side of an assignment, for example. The **yield** statement specifies the value that is produced by a **switch** expression. It is now possible to specify more than one **case** constant in a **case** statement through the use of a list of **case** constants. A second form of **case** has been added that uses an arrow (- >) instead of a colon. The arrow gives **case** new capabilities. In the sections that follow, each new **switch** feature is described in detail.

Ask the Expert

Q: The expanded switch features appear to involve very significant changes and additions to the original switch statement. Am I understanding this correctly?

A: Yes. Collectively, the enhancements to **switch** represent a significant change to the original **switch**, and to the Java language in general. Not only do they provide new capabilities, but in some situations, they also offer superior alternatives to the traditional approaches. Frankly, in the years to come, the expanded **switch** features will change the way you craft solutions and code your programs. Because of this, a solid understanding of the "how" and "why" behind the **switch** enhancements is important. The "new" **switch** really is that important.

Perhaps the best way to understand the **switch** enhancements is to start with an example that uses a traditional **switch** and then gradually incorporate each new feature. This way, the use and benefit of the enhancements will be clearly apparent. To begin, imagine a distribution center that ships various products, each identified by an ID number. Most products are shipped in the standard way, but a few require special handling. Here is a program that uses a traditional **switch** to supply the shipping method for a given product ID:

```java
// Use a traditional switch to obtain the shipping method
// associated with a product ID. Most products use standard
// shipping, but a few require special handling.
class TraditionalSwitch {

  enum ShipMethod { STANDARD, TRUCK, AIR, OVERNIGHT }

  public static void main(String[] args) {
    ShipMethod shipBy;

    int productID = 5099;

    // Here, a traditional switch is used to obtain the
    // shipping method. Notice that case stacking is used.
    switch(productID) {
      case 1774:
      case 8708:
      case 6709:
        shipBy = ShipMethod.TRUCK;
        break;
      case 4657:
      case 2195:
      case 3621:
      case 1887:
        shipBy = ShipMethod.AIR;
        break;
      case 2907:
      case 5099:
        shipBy = ShipMethod.OVERNIGHT;
        break;
      default:
        shipBy = ShipMethod.STANDARD;
    }

    System.out.println("Shipping method for product number " +
                       productID + " is " + shipBy);
  }
}
```

The output is shown here:

```
Shipping method for product number 5099 is OVERNIGHT
```

There is certainly nothing wrong with using a traditional **switch** as shown in the program, and this is the way Java code has been written for more than two decades. However, as the following sections will show, in many cases, the traditional **switch** can be improved by use of the enhanced **switch** features.

Use a List of case Constants

We begin with one of the easiest ways to modernize a traditional **switch**: by use of a list of **case** constants. In the past, when two or more constants were both handled by the same code sequence, *case stacking* was employed, and this is the approach used by the preceding program. For example, here are how the **case**s for product IDs 1774, 8708, and 6709 are handled:

```
case 1774:
case 8708:
case 6709:
  shipBy = ShipMethod.TRUCK;
  break;
```

The stacking of **case** statements enables all three **case** statements to use the same code sequence that sets the shipping method. As explained in Chapter 3, in a traditional-style **switch**, the stacking of **case**s is made possible because execution falls through each **case** until a **break** is encountered. Although this approach works, a more elegant solution can be achieved by use of a *case constant list*.

Beginning with JDK 14, you can specify more than one **case** constant in a single **case**. To do so, simply separate each constant with a comma. For example, here is a more compact way to code the **case** for product IDs 1774, 8708, and 6709:

```
case 1774, 8708, 6709:
  shipBy = ShipMethod.TRUCK;
  break;
```

Here, one **case** statement replaces what previously took three. If the **switch** matches any of the three constants in the **case** statement, **shipBy** is set to **ShipMethod.TRUCK**. Thus, a **case** constant list streamlines the code. Because of the ease with which you can incorporate a list of **case** constants into existing code, it is a feature that you will want to put to work immediately.

Introducing the switch Expression and the yield Statement

The **switch** enhancement that will have the most profound impact on the way you write code is the *switch expression*. A **switch** expression is, essentially, a **switch** that returns a value. Thus, it has all of the capabilities of a traditional **switch** statement, plus the ability to produce a result. This added capability can significantly improve the way that certain uses of **switch** are coded.

One way to supply the value of a **switch** expression is with the **yield** statement. It has this general form:

yield *value*;

Here, *value* is the value produced by the **switch**, and it can be any expression compatible with the type of value required. A key point to understand about **yield** is that it immediately terminates the **switch**. Thus, it works somewhat like **break**, with the added capability of supplying a value. It is important to point out that **yield** is a context-sensitive keyword. This means that outside its use in a **switch** expression, **yield** is simply an identifier with no special meaning. However, if you use a method called **yield( )**, it must be qualified. For example, if **yield( )** is a non-**static** method within its class, you must use **this.yield( )**.

It is very easy to specify a **switch** expression. Simply use the **switch** in a context in which a value is required, such as on the right side of an assignment statement, an argument to a method, or a return value. For example, this line indicates that a **switch** expression is being employed:

```
int result = switch(what) { // ...
```

Here, the **switch** result is being assigned to the **result** variable. A key point about using a **switch** expression is that each **case** (plus **default**) must produce a value (unless it throws an exception). In other words, each path through a **switch** expression must produce a result.

The addition of the **switch** expression simplifies the coding of situations in which each **case** sets the value of some variable. Such situations can occur in a number of different ways. For example, each **case** could set a **boolean** variable that indicates the success or failure of some action taken by the **switch**. Often, however, the setting of a variable is the *primary purpose* of the **switch**, as is the case with the **switch** used by the preceding program. Its job is to obtain the shipping method associated with a product ID. With a traditional **switch** statement, each **case** statement must individually assign a value to the variable, and this variable becomes the de facto result of the **switch**. This is the approach used by the preceding program, in which the value of the variable **shipBy** is set by each **case**. Although this approach has been used in Java programs for decades, the **switch** expression offers a better solution because the desired value is produced by the **switch** itself.

The following version of the program puts the preceding discussion into action by changing the **switch** statement into a **switch** expression. It also uses **case** constant lists.

```
// Convert a switch statement into a switch expression.
class SwitchExprDemo {

  enum ShipMethod { STANDARD, TRUCK, AIR, OVERNIGHT }

  public static void main(String[] args) {

    int productID = 5099;

    // This is a switch expression. The value produced by
    // the yield statement in the case that matches productID
    // is assigned to the shipBy variable.
    ShipMethod shipBy = switch(productID) {
```

```
      case 1774, 8708, 6709:
        yield ShipMethod.TRUCK;
      case 4657, 2195, 1887, 3621:
        yield ShipMethod.AIR;
      case 2907, 5099:                          Use yield to produce a value.
        yield ShipMethod.OVERNIGHT;
      default:
        yield ShipMethod.STANDARD;
    };  // Notice that a semicolon is required here.

    System.out.println("Shipping method for product number " +
                       productID + " is " + shipBy);
  }
}
```

Look closely at the **switch** in the program. Notice that it differs in important ways from the one used in the previous example. Instead of each **case** assigning a value to **shipBy** individually, this version assigns the outcome of the **switch** itself to the **shipBy** variable. Thus, only one assignment to **shipBy** is required, and the length of the **switch** is reduced. Using a **switch** expression also ensures that each **case** yields a value, thus avoiding the possibility of forgetting to give **shipBy** a value in one of the **cases**. Notice that the value of the **switch** is produced by the **yield** statement inside each **case**. As explained, **yield** causes immediate termination of the **switch**, so no fall through from **case** to **case** will occur. Thus, no **break** statement is required, or allowed. One other thing to notice is the semicolon after the closing brace of the **switch**. Because this **switch** is used in an assignment, it must be terminated by a semicolon.

There is an important restriction that applies to a **switch** expression: the **case** statements must handle all of the values that might occur. Thus, a **switch** expression must be *exhaustive*. For example, if its controlling expression is of type **int**, then all **int** values must be handled by the **switch**. This would, of course, constitute a very large number of **case** statements! For this reason, most **switch** expressions will have a **default** statement. The exception to this rule is when an enumeration is used, and each value of the enumeration is matched by a **case**.

Introducing the Arrow in a case Statement

Although the use of **yield** in the preceding program is a perfectly valid way to specify a value for a **switch** expression, it is not the only way to do so. In many situations, an easier way to supply a value is through the use of a new form of the **case** that substitutes **->** for the colon in a **case**. For example, this line:

```
case 'X': // ...
```

can be rewritten using the arrow like this:

```
case 'X' -> // ...
```

To avoid confusion, in this discussion we will refer to a **case** with an arrow as an *arrow case* and the traditional, colon-based form as a *colon case*. Although both forms will match the character X, the precise action of each style of **case** statement differs in three very important ways.

First, one arrow **case** *does not* fall through to the next **case**. Thus, there is no need to use **break**. Execution simply terminates at the end of an arrow **case**. Although the fall-through nature of a traditional colon **case** has always been part of Java, fall through has been criticized because it can be a source for bugs, such as when the programmer forgets to add a **break** statement to prevent fall through when fall through is not desired. The arrow **case** avoids this situation. Second, the arrow **case** provides a "shorthand" way to supply a value when used in a **switch** expression. For this reason, the arrow **case** is often used in **switch** expressions. Third, the target of an arrow **case** must be either an expression, a block, or throw an exception. It cannot be a statement sequence, as is allowed with a traditional **case**. Thus, the arrow case will have one of these general forms:

case *constant* -> *expression*;
case *constant* -> { *block-of-statements* }
case *constant* -> throw …

Of course, the first two forms represent the primary uses.

Arguably, the most common use of an arrow **case** is in a **switch** expression, and the most common target of the arrow **case** is an expression. Thus, it is here that we will begin. When the target of an arrow **case** is an expression, the value of that expression becomes the value of the **switch** when that **case** is matched. As such, it provides a very efficient alternative to the **yield** statement in many situations. For example, here is the first **case** in the preceding example rewritten to use an arrow **case**:

```
case 1774, 8708, 6709 -> ShipMethod.TRUCK;
```

Here, the value of the expression (which is **ShipMethod.TRUCK**) automatically becomes the value produced by the **switch** when this **case** is matched. In other words, the expression becomes the value yielded by the **switch**. Notice that this statement is quite compact, yet clearly expresses the intent to supply a value.

In the following program, the entire **switch** expression has been completely rewritten to use the arrow **case**:

```
// Use the arrow case "shorthand" to supply the shipping method.
class SwitchExprDemo2 {

  enum ShipMethod { STANDARD, TRUCK, AIR, OVERNIGHT }

  public static void main(String[] args) {

    int productID = 5099;

    // In this switch expression, the value is supplied
    // by use of an arrow case, rather than a yield statement.
    // Notice that no break statements are required because
    // arrow cases do not fall through.
    ShipMethod shipBy = switch(productID) {
      case 1774, 8708, 6709 -> ShipMethod.TRUCK;
      case 4657, 2195, 1887, 3621 -> ShipMethod.AIR;
```

Notice the use of the arrow and that no **break** is required.

```
        case 2907, 5099 -> ShipMethod.OVERNIGHT;
        default -> ShipMethod.STANDARD;
    };

    System.out.println("Shipping method for product number " +
                        productID + " is " + shipBy);
  }
}
```

This produces the same output as before, but the **switch** is more compact and eliminates the need for a separate **yield** statement. Because the arrow **case** does not fall through, there is no need for a **break** statement. Each **case** terminates by yielding the value of its expression. Furthermore, if you compare this final version of the **switch** to the original, traditional **switch** shown at the start of this discussion, it is readily apparent how streamlined and expressive this version is. In combination, the **switch** enhancements offer a truly impressive way to improve the clarity and resiliency of your code.

A Closer Look at the Arrow case

As mentioned, the target of the -> can also be a block of code. You will need to use a block as the target of an arrow **case** whenever you need more than a single expression. For example, in addition to yielding the shipping method, each **case** in this version of the previous program sets a variable called **extraCharge** to indicate if an extra shipping charge is required. Therefore, a block of code is required.

```
// Use blocks with the arrow case.
class BlockArrowCaseDemo {

  enum ShipMethod { STANDARD, TRUCK, AIR, OVERNIGHT }

  public static void main(String[] args) {

    int productID = 5099;

    boolean extraCharge;

    // Use code blocks with an arrow case. Because
    // the target of the arrow is a block, yield must
    // be used to supply the value. As before, no break
    // statements are needed (or legal) because no fall
    // through occurs with the arrow.
    ShipMethod shipBy = switch(productID) {
      case 1774, 8708, 6709 -> {
            extraCharge = true;
            yield ShipMethod.TRUCK;      ◄———— In a block of code, yield
      }                                        is used to supply a value.
      case 4657, 2195, 1887, 3621 -> {
            extraCharge = false;
            yield ShipMethod.AIR;
      }
```

```
        case 2907, 5099 -> {
                extraCharge = true;
                yield ShipMethod.OVERNIGHT;
            }
        default -> {
                extraCharge = false;
                yield ShipMethod.STANDARD;
            }
    };

    System.out.println("Shipping method for product number " +
                        productID + " is " + shipBy);
    if(extraCharge) System.out.println("Extra charge required.");
    }
}
```

Here is the output:

```
Shipping method for product number 5099 is OVERNIGHT
Extra charge required.
```

As this example shows, when using a block, you must use **yield** to supply a value to a **switch** expression. Furthermore, even though block targets are used, each path through the **switch** expression must still provide a value.

Although the arrow **case** is very helpful in a **switch** expression, it is important to emphasize that it is not limited to that use. The arrow **case** can also be used in a **switch** statement, which enables you to write **switch**es in which no **case** fall through can occur. In this situation, no **yield** statement is required (or allowed), and no value is produced by the **switch**. In essence, it works much like a traditional **switch** but without the fall through. Here is an example. Assume a situation

Ask the Expert

Q: In the BlockArrowCaseDemo **program, the shipping method is returned by the** switch, **but the** extraCharge **variable is still set explicitly within each** case **block. Is there a way for the** switch **to efficiently yield more than one value?**

A: Yes. Although this program was designed to provide a simple illustration of a block target of an arrow **case**, it can be improved through the new **record** feature. Added by JDK 16 and described later in this chapter, a **record** offers a convenient and efficient way to link two or more values in a single logical unit. In this case, the **record** would hold both the shipping method and the **extraCharge** value, and this **record** can be supplied by the **switch** as a unit. Thus, using a **record**, a **switch** could yield more than a single value. Reworking this program to use a **record** is the subject of Exercise 14.

in which a factory has three production lines and a count of the number of units produced by each line is needed. The following program simulates this situation. It uses a **switch** to increase the unit count for each production line as units are produced.

```
// Use case arrows with a switch statement. This example
// uses a switch to count the number of units produced
// by three simulated production lines.
class StatementSwitchWithArrows {

  public static void main(String[] args) {
    // Production line counters.
    int line1count = 0;
    int line2count = 0;
    int line3count = 0;

    // Production line number.
    int productionLine;

    for(int i=1; i < 10; i++) {
      // Simulate production line output.
      productionLine = (i % 3) + 1;

      // Use arrows with a switch statement. Notice that
      // no value is yielded. Instead, a line counter
      // is updated based on which line produced the unit and
      // a message indicating the unit is displayed.
      switch(productionLine) {
        case 1 -> { line1count++;
                    System.out.println("Line 1 produced a unit.");
                  }
        case 2 -> { line2count++;
                    System.out.println("Line 2 produced a unit.");
                  }
        case 3 -> { line3count++;
                    System.out.println("Line 3 produced a unit.");
                  }
      }
    }

    System.out.println("Total counts for Lines 1, 2, and 3: " + line1count +
                       ", " + line2count + ", " + line3count);
  }
}
```

In this program, the **switch** is a statement, not an expression, because no value is produced. Also, this **switch** does not have a **default** clause. If this **switch** were an expression, then **default** would be needed because a **switch** expression is required to be exhaustive, but a **switch** statement is not. Because no fall through occurs with an arrow **case**, no **break** statements are needed (or allowed). Instead, the matching line counter is updated, a message is displayed, and then the **switch** ends. As a point of interest, because each **case** increases the value of a different variable, it would not be possible to transform this **switch** into an expression. What value would it produce? All three **case**s set a different variable.

One last point: you cannot mix arrow **case**s with traditional, colon **case**s in the same **switch**. You must choose one or the other. For example, this sequence is invalid:

```java
// This won't work! You cannot mix a colon case with an arrow case.
switch(productionLine) {
  case 1 -> { line1count++;
              System.out.println("Line 1 produced a unit.");
            }
  case 2: line2count++; // Wrong! Can't mix case styles!
          System.out.println("Line 2 produced a unit.");
          break;
  case 3 -> { line3count++;
              System.out.println("Line 3 produced a unit.");
            }
}
```

Ask the Expert

Q: In the foregoing switch **expression examples, constant values have been returned via** yield **or as the target of an arrow** case. **Can the result of a more complex expression be used?**

A: Yes. Any valid expression can be used as the target of the **->** or **yield** as long as it is compatible with the type required by the **switch**. For example, assuming that **getErrorCode( )** returns an **int** value and that the **switch** is expected to produce an **int** result, then the following is a valid **case** statement:

```java
case -1 -> getErrorCode();
```

Here, the result of the call to **getErrorCode( )** becomes the value of the **switch** expression. Here is another example:

```java
case 0 -> normalCompletion = true;
```

In this case, the result of the assignment, which is **true**, becomes the value produced. Of course, the context in which the **switch** is used must be expecting a **boolean** result. In the next example, the uppercase version of **str** is yielded:

```java
case UPCASE -> str.toUpperCase();
```

Here, **toUpperCase( )** returns a string, which means that the **switch** must be expected to produce a **String** result. The key point is that the type of expression must be compatible with the type required by the **switch**.

Use a switch Expression to Obtain a City's Time Zone

CityTZDemo.java

Imagine that you are a programmer working for a large company that has offices in several major cities located within the continental United States. Furthermore, these offices are spread out across the country, resulting in offices in four different time zones. (For the purposes of this project, it is not necessary to distinguish between daylight saving time and standard time.) The company offices and time zones are shown here.

Time Zone	Offices
Eastern	New York, Boston, Miami
Central	Chicago, St. Louis, Des Moines
Mountain	Denver, Albuquerque
Pacific	Seattle, San Francisco, Los Angeles, Portland

In this project, you will create a **switch** expression that yields the time zone of a city given the city's name and demonstrate the **switch** in a short program. You will use arrow **case**s and **case** constant lists to streamline the **switch**.

1. This program involves several pieces. To begin, declare a class called **CityTZDemo**. Inside that class, create an enumeration called **TZ** that enumerates the four time zones in the continental United States, plus a value that indicates that an office is outside the continental United States. The enumeration is shown here:

```
enum TZ { Eastern, Central, Mountain, Pacific, Other }
```

2. Declare the **main( )** method and then add the following array, which contains the names of cities in which the company has offices:

```
String[] cities = {
                "New York", "Boston", "Miami", "Chicago",
                "St. Louis", "Des Moines", "Denver",
                "Albuquerque", "Seattle", "San Francisco",
                "Los Angeles", "Portland"
             };
```

You will use this array to demonstrate the **switch** expression.

3. Add the **switch** expression that returns the time zone for a city. It is shown here:

```
TZ zone = switch(city) {
  case "New York", "Boston", "Miami" -> TZ.Eastern;
  case "Chicago", "St. Louis", "Des Moines" -> TZ.Central;
  case "Albuquerque", "Denver" -> TZ.Mountain;
  case "Seattle", "San Francisco", "Los Angeles",
            "Portland" -> TZ.Pacific;
  default -> TZ.Other;
};
```

(continued)

Because a **switch** expression must handle all possible values of its controlling expression, the **default** clause is required. Notice that this **switch** uses the arrow to produce a value. Thus, no **break** statements are needed because fall through does not occur. Also, notice the use of **case** constant lists. In combination, these features substantially shorten the **switch** and make it more resilient.

4. To cycle through the **cities** array and display the time zones, put the **switch** inside a for-each style **for** loop, as shown here:

```java
// Display the time zone for each city in the array.
for(String city: cities) {

  // This expression switch yields an enumeration value
  // that indicates the time zone of a city.
  TZ zone = switch(city) {
    case "New York", "Boston", "Miami" -> TZ.Eastern;
    case "Chicago", "St. Louis", "Des Moines" -> TZ.Central;
    case "Albuquerque", "Denver" -> TZ.Mountain;
    case "Seattle", "San Francisco", "Los Angeles",
         "Portland" -> TZ.Pacific;
    default -> TZ.Other;
  };

  if(zone == TZ.Other)
    System.out.println(city + " is outside the Continental US");
  else
    System.out.println(city + " is in the " + zone + " time zone");
}
```

5. Here is the entire demonstration program, with all of the pieces assembled:

```java
// Use a switch expression to obtain the time zone for selected
// cities in continental US.
class CityTZDemo {

  // Use an enumeration to describe the time zones.
  enum TZ { Eastern, Central, Mountain, Pacific, Other }

  public static void main(String[] args) {

    // An array of various cities in North America.
    String[] cities = {
                        "New York", "Boston", "Miami", "Chicago",
                        "St. Louis", "Des Moines", "Denver",
                        "Albuquerque", "Seattle", "San Francisco",
                        "Los Angeles", "Portland"
                      };

    // Display the time zone for each city in the array.
    for(String city: cities) {
```

```
      // This switch expression yields an enumeration value
      // that indicates the time zone of a city.
      TZ zone = switch(city) {
        case "New York", "Boston", "Miami" -> TZ.Eastern;
        case "Chicago", "St. Louis", "Des Moines" -> TZ.Central;
        case "Albuquerque", "Denver" -> TZ.Mountain;
        case "Seattle", "San Francisco", "Los Angeles",
             "Portland" -> TZ.Pacific;
        default -> TZ.Other;
      };

      if(zone == TZ.Other)
        System.out.println(city + " is outside the Continental US");
      else
        System.out.println(city + " is in the " + zone + " time zone");
    }
  }
}
```

The output is shown here:

```
New York is in the Eastern time zone
Boston is in the Eastern time zone
Miami is in the Eastern time zone
Chicago is in the Central time zone
St. Louis is in the Central time zone
Des Moines is in the Central time zone
Denver is in the Mountain time zone
Albuquerque is in the Mountain time zone
Seattle is in the Pacific time zone
San Francisco is in the Pacific time zone
Los Angeles is in the Pacific time zone
Portland is in the Pacific time zone
```

6. As an experiment, try adding a city called Honolulu to the **cities** array, but do not include it in the **switch**. After making that change, when you run the program, Honolulu will be handled by the **default** clause. This demonstrates that this **switch** expression is complete and handles all possible values.

Records

Beginning with JDK 16, Java supports a special-purpose class called a *record*. A record is designed to provide an efficient, easy-to-use way to hold a group of values. For example, you might use a record to hold a set of coordinates; bank account numbers and balances; the length, width, and height of a shipping container; and so on. Because it holds a group of values, a record is commonly referred to as an *aggregate* type. However, the record is more than simply a means of grouping data because records also have some of the capabilities of a class. In addition, a record has unique features that simplify its declaration and streamline access to its values. As a result, records make it much easier to work with groups of related data.

One of the central motivations for records is the reduction of the effort required to create a class whose primary purpose is to organize two or more values into a single unit. Although it has always been possible to use **class** for this purpose, doing so can entail writing a number of lines of code for constructors, getter methods, and possibly (depending on use) overriding one or more of the methods inherited from **Object**. As you will see, by creating a data aggregate by use of **record**, these elements are handled automatically for you, greatly simplifying your code. Another reason for the addition of records is to enable a program to clearly indicate that the intended purpose of a class is to hold a grouping of data, rather than to act as a full-featured class. Because of these advantages, records are a much welcomed addition to Java.

Record Basics

As stated, a record is a narrowly focused, specialized class. It is declared by use of the **record** context-sensitive keyword. As such, **record** is a keyword only in the context of a **record** declaration. Otherwise, it is treated as a user-defined identifier with no special meaning. Thus, the addition of **record** does not impact or break existing code.

The general form of a basic **record** declaration is shown here:

record *recordName*(*component-list*) {
    // *optional body statements*
}

As the general form shows, a **record** declaration has significant differences from a **class** declaration. First, notice that the record name is immediately followed by a comma-separated list of parameter declarations called a *component list*. This list defines the data that the record will hold. Second, notice that the body is optional. This is made possible because the compiler will automatically provide the elements necessary to store the data; construct a record; create *getter methods* to access the data; and override **toString( )**, **equals( )**, and **hashCode( )** inherited from **Object**. As a result, for many uses of a record, no body is required because the **record** declaration itself fully defines the record.

Here is an example of a simple **record** declaration:

```
record Item(String name, int itemNum, double price) { }
```

The record name is **Item** and it has three components: the string **name**, the integer **itemNum**, and the double **price**. It specifies no statements in its body, so its body is empty. As the names imply, the record aggregates three pieces of information about some item: its name, its item identification number, and the item's price. Such a record could be used to describe an entry in an online retailer's catalog, for example.

Given the **Item** declaration just shown, a number of elements are automatically created. First, private final fields for **name**, **itemNum**, and **price** are declared, with types **String**, **int**, and **double**, respectively. Second, **public** read-only accessor methods (getter methods) that have the same names and types as the record components **name**, **itemNum**, and **price** are provided.

Therefore, these getter methods are called **name( )**, **itemNum( )**, and **price( )**. In general, each record component will have a corresponding private final field and a read-only public getter method automatically created by the compiler.

Another element created automatically by the compiler will be the record's *canonical constructor*. This constructor has a parameter list that contains the same elements, in the same order, as the component list in the record declaration. The values passed to the constructor are automatically assigned to the corresponding fields in the record. In a **record**, the canonical constructor takes the place of the default constructor used by a **class**.

A **record** is instantiated by use of **new**, just the way you create an instance of a **class**. For example, this creates a new **Item** object, with the name "Hammer", the item identification number 257, and a price of 10.99:

```
Item myItem = new Item("Hammer", 257, 10.99);
```

After this declaration executes, the private fields **name**, **itemNum**, and **price** for **myItem** will contain the values "Hammer", 257, and 10.99, respectively. Therefore, you can use the following statement to display the information associated with **myItem**:

```
System.out.println(myItem.name() + ", Item Number " + myItem.itemNum() + ", " +
                   " Price: " + myItem.price());
```

The resulting output is shown here:

```
Hammer, Item Number 257,  Price: 10.99
```

A key point about a record is that its data is held in private final fields and only getter methods are provided. Thus, a record is immutable. In other words, once you construct a record, its contents cannot be changed. However, if a record holds a reference to some object, you can make a change to that object, but you cannot change to what object the reference refers. Thus, in Java terms, records are said to be *shallowly immutable*.

The following program puts the preceding discussion into action. It creates a small array of **Item** records. It then cycles through the array, displaying the contents of each record.

```
// A simple Record example.

// Declare an Item record. This automatically creates
// a record class with private, final fields called name, itemNum,
// and price, and with read-only accessors called name(), itemNum(),
// and price().
record Item(String name, int itemNum, double price) {}          ◄——— Declare a record.

class RecordDemo {
  public static void main(String[] args) {
    // Create an array of Item records.
    Item[] items = new Item[4];

    // Fill the array with items.
    // Notice how each record is constructed. The arguments
    // are automatically assigned to the name, itemNum, and
    // price fields in the record that is being created.
    items[0] = new Item("Hammer", 257, 10.99);          ◄——— Construct a record instance.
```

```
        items[1] = new Item("Wrench", 18, 19.29);
        items[2] = new Item("Drill", 903, 22.25);
        items[3] = new Item("Saw", 27, 34.59);

        // Use the record accessors to display the list of items
        for(Item i: items) {
          System.out.println(i.name() + ", Item Number " + i.itemNum() + ", " +
                             " Price: " +  i.price());
        }
      }
    }
```

Use accessor methods to obtain a record's data.

The output is shown here:

```
Hammer, Item Number 257,   Price: 10.99
Wrench, Item Number 18,   Price: 19.29
Drill, Item Number 903,   Price: 22.25
Saw, Item Number 27,   Price: 34.59
```

Before continuing, it is important to mention some key points related to records. First, a **record** cannot inherit another class. However, a **record** implicitly inherits **java.lang.Record**, which specifies abstract overrides of the **equals( )**, **hashCode( )**, and **toString( )** methods declared by **Object**. Implicit implementations of these methods are automatically created, based on the record declaration. A **record** type cannot be extended. Thus, all **record** declarations are considered final. Although a **record** cannot extend another class, it can implement one or more interfaces. With the exception of **equals**, you cannot use the names of methods defined by **Object** as names for a **record**'s components. Aside from the fields associated with a **record**'s components, any other fields must be **static**. Finally, a **record** can be generic.

Create Record Constructors

Although you will often find that the automatically supplied canonical constructor is precisely what you want, you can also declare one or more of your own constructors. You can also define your own implementation of the canonical constructor. You might want to declare a **record** constructor for a number of reasons. For example, the constructor could check that a value is within a required range, ensure that an object is in the proper format, or confirm that an argument is not **null**. For a **record**, there are two general types of constructors that you can explicitly create: canonical and non-canonical, and there are some differences between the two. The creation of each type is examined here, beginning with defining your own implementation of the canonical constructor.

Declare a Canonical Constructor

Although the canonical constructor has a specific, predefined form, there are two ways that you can code your own implementation. First, you can explicitly declare the full form of the canonical constructor. Second, you can use what is called a *compact canonical constructor*. Each approach is examined here, beginning with the full form.

To define your own implementation of a canonical constructor, simply do so as you would with any other constructor, specifying the record's name and its parameter list. It is important to emphasize that for the canonical constructor, the types and parameter names must be the same as those specified by the **record** declaration. This is because the parameter names are linked to the automatically created fields and accessor methods defined by the **record** declaration. Thus, they must agree in both type and name. Furthermore, each component must be fully initialized upon completion of the constructor. The following restrictions also apply: the constructor must be at least as accessible as its **record** declaration. Thus, if the access modifier for the **record** is **public**, the constructor must also be specified **public**. A constructor cannot be generic, and it cannot include a **throws** clause. It also cannot invoke another constructor defined for the record.

Here is an example of the **Item** record that explicitly defines the canonical constructor. It uses the constructor to remove any leading and/or trailing spaces from the item's name, thus ensuring that the name is in a standardized form. This would make it easier for an item to be found when searched for by name, for example.

```
// An explicitly declared canonical constructor for Item.
public Item(String name, int itemNum, double price) {
  // Remove leading and trailing spaces by use of the
  // trim() method defined by the String class.
  this.name = name.trim();

  // Set the other fields in Item.
  this.itemNum = itemNum;
  this.price = price;
}
```

In the constructor, any leading and/or trailing spaces in the string passed to **name** are removed. This is done by a call to **trim( )**, which is a method defined by the **String** class. It removes leading and trailing space from the string on which it is called and returns the result. The resulting string is assigned to the field **this.name**. Next, the values passed to the parameters **itemNum** and **price** are assigned to their corresponding fields. Because the parameters **name**, **itemNum**, and **price** are the same as their corresponding fields in **Item**, the field names must be qualified by **this**.

Although there is certainly nothing wrong with creating a canonical constructor as just shown, there is often an easier way: through the use of a *compact canonical constructor*. A compact canonical constructor is declared by specifying the name of the record, but without parameters. The compact constructor implicitly has parameters that are the same as the record's components, and its components are automatically assigned the values of the arguments passed to the constructor. Within the compact constructor you can, however, alter one or more of the arguments prior to their value being assigned to the components. You could also throw an exception if an error condition is encountered, or perform some other procedure.

The following example converts the previous canonical constructor into its compact form:

```
// A compact canonical constructor
public Item {
  // Remove leading and trailing spaces by use of the
  // trim() method defined by the String class.
  name = name.trim();
```

A compact canonical constructor does not specify parameters.

```
    // The name, itemNum, and price fields are automatically
    // assigned the values of their corresponding parameters when
    // the constructor ends.
  }
```

Here, the trimmed **name** string is assigned back to the **name** parameter. The **name**, **itemNum**, and **price** fields are automatically set to the values of their corresponding parameters when the constructor ends. There is no need to initialize these fields inside the compact constructor. Moreover, it would not be legal to do so.

Here is a reworked version of the previous program that demonstrates the compact canonical constructor:

```java
// Use a compact canonical constructor for Item.
record Item(String name, int itemNum, double price) {

  public Item {
    // Remove leading and trailing spaces by use of the
    // trim() method defined by the String class.
    name = name.trim();

    // The name, itemNum, and price fields are automatically
    // assigned the values of their corresponding parameters when
    // the constructor ends.
  }
}

class RecordDemo2 {
  public static void main(String[] args) {
    // Create an array of Item records.
    Item[] items = new Item[4];

    // Notice how each record is constructed. Here, no leading
    // or trailing spaces are present in the name.
    items[0] = new Item("Hammer", 257, 10.99);

    // These entries have leading and/or trailing spaces in their
    // names. The canonical constructor will remove the spaces.
    items[1] = new Item("  Wrench", 18, 19.29);
    items[2] = new Item("Drill   ", 903, 22.25);
    items[3] = new Item("  Saw   ", 27, 34.59);

    // Use the record accessors to display the list of items
    for(Item i: items) {
      System.out.println(i.name() + ", Item Number " + i.itemNum() + ", " +
                         " Price: " +  i.price());
    }
  }
}
```

The output is shown here:

```
Hammer, Item Number 257,  Price: 10.99
Wrench, Item Number 18,  Price: 19.29
Drill, Item Number 903,  Price: 22.25
Saw, Item Number 27,  Price: 34.59
```

As you can see, the leading and trailing spaces have been removed from the names. As an experiment, try commenting-out the line in the constructor that trims the spaces. This will result in the leading and trailing spaces remaining in the names.

Declare a Non-Canonical Constructor

Although the canonical constructor will often be sufficient, you can declare other constructors. The key requirement is that any non-canonical constructor must first call another constructor in the record via **this( )**. (See Appendix E for a discussion of this technique.) The constructor invoked will often be the canonical constructor. Doing this ultimately ensures that all fields are assigned. Declaring a non-canonical constructor enables you to handle special-case situations. For example, you might use such a constructor to create a record in which one or more of the components is given a default, place-holder value. Another use is when an argument is not in a form compatible with the canonical constructor. Whatever the reason, a non-canonical constructor gives you added flexibility when constructing records.

The following program declares a non-canonical constructor for **Item** that handles a situation in which the price of an item is given as a numeric string, rather than as a numeric value. For example, "88.29" is used instead of the **double** value 88.29. The non-canonical constructor shown here converts the numeric string into its **double** equivalent. To do so, it uses the **parseDouble( )** method supplied by the **Double** wrapper class, as described in Chapter 10.

```
// Use a non-canonical constructor.

// Declare a record that holds items.
record Item(String name, int itemNum, double price) {
  // Use a static field in a record.
  static double pricePending = -1;

  // This is a non-canonical constructor.
  // It creates a record in which the price of the item
  // is passed as a string instead of a double. Thus, it
  // must be converted to a double when passed to the
  // canonical constructor.
  public Item(String name, int itemNum, String price) {
    this(name, itemNum, Double.parseDouble(price));
  }
}

class RecordDemo3 {
  public static void main(String[] args) {
    // Create an array of Item records.
    Item[] items = new Item[4];

    // Create some item entries. These will use the implicit
    // canonical constructor.
    items[0] = new Item("Hammer", 257, 10.99);
    items[1] = new Item("Wrench", 18, 19.29);

    // These will use the non-canonical constructor because
    // the price is passed as a string, not a double.
    items[2] = new Item("Drill", 903, "22.25");
    items[3] = new Item("Saw", 27, "34.59");
```

```
     // Use the record accessors to display the list of items
     for(Item i: items)
       System.out.println(i.name() + ", Item Number " + i.itemNum() + ", " +
                          " Price: " + i.price());
   }
 }
```

Pay special attention to the way that the records for Drill and Saw are created by use of the non-canonical constructor. In the call to **this**, the constructor passes the **name** and **itemNum** arguments as-is, but converts the numeric string in **price** to its **double** form. Thus, the non-canonical constructor converts the price into its required form. The output is the same as before.

Ask the Expert

Q: Can a record have both a user-defined canonical and non-canonical constructor?

A: Yes, you can declare both a canonical constructor and one or more non-canonical constructors. For example, here is a version of **Item** in which the compact canonical constructor throws an exception if the **name** component is an empty string. It also declares the non-canonical constructor used by the preceding example.

```
// Declare an Item record that explicitly declares both
// a canonical and non-canonical constructor.
record Item(String name, int itemNum, double price) {

  // This compact canonical constructor throws an exception
  // if the name parameter is empty.
  public Item {
    if(name.length() == 0)
      throw new IllegalArgumentException("Item name is empty.");
  }

  // This is a non-canonical constructor.
  // It creates a record in which the price of the item
  // is passed as a string instead of a double. Thus, it
  // must be converted to a double when passed to the
  // canonical constructor.
  public Item(String name, int itemNum, String price) {
    this(name, itemNum, Double.parseDouble(price));
  }
}
```

In general, and within the constraints of a **record**, you can create constructors to meet the needs of your program. As you work more with records, you will find that they are a powerful, yet flexible feature.

A Closer Look at Record Getter Methods

Although it is seldom necessary, it is possible to create your own implementation of a getter method. When you declare the getter, the implicit version is no longer supplied. One possible reason you might want to declare your own getter is to throw an exception if some condition is not met. For example, if a record holds a filename and a URL, the getter for the filename might throw a **FileNotFoundException** if the file is not present at the URL. There is a very important requirement, however, that applies to creating your getters: they must adhere to the principle that a record is immutable. Thus, a getter that returns an altered value is semantically questionable and should be avoided even though such code would be syntactically correct.

If you do declare your own version of a built-in getter, there are a number of rules that apply. A getter must have the same return type and name as the component that it obtains. It must also be explicitly declared public. (Thus, default accessibility is not sufficient for a getter declaration in a **record**.) No **throws** clause is allowed in a getter declaration. Finally, a getter must be non-generic and non-static.

Often, rather than overriding a built-in getter, it is a better idea to simply create a method that returns the value that you desire. For example, this version of **Item** defines a method called **discountPrice( )** that returns the price of an item discounted by a specified percentage. With this approach the built-in **price( )** getter is unchanged, thus preserving immutability. If a discounted value is needed, **discountPrice( )** is called instead.

```
record Item(String name, int itemNum, double price) {
  // ...

  double discountPrice(double percentage) {
    return price - (price * percentage / 100.0);
  }
}
```

Pattern Matching with instanceof

The traditional form of the **instanceof** operator was introduced in Chapter 7. As you learned there, **instanceof** evaluates to **true** if and only if an object is of a specified type or can be cast to that type. Beginning with JDK 16, a second form of **instanceof** has been added to Java that supports the new *pattern matching* feature. In general terms, *pattern matching* defines a mechanism that determines if a value fits a general form. As it relates to **instanceof**, pattern matching is used to test the type of a value (which must be a reference type) against a specified type. This kind of pattern is called a *type pattern*. If the pattern matches, a *pattern variable* will receive a reference to the object matched by the pattern.

The pattern matching form of **instanceof** is shown here:

objref instanceof *type pattern-var*

If **instanceof** succeeds, *pattern-var* will be created and contain a reference to the object that matches the pattern. If it fails, *pattern-var* is never created. This form of **instanceof** succeeds if the object referred to by *objref* can be cast to *type* and the static type of *objref* is not a subtype of *type*.

For example, the following fragment creates a **Number** reference called **myOb** that refers to an **Integer** object. (Recall that **Number** is a superclass of all numeric primitive type wrappers.) It then uses the **instanceof** operator to check if the object referred to by **myOb** is an **Integer**, which it will be in this example. This results in an object called **iObj** of type **Integer** being instantiated that contains the matched value.

```
Number myOb = Integer.valueOf(27);

// Use the pattern matching version of instanceof.
if(myOb instanceof Integer iObj) {        iObj is created only if
  // iObj is known and in scope here.      instanceof succeeds.
  System.out.println("iObj refers to an integer: " + iObj);
}
// iObj does not exist here
```

As the comments indicate, **iObj** is known only within the scope of the **if** clause. It is not known outside of the **if**. It also would not be known within an **else** clause, should one have been included. It is crucial to understand that the pattern variable **iObj** is created only if the pattern matching succeeds.

The primary advantage of the pattern matching form of **instanceof** is that it reduces the amount of code that was typically needed by its traditional form. For example, consider this functionally equivalent version of the preceding example that uses the traditional approach:

```
// Use a traditional instanceof.
if(myOb instanceof Integer) {
  // Use an explicit cast to obtain iObj.
  Integer iObj = (Integer) myOb;
  System.out.println("iObj refers to an integer: " + iObj);
}
```

With the traditional form, a separate declaration statement and explicit cast are required to create the **iObj** variable. The pattern matching form of **instanceof** streamlines the process.

Ask the Expert

Q: Can the pattern form of instanceof be used as part of an AND logical expression where the pattern variable is also used within another part of the expression?

A: Yes, as long as you understand that the pattern variable is in scope only if the **instanceof** expression succeeds. Consider the following example. The **if** succeeds only when **myOb** refers to an **Integer** and its value is between 1 and 10, inclusive. Pay special attention to the expression in the **if**:

```
if((myOb instanceof Integer iObj) && ((iObj > 0) && (iObj < 11))) {
  // myOb is both an Integer and between 1 and 10, inclusive.
  // ...
}
```

(continued)

The **iObj** pattern variable is created only if the left side of the first **&&** (the part that contains the **instanceof** operator) is true. However, notice that **iObj** is used by the right side. This is possible because the short-circuit form of the AND logical operator is used, and the right side is evaluated only if the left succeeds. Thus, if the right side operand is evaluated, **iObj** will be in scope. However, if you tried to write the preceding **if** using the **&** operator like this:

```
if((myOb instanceof Integer iObj) & ((iObj > 0) && (iObj < 11))) { // Wrong!
```

a compilation error would occur because **iObj** will not be in scope if the left side fails. Recall that the **&** operator causes both sides of the expression to be evaluated, but **iObj** is only in scope if the left side is true.

One other point: a logical expression cannot introduce the same pattern variable more than once. For example, in a logical AND, it is an error if both operands create the same pattern variable.

Sealed Classes and Interfaces

Beginning with JDK 17, it is possible to declare a class that can be inherited by only specific subclasses. Such a class is called *sealed*. Prior to sealed classes, inheritance was an "all or nothing" situation. A class could either be extended by any subclass or marked as **final**, which prevented its inheritance entirely. Sealed classes fall between these two extremes because they enable you to specify precisely what subclasses a superclass will allow. In a similar fashion, it is also possible to declare a sealed interface in which you specify only those classes that implement the interface and/or those interfaces that extend the sealed interface. Together, sealed classes and interfaces give you significantly greater control over inheritance.

It is important to state at the outset that sealed classes and interfaces constitute a rather specialized feature. Arguably, their primary use is found when designing class libraries. It is, therefore, a feature that not all Java programmers will use in their day-to-day coding. That said, it is important to have a general understanding because sealed classes and interfaces represent a significant addition to the Java language.

Sealed Classes

To declare a sealed class, precede the declaration with **sealed**. Then, after the class name, include a **permits** clause that specifies the allowed subclasses. Both **sealed** and **permits** are context-sensitive keywords and have special meaning only in a class or interface declaration. Outside of a class or interface declaration, **sealed** and **permits** are unrestricted. Thus, no preexisting code was broken by the addition of these two keywords.

Here is a simple example of a sealed class:

```
public sealed class Fruit permits Apple, Pear, Grape {
  // ...
}
```

Here, the sealed class is called **Fruit**. It allows only three subclasses: **Apple**, **Pear**, and **Grape**. If any other class attempts to inherit **Fruit**, a compile-time error will occur.

Here is an important point: a subclass of a sealed class must be declared as either **final**, **sealed**, or **non-sealed**. Let's look at each option in turn. Here, the **Apple** subclass is specified **final**:

```
public final class Apple extends Fruit {
  // ...
}
```

As you learned in Chapter 7, specifying a class **final** prevents it from having any subclasses. In this case, it means that no subclasses of **Apple** are allowed.

To indicate that a subclass is itself sealed, it must be declared **sealed** and its permitted subclasses must be specified. For example, this version of **Apple** permits two subclasses called **Fuji** and **Jonathan**:

```
public sealed class Apple extends Fruit permits Fuji, Jonathan {
  // ...
}
```

Of course, the classes **Fuji** and **Jonathan** must then be declared either **sealed**, **final**, or **non-sealed**.

By declaring a subclass **non-sealed** you can unseal a subclass of a sealed class. The context-sensitive keyword **non-sealed** was added by JDK 17. It unlocks the subclass, enabling it to be inherited by any other class. For example, **Apple** could be coded like this:

```
public non-sealed class Apple extends Fruit {
  // ...
}
```

Now, any class may inherit **Apple**. It is important to understand that even though **Apple** has been declared **non-sealed**, the only direct subclasses of **Fruit** remain **Apple**, **Pear**, and **Grape**. A primary reason for **non-sealed** is to enable a superclass to specify a limited set of direct subclasses that provide a baseline of well-defined functionality, but allow those subclasses to be freely extended.

If a class is specified in a **permits** clause for a sealed class, then that class *must* directly extend the sealed class. Otherwise, a compile-time error will result. Thus, a sealed class and its subclasses define a mutually dependent logical unit. Additionally, it is illegal to declare a class that does not extend a sealed class as **non-sealed**.

A key requirement of a sealed class is that every subclass that it permits must be accessible. Furthermore, if a sealed class is contained in a named module, then each subclass must also be in the same named module. In this case, a subclass can be in a different package from the sealed class. If the sealed class is in the unnamed module, then the sealed class and all permitted subclasses must be in the same package.

In the preceding discussion, notice that the superclass **Fruit** and its subclasses **Apple**, **Pear**, and **Grape** are all public classes. Thus, each would have been stored in its own separate file (formally, a compilation unit). However, if the subclasses have default package access (rather than public access), it is possible for a sealed class and its subclasses to be stored in a single file. In cases such as this, no **permits** clause is required for a sealed class. For example, here all three subclassess are in the same file as **Fruit**:

```
// Because this is all in one file, Fruit does not require
// a permits clause.
public sealed class Fruit {
  // ...
}

final class Apple extends Fruit {
  // ...
}

final class Pear extends Fruit {
  // ...
}

final class Grape extends Fruit {
  // ...
}
```

In this case, the classes **Apple**, **Pear**, and **Grape** are implicitly permitted.

Ask the Expert

Q: Can an abstract class be sealed?

A: Yes. There is no restriction in this regard. For example, here **Fruit** is declared as an abstract class:

```
public sealed abstract class Fruit permits Apple, Pear, Grape {
  // ...
}
```

Of course, at some point a concrete subclass of **Fruit** must occur. For example, assuming that **Apple** is declared **final**, then **Apple** must be a concrete class.

Sealed Interfaces

A sealed interface is declared in the same way as a sealed class, by the use of **sealed**.
A sealed interface uses its **permits** clause to specify the classes allowed to implement it
and/or the interfaces allowed to extend it. Thus, a class that is not part of the **permits** clause
cannot implement a sealed interface, and an interface not included in the **permits** clause
cannot extend it.

Here is a simple example of a sealed interface that permits only the classes **Apple**, **Pear**,
and **Grape** to implement it:

```
public sealed interface FruitIF permits Apple, Pear, Grape {
  String type();
}
```

A class that implements a sealed interface must, itself, be specified as either **final**, **sealed**, or
non-sealed. For example, here **Apple** is marked **non-sealed**, **Pear** is specified as **final**, and
Grape is declared **sealed**:

```
public non-sealed class Apple implements FruitIF {
  public String type() { return "Apple is a tree fruit"; }
  // ...
}

public final class Pear implements FruitIF {
  public String type() { return "Pear is a tree fruit"; }
  // ...
}

public sealed class Grape implements FruitIF permits Concord {
  public String type() { return "Grape is a vine fruit"; }
  // ...
}
```

Here is a key point: any class specified in a sealed interface's **permits** clause *must* implement
the interface. Therefore, a sealed interface and its implementing classes form a logical unit.

A sealed interface can also specify which other interfaces can extend the sealed interface.
For example, here the sealed **PlantIF** interface permits **FruitIF**:

```
public sealed interface PlantIF permits FruitIF {
  // ...
}
```

Because **FruitIF** is a permitted subinterface of **PlantIF**, it must extend **PlantIF**. For example,

```
public sealed interface FruitIF extends PlantIF permits Apple, Pear, Grape {
  // ...
}
```

Because **FruitIF** is permitted to extend **PlantIF**, the permitted subclasses of **FruitIF** also have
access to **PlantIF**.

Three more points: First, an interface specified in a **permits** clause must extend the permitting interface. Second, the permitted interface must be declared either **non-sealed** or **sealed**. Third, it is possible for a class to inherit a sealed class *and* implement a sealed interface, as does **Grape** in the foregoing example. There is no restriction in this regard.

In the preceding interface examples, each class and interface are declared **public**. Thus, each one is in its own file. However, as is the case with sealed classes, it is also possible for a sealed interface and its implementing classes (and extending interfaces) to be stored in a single file as long as the classes and interfaces have default, package access. In cases such as this, no **permits** clause is required for a sealed interface.

Future Directions

Beginning with JDK 12, Java releases may, and often do, include preview features. As explained in Chapter 1, a preview feature is a new, fully developed enhancement to Java. However, a preview feature is *not yet* formally part of Java. Instead, a feature is previewed to allow programmers time to experiment with the feature and, if desired, communicate their thoughts and opinions prior to the feature being made permanent. This process enables a new feature to be improved or optimized based on actual developer use. As a result, a preview feature is *subject to change*. It can even be withdrawn. This means that a preview feature should not be used for code that you intend to publicly release That said, it is expected that most preview features will ultimately become part of Java, possibly after a period of refinement. Preview features chart the course of Java's future direction.

JDK 17 includes one preview feature: Pattern Matching for **switch** (JEP 406). It adds pattern matching capabilities to **switch**. As described earlier in this chapter, pattern matching was first introduced by the enhancement of **instanceof** in JDK 16. Adding pattern matching to **switch** continues the process. Because this is a preview feature that is subject to change, it is not discussed further in this book.

Java releases may also include *incubator modules*, which preview a new API or tool that is undergoing development. Like a preview feature, an incubator feature is subject to change. Furthermore, an incubator feature can be removed in the future. Thus, there is no guarantee that an incubating module will formally become part of Java in the future. Incubator features give developers an opportunity to experiment with the API or tool and possibly supply feedback. JDK 17 includes two incubator modules The first is Foreign Function and Memory API (JEP 412). The second is Vector API (JEP 414).

It is important to emphasize that preview features and incubator modules can be introduced in any Java release. Therefore, you will want to watch for them in each new version of Java. They give you a chance to try a new enhancement before it potentially becomes a formal part of Java. Perhaps more importantly, preview features and incubator modules give you advance information on where Java's development is headed.

Chapter 16 Self Test

1. Rewrite the following sequence so that it uses a constant list:

```
case 3: prime = true;
          break;
case 5: prime = true;
          break;
case 7: prime = true;
          break;
```

2. When using an arrow **case**, does execution fall through to the next **case**?

3. Given this **switch**, show the **yield** statement that returns the value 98.6:

```
double val = switch(x) {
  case "temp":  // produce the value 98.6
// ...
```

4. Assuming the **switch** in Question 3, show how to use an arrow **case** to yield the value 98.6.

5. Can you mix an arrow **case** and a colon **case** in the same **switch**?

6. Can the target of an arrow **case** be a block?

7. A **record** is commonly referred to as a/an _____ type.

8. Given this record declaration, what are its components? What elements are implicitly created?

```
record MyRec(Double highTemp, Double lowTemp, String location) { }
```

9. Does a **record** have a default constructor? If not, what type of constructor does a **record** automatically have?

10. Given **MyRec** from Question 8, show the compact canonical constructor that removes leading and trailing spaces from the **location** string.

11. If you were to override a **record** getter method, in what way would you need to be very careful?

12. In Try This 13-1 you created a generic queue class. Can this class be used to store **record** objects without any changes? If so, demonstrate its use to store the **Item** records used in the record examples.

13. Rework the **Item** record so that the **price** component is generic, with an upper bound of **Number**.

14. In the **BlockArrowCaseDemo** program, the **switch** expression yields the shipping method, but the variable **extraCharge** is set separately inside each **case**. This program can be improved by having the **switch** yield a **record** that contains both the shipping method and the **extraCharge** value. In essence, the use of a **record** enables the **switch** to yield two or more values when it returns its result. Rework the **BlockArrowCaseDemo** program to demonstrate this approach.

15. Show the general form of **instanceof** when using pattern matching.

16. Given

```
Object myOb = "A test string";
```

fill in the blank in the following **if** statement that uses **instanceof** to determine whether **myOb** refers to a **String**.

```
if(myObj instanceof _____) System.out.println("Is a string: " + str);
```

17. A **sealed** class explicitly specifies the subclasses that can inherit it. True or false?

18. Given the following:

```
public sealed class MyClass permits Alpha, Beta, Gamma { // ...
```

which of the follow declarations are legal?

A. `public final class Alpha extends MyClass { // ...`

B. `public final class Beta { // ...`

C. `public class Gamma extends MyClass { // ...`

D. `public non-sealed SomeOtherClass extends MyClass { // ...`

19. Can an interface be sealed? If so, what effect does sealing an interface have?

20. A preview feature is a new feature that is fully developed, but not yet formally part of Java. True or False?

21. A preview feature is subject to change or may even be withdrawn. True or False?

Chapter 17

Introducing Swing

Key Skills & Concepts

- Know the origins and design philosophy of Swing

- Understand Swing components and containers

- Know layout manager basics

- Create, compile, and run a simple Swing application

- Learn event handling fundamentals

- Use **JButton**

- Work with **JTextField**

- Create a **JCheckBox**

- Work with **JList**

- Use anonymous inner classes or lambda expressions to handle events

So far, all of the programs in this book have been console-based. This means that they do not make use of a graphical user interface (GUI). Although console-based programs are excellent for teaching the basics of Java and for some types of programs, such as server-side code, most real-world client applications will be GUI-based. At the time of this writing, the most widely used Java GUI is Swing.

Swing defines a collection of classes and interfaces that support a rich set of visual components, such as buttons, text fields, scroll panes, check boxes, trees, and tables, to name a few. Collectively, these controls can be used to construct powerful, yet easy-to-use graphical interfaces. Because of its widespread use, Swing is something with which all Java programmers should be familiar. Therefore, this chapter provides an introduction to this important GUI framework.

It is important to state at the outset that Swing is a very large topic that requires an entire book of its own. This chapter can only scratch its surface. However, the material presented here will give you a general understanding of Swing, including its history, basic concepts, and design philosophy. It then introduces five commonly used Swing components: the label, push button, text field, check box, and list. Although this chapter describes only a small part of Swing's features, after completing it, you will be able to begin writing simple GUI-based programs. You will also have a foundation upon which to continue your study of Swing.

NOTE

For a comprehensive introduction to Swing, see *Swing: A Beginner's Guide*
(McGraw Hill, 2007).

The Origins and Design Philosophy of Swing

Swing did not exist in the early days of Java. Rather, it was a response to deficiencies present
in Java's original GUI subsystem: the Abstract Window Toolkit (AWT). The AWT defines
a basic set of components that support a usable, but limited, graphical interface. One reason
for the limited nature of the AWT is that it translates its various visual components into their
corresponding, platform-specific equivalents, or *peers*. This means that the look and feel of an
AWT component is defined by the platform, not by Java. Because the AWT components use
native code resources, they are referred to as *heavyweight*.

The use of native peers led to several problems. First, because of differences between
operating systems, a component might look, or even act, differently on different platforms.
This potential variability threatened the overarching philosophy of Java: write once, run
anywhere. Second, the look and feel of each component was fixed (because it is defined by the
platform) and could not be (easily) changed. Third, the use of heavyweight components caused
some frustrating restrictions. For example, a heavyweight component was always opaque.

Not long after Java's original release, it became apparent that the limitations and
restrictions present in the AWT were sufficiently serious that a better approach was
needed. The solution was Swing. Introduced in 1997, Swing was included as part of the
Java Foundation Classes (JFC). Swing was initially available for use with Java 1.1 as a
separate library. However, beginning with Java 1.2, Swing (and the rest of JFC) was fully
integrated into Java.

Swing addresses the limitations associated with the AWT's components through the use
of two key features: *lightweight components* and a *pluggable look and feel*. Although they
are largely transparent to the programmer, these two features are at the foundation of Swing's
design philosophy and the reason for much of its power and flexibility. Let's look at each.

With very few exceptions, Swing components are *lightweight*. This means that a component
is written entirely in Java. They do not rely on platform-specific peers. Lightweight components
have some important advantages, including efficiency and flexibility. Furthermore, because
lightweight components do not translate into platform-specific peers, the look and feel of each
component is determined by Swing, not by the underlying operating system. This means that
each component can work in a consistent manner across all platforms.

Because each Swing component is rendered by Java code rather than by platform-specific
peers, it is possible to separate the look and feel of a component from the logic of the component,
and this is what Swing does. Separating out the look and feel provides a significant advantage:
it becomes possible to change the way that a component is rendered without affecting any

of its other aspects. In other words, it is possible to "plug in" a new look and feel for any given component without creating any side effects in the code that uses that component.

Java provides look-and-feels, such as metal and Nimbus, that are available to all Swing users. The metal look and feel is also called the *Java look and feel.* It is a platform-independent look and feel that is available in all Java execution environments. It is also the default look and feel. For this reason, the default Java look and feel (metal) is used by the examples in this chapter.

Swing's pluggable look and feel is made possible because Swing uses a modified version of the classic *model-view-controller (MVC)* architecture. In MVC terminology, the *model* corresponds to the state information associated with the component. For example, in the case of a check box, the *model* contains a field that indicates if the box is checked or unchecked. The *view* determines how the component is displayed on the screen, including any aspects of the view that are affected by the current state of the model. The *controller* determines how the component reacts to the user. For example, when the user clicks a check box, the controller reacts by changing the model to reflect the user's choice (checked or unchecked). This then results in the view being updated. By separating a component into a model, a view, and a controller, the specific implementation of each can be changed without affecting the other two. For instance, different view implementations can render the same component in different ways without affecting the model or the controller.

Although the MVC architecture and the principles behind it are conceptually sound, the high level of separation between the view and the controller was not beneficial for Swing components. Instead, Swing uses a modified version of MVC that combines the view and the controller into a single logical entity called the *UI delegate.* For this reason, Swing's approach is called either the *model-delegate* architecture or the *separable model* architecture. Therefore, although Swing's component architecture is based on MVC, it does not use a classical implementation of it. Although you won't work directly with models or UI delegates in this chapter, they are, nevertheless, present behind the scene.

As you work through this chapter, you will see that even though Swing embodies very sophisticated design concepts, it is easy to use. In fact, one could argue that Swing's ease of use is its most important advantage. Simply stated, Swing makes manageable the often difficult task of developing your program's user interface. This lets you concentrate on the GUI itself, rather than on implementation details.

Ask the Expert

Q: You say that Swing defines a GUI that is superior to the AWT. Does this mean that Swing replaces the AWT?

A: No, Swing does not replace the AWT. Rather, Swing builds upon the foundation provided by aspects of the AWT. Thus, portions of the AWT are still a crucial part of Java. Although knowledge of the AWT is not required by this chapter, you need a solid understanding of its structure and features if you seek full Swing mastery.

Components and Containers

A Swing GUI consists of two key items: *components* and *containers*. However, this distinction is mostly conceptual because all containers are also components. The difference between the two is found in their intended purpose: As the term is commonly used, a component is an independent visual control, such as a push button or text field. A container holds a group of components. Thus, a container is a special type of component that is designed to hold other components. Furthermore, in order for a component to be displayed, it must be held within a container. Thus, all Swing GUIs will have at least one container. Because containers are components, a container can also hold other containers. This enables Swing to define what is called a *containment hierarchy,* at the top of which must be a *top-level container.*

Components

In general, Swing components are derived from the **JComponent** class. (The only exceptions to this are the four top-level containers, described in the next section.) **JComponent** provides the functionality that is common to all components. For example, **JComponent** supports the pluggable look and feel. **JComponent** inherits the AWT classes **Container** and **Component**. Thus, a Swing component is built on and compatible with an AWT component.

All of Swing's components are represented by classes defined within the package **javax.swing.** The following table shows the class names for Swing components (including those used as containers):

JApplet (deprecated)	JButton	JCheckBox	JCheckBoxMenuItem
JColorChooser	JComboBox	JComponent	JDesktopPane
JDialog	JEditorPane	JFileChooser	JFormattedTextField
JFrame	JInternalFrame	JLabel	JLayer
JLayeredPane	JList	JMenu	JMenuBar
JMenuItem	JOptionPane	JPanel	JPasswordField
JPopupMenu	JProgressBar	JRadioButton	JRadioButtonMenuItem
JRootPane	JScrollBar	JScrollPane	JSeparator
JSlider	JSpinner	JSplitPane	JTabbedPane
JTable	JTextArea	JTextField	JTextPane
JTogglebutton	JToolBar	JToolTip	JTree
JViewport	JWindow		

Notice that all component classes begin with the letter **J**. For example, the class for a label is **JLabel**, the class for a push button is **JButton**, and the class for a check box is **JCheckBox**.

This chapter introduces five commonly used components: **JLabel**, **JButton**, **JTextField**, **JCheckBox**, and **JList**. Once you understand their basic operation, it will be easy for you to learn to use the others.

Containers

Swing defines two types of containers. The first are top-level containers: **JFrame**, **JApplet**, **JWindow**, and **JDialog**. (**JApplet**, which supports Swing-based applets, has been deprecated since JDK 9, and is now deprecated for removal.) These containers do not inherit **JComponent**. They do, however, inherit the AWT classes **Component** and **Container**. Unlike Swing's other components, which are lightweight, the top-level containers are heavyweight. This makes the top-level containers a special case in the Swing component library.

As the name implies, a top-level container must be at the top of a containment hierarchy. A top-level container is not contained within any other container. Furthermore, every containment hierarchy must begin with a top-level container. The one most commonly used for applications is **JFrame**.

The second type of container supported by Swing is the lightweight container. Lightweight containers *do* inherit **JComponent**. Examples of lightweight containers are **JPanel**, **JScrollPane**, and **JRootPane**. Lightweight containers are often used to collectively organize and manage groups of related components because a lightweight container can be contained within another container. Thus, you can use lightweight containers to create subgroups of related controls that are contained within an outer container.

The Top-Level Container Panes

Each top-level container defines a set of *panes*. At the top of the hierarchy is an instance of **JRootPane**. **JRootPane** is a lightweight container whose purpose is to manage the other panes. It also helps manage the optional menu bar. The panes that compose the root pane are called the *glass pane,* the *content pane,* and the *layered pane.*

The glass pane is the top-level pane. It sits above and completely covers all other panes. The glass pane enables you to manage mouse events that affect the entire container (rather than an individual control) or to paint over any other component, for example. In most cases, you won't need to use the glass pane directly. The layered pane allows components to be given a depth value. This value determines which component overlays another. (Thus, the layered pane lets you specify a Z-order for a component, although this is not something that you will usually need to do.) The layered pane holds the content pane and the (optional) menu bar. Although the glass pane and the layered panes are integral to the operation of a top-level container and serve important purposes, much of what they provide occurs behind the scene.

The pane with which your application will interact the most is the content pane, because this is the pane to which you will add visual components. In other words, when you add a component, such as a button, to a top-level container, you will add it to the content pane. Therefore, the content pane holds the components that the user interacts with.

Layout Managers

Before you begin writing a Swing program, there is one more thing that you need to be aware of: the *layout manager*. The layout manager controls the position of components within a container. Java offers several layout managers. Most are provided by the AWT (within **java.awt**), but Swing adds a few of its own. All layout managers are instances of a class that implements the **LayoutManager** interface. (Some will also implement the **LayoutManager2** interface.) Here is a list of a few of the layout managers available to the Swing programmer:

FlowLayout	A simple layout that positions components left-to-right, top-to-bottom. (Positions components right-to-left for some cultural settings.)
BorderLayout	Positions components within the center or the borders of the container. This is the default layout for a content pane.
GridLayout	Lays out components within a grid.
GridBagLayout	Lays out different size components within a flexible grid.
BoxLayout	Lays out components vertically or horizontally within a box.
SpringLayout	Lays out components subject to a set of constraints.

Frankly, the topic of layout managers is quite large, and it is not possible to examine it in detail in this book. Fortunately, this chapter uses only two layout managers—**BorderLayout** and **FlowLayout**—and both are very easy to use.

BorderLayout is the default layout manager for the content pane. It implements a layout style that defines five locations to which a component can be added. The first is the center. The other four are the sides (i.e., borders), which are called north, south, east, and west. By default, when you add a component to the content pane, you are adding the component to the center. To add a component to one of the other regions, specify its name.

Although a border layout is useful in some situations, often another, more flexible layout manager is needed. One of the simplest is **FlowLayout**. A flow layout lays out components one row at a time, top to bottom. When one row is full, layout advances to the next row. Although this scheme gives you little control over the placement of components, it is quite simple to use. However, be aware that if you resize the frame, the position of the components will change.

A First Simple Swing Program

Swing programs differ from the console-based programs shown earlier in this book. Not only do Swing programs use the Swing component set to handle user interaction, but they also have special requirements that relate to threading. The best way to understand the structure of a Swing program is to work through an example.

NOTE

The type of Swing programs shown in this chapter are desktop applications. In the past, Swing was also used to create applets. However, applets have been deprecated since JDK 9 and are not recommended for new code. For this reason, they are not discussed in this book.

Although quite short, the following program shows one way to write a Swing application. In the process it demonstrates several key features of Swing. It uses two Swing components: **JFrame** and **JLabel**. **JFrame** is the top-level container that is commonly used for Swing applications. **JLabel** is the Swing component that creates a label, which is a component that displays information. The label is Swing's simplest component because it is passive. That is, a label does not respond to user input. It just displays output. The program uses a **JFrame** container to hold an instance of a **JLabel**. The label displays a short text message.

```java
// A simple Swing program.

import javax.swing.*;                    Swing programs must import javax.swing.

public class SwingDemo {

  SwingDemo() {                                          Create a container.

    // Create a new JFrame container.
    JFrame jfrm = new JFrame("A Simple Swing Application");

    // Give the frame an initial size.
    jfrm.setSize(275, 100);              Set the dimensions of the frame.

    // Terminate the program when the user closes the application.
    jfrm.setDefaultCloseOperation(JFrame.EXIT_ON_CLOSE);     Terminate
                                                             on close.
    // Create a text-based label.
    JLabel jlab = new JLabel(" GUI programming with Swing.");

    // Add the label to the content pane.                Create a Swing label.
    jfrm.add(jlab);         Add the label to the content pane.

    // Display the frame.
    jfrm.setVisible(true);      Make the frame visible.
  }

  public static void main(String[] args) {
    // Create the frame on the event dispatching thread.
    SwingUtilities.invokeLater(new Runnable() {
      public void run() {
        new SwingDemo();         SwingDemo must be created on the event
      }                          dispatching thread.
    });
  }
}
```

Swing programs are compiled and run in the same way as other Java applications. Thus, to compile this program, you can use this command line:

```
javac SwingDemo.java
```

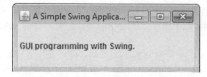

Figure 17-1 The window produced by the **SwingDemo** program

To run the program, use this command line:

```
java SwingDemo
```

When the program is run, it will produce the window shown in Figure 17-1.

The First Swing Example Line by Line

Because the **SwingDemo** program illustrates several key Swing concepts, we will examine it carefully, line by line. The program begins by importing the following package:

```
import javax.swing.*;
```

This **javax.swing** package contains the components and models defined by Swing. For example, it defines classes that implement labels, buttons, edit controls, and menus. This package will be included in all programs that use Swing. Beginning with JDK 9, **javax.swing** is in the **java.desktop** module.

Next, the program declares the **SwingDemo** class and a constructor for that class. The constructor is where most of the action of the program occurs. It begins by creating a **JFrame**, using this line of code:

```
JFrame jfrm = new JFrame("A Simple Swing Application.");
```

This creates a container called **jfrm** that defines a rectangular window complete with a title bar; close, minimize, maximize, and restore buttons; and a system menu. Thus, it creates a standard, top-level window. The title of the window is passed to the constructor.

Next, the window is sized using this statement:

```
jfrm.setSize(275, 100);
```

The **setSize( )** method sets the dimensions of the window, which are specified in pixels. Its general form is shown here:

void setSize(int *width*, int *height*)

In this example, the width of the window is set to 275 and the height is set to 100.

By default, when a top-level window is closed (such as when the user clicks the close box), the window is removed from the screen, but the application is not terminated. While this default behavior is useful in some situations, it is not what is needed for most applications. Instead, you will usually want the entire application to terminate when its top-level window is closed. There are a couple of ways to achieve this. The easiest way is to call **setDefaultCloseOperation( )**, as the program does:

```
jfrm.setDefaultCloseOperation(JFrame.EXIT_ON_CLOSE);
```

After this call executes, closing the window causes the entire application to terminate. The general form of **setDefaultCloseOperation( )** is shown here:

void setDefaultCloseOperation(int *what*)

The value passed in *what* determines what happens when the window is closed. There are several other options in addition to **JFrame.EXIT_ON_CLOSE**. They are shown here:

JFrame.DISPOSE_ON_CLOSE

JFrame.HIDE_ON_CLOSE

JFrame.DO_NOTHING_ON_CLOSE

Their names reflect their actions. These constants are declared in **WindowConstants**, which is an interface declared in **javax.swing** that is implemented by **JFrame**.

The next line of code creates a **JLabel** component:

```
JLabel jlab = new JLabel(" GUI programming with Swing.");
```

JLabel is the easiest-to-use Swing component because it does not accept user input. It simply displays information, which can consist of text, an icon, or a combination of the two. The label created by the program contains only text, which is passed to its constructor.

The next line of code adds the label to the content pane of the frame:

```
jfrm.add(jlab);
```

As explained earlier, all top-level containers have a content pane in which components are stored. Thus, to add a component to a frame, you must add it to the frame's content pane. This is accomplished by calling **add( )** on the **JFrame** reference (**jfrm** in this case). The **add( )** method has several versions. The general form of the one used by the program is shown here:

Component add(Component *comp*)

By default, the content pane associated with a **JFrame** uses a border layout. This version of **add( )** adds the component (in this case, a label) to the center location. Other versions of **add( )** enable you to specify one of the border regions. When a component is added to the center, its size is automatically adjusted to fit the size of the center.

The last statement in the **SwingDemo** constructor causes the window to become visible.

```
jfrm.setVisible(true);
```

The **setVisible( )** method has this general form:

void setVisible(boolean *flag*)

If *flag* is **true**, the window will be displayed. Otherwise, it will be hidden. By default, a **JFrame** is invisible, so **setVisible(true)** must be called to show it.

Inside **main( )**, a **SwingDemo** object is created, which causes the window and the label to be displayed. Notice that the **SwingDemo** constructor is invoked using these lines of code:

```
SwingUtilities.invokeLater(new Runnable() {
  public void run() {
    new SwingDemo();
  }
});
```

This sequence causes a **SwingDemo** object to be created on the *event-dispatching thread* rather than on the main thread of the application. Here's why. In general, Swing programs are event-driven. For example, when a user interacts with a component, an event is generated. An event is passed to the application by calling an event handler defined by the application. However, the handler is executed on the event-dispatching thread provided by Swing and not on the main thread of the application. Thus, although event handlers are defined by your program, they are called on a thread that was not created by your program. To avoid problems (such as two different threads trying to update the same component at the same time), all Swing GUI components must be created and updated from the event-dispatching thread, not the main thread of the application. However, **main( )** is executed on the main thread. Thus, it cannot directly instantiate a **SwingDemo** object. Instead, it must create a **Runnable** object that executes on the event-dispatching thread, and have this object create the GUI.

To enable the GUI code to be created on the event-dispatching thread, you must use one of two methods that are defined by the **SwingUtilities** class. These methods are **invokeLater( )** and **invokeAndWait( )**. They are shown here:

static void invokeLater(Runnable *obj*)

static void invokeAndWait(Runnable *obj*)
    throws InterruptedException, InvocationTargetException

Here, *obj* is a **Runnable** object that will have its **run( )** method called by the event-dispatching thread. The difference between the two methods is that **invokeLater( )** returns immediately, but **invokeAndWait( )** waits until *obj*.**run( )** returns. You can use these methods to call a method that constructs the GUI for your Swing application, or whenever you need to modify the state of the GUI from code not executed by the event-dispatching thread. For the types of programs shown in this chapter, you will normally want to use **invokeLater( )**, as the preceding program does.

One more point: The preceding program does not respond to any events, because **JLabel** is a passive component. In other words, a **JLabel** does not generate any events. Therefore, the preceding program does not include any event handlers. However, all other components generate events to which your program must respond, as the subsequent examples in this chapter show.

Ask the Expert

Q: You state that it is possible to add a component to the other regions of a border layout by using an overloaded version of add( ). Can you explain?

A: As explained, **BorderLayout** implements a layout style that defines five locations to which a component can be added. The first is the center. The other four are the sides (i.e., borders), which are called north, south, east, and west. By default, when you add a component to the content pane, you are adding the component to the center. To specify one of the other locations, use this form of **add( )**:

void add(Component *comp*, Object *loc*)

Here, *comp* is the component to add and *loc* specifies the location to which it is added. The *loc* value is typically one of the following:

BorderLayout.CENTER	BorderLayout.EAST	BorderLayout.NORTH
BorderLayout.SOUTH	BorderLayout.WEST	

In general, **BorderLayout** is most useful when you are creating a **JFrame** that contains a centered component (which might be a group of components held within one of Swing's lightweight containers) that has a header and/or footer component associated with it. In other situations, one of Java's other layout managers will be more appropriate.

Swing Event Handling

As just explained, in general, Swing programs are event driven, with components interacting with the program through events. For example, an event is generated when the user clicks a button, moves the mouse, types a key, or selects an item from a list. Events can also be generated in other ways. For example, an event is generated when a timer goes off. When an event is sent to a program, the program responds to the event by use of an event handler. Thus, event handling is an important part of nearly all Swing applications.

The event handling mechanism used by Swing is called the *delegation event model*. Its concept is quite simple. An event *source* generates an event and sends it to one or more *listeners*. With this approach, the listener simply waits until it receives an event. Once an event arrives, the listener processes the event and then returns. The advantage of this design is that the application logic that processes events is cleanly separated from the user interface logic that generates the events. Therefore, a user interface element is able to "delegate" the handling of an event to a separate piece of code. In the delegation event model, a listener must register with a source in order to receive an event.

Let's look at events, event sources, and listeners a bit more closely.

Events

In Java, an *event* is an object that describes a state change in an event source. It can be generated as a consequence of a person interacting with an element in a graphical user interface or generated under program control. The superclass for all events is **java.util.EventObject**. Many events are declared in **java.awt.event**. Events specifically related to Swing are found in **javax.swing.event**.

Event Sources

An event source is an object that generates an event. When a source generates an event, it sends that event to all registered listeners. Therefore, in order for a listener to receive an event, it must register with the source of that event. In Swing, listeners register with a source by calling a method on the event source object. Each type of event has its own registration method. Typically, events use the following naming convention:

public void add*Type*Listener(*Type*Listener *el*)

Here, *Type* is the name of the event and *el* is a reference to the event listener. For example, the method that registers a keyboard event listener is called **addKeyListener( )**. The method that registers a mouse motion listener is called **addMouseMotionListener( )**. When an event occurs, the event is passed to all registered listeners.

A source must also provide a method that allows a listener to unregister an interest in a specific type of event. In Swing, the naming convention of such a method is this:

public void remove*Type*Listener(*Type*Listener *el*)

Again, *Type* is the name of the event and *el* is a reference to the event listener. For example, to remove a keyboard listener, you would call **removeKeyListener( )**.

The methods that add or remove listeners are provided by the source that generates events. For example, as you will soon see, the **JButton** class is a source of **ActionEvent**s, which are events that indicate that some action, such as a button press, has occurred. Thus, **JButton** provides methods to add or remove an action listener.

Event Listeners

A *listener* is an object that is notified when an event occurs. It has two major requirements. First, it must have registered with one or more sources to receive a specific type of event. Second, it must implement a method to receive and process that event.

The methods that receive and process events applicable to Swing are defined in a set of interfaces, such as those found in **java.awt.event** and **javax.swing.event**. For example, the **ActionListener** interface defines a method that handles an **ActionEvent**. Any object may receive and process this event if it provides an implementation of the **ActionListener** interface.

There is an important general principle that must be stated now. An event handler should do its job quickly and then return. In most cases, it should not engage in a long operation because doing so will slow down the entire application. If a time-consuming operation is required, then a separate thread should be created for this purpose.

Event Classes and Listener Interfaces

The classes that represent events are at the core of Swing's event handling mechanism. At the root of the event class hierarchy is **EventObject**, which is in **java.util**. It is the superclass for all events in Java. The class **AWTEvent**, declared in the **java.awt** package, is a subclass of **EventObject**. It is the superclass (either directly or indirectly) of all AWT-based events used by the delegation event model. Although Swing uses the AWT events, it also adds several of its own. As mentioned, these are in **javax.swing.event**. Thus, Swing supports a large number of events. However, in this chapter only three are used. They are shown here, along with their corresponding listener.

Event Class	Description	Corresponding Event Listener
ActionEvent	Generated when an action occurs within a control, such as when a button is clicked.	ActionListener
ItemEvent	Generated when an item is selected, such as when a check box is clicked.	ItemListener
ListSelectionEvent	Generated when a list selection changes.	ListSelectionListener

The examples that follow illustrate the general procedures that you will use to these handle events. However, the same basic mechanism applies to Swing event handling in general. As you will see, the process is both streamlined and easy to use.

Use JButton

One of the most commonly used Swing controls is the push button. A push button is an instance of **JButton**. **JButton** inherits the abstract class **AbstractButton**, which defines the functionality common to all buttons. Swing push buttons can contain text, an image, or both, but this book uses only text-based buttons.

JButton supplies several constructors. The one used here is

JButton(String *msg*)

Here, *msg* specifies the string that will be displayed inside the button.

When a push button is pressed, it generates an **ActionEvent**. **JButton** provides the following methods, which are used to add or remove an action listener:

void addActionListener(ActionListener *al*)

void removeActionListener(ActionListener *al*)

Here, *al* specifies an object that will receive event notifications. This object must be an instance of a class that implements the **ActionListener** interface.

The **ActionListener** interface defines only one method: **actionPerformed( )**. It is shown here:

void actionPerformed(ActionEvent *ae*)

This method is called when a button is pressed. In other words, it is the event handler that is called when a button press event has occurred. Your implementation of **actionPerformed( )** must quickly respond to that event and return. As explained earlier, as a general rule, event handlers must not engage in long operations, because doing so will slow down the entire application.

Using the **ActionEvent** object passed to **actionPerformed( )**, you can obtain several useful pieces of information relating to the button-press event. The one used by this chapter is the *action command* string associated with the button. By default, this is the string displayed inside the button. The action command is obtained by calling **getActionCommand( )** on the event object. It is declared like this:

String getActionCommand( )

The action command identifies the button. Thus, when using two or more buttons within the same application, the action command gives you an easy way to determine which button was pressed.

The following program demonstrates how to create a push button and respond to button-press events. Figure 17-2 shows how the example appears on the screen.

```java
// Demonstrate a push button and handle action events.

import java.awt.*;
import java.awt.event.*;
import javax.swing.*;

public class ButtonDemo implements ActionListener {

  JLabel jlab;

  ButtonDemo() {

    // Create a new JFrame container.
    JFrame jfrm = new JFrame("A Button Example");

    // Specify FlowLayout for the layout manager.
    jfrm.setLayout(new FlowLayout());

    // Give the frame an initial size.
    jfrm.setSize(220, 90);

    // Terminate the program when the user closes the application.
    jfrm.setDefaultCloseOperation(JFrame.EXIT_ON_CLOSE);

    // Make two buttons.
    JButton jbtnUp = new JButton("Up");          // ─────┐ Create two push buttons.
    JButton jbtnDown = new JButton("Down");      // ─────┘

    // Add action listeners.
    jbtnUp.addActionListener(this);              // ─────┐ Add action listeners for the buttons.
    jbtnDown.addActionListener(this);            // ─────┘
```

```
    // Add the buttons to the content pane.
    jfrm.add(jbtnUp);
    jfrm.add(jbtnDown);
```
Add the buttons to the content pane.

```
    // Create a label.
    jlab = new JLabel("Press a button.");

    // Add the label to the frame.
    jfrm.add(jlab);

    // Display the frame.
    jfrm.setVisible(true);
  }
```

Handle button events.

```
  // Handle button events.
  public void actionPerformed(ActionEvent ae) {
    if(ae.getActionCommand().equals("Up"))
      jlab.setText("You pressed Up.");
    else
      jlab.setText("You pressed down. ");
  }
```

Use the action command to determine which button was pressed.

```
  public static void main(String[] args) {
    // Create the frame on the event dispatching thread.
    SwingUtilities.invokeLater(new Runnable() {
      public void run() {
        new ButtonDemo();
      }
    });
  }
}
```

Let's take a close look at the new things in this program. First, notice that the program now imports both the **java.awt** and **java.awt.event** packages. The **java.awt** package is needed because it contains the **FlowLayout** class, which supports the flow layout manager. The **java.awt.event** package is needed because it defines the **ActionListener** interface and the **ActionEvent** class. Beginning with JDK 9, both packages are in the **java.desktop** module.

Next, the class **ButtonDemo** is declared. Notice that it implements **ActionListener**. This means that **ButtonDemo** objects can be used to receive action events. Next, a **JLabel** reference

Figure 17-2 Output from the **ButtonDemo** program

is declared. This reference will be used within the **actionPerformed( )** method to display which button has been pressed.

The **ButtonDemo** constructor begins by creating a **JFrame** called **jfrm**. It then sets the layout manager for the content pane of **jfrm** to **FlowLayout**, as shown here:

```
jfrm.setLayout(new FlowLayout());
```

As explained earlier, by default, the content pane uses **BorderLayout** as its layout manager, but for many applications, **FlowLayout** is more convenient. Recall that a flow layout lays out components one row at a time, top to bottom. When one row is full, layout advances to the next row. Although this scheme gives you little control over the placement of components, it is quite simple to use. However, be aware that if you resize the frame, the position of the components will change.

After setting the size and the default close operation, **ButtonDemo( )** creates two buttons, as shown here:

```
JButton jbtnUp = new JButton("Up");
JButton jbtnDown = new JButton("Down");
```

The first button will contain the text "Up", and the second will contain "Down".

Next, the instance of **ButtonDemo** referred to via **this** is added as an action listener for the buttons by these two lines:

```
jbtnUp.addActionListener(this);
jbtnDown.addActionListener(this);
```

This approach means that the object that creates the buttons will also receive notifications when a button is pressed.

Each time a button is pressed, it generates an action event and all registered listeners are notified by calling the **actionPerformed( )** method. The **ActionEvent** object representing the button event is passed as a parameter. In the case of **ButtonDemo**, this event is passed to this implementation of **actionPerformed( )**:

```
// Handle button events.
public void actionPerformed(ActionEvent ae) {
  if(ae.getActionCommand().equals("Up"))
    jlab.setText("You pressed Up.");
  else
    jlab.setText("You pressed down. ");
}
```

The event that occurred is passed via **ae**. Inside the method, the action command associated with the button that generated the event is obtained by calling **getActionCommand( )**. (Recall that, by default, the action command is the same as the text displayed by the button.) Based on the contents of that string, the text in the label is set to show which button was pressed.

One last point: Remember that **actionPerformed( )** is called on the event-dispatching thread as explained earlier. It must return quickly in order to avoid slowing down the application.

Work with JTextField

Another commonly used control is **JTextField**. It enables the user to enter a line of text. **JTextField** inherits the abstract class **JTextComponent**, which is the superclass of all text components. **JTextField** defines several constructors. The one we will use is shown here:

JTextField(int *cols*)

Here, *cols* specifies the width of the text field in columns. It is important to understand that you can enter a string that is longer than the number of columns. It's just that the physical size of the text field on the screen will be *cols* columns wide.

When you press ENTER when inputting into a text field, an **ActionEvent** is generated. Therefore, **JTextField** provides the **addActionListener( )** and **removeActionListener( )** methods. To handle action events, you must implement the **actionPerformed( )** method defined by the **ActionListener** interface. The process is similar to handling action events generated by a button, as described earlier.

Like a **JButton**, a **JTextField** has an action command string associated with it. By default, the action command is the current content of the text field. However, this default is seldom used. Instead, you will usually set the action command to a fixed value of your own choosing by calling the **setActionCommand( )** method, shown here:

void setActionCommand(String *cmd*)

The string passed in *cmd* becomes the new action command. The text in the text field is unaffected. Once you set the action command string, it remains the same no matter what is entered into the text field. One reason that you might want to explicitly set the action command is to provide a way to recognize the text field as the source of an action event. This is especially important when another control in the same frame also generates action events and you want to use the same event handler to process both events. Setting the action command gives you a way to tell them apart. Also, if you don't set the action command associated with a text field, then by happenstance the contents of the text field might match the action command of another component.

Ask the Expert

Q: You explained that the action command associated with a text field can be set by calling setActionCommand( ). **Can I use this method to set the action command associated with a push button?**

A: Yes. As explained, by default the action command associated with a push button is the name of the button. To set the action command to a different value, you can use the **setActionCommand( )** method. It works the same for **JButton** as it does for **JTextField**.

To obtain the string that is currently displayed in the text field, call **getText( )** on the **JTextField** instance. It is declared as shown here:

String getText( )

You can set the text in a **JTextField** by calling **setText( )**, shown next:

void setText(String *text*)

Here, *text* is the string that will be put into the text field.

The following program demonstrates **JTextField**. It contains one text field, one push button, and two labels. One label prompts the user to enter text into the text field. When the user presses ENTER while focus is within the text field, the contents of the text field are obtained and displayed within a second label. The push button is called Reverse. When pressed, it reverses the contents of the text field. Sample output is shown in Figure 17-3.

```java
// Use a text field.

import java.awt.*;
import java.awt.event.*;
import javax.swing.*;

public class TFDemo implements ActionListener {

  JTextField jtf;
  JButton jbtnRev;
  JLabel jlabPrompt, jlabContents;

  TFDemo() {

    // Create a new JFrame container.
    JFrame jfrm = new JFrame("Use a Text Field");

    // Specify FlowLayout for the layout manager.
    jfrm.setLayout(new FlowLayout());

    // Give the frame an initial size.
    jfrm.setSize(240, 120);

    // Terminate the program when the user closes the application.
    jfrm.setDefaultCloseOperation(JFrame.EXIT_ON_CLOSE);

    // Create a text field.
    jtf = new JTextField(10);              // ◄──── Create a text field that is 10 columns wide.

    // Set the action commands for the text field.
    jtf.setActionCommand("myTF");          // ◄──── Set the action command for the text field.

    // Create the Reverse button.
    JButton jbtnRev = new JButton("Reverse");
```

```
        // Add action listeners.
        jtf.addActionListener(this);
        jbtnRev.addActionListener(this);
```
Add action listeners for both the text field and the button.

```
        // Create the labels.
        jlabPrompt = new JLabel("Enter text: ");
        jlabContents = new JLabel("");

        // Add the components to the content pane.
        jfrm.add(jlabPrompt);
        jfrm.add(jtf);
        jfrm.add(jbtnRev);
        jfrm.add(jlabContents);

        // Display the frame.
        jfrm.setVisible(true);
    }

    // Handle action events.
    public void actionPerformed(ActionEvent ae) {
```
This method handles both button and text field events.

```
        if(ae.getActionCommand().equals("Reverse")) {
```
Use the action command to determine which component generated the event.

```
            // The Reverse button was pressed.
            String orgStr = jtf.getText();
            String resStr = "";

            // Reverse the string in the text field.
            for(int i=orgStr.length()-1; i >=0; i--)
                resStr += orgStr.charAt(i);

            // Store the reversed string in the text field.
            jtf.setText(resStr);
        } else
            // Enter was pressed while focus was in the
            // text field.
            jlabContents.setText("You pressed ENTER. Text is: " +
                                 jtf.getText());

    }

    public static void main(String[] args) {
        // Create the frame on the event dispatching thread.
        SwingUtilities.invokeLater(new Runnable() {
            public void run() {
                new TFDemo();
            }
        });
    }
}
```

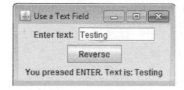

Figure 17-3 Sample output from the **TFDemo** program

Much of the program will be familiar, but a few parts warrant special attention. First, notice that the action command associated with the text field is set to "myTF" by the following line:

```
jtf.setActionCommand("myTF");
```

After this line executes, the action command string will always be "myTF" no matter what text is currently held in the text field. Therefore, the action command generated by **jtf** will not accidentally conflict with the action command associated with the Reverse push button. The **actionPerformed( )** method makes use of this fact to determine what event has occurred. If the action command string is "Reverse", it can mean only one thing: that the Reverse push button has been pressed. Otherwise, the action command was generated by the user pressing ENTER while the text field had input focus.

Finally, notice this line from within the **actionPerformed( )** method:

```
jlabContents.setText("You pressed ENTER. Text is: " +
                     jtf.getText());
```

As explained, when the user presses ENTER while focus is inside the text field, an **ActionEvent** is generated and sent to all registered action listeners, through the **actionPerformed( )** method. For **TFDemo**, this method simply obtains the text currently held in the text field by calling **getText( )** on **jtf**. It then displays the text through the label referred to by **jlabContents**.

Create a JCheckBox

After the push button, perhaps the next most widely used control is the check box. In Swing, a check box is an object of type **JCheckBox**. **JCheckBox** inherits **AbstractButton** and **JToggleButton**. Thus, a check box is, essentially, a special type of button.

JCheckBox defines several constructors. The one used here is

JCheckBox(String *str*)

It creates a check box that has the text specified by *str* as a label.

When a check box is selected or deselected (that is, checked or unchecked), an item event is generated. Item events are represented by the **ItemEvent** class. Item events are handled by classes that implement the **ItemListener** interface. This interface specifies only one method, **itemStateChanged( )**, which is shown here:

void itemStateChanged(ItemEvent *ie*)

The item event is received in *ie*.

To obtain a reference to the item that changed, call **getItem( )** on the **ItemEvent** object. This method is shown here:

Object getItem( )

The reference returned must be cast to the component class being handled, which in this case is **JCheckBox**.

You can obtain the text associated with a check box by calling **getText( )**. You can set the text after a check box is created by calling **setText( )**. These methods work the same as they do for **JButton**, described earlier.

The easiest way to determine the state of a check box is to call the **isSelected( )** method. It is shown here:

boolean isSelected( )

It returns true if the check box is selected and false otherwise.

The following program demonstrates check boxes. It creates three check boxes called Alpha, Beta, and Gamma. Each time the state of a box is changed, the current action is displayed. Also, the list of all currently selected check boxes is displayed. Sample output is shown in Figure 17-4.

```java
// Demonstrate check boxes.

import java.awt.*;
import java.awt.event.*;
import javax.swing.*;

public class CBDemo implements ItemListener {

  JLabel jlabSelected;
  JLabel jlabChanged;
  JCheckBox jcbAlpha;
  JCheckBox jcbBeta;
  JCheckBox jcbGamma;

  CBDemo() {
    // Create a new JFrame container.
    JFrame jfrm = new JFrame("Demonstrate Check Boxes");

    // Specify FlowLayout for the layout manager.
    jfrm.setLayout(new FlowLayout());
```

```
     // Give the frame an initial size.
     jfrm.setSize(280, 120);

     // Terminate the program when the user closes the application.
     jfrm.setDefaultCloseOperation(JFrame.EXIT_ON_CLOSE);

     // Create empty labels.
     jlabSelected = new JLabel("");
     jlabChanged = new JLabel("");

     // Make check boxes.
     jcbAlpha = new JCheckBox("Alpha");
     jcbBeta = new JCheckBox("Beta");                          Create the check boxes.
     jcbGamma = new JCheckBox("Gamma");

     // Events generated by the check boxes
     // are handled in common by the itemStateChanged()
     // method implemented by CBDemo.
     jcbAlpha.addItemListener(this);
     jcbBeta.addItemListener(this);
     jcbGamma.addItemListener(this);

     // Add check boxes and labels to the content pane.
     jfrm.add(jcbAlpha);
     jfrm.add(jcbBeta);
     jfrm.add(jcbGamma);
     jfrm.add(jlabChanged);
     jfrm.add(jlabSelected);

     // Display the frame.
     jfrm.setVisible(true);
   }

   // This is the handler for the check boxes.
   public void itemStateChanged(ItemEvent ie) {          Handle check box item events.
     String str = "";

     // Obtain a reference to the check box that
     // caused the event.
     JCheckBox cb = (JCheckBox) ie.getItem();          Get a reference to the check
                                                        box that changed.

     // Report what check box changed.
     if(cb.isSelected())          Determine what happened.
       jlabChanged.setText(cb.getText() + " was just selected.");
     else
       jlabChanged.setText(cb.getText() + " was just cleared.");
```

```
                  // Report all selected boxes.
                  if(jcbAlpha.isSelected()) {
                    str += "Alpha ";
                  }
                  if(jcbBeta.isSelected()) {
                    str += "Beta ";
                  }
                  if(jcbGamma.isSelected()) {
                    str += "Gamma";
                  }

                  jlabSelected.setText("Selected check boxes: " + str);
              }

              public static void main(String[] args) {
                // Create the frame on the event dispatching thread.
                SwingUtilities.invokeLater(new Runnable() {
                  public void run() {
                    new CBDemo();
                  }
                });
              }
            }
```

The main point of interest in this program is the item event handler, **itemStateChanged( )**. It performs two functions. First, it reports whether the check box has been selected or cleared. Second, it displays all selected check boxes. It begins by obtaining a reference to the check box that generated the **ItemEvent**, as shown here:

```
JCheckBox cb = (JCheckBox) ie.getItem();
```

The cast to **JCheckBox** is necessary because **getItem( )** returns a reference of type **Object**. Next, **itemStateChanged( )** calls **isSelected( )** on **cb** to determine the current state of the check box. If **isSelected( )** returns true, it means that the user selected the check box. Otherwise, the check box was cleared. It then sets the **jlabChanged** label to reflect what happened.

Finally, **itemStateChanged( )** checks the selected state of each check box, building a string that contains the names of those that are selected. It displays this string in the **jlabSelected** label.

Figure 17-4   Sample output from the **CBDemo** program

Work with JList

The last component that we will examine is **JList**. This is Swing's basic list class. It supports the selection of one or more items from a list. Although often the list consists of strings, it is possible to create a list of just about any object that can be displayed. **JList** is so widely used in Java that it is highly unlikely that you have not seen one before.

JList is a generic class that is declared as shown here:

class JList<E>

Here, **E** represents the type of the items in the list.

JList provides several constructors. The one used here is

JList(E[ ] *items*)

This creates a **JList** that contains the items in the array specified by *items*.

Although a **JList** will work properly by itself, most of the time you will wrap a **JList** inside a **JScrollPane**, which is a container that automatically provides scrolling for its contents. Here is the constructor that we will use:

JScrollPane(Component *comp*)

Here, *comp* specifies the component to be scrolled, which in this case will be a **JList**. When you wrap a **JList** in a **JScrollPane**, long lists will automatically be scrollable. This simplifies GUI design. It also makes it easy to change the number of entries in a list without having to change the size of the **JList** component.

A **JList** generates a **ListSelectionEvent** when the user makes or changes a selection. This event is also generated when the user deselects an item. It is handled by implementing **ListSelectionListener**, which is packaged in **javax.swing.event**. This listener specifies only one method, called **valueChanged( )**, which is shown here:

void valueChanged(ListSelectionEvent *le*)

Here, *le* is a reference to the object that generated the event. Although **ListSelectionEvent** does provide some methods of its own, often you will interrogate the **JList** object itself to determine what has occurred. **ListSelectionEvent** is also packaged in **javax.swing.event**.

By default, a **JList** allows the user to select multiple ranges of items within the list, but you can change this behavior by calling **setSelectionMode( )**, which is defined by **JList**. It is shown here:

void setSelectionMode(int *mode*)

Here, *mode* specifies the selection mode. It must be one of these values defined by the **ListSelectionModel** interface (which is packaged in **javax.swing**):

SINGLE_SELECTION

SINGLE_INTERVAL_SELECTION

MULTIPLE_INTERVAL_SELECTION

The default, multiple-interval selection lets the user select multiple ranges of items within a list. With single-interval selection, the user can select one range of items. With single selection, the user can select only a single item. Of course, a single item can be selected in the other two modes, too. It's just that they also allow a range to be selected.

You can obtain the index of the first item selected, which will also be the index of the only selected item when using single-selection mode, by calling **getSelectedIndex( )**, shown here:

int getSelectedIndex( )

Indexing begins at zero. So, if the first item is selected, this method will return 0. If no item is selected, −1 is returned.

You can obtain an array containing all selected items by calling **getSelectedIndices( )**, shown next:

int[ ] getSelectedIndices( )

In the returned array, the indices are ordered from smallest to largest. If a zero-length array is returned, it means that no items are selected.

The following program demonstrates a simple **JList**, which holds a list of names. Each time a name is selected in the list, a **ListSelectionEvent** is generated, which is handled by the **valueChanged( )** method defined by **ListSelectionListener**. It responds by obtaining the index of the selected item and displaying the corresponding name. Sample output is shown in Figure 17-5.

```
// Demonstrate a simple JList.

import javax.swing.*;
import javax.swing.event.*;
import java.awt.*;
import java.awt.event.*;

public class ListDemo implements ListSelectionListener {

  JList<String> jlst;
  JLabel jlab;
  JScrollPane jscrlp;
```

```
// Create an array of names.
String[] names = { "Sherry", "Jon", "Rachel",
                   "Sasha", "Josselyn", "Randy",
                   "Tom", "Mary", "Ken",
                   "Andrew", "Matt", "Todd" };
```

This array will be displayed in a **JList**.

```
ListDemo() {
  // Create a new JFrame container.
  JFrame jfrm = new JFrame("JList Demo");

  // Specify a flow Layout.
  jfrm.setLayout(new FlowLayout());

  // Give the frame an initial size.
  jfrm.setSize(200, 160);

  // Terminate the program when the user closes the application.
  jfrm.setDefaultCloseOperation(JFrame.EXIT_ON_CLOSE);

  // Create a JList.
  jlst = new JList<String>(names);
```
Create the list.
```
  // Set the list selection mode to single selection.
  jlst.setSelectionMode(ListSelectionModel.SINGLE_SELECTION);
```
Switch to single-selection mode.
```
  // Add list to a scroll pane.
  jscrlp = new JScrollPane(jlst);
```
Wrap the list in a scroll pane.
```
  // Set the preferred size of the scroll pane.
  jscrlp.setPreferredSize(new Dimension(120, 90));

  // Make a label that displays the selection.
  jlab = new JLabel("Please choose a name");

  // Add list selection handler.
  jlst.addListSelectionListener(this);
```
Listen for list selection events.
```
  // Add the list and label to the content pane.
  jfrm.add(jscrlp);
  jfrm.add(jlab);

  // Display the frame.
  jfrm.setVisible(true);
}
```
Handle list selection events.
```
// Handle list selection events.
public void valueChanged(ListSelectionEvent le) {
  // Get the index of the changed item.
  int idx = jlst.getSelectedIndex();
```
Get the index of the selected/ deselected item.

```
    // Display selection, if item was selected.
    if(idx != -1)
      jlab.setText("Current selection: " + names[idx]);
    else // Otherwise, reprompt.
      jlab.setText("Please choose a name");
  }

  public static void main(String[] args) {
    // Create the frame on the event dispatching thread.
    SwingUtilities.invokeLater(new Runnable() {
      public void run() {
        new ListDemo();
      }
    });
  }
}
```

Let's look closely at this program. First, notice the **names** array near the top of the program. It is initialized to a list of strings that contain various names. Inside **ListDemo( )**, a **JList** called **jlst** is constructed using the **names** array. As mentioned, when the array constructor is used (as it is in this case), a **JList** instance is automatically created that contains the contents of the array. Thus, the list will contain the names in **names**.

Next, the selection mode is set to single selection. This means that only one item in this list can be selected at any one time. Then, **jlst** is wrapped inside a **JScrollPane**, and the preferred size of the scroll pane is set to 120 by 90. This makes for a compact, but easy-to-use scroll pane. In Swing, the **setPreferredSize( )** method sets the ideal size of a component. Be aware that some layout managers are free to ignore this request, but most often the preferred size determines the size of the component.

A list selection event occurs whenever the user selects an item or changes the item selected. Inside the **valueChanged( )** event handler, the index of the item selected is obtained by calling **getSelectedIndex( )**. Because the list has been set to single-selection mode, this is also the index of the only item selected. This index is then used to index the **names** array to obtain the selected name. Notice that this index value is tested against –1. Recall that this is the value returned if no item has been selected. This will be the case when the selection event handler is called if the user has deselected an item. Remember: A selection event is generated when the user selects or deselects an item.

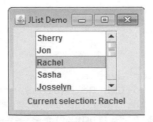

Figure 17-5 Output from the **ListDemo** program

Try This 17-1   A Swing-Based File Comparison Utility

SwingFC.java   Although you know only a small amount about Swing, you can still put it to use
to create a practical application. In Try This 10-1, you created a console-based file
comparison utility. This project creates a Swing-based version of the program. As you will see,
giving this application a Swing-based user interface substantially improves its appearance and
makes it easier to use. Here is how the Swing version looks:

Because Swing streamlines the creation of GUI-based programs, you might be surprised
by how easy it is to create this program.

1. Begin by creating a file called **SwingFC.java** and then enter the following comment and
import statements:

```
/*
    Try This 17-1

    A Swing-based file comparison utility.

*/

import java.awt.*;
import java.awt.event.*;
import javax.swing.*;
import java.io.*;
```

2. Next, begin the **SwingFC** class, as shown here:

```
public class SwingFC implements ActionListener {

    JTextField jtfFirst; // holds the first file name
    JTextField jtfSecond; // holds the second file name

    JButton jbtnComp; // button to compare the files

    JLabel jlabFirst, jlabSecond; // displays prompts
    JLabel jlabResult; // displays results and error messages
```

(continued)

The names of the files to compare are entered into the text fields defined by **jtfFirst** and **jtfSecond**. To compare the files, the user presses the **jbtnComp** button. Prompting messages are displayed in **jlabFirst** and **jlabSecond**. The results of the comparison, or any error messages, are displayed in **jlabResult**.

3. Code the **SwingFC** constructor like this:

```
SwingFC() {

  // Create a new JFrame container.
  JFrame jfrm = new JFrame("Compare Files");

  // Specify FlowLayout for the layout manager.
  jfrm.setLayout(new FlowLayout());

  // Give the frame an initial size.
  jfrm.setSize(200, 190);

  // Terminate the program when the user closes the application.
  jfrm.setDefaultCloseOperation(JFrame.EXIT_ON_CLOSE);

  // Create the text fields for the file names.
  jtfFirst = new JTextField(14);
  jtfSecond = new JTextField(14);

  // Set the action commands for the text fields.
  jtfFirst.setActionCommand("fileA");
  jtfSecond.setActionCommand("fileB");

  // Create the Compare button.
  JButton jbtnComp = new JButton("Compare");

  // Add action listener for the Compare button.
  jbtnComp.addActionListener(this);

  // Create the labels.
  jlabFirst = new JLabel("First file: ");
  jlabSecond = new JLabel("Second file: ");
  jlabResult = new JLabel("");

  // Add the components to the content pane.
  jfrm.add(jlabFirst);
  jfrm.add(jtfFirst);
  jfrm.add(jlabSecond);
  jfrm.add(jtfSecond);
  jfrm.add(jbtnComp);
  jfrm.add(jlabResult);
```

```
   // Display the frame.
   jfrm.setVisible(true);
 }
```

Most of the code in this constructor should be familiar to you. However, notice one thing: an action listener is added only to the push button **jbtnCompare**. Action listeners are not added to the text fields. Here's why: the contents of the text fields are needed only when the Compare button is pushed. At no other time are their contents required. Thus, there is no reason to respond to any text field events. As you begin to write more Swing programs, you will find that this is often the case when using a text field.

4. Begin creating the **actionPerformed( )** event handler, as shown next. This method is called when the Compare button is pressed.

```
// Compare the files when the Compare button is pressed.
public void actionPerformed(ActionEvent ae) {
  int i=0, j=0;

  // First, confirm that both file names have
  // been entered.
  if(jtfFirst.getText().equals("")) {
    jlabResult.setText("First file name missing.");
    return;
  }
  if(jtfSecond.getText().equals("")) {
    jlabResult.setText("Second file name missing.");
    return;
  }
```

The method begins by confirming that the user has entered a file name into each of the text fields. If this is not the case, the missing file name is reported and the handler returns.

5. Now, finish **actionPerformed( )** by adding the code that actually opens the files and then compares them.

```
// Compare files. Use try-with-resources to manage the files.
try (FileInputStream f1 = new FileInputStream(jtfFirst.getText());
     FileInputStream f2 = new FileInputStream(jtfSecond.getText()))
{

  // Check the contents of each file.
  do {
    i = f1.read();
    j = f2.read();
    if(i != j) break;
  } while(i != -1 && j != -1);

  if(i != j)
    jlabResult.setText("Files are not the same.");
```

(continued)

```
      else
        jlabResult.setText("Files compare equal.");
    } catch(IOException exc) {
      jlabResult.setText("File Error");
    }
  }
}
```

6. Finish **SwingFC** by adding the following **main( )** method.

```
public static void main(String[] args) {
  // Create the frame on the event dispatching thread.
  SwingUtilities.invokeLater(new Runnable() {
    public void run() {
    new SwingFC();
    }
  });
  }
}
```

7. The entire Swing-based file comparison program is shown here:

```
/*
    Try This 17-1

    A Swing-based file comparison utility.

*/

import java.awt.*;
import java.awt.event.*;
import javax.swing.*;
import java.io.*;

public class SwingFC implements ActionListener {

  JTextField jtfFirst;  // holds the first file name
  JTextField jtfSecond; // holds the second file name

  JButton jbtnComp; // button to compare the files

  JLabel jlabFirst, jlabSecond; // displays prompts
  JLabel jlabResult; // displays results and error messages

  SwingFC() {

    // Create a new JFrame container.
    JFrame jfrm = new JFrame("Compare Files");

    // Specify FlowLayout for the layout manager.
    jfrm.setLayout(new FlowLayout());

    // Give the frame an initial size.
    jfrm.setSize(200, 190);
```

```
    // Terminate the program when the user closes the application.
    jfrm.setDefaultCloseOperation(JFrame.EXIT_ON_CLOSE);

    // Create the text fields for the file names.
    jtfFirst = new JTextField(14);
    jtfSecond = new JTextField(14);

    // Set the action commands for the text fields.
    jtfFirst.setActionCommand("fileA");
    jtfSecond.setActionCommand("fileB");

    // Create the Compare button.
    JButton jbtnComp = new JButton("Compare");

    // Add action listener for the Compare button.
    jbtnComp.addActionListener(this);

    // Create the labels.
    jlabFirst = new JLabel("First file: ");
    jlabSecond = new JLabel("Second file: ");
    jlabResult = new JLabel("");

    // Add the components to the content pane.
    jfrm.add(jlabFirst);
    jfrm.add(jtfFirst);
    jfrm.add(jlabSecond);
    jfrm.add(jtfSecond);
    jfrm.add(jbtnComp);
    jfrm.add(jlabResult);

    // Display the frame.
    jfrm.setVisible(true);
  }

  // Compare the files when the Compare button is pressed.
  public void actionPerformed(ActionEvent ae) {
    int i=0, j=0;

    // First, confirm that both file names have
    // been entered.
    if(jtfFirst.getText().equals("")) {
      jlabResult.setText("First file name missing.");
      return;
    }
    if(jtfSecond.getText().equals("")) {
      jlabResult.setText("Second file name missing.");
      return;
    }
```

(continued)

```
    // Compare files. Use try-with-resources to manage the files.
    try (FileInputStream f1 = new FileInputStream(jtfFirst.getText());
         FileInputStream f2 = new FileInputStream(jtfSecond.getText()))
    {
      // Check the contents of each file.
      do {
        i = f1.read();
        j = f2.read();
        if(i != j) break;
      } while(i != -1 && j != -1);

      if(i != j)
        jlabResult.setText("Files are not the same.");
      else
        jlabResult.setText("Files compare equal.");
    } catch(IOException exc) {
      jlabResult.setText("File Error");
    }
  }

  public static void main(String[] args) {
    // Create the frame on the event dispatching thread.
    SwingUtilities.invokeLater(new Runnable() {
      public void run() {
        new SwingFC();
      }
    });
  }
}
```

Use Anonymous Inner Classes
or Lambda Expressions to Handle Events

Up to this point, the programs in this chapter have used a simple, straightforward approach
to handling events in which the main class of the application has implemented the listener
interface itself and all events are sent to an instance of that class. While this is perfectly
acceptable, it is not the only way to handle events. For example, you could use separate listener
classes. Thus, different classes could handle different events and these classes would be
separate from the main class of the application. However, two other approaches offer powerful
alternatives. First, you can implement listeners through the use of *anonymous inner classes*.
Second, in some cases, you can use a lambda expression to handle an event. Let's look at each
approach.

Anonymous inner classes are inner classes that don't have a name. Instead, an instance
of the class is simply generated "on the fly" as needed. Anonymous inner classes make

implementing some types of event handlers much easier. For example, given a **JButton** called **jbtn**, you could implement an action listener for it like this:

```
jbtn.addActionListener(new ActionListener() {
  public void actionPerformed(ActionEvent ae) {
    // Handle action event here.
  }
});
```

Here, an anonymous inner class is created that implements the **ActionListener** interface. Pay special attention to the syntax. The body of the inner class begins after the **{** that follows **new ActionListener( )**. Also notice that the call to **addActionListener( )** ends with a **)** and a **;** just like normal. The same basic syntax and approach is used to create an anonymous inner class for any event handler. Of course, for different events, you specify different event listeners and implement different methods.

One advantage to using an anonymous inner class is that the component that invokes the class' methods is already known. For instance, in the preceding example, there is no need to call **getActionCommand( )** to determine what component generated the event, because this implementation of **actionPerformed( )** will only be called by events generated by **jbtn**.

In the case of an event whose listener defines a functional interface, you can handle the event by use of a lambda expression. For example, action events can be handled with a lambda expression because **ActionListener** defines only one abstract method, **actionPerformed( )** Using a lambda expression to implement **ActionListener** provides a compact alternative to explicitly declaring an anonymous inner class. For example, again assuming a **JButton** called **jbtn**, you could implement the action listener like this:

```
jbtn.addActionListener((ae) -> {
  // Handle action event here.
});
```

As was the case with the anonymous inner class approach, the object that generates the event is known. In this case, the lambda expression applies only to the **jbtn** button.

Of course, in cases in which an event can be handled by use of a single expression, it is not necessary to use a block lambda. For example, here is an action event handler for the Up button in the **ButtonDemo** program shown earlier. It requires only an expression lambda.

```
jbtnUp.addActionListener((ae) -> jlab.setText("You pressed Up."));
```

Notice how much shorter this code is compared with the original approach. It is also shorter than it would be if you explicitly used an anonymous inner class.

In general, you can use a lambda expression to handle an event when its listener defines a functional interface. For example, **ItemListener** is also a functional interface. Of course, whether you use the traditional approach, an anonymous inner class, or a lambda expression will be determined by the precise nature of your application. To gain experience with each, try converting the event handlers in the foregoing examples to lambda expressions or anonymous inner classes.

Chapter 17 Self Test

1. In general, AWT components are heavyweight and Swing components are _____.

2. Can the look and feel of a Swing component be changed? If so, what feature enables this?

3. What is the most commonly used top-level container for an application?

4. Top-level containers have several panes. To what pane are components added?

5. Show how to construct a label that contains the message "Select an entry from the list".

6. All interaction with GUI components must take place on what thread?

7. What is the default action command associated with a **JButton**? How can the action command be changed?

8. What event is generated when a push button is pressed?

9. Show how to create a text field that has 32 columns.

10. Can a **JTextField** have its action command set? If so, how?

11. What Swing component creates a check box? What event is generated when a check box is selected or deselected?

12. **JList** displays a list of items from which the user can select. True or False?

13. What event is generated when the user selects or deselects an item in a **JList**?

14. What method sets the selection mode of a **JList**? What method obtains the index of the first selected item?

15. Add a check box to the file comparer developed in Try This 17-1 that has the following text: Show position of mismatch. When this box is checked, have the program display the location of the first point in the files at which a mismatch occurs.

16. Change the **ListDemo** program so that it allows multiple items in the list to be selected.

17. Bonus challenge: Convert the Help class developed in Try This 4-1 into a Swing-based GUI program. Display the keywords (**for**, **while**, **switch**, and so on) in a **JList**. When the user selects one, display the keyword's syntax. To display multiple lines of text within a label, you can use HTML. When doing so, you must begin the text with the sequence **<html>**. When this is done, the text is automatically formatted as described by the markup. In addition to other benefits, using HTML enables you to create labels that span two or more lines. For example, this creates a label that displays two lines of text, with the string "Top" over the string "Bottom".

```
JLabel jlabhtml = new JLabel("<html>Top<br>Bottom</html>");
```

No answer is shown for this exercise. You have reached the point where you are ready to apply your Java skills on your own!

Appendix A

Answers to Self Tests

Chapter 1: Java Fundamentals

1. What is bytecode and why is it important to Java's use for Internet programming?

Bytecode is a highly optimized set of instructions that is executed by the Java Virtual Machine. Bytecode helps Java achieve both portability and security.

2. What are the three main principles of object-oriented programming?

Encapsulation, polymorphism, and inheritance.

3. Where do Java programs begin execution?

Java programs begin execution at **main( )**.

4. What is a variable?

A variable is a named memory location. The contents of a variable can be changed during the execution of a program.

5. Which of the following variable names is invalid?

The invalid variable is **D**. Variable names cannot begin with a digit.

6. How do you create a single-line comment? How do you create a multiline comment?

A single-line comment begins with **//** and ends at the end of the line. A multiline comment begins with **/*** and ends with ***/**.

7. Show the general form of the **if** statement. Show the general form of the **for** loop.

The general form of the **if**:

if(*condition*) *statement;*

The general form of the **for**:

for(*initialization; condition; iteration*) *statement;*

8. How do you create a block of code?

A block of code is started with a **{** and ended with a **}**.

9. The moon's gravity is about 17 percent that of the earth's. Write a program that computes your effective weight on the moon.

```
/*
   Compute your weight on the moon.

   Call this file Moon.java.
*/
class Moon {
  public static void main(String[] args) {
    double earthweight; // weight on earth
    double moonweight; // weight on moon
```

```
        earthweight = 165;

        moonweight = earthweight * 0.17;

        System.out.println(earthweight +
                        " earth-pounds is equivalent to " +
                        moonweight + " moon-pounds.");

    }
}
```

10. Adapt Try This 1-2 so that it prints a conversion table of inches to meters. Display 12 feet of conversions, inch by inch. Output a blank line every 12 inches. (One meter equals approximately 39.37 inches.)

```
/*
    This program displays a conversion
    table of inches to meters.

    Call this program InchToMeterTable.java.
*/
class InchToMeterTable {
  public static void main(String[] args) {
     double inches, meters;
     int counter;

     counter = 0;
     for(inches = 1; inches <= 144; inches++) {
       meters = inches / 39.37; // convert to meters
       System.out.println(inches + " inches is " +
                        meters + " meters.");

       counter++;
       // every 12th line, print a blank line
       if(counter == 12) {
         System.out.println();
         counter = 0; // reset the line counter
       }
     }
  }
}
```

11. If you make a typing mistake when entering your program, what sort of error will result?

    A syntax error.

12. Does it matter where on a line you put a statement?

    No, Java is a free-form language.

Chapter 2: Introducing Data Types and Operators

1. Why does Java strictly specify the range and behavior of its primitive types?

Java strictly specifies the range and behavior of its primitive types to ensure portability across platforms.

2. What is Java's character type, and how does it differ from the character type used by some other programming languages?

Java's character type is **char**. Java characters are Unicode rather than ASCII, which is used by some other computer languages.

3. A **boolean** value can have any value you like because any non-zero value is true. True or False?

False. A boolean value must be either **true** or **false**.

4. Given this output,

```
One
Two
Three
```

use a single string to show the **println( )** statement that produced it.

```
System.out.println("One\nTwo\nThree");
```

5. What is wrong with this fragment?

```
for(i = 0; i < 10; i++) {
  int sum;

  sum = sum + i;
}
System.out.println("Sum is: " + sum);
```

There are two fundamental flaws in the fragment. First, **sum** is created each time the block defined by the **for** loop is entered and destroyed on exit. Thus, it will not hold its value between iterations. Attempting to use **sum** to hold a running sum of the iterations is pointless. Second, **sum** will not be known outside of the block in which it is declared. Thus, the reference to it in the **println( )** statement is invalid.

6. Explain the difference between the prefix and postfix forms of the increment operator.

When the increment operator precedes its operand, Java will perform the increment prior to obtaining the operand's value for use by the rest of the expression. If the operator follows its operand, then Java will obtain the operand's value before incrementing.

7. Show how a short-circuit AND can be used to prevent a divide-by-zero error.

```
if((b != 0) && (val / b)) ...
```

8. In an expression, what type are **byte** and **short** promoted to?

In an expression, **byte** and **short** are promoted to **int**.

9. In general, when is a cast needed?

A cast is needed when converting between incompatible types or when a narrowing conversion is occurring.

10. Write a program that finds all of the prime numbers between 2 and 100.

```
// Find prime numbers between 2 and 100.
class Prime {
  public static void main(String[] args) {
    int i, j;
    boolean isprime;

    for(i=2; i < 100; i++) {
      isprime = true;

      // see if the number is evenly divisible
      for(j=2; j <= i/j; j++)
        // if it is, then it's not prime
        if((i%j) == 0) isprime = false;

      if(isprime)
        System.out.println(i + " is prime.");
    }
  }
}
```

11. Does the use of redundant parentheses affect program performance?

No.

12. Does a block define a scope?

Yes.

Chapter 3: Program Control Statements

1. Write a program that reads characters from the keyboard until a period is received. Have the program count the number of spaces. Report the total at the end of the program.

```
// Count spaces.
class Spaces {
  public static void main(String[] args)
    throws java.io.IOException {

    char ch;
    int spaces = 0;



    System.out.println("Enter a period to stop.");
```

```
        do {
          ch = (char) System.in.read();
          if(ch == ' ') spaces++;
        } while(ch != '.');

        System.out.println("Spaces: " + spaces);
      }
    }
```

2. Show the general form of the **if**-**else**-**if** ladder.

```
if(condition)
   statement;
else if(condition)
   statement;
else if(condition)
   statement;
   .
   .
   .
else
   statement;
```

3. Given

```
if(x < 10)
  if(y > 100) {
     if(!done) x = z;
     else y = z;
  }
else System.out.println("error"); // what if?
```

to what **if** does the last **else** associate?

The last **else** associates with **if(y > 100)**.

4. Show the **for** statement for a loop that counts from 1000 to 0 by –2.

```
for(int i = 1000; i >= 0; i -= 2) // ...
```

5. Is the following fragment valid?

```
for(int i = 0; i < num; i++)
   sum += i;

count = i;
```

No; **i** is not known outside of the **for** loop in which it is declared.

6. Explain what **break** does. Be sure to explain both of its forms.

A **break** without a label causes termination of its immediately enclosing loop or **switch** statement. A **break** with a label causes control to transfer to the end of the labeled block.

7. In the following fragment, after the **break** statement executes, what is displayed?

```
for(i = 0; i < 10; i++) {
  while(running) {
    if(x<y) break;
    // ...
  }
  System.out.println("after while");
}
System.out.println("After for");
```

After **break** executes, "after while" is displayed.

8. What does the following fragment print?

```
for(int i = 0; i<10; i++) {
  System.out.print(i + " ");
  if((i%2) -- 0) continue;
  System.out.println();
}
```

Here is the answer.

```
0 1
2 3
4 5
6 7
8 9
```

9. The iteration expression in a **for** loop need not always alter the loop control variable by a fixed amount. Instead, the loop control variable can change in any arbitrary way. Using this concept, write a program that uses a **for** loop to generate and display the progression 1, 2, 4, 8, 16, 32, and so on.

```
/* Use a for loop to generate the progression

     1 2 4 8 16, ...
*/
class Progress {
  public static void main(String[] args) {

    for(int i = 1; i < 100; i += i)
      System.out.print(i + " ");

  }
}
```

10. The ASCII lowercase letters are separated from the uppercase letters by 32. Thus, to convert a lowercase letter to uppercase, subtract 32 from it. Use this information to write a program that reads characters from the keyboard. Have it convert all lowercase letters to uppercase, and all uppercase letters to lowercase, displaying the result. Make no changes to any other character. Have the program stop when the user enters a period. At the end, have the program display the number of case changes that have taken place.

```java
// Change case.
class CaseChg {
  public static void main(String[] args)
    throws java.io.IOException {
    char ch;
    int changes = 0;

    System.out.println("Enter period to stop.");

    do {
      ch = (char) System.in.read();
      if(ch >= 'a' & ch <= 'z') {
        ch -= 32;
        changes++;
        System.out.println(ch);
      }
      else if(ch >= 'A' & ch <= 'Z') {
        ch += 32;
        changes++;
        System.out.println(ch);
      }
    } while(ch != '.');
    System.out.println("Case changes: " + changes);
  }
}
```

11. What is an infinite loop?

An infinite loop is a loop that runs indefinitely.

12. When using **break** with a label, must the label be on a block that contains the **break**?

Yes.

Chapter 4: Introducing Classes, Objects, and Methods

1. What is the difference between a class and an object?

A class is a logical abstraction that describes the form and behavior of an object. An object is a physical instance of the class.

2. How is a class defined?

A class is defined by using the keyword **class**. Inside the **class** statement, you specify the code and data that comprise the class.

3. What does each object have its own copy of?

   Each object of a class has its own copy of the class' instance variables.

4. Using two separate statements, show how to declare an object called **counter** of a class called **MyCounter**.

   ```
   MyCounter counter;
   counter = new MyCounter();
   ```

5. Show how a method called **myMeth( )** is declared if it has a return type of **double** and has two **int** parameters called **a** and **b**.

   ```
   double myMeth(int a, int b) { // ...
   ```

6. How must a method return if it returns a value?

   A method that returns a value must return via the **return** statement, passing back the return value in the process.

7. What name does a constructor have?

   A constructor has the same name as its class.

8. What does **new** do?

   The **new** operator allocates memory for an object and initializes it using the object's constructor.

9. What is garbage collection and how does it work?

   Garbage collection is the mechanism that recycles unused objects so that their memory can be reused.

10. What is **this**?

    The **this** keyword is a reference to the object on which a method is invoked. It is automatically passed to a method.

11. Can a constructor have one or more parameters?

    Yes.

12. If a method returns no value, what must its return type be?

    void

Chapter 5: More Data Types and Operators

1. Show two ways to declare a one-dimensional array of 12 **doubles**.

   ```
   double x[] = new double[12];
   double[] x = new double[12];
   ```

2. Show how to initialize a one-dimensional array of integers to the values 1 through 5.

   ```
   int[] x = { 1, 2, 3, 4, 5 };
   ```

3. Write a program that uses an array to find the average of ten **double** values. Use any ten values you like.

```
// Average 10 double values.
class Avg {
  public static void main(String[] args) {
    double[] nums = { 1.1, 2.2, 3.3, 4.4, 5.5,
                      6.6, 7.7, 8.8, 9.9, 10.1 };
    double sum = 0;

    for(int i=0; i < nums.length; i++)
      sum += nums[i];

    System.out.println("Average: " + sum / nums.length);
  }
}
```

4. Change the sort in Try This 5-1 so that it sorts an array of strings. Demonstrate that it works.

```
// Demonstrate the Bubble sort with strings.
class StrBubble {
  public static void main(String[] args) {
    String[] strs = {
                      "this", "is", "a", "test",
                      "of", "a", "string", "sort"
                    };
    int a, b;
    String t;
    int size;

    size = strs.length; // number of elements to sort

    // display original array
    System.out.print("Original array is:");
    for(int i=0; i < size; i++)
      System.out.print(" " + strs[i]);
    System.out.println();

    // This is the bubble sort for strings.
    for(a=1; a < size; a++)
      for(b=size-1; b >= a; b--) {
        if(strs[b-1].compareTo(strs[b]) > 0) { // if out of order
          // exchange elements
          t = strs[b-1];
          strs[b-1] = strs[b];
          strs[b] = t;
        }
      }
```

```
    // display sorted array
    System.out.print("Sorted array is:");
    for(int i=0; i < size; i++)
      System.out.print(" " + strs[i]);
    System.out.println();
  }
}
```

5. What is the difference between the **String** methods **indexOf( )** and **lastIndexOf( )**?

   The **indexOf( )** method finds the first occurrence of the specified substring. **lastIndexOf( )** finds the last occurrence.

6. Since all strings are objects of type **String**, show how you can call the **length( )** and **charAt( )** methods on this string literal: "I like Java".

   As strange as it may look, this is a valid call to **length( )**:

   ```
   System.out.println("I like Java".length());
   ```

   The output displayed is 11. **charAt( )** is called in a similar fashion.

7. Expanding on the **Encode** cipher class, modify it so that it uses an eight-character string as the key.

   ```
   // An improved XOR cipher.
   class Encode {
     public static void main(String[] args) {
       String msg = "This is a test";
       String encmsg = "";
       String decmsg = "";
       String key = "abcdefgi";
       int j;

       System.out.print("Original message: ");
       System.out.println(msg);

       // encode the message
       j = 0;
       for(int i=0; i < msg.length(); i++) {
         encmsg = encmsg + (char) (msg.charAt(i) ^ key.charAt(j));
         j++;
         if(j==8) j = 0;
       }

       System.out.print("Encoded message: ");
       System.out.println(encmsg);

       // decode the message
       j = 0;
       for(int i=0; i < msg.length(); i++) {
         decmsg = decmsg + (char) (encmsg.charAt(i) ^ key.charAt(j));
   ```

```
          j++;
          if(j==8) j = 0;
        }

        System.out.print("Decoded message: ");
        System.out.println(decmsg);
      }
    }
```

8. Can the bitwise operators be applied to the **double** type?

No.

9. Show how this sequence can be rewritten using the **?** operator.

```
if(x < 0) y = 10;
else y = 20;
```

Here is the answer:

```
y = x < 0 ? 10 : 20;
```

10. In the following fragment, is the **&** a bitwise or logical operator? Why?

```
boolean a, b;
// ...
if(a & b) ...
```

It is a logical operator because the operands are of type **boolean**.

11. Is it an error to overrun the end of an array?

Yes.

Is it an error to index an array with a negative value?

Yes. All array indexes start at zero.

12. What is the unsigned right-shift operator?

```
>>>
```

13. Rewrite the **MinMax** class shown earlier in this chapter so that it uses a for-each style **for** loop.

```
// Find the minimum and maximum values in an array.
class MinMax {
  public static void main(String[] args) {
    int[] nums = new int[10];
    int min, max;

    nums[0] = 99;
    nums[1] = -10;
    nums[2] = 100123;
    nums[3] = 18;
    nums[4] = -978;
    nums[5] = 5623;
```

```
     nums[6] = 463;
     nums[7] = -9;
     nums[8] = 287;
     nums[9] = 49;

     min = max = nums[0];
     for(int v : nums) {
       if(v < min) min = v;
       if(v > max) max = v;
     }
     System.out.println("min and max: " + min + " " + max);
   }
 }
```

14. Can the **for** loops that perform sorting in the **Bubble** class shown in Try This 5-1 be converted into for-each style loops? If not, why not?

No, the **for** loops in the **Bubble** class that perform the sort cannot be converted into for-each style loops. In the case of the outer loop, the current value of its loop counter is needed by the inner loop. In the case of the inner loop, out-of-order values must be exchanged, which implies assignments. Assignments to the underlying array cannot take place when using a for-each style loop.

15. Can a **String** control a **switch** statement?

Yes.

16. What keyword is reserved for use with local variable type inference?

The context-sensitive keyword **var** is reserved for use with local variable type inference.

17. Show how to use local variable type inference to declare a **boolean** variable called **done** that has an initial value of **false**.

```
var done = false;
```

18. Can **var** be the name of a variable? Can **var** be the name of a class?

Yes, **var** can be the name of a variable. No, **var** cannot be the name of a class.

19. Is the following declaration valid? If not, why not.

```
var[] avgTemps = new double[7];
```

No, it is not valid because array brackets are not allowed after **var**. Remember, the complete type is inferred from the initializer.

20. Is the following declaration valid? If not, why not?

```
var alpha = 10, beta = 20;
```

No, only one variable at a time can be declared when type inference is used.

21. In the **show( )** method of the **ShowBits** class developed in Try This 5-3, the local variable **mask** is declared as shown here:

```
long mask = 1;
```

Change this declaration so that it uses local variable type inference. When doing so, be sure that **mask** is of type **long** (as it is here), and not of type **int**.

```
var mask = 1L; // Notice that the initial value is explicitly
               // specified as long so that mask will be inferred to
               // be long.
```

Chapter 6: A Closer Look at Methods and Classes

1. Given this fragment,

```
class X {
  private int count;
```

is the following fragment correct?

```
class Y {
  public static void main(String[] args) {
    X ob = new X();

    ob.count = 10;
```

No; a **private** member cannot be accessed outside of its class.

2. An access modifier must _____ a member's declaration.

precede

3. The complement of a queue is a stack. It uses first-in, last-out accessing and is often likened to a stack of plates. The first plate put on the table is the last plate used. Create a stack class called **Stack** that can hold characters. Call the methods that access the stack **push( )** and **pop( )**. Allow the user to specify the size of the stack when it is created. Keep all other members of the **Stack** class private. (Hint: You can use the **Queue** class as a model; just change the way that the data is accessed.)

```
// A stack class for characters.
class Stack {
  private char[] stck; // this array holds the stack
  private int tos; // top of stack

  // Construct an empty Stack given its size.
  Stack(int size) {
    stck = new char[size]; // allocate memory for stack
    tos = 0;
  }

  // Construct a Stack from a Stack.
  Stack(Stack ob) {
    tos = ob.tos;
    stck = new char[ob.stck.length];
```

```
      // copy elements
      for(int i=0; i < tos; i++)
        stck[i] = ob.stck[i];
    }

    // Construct a stack with initial values.
    Stack(char[] a) {
      stck = new char[a.length];

      for(int i = 0; i < a.length; i++) {
        push(a[i]);
      }
    }

    // Push characters onto the stack.
    void push(char ch) {
      if(tos==stck.length) {
        System.out.println(" -- Stack is full.");
        return;
      }

      stck[tos] = ch;
      tos++;
    }

    // Pop a character from the stack.
    char pop() {
      if(tos==0) {
        System.out.println(" -- Stack is empty.");
        return (char) 0;
      }

      tos--;
      return stck[tos];
    }
  }

  // Demonstrate the Stack class.
  class SDemo {
    public static void main(String[] args) {
      // construct 10-element empty stack
      Stack stk1 = new Stack(10);

      char[] name = {'T', 'o', 'm'};

      // construct stack from array
      Stack stk2 = new Stack(name);
```

```
      char ch;
      int i;

      // put some characters into stk1
      for(i=0; i < 10; i++)
        stk1.push((char) ('A' + i));

      // construct stack from another stack
      Stack stk3 = new Stack(stk1);

      // show the stacks.
      System.out.print("Contents of stk1: ");
      for(i=0; i < 10; i++) {
        ch = stk1.pop();
        System.out.print(ch);
      }

      System.out.println("\n");

      System.out.print("Contents of stk2: ");
      for(i=0; i < 3; i++) {
        ch = stk2.pop();
        System.out.print(ch);
      }

      System.out.println("\n");
      System.out.print("Contents of stk3: ");
      for(i=0; i < 10; i++) {
        ch = stk3.pop();
        System.out.print(ch);
      }
    }
  }
```

Here is the output from the program:

```
Contents of stk1: JIHGFEDCBA
Contents of stk2: moT
Contents of stk3: JIHGFEDCBA
```

4. Given this class,

```
class Test {
  int a;
  Test(int i) { a = i; }
}
```

write a method called **swap( )** that exchanges the contents of the objects referred to by two **Test** object references.

```
void swap(Test ob1, Test ob2) {
   int t;

   t = ob1.a;
   ob1.a = ob2.a;
   ob2.a = t;
}
```

5. Is the following fragment correct?

```
class X {
   int meth(int a, int b) { ... }
   String meth(int a, int b) { ... }
```

No. Overloaded methods can have different return types, but they do not play a role in overload resolution. Overloaded methods *must* have different parameter lists.

6. Write a recursive method that displays the contents of a string backwards.

```
// Display a string backwards using recursion.
class Backwards {
   String str;

   Backwards(String s) {
      str = s;
   }

   void backward(int idx) {
      if(idx != str.length()-1) backward(idx+1);

      System.out.print(str.charAt(idx));
   }
}

class BWDemo {
   public static void main(String[] args) {
      Backwards s = new Backwards("This is a test");

      s.backward(0);
   }
}
```

7. If all objects of a class need to share the same variable, how must you declare that variable?

Shared variables are declared as **static**.

8. Why might you need to use a **static** block?

A **static** block is used to perform any initializations related to the class, before any objects are created.

9. What is an inner class?

An inner class is a nonstatic nested class.

10. To make a member accessible by only other members of its class, what access modifier must be used?

private

11. The name of a method plus its parameter list constitutes the method's _____.

signature

12. An **int** argument is passed to a method by using call-by-_____.

value

13. Create a varargs method called **sum( )** that sums the **int** values passed to it. Have it return the result. Demonstrate its use.

There are many ways to craft the solution. Here is one:

```
class SumIt {
  int sum(int ... n) {
    int result = 0;

    for(int i = 0; i < n.length; i++)
      result += n[i];

    return result;
  }
}

class SumDemo {
  public static void main(String[] args) {

    SumIt siObj = new SumIt();

    int total = siObj.sum(1, 2, 3);
    System.out.println("Sum is " + total);

    total = siObj.sum(1, 2, 3, 4, 5);
    System.out.println("Sum is " + total);
  }
}
```

14. Can a varargs method be overloaded?

Yes.

15. Show an example of an overloaded varargs method that is ambiguous.

Here is one example of an overloaded varargs method that is ambiguous:

```
double myMeth(double ... v ) { // ...

double myMeth(double d, double ... v) { // ...
```

If you try to call **myMeth( )** with one argument, like this,

```
myMeth(1.1);
```

the compiler can't determine which version of the method to invoke.

Chapter 7: Inheritance

1. Does a superclass have access to the members of a subclass? Does a subclass have access to the members of a superclass?

No, a superclass has no knowledge of its subclasses. Yes, a subclass has access to all nonprivate members of its superclass.

2. Create a subclass of **TwoDShape** called **Circle**. Include an **area( )** method that computes the area of the circle and a constructor that uses **super** to initialize the **TwoDShape** portion.

```
// A subclass of TwoDShape for circles.
class Circle extends TwoDShape {
  // A default constructor.
  Circle() {
    super();
  }

  // Construct Circle
  Circle(double x) {
    super(x, "circle"); // call superclass constructor
  }

  // Construct an object from an object.
  Circle(Circle ob) {
    super(ob); // pass object to TwoDShape constructor
  }

  double area() {
    return (getWidth() / 2) * (getWidth() / 2) * 3.1416;
  }
}
```

3. How do you prevent a subclass from having access to a member of a superclass?

To prevent a subclass from having access to a superclass member, declare that member as **private**.

4. Describe the purpose and use of the two versions of **super** described in this chapter.

The **super** keyword has two forms. The first is used to call a superclass constructor. The general form of this usage is

> super (*param-list*);

The second form of **super** is used to access a superclass member. It has this general form:

> super.*member*

5. Given the following hierarchy, in what order do the constructors for these classes complete their execution when a **Gamma** object is instantiated?

```
class Alpha { ...

class Beta extends Alpha { ...

Class Gamma extends Beta { ...
```

Constructors complete their execution in order of derivation. Thus, when a **Gamma** object is created, the order is **Alpha**, **Beta**, **Gamma**.

6. A superclass reference can refer to a subclass object. Explain why this is important as it is related to method overriding.

When an overridden method is called through a superclass reference, it is the type of the object being referred to that determines which version of the method is called.

7. What is an abstract class?

An abstract class contains at least one abstract method.

8. How do you prevent a method from being overridden? How do you prevent a class from being inherited?

To prevent a method from being overridden, declare it as **final**. To prevent a class from being inherited, declare it as **final**.

9. Explain how inheritance, method overriding, and abstract classes are used to support polymorphism.

Inheritance, method overriding, and abstract classes support polymorphism by enabling you to create a generalized class structure that can be implemented by a variety of classes. Thus, the abstract class defines a consistent interface that is shared by all implementing classes. This embodies the concept of "one interface, multiple methods."

10. What class is a superclass of every other class?

The **Object** class.

11. A class that contains at least one abstract method must, itself, be declared abstract. True or False?

True.

12. What keyword is used to create a named constant?

final

13. Assume that class **B** inherits class **A**. Further, assume a method called **makeObj( )** that is declared as shown here:

```
A makeObj(int which ) {
   if(which == 0) return new A();
   else return new B();
}
```

Notice that **makeObj( )** returns a reference to an object of either type **A** or **B**, depending on the value of **which**. Notice, however, that the return type of **makeObj( )** is **A**. (Recall that a superclass reference can refer to a subclass object.) Given this situation and assuming that you are using JDK 10 or later, what is the type of **myRef** in the following declaration and why?

```
var myRef = makeObj(1);
```

Even though a **B** object is created, the type of **myRef** will be **A** because that is the declared return type of **makeObj( )**. When using local variable type inference, the inferred type of a variable is based on the *declared type* of its initializer. Therefore, if the initializer is of a superclass type (which is **A** in this case), that will be the type of the variable. It does not matter if the actual object being referred to by the initializer is an instance of a derived class.

14. Assuming the situation described in Question 13, what will the type of **myRef** be given this statement?

```
var myRef = (B) makeObj(1);
```

In this case, the cast to **B** specifies the type of the initializer, and **myRef** is of type **B**.

Chapter 8: Packages and Interfaces

1. Using the code from Try This 8-1, put the **ICharQ** interface and its three implementations into a package called **qpack**. Keeping the queue demonstration class **IQDemo** in the default package, show how to import and use the classes in **qpack**.

To put **ICharQ** and its implementations into the **qpack** package, you must separate each into its own file, make each implementation class **public**, and add this statement to the top of each file.

```
package qpack;
```

Once this has been done, you can use **qpack** by adding this **import** statement to **IQDemo**.

```
import qpack.*;
```

2. What is a namespace? Why is it important that Java allows you to partition the namespace?

A namespace is a declarative region. By partitioning the namespace, you can prevent name collisions.

3. Typically, packages are stored in _____.

directories

4. Explain the difference between **protected** and default access.

A member with **protected** access can be used within its package and by a subclass in other packages.

A member with default access can be used only within its package.

5. Explain the two ways that the members of a package can be used by other packages.

To use a member of a package, you can either fully qualify its name, or you can import it using **import**.

6. "One interface, multiple methods" is a key tenet of Java. What feature best exemplifies it?

The interface best exemplifies the one interface, multiple methods principle of OOP.

7. How many classes can implement an interface? How many interfaces can a class implement?

An interface can be implemented by an unlimited number of classes. A class can implement as many interfaces as it chooses.

8. Can interfaces be extended?

Yes, interfaces can be extended.

9. Create an interface for the **Vehicle** class from Chapter 7. Call the interface **IVehicle**.

```
interface IVehicle {

  // Return the range.
  int range();

  // Compute fuel needed for a given distance.
  double fuelneeded(int miles);

  // Access methods for instance variables.
  int getPassengers();
  void setPassengers(int p);
  int getFuelcap();
  void setFuelcap(int f);
  int getMpg();
  void setMpg(int m);
}
```

10. Variables declared in an interface are implicitly **static** and **final**. Can they be shared with other parts of a program?

Yes, interface variables can be used as named constants that are shared by all files in a program. They are brought into view by implementing their interface.

11. A package is, in essence, a container for classes. True or False?

True.

12. What standard Java package is automatically imported into a program?

java.lang

13. What keyword is used to declare a default **interface** method?

default

14. Is it possible to define a **static** method in an **interface**?

Yes

15. Assume that the **ICharQ** interface shown in Try This 8-1 has been in widespread use for several years. Now, you want to add a method to it called **reset( )**, which will be used to reset the queue to its empty, starting condition. How can this be accomplished without breaking preexisting code?

To avoid breaking preexisting code, you must use a default interface method. Because you can't know how to reset each queue implementation, the default **reset( )** implementation will need to report an error that indicates that it is not implemented. (The best way to do this is to use an exception. Exceptions are examined in the following chapter.) Fortunately, since no preexisting code assumes that **ICharQ** defines a **reset( )** method, no preexisting code will encounter that error, and no preexisting code will be broken.

16. How is a **static** method in an interface called?

A **static** interface method is called through its interface name, by use of the dot operator.

17. Can an interface have a private method?

Yes.

Chapter 9: Exception Handling

1. What class is at the top of the exception hierarchy?

Throwable is at the top of the exception hierarchy.

2. Briefly explain how to use **try** and **catch**.

The **try** and **catch** statements work together. Program statements that you want to monitor for exceptions are contained within a **try** block. An exception is caught using **catch**.

3. What is wrong with this fragment?

```
// ...
vals[18] = 10;
catch (ArrayIndexOutOfBoundsException exc) {
  // handle error
}
```

There is no **try** block preceding the **catch** statement.

4. What happens if an exception is not caught?

If an exception is not caught, abnormal program termination results.

5. What is wrong with this fragment?

```
class A extends Exception { ...

class B extends A { ...

// ...

try {
  // ...
}
catch (A exc) { ... }
catch (B exc) { ... }
```

In the fragment, a superclass **catch** precedes a subclass **catch**. Since the superclass **catch** will catch all subclasses too, unreachable code is created.

6. Can an inner **catch** rethrow an exception to an outer **catch**?

Yes, an exception can be rethrown.

7. The **finally** block is the last bit of code executed before your program ends. True or False? Explain your answer.

False. The **finally** block is the code executed when a **try** block ends.

8. What type of exceptions must be explicitly declared in a **throws** clause of a method?

All exceptions except those of type **RuntimeException** and **Error** must be declared in a **throws** clause.

9. What is wrong with this fragment?

```
class MyClass { // ... }
// ...
throw new MyClass();
```

MyClass does not extend **Throwable**. Only subclasses of **Throwable** can be thrown by **throw**.

10. In question 3 of the Chapter 6 Self Test, you created a **Stack** class. Add custom exceptions to your class that report stack full and stack empty conditions.

```
// An exception for stack-full errors.
class StackFullException extends Exception {
  int size;

  StackFullException(int s) { size = s; }

  public String toString() {
   return "\nStack is full. Maximum size is " +
          size;
  }
}
```

```java
// An exception for stack-empty errors.
class StackEmptyException extends Exception {

  public String toString() {
    return "\nStack is empty.";
  }
}

// A stack class for characters.
class Stack {
  private char[] stck; // this array holds the stack
  private int tos; // top of stack

  // Construct an empty Stack given its size.
  Stack(int size) {
    stck = new char[size]; // allocate memory for stack
    tos = 0;
  }

  // Construct a Stack from a Stack.
  Stack(Stack ob) {
    tos = ob.tos;
    stck = new char[ob.stck.length];

    // copy elements
    for(int i=0; i < tos; i++)
      stck[i] = ob.stck[i];
  }

  // Construct a stack with initial values.
  Stack(char[] a) {
    stck = new char[a.length];

    for(int i = 0; i < a.length; i++) {
      try {
        push(a[i]);
      }
      catch(StackFullException exc) {
        System.out.println(exc);
      }
    }
  }

  // Push characters onto the stack.
  void push(char ch) throws StackFullException {
    if(tos==stck.length)
      throw new StackFullException(stck.length);
```

```
      stck[tos] = ch;
      tos++;
    }

    // Pop a character from the stack.
    char pop() throws StackEmptyException {
      if(tos==0)
        throw new StackEmptyException();
      tos--;
      return stck[tos];
    }
  }
```

11. What are the three ways that an exception can be generated?

An exception can be generated by an error in the JVM, by an error in your program, or explicitly via a **throw** statement.

12. What are the two direct subclasses of **Throwable**?

Error and **Exception**

13. What is the multi-catch feature?

The multi-catch feature allows one **catch** clause to catch two or more exceptions.

14. Should your code typically catch exceptions of type **Error**?

No.

Chapter 10: Using I/O

1. Why does Java define both byte and character streams?

The byte streams are the original streams defined by Java. They are especially useful for binary I/O, and they support random-access files. The character streams are optimized for Unicode.

2. Even though console input and output is text-based, why does Java still use byte streams for this purpose?

The predefined streams, **System.in**, **System.out**, and **System.err**, were defined before Java added the character streams.

3. Show how to open a file for reading bytes.

Here is one way to open a file for **byte** input:

```
FileInputStream fin = new FileInputStream("test");
```

4. Show how to open a file for reading characters.

Here is one way to open a file for reading characters:

```
FileReader fr = new FileReader("test");
```

5. Show how to open a file for random-access I/O.

Here is one way to open a file for random access:

```
randfile = new RandomAccessFile("test", "rw");
```

6. How do you convert a numeric string such as "123.23" into its binary equivalent?

To convert numeric strings into their binary equivalents, use the parsing methods defined by the type wrappers, such as **Integer** or **Double**.

7. Write a program that copies a text file. In the process, have it convert all spaces into hyphens. Use the byte stream file classes. Use the traditional approach to closing a file by explicitly calling **close( )**.

```java
/* Copy a text file, substituting hyphens for spaces.

   This version uses byte streams.

   To use this program, specify the name
   of the source file and the destination file.
   For example,

   java Hyphen source target
*/

import java.io.*;

class Hyphen {
  public static void main(String[] args)
  {
    int i;
    FileInputStream fin = null;
    FileOutputStream fout = null;

    // First make sure that both files have been specified.
    if(args.length !=2 ) {
      System.out.println("Usage: Hyphen From To");
      return;
    }

    // Copy file and substitute hyphens.
    try {
      fin = new FileInputStream(args[0]);
      fout = new FileOutputStream(args[1]);

      do {
        i = fin.read();

        // convert space to a hyphen
        if((char)i == ' ') i = '-';
```

```
            if(i != -1) fout.write(i);
          } while(i != -1);
        } catch(IOException exc) {
          System.out.println("I/O Error: " + exc);
        } finally {
          try {
            if(fin != null) fin.close();
          } catch(IOException exc) {
            System.out.println("Error closing input file.");
          }

          try {
            if(fin != null) fout.close();
          } catch(IOException exc) {
            System.out.println("Error closing output file.");
          }
        }
      }
    }
  }
```

8. Rewrite the program in question 7 so that it uses the character stream classes. This time, use the **try**-with-resources statement to automatically close the file.

```
/* Copy a text file, substituting hyphens for spaces.

   This version uses character streams.

   To use this program, specify the name
   of the source file and the destination file.
   For example,

   java Hyphen2 source target

*/

import java.io.*;

class Hyphen2 {
  public static void main(String[] args)
    throws IOException
  {
    int i;

    // First make sure that both files have been specified.
    if(args.length !=2 ) {
      System.out.println("Usage: CopyFile From To");
      return;
    }
```

```
// Copy file and substitute hyphens.
// Use the try-with-resources statement.
try (FileReader fin = new FileReader(args[0]);
     FileWriter fout = new FileWriter(args[1]))
{
  do {
    i = fin.read();

    // convert space to a hyphen
    if((char)i == ' ') i = '-';

    if(i != -1) fout.write(i);
  } while(i != -1);
} catch(IOException exc) {
  System.out.println("I/O Error: " + exc);
}
  }
}
```

9. What type of stream is **System.in**?

   InputStream

10. What does the **read( )** method of **InputStream** return when an attempt is made to read at the end of the stream?

    –1

11. What type of stream is used to read binary data?

    DataInputStream

12. **Reader** and **Writer** are at the top of the _____ class hierarchies.

    character-based I/O

13. The **try**-with-resources statement is used for _____ _____ _____.

    automatic resource management

14. If you are using the traditional method of closing a file, then closing a file within a **finally** block is generally a good approach. True or False?

    True

15. Can local variable type inference be used when declaring the resource in a **try**-with-resources statement?

    Yes.

Chapter 11: Multithreaded Programming

1. How does Java's multithreading capability enable you to write more efficient programs?

   Multithreading allows you to take advantage of the idle time that is present in nearly all programs. When one thread can't run, another can. In multicore systems, two or more threads can execute simultaneously.

2. Multithreading is supported by the _____ class and the _____ interface.

   Multithreading is supported by the **Thread** class and the **Runnable** interface.

3. When creating a runnable object, why might you want to extend **Thread** rather than implement **Runnable**?

   You will extend **Thread** when you want to override one or more of **Thread**'s methods other than **run( )**.

4. Show how to use **join( )** to wait for a thread object called **MyThrd** to end.

   ```
   MyThrd.join();
   ```

5. Show how to set a thread called **MyThrd** to three levels above normal priority.

   ```
   MyThrd.setPriority(Thread.NORM_PRIORITY+3);
   ```

6. What is the effect of adding the **synchronized** keyword to a method?

   Adding **synchronized** to a method allows only one thread at a time to use the method for any given object of its class.

7. The **wait( )** and **notify( )** methods are used to perform _____.

   interthread communication

8. Change the **TickTock** class so that it actually keeps time. That is, have each tick take one half second, and each tock take one half second. Thus, each tick-tock will take one second. (Don't worry about the time it takes to switch tasks, etc.)

   To make the **TickTock** class actually keep time, simply add calls to **sleep( )**, as shown here:

   ```
   // Make the TickTock class actually keep time.

   class TickTock {

     String state; // contains the state of the clock

     synchronized void tick(boolean running) {
       if(!running) { // stop the clock
         state = "ticked";
         notify(); // notify any waiting threads
         return;
       }
   ```

```java
      System.out.print("Tick ");

      // wait 1/2 second
      try {
        Thread.sleep(500);
      } catch(InterruptedException exc) {
        System.out.println("Thread interrupted.");
      }

      state = "ticked"; // set the current state to ticked

      notify(); // let tock() run
      try {
        while(!state.equals("tocked"))
          wait(); // wait for tock() to complete
      }
      catch(InterruptedException exc) {
        System.out.println("Thread interrupted.");
      }
    }

    synchronized void tock(boolean running) {
      if(!running) { // stop the clock
        state = "tocked";
        notify(); // notify any waiting threads
        return;
      }

      System.out.println("Tock");

      // wait 1/2 second
      try {
        Thread.sleep(500);
      } catch(InterruptedException exc) {
        System.out.println("Thread interrupted.");
      }

      state = "tocked"; // set the current state to tocked

      notify(); // let tick() run
      try {
        while(!state.equals("ticked"))
          wait(); // wait for tick to complete
      }
      catch(InterruptedException exc) {
        System.out.println("Thread interrupted.");
      }
    }
  }
```

9. Why can't you use **suspend( )**, **resume( )**, and **stop( )** for new programs?

   The **suspend( )**, **resume( )**, and **stop( )** methods have been deprecated because they can cause serious run-time problems.

10. What method defined by **Thread** obtains the name of a thread?

    getName( )

11. What does **isAlive( )** return?

    It returns **true** if the invoking thread is still running, and **false** if it has been terminated.

Chapter 12: Enumerations, Autoboxing, Annotations, and More

1. Enumeration constants are said to be *self-typed*. What does this mean?

   In the term *self-typed,* the "self" refers to the type of the enumeration in which the constant is defined. Thus, an enumeration constant is an object of the enumeration of which it is a part.

2. What class do all enumerations automatically inherit?

   The **Enum** class is automatically inherited by all enumerations.

3. Given the following enumeration, write a program that uses **values( )** to show a list of the constants and their ordinal values.

```
enum Tools {
   SCREWDRIVER, WRENCH, HAMMER, PLIERS
}
```

   The solution is

```
enum Tools {
   SCREWDRIVER, WRENCH, HAMMER, PLIERS
}

class ShowEnum {
  public static void main(String[] args) {
    for(Tools d : Tools.values())
      System.out.print(d + " has ordinal value of " +
                       d.ordinal() + '\n');
  }
}
```

4. The traffic light simulation developed in Try This 12-1 can be improved with a few simple changes that take advantage of an enumeration's class features. In the version shown, the duration of each color was controlled by the **TrafficLightSimulator** class by hard-coding these values into the **run( )** method. Change this so that the duration of each color is stored by the constants in the **TrafficLightColor** enumeration. To do this, you will need to add a constructor, a private instance variable, and a method called **getDelay( )**. After making

these changes, what improvements do you see? On your own, can you think of other improvements? (Hint: Try using ordinal values to switch light colors rather than relying on a **switch** statement.)

The improved version of the traffic light simulation is shown here. There are two major improvements. First, a light's delay is now linked with its enumeration value, which gives more structure to the code. Second, the **run( )** method no longer needs to use a **switch** statement to determine the length of the delay. Instead, **sleep( )** is passed **tlc.getDelay( )**, which causes the delay associated with the current color to be used automatically.

```java
// An improved version of the traffic light simulation that
// stores the light delay in TrafficLightColor.

// An enumeration of the colors of a traffic light.
enum TrafficLightColor {
  RED(12000), GREEN(10000), YELLOW(2000);

  private int delay;

  TrafficLightColor(int d) {
    delay = d;
  }

  int getDelay() { return delay; }
}

// A computerized traffic light.
class TrafficLightSimulator implements Runnable {
  private TrafficLightColor tlc; // holds the current traffic light color
  private boolean stop = false; // set to true to stop the simulation
  private boolean changed = false; // true when the light has changed

  TrafficLightSimulator(TrafficLightColor init) {
    tlc = init;
  }

  TrafficLightSimulator() {
    tlc = TrafficLightColor.RED;
  }

  // Start up the light.
  public void run() {
    while(!stop) {
      // Notice how this code has been simplified!
      try {
        Thread.sleep(tlc.getDelay());
      } catch(InterruptedException exc) {
        System.out.println(exc);
      }

      changeColor();
    }
  }
}
```

```java
// Change color.
synchronized void changeColor() {
  switch(tlc) {
    case RED:
      tlc = TrafficLightColor.GREEN;
      break;
    case YELLOW:
      tlc = TrafficLightColor.RED;
      break;
    case GREEN:
      tlc = TrafficLightColor.YELLOW;
  }

  changed = true;
  notify(); // signal that the light has changed
}

// Wait until a light change occurs.
synchronized void waitForChange() {
  try {
    while(!changed)
      wait(); // wait for light to change
    changed = false;
  } catch(InterruptedException exc) {
    System.out.println(exc);
  }
}

// Return current color.
synchronized TrafficLightColor getColor() {
  return tlc;
}

// Stop the traffic light.
synchronized void cancel() {
  stop = true;
}
}

class TrafficLightDemo {
  public static void main(String[] args) {
    TrafficLightSimulator tl =
      new TrafficLightSimulator(TrafficLightColor.GREEN);

    Thread thrd = new Thread(tl);
    thrd.start();
    for(int i=0; i < 9; i++) {
      System.out.println(tl.getColor());
      tl.waitForChange();
    }

    tl.cancel();
  }
}
```

5. Define boxing and unboxing. How does autoboxing/unboxing affect these actions?

Boxing is the process of storing a primitive value in a type wrapper object. Unboxing is the process of retrieving the primitive value from the type wrapper. Autoboxing automatically boxes a primitive value without having to explicitly construct an object. Auto-unboxing automatically retrieves the primitive value from a type wrapper without having to explicitly call a method, such as **intValue( )**.

6. Change the following fragment so that it uses autoboxing.

```
Double val = Double.valueOf(123.0);
```

The solution is

```
Double val = 123.0;
```

7. In your own words, what does static import do?

Static import brings into the current namespace the static members of a class or interface. This means that static members can be used without having to be qualified by their class or interface name.

8. What does this statement do?

```
import static java.lang.Integer.parseInt;
```

The statement brings into the current namespace the **parseInt( )** method of the type wrapper **Integer**.

9. Is static import designed for special-case situations, or is it good practice to bring all static members of all classes into view?

Static import is designed for special cases. Bringing many static members into view will lead to namespace collisions and destructure your code.

10. An annotation is syntactically based on a/an _____

interface

11. What is a marker annotation?

A marker annotation is one that does not take arguments.

12. An annotation can be applied only to methods. True or False?

False. Any type of declaration can have an annotation. Beginning with JDK 8, a type use can also have an annotation.

13. What operator determines if an object is of a specified type?

instanceof

14. Will an invalid cast that occurs at run time result in an exception?

Yes

Chapter 13: Generics

1. Generics are important to Java because they enable the creation of code that is

   A. Type-safe

   B. Reusable

   C. Reliable

   D. All of the above

   D. All of the above

2. Can a primitive type be used as a type argument?

   No, type arguments must be object types.

3. Show how to declare a class called **FlightSched** that takes two generic parameters.

   The solution is

   ```
   class FlightSched<T, V> {
   ```

4. Beginning with your answer to question 3, change **FlightSched**'s second type parameter so that it must extend **Thread**.

   The solution is

   ```
   class FlightSched<T, V extends Thread> {
   ```

5. Now, change **FlightSched** so that its second type parameter must be a subclass of its first type parameter.

   The solution is

   ```
   class FlightSched<T, V extends T> {
   ```

6. As it relates to generics, what is the **?** and what does it do?

   The **?** is the wildcard argument. It matches any valid type.

7. Can the wildcard argument be bounded?

   Yes, a wildcard can have either an upper or lower bound.

8. A generic method called **MyGen( )** has one type parameter. Furthermore, **MyGen( )** has one parameter whose type is that of the type parameter. It also returns an object of that type parameter. Show how to declare **MyGen( )**.

   The solution is

   ```
   <T> T MyGen(T o) { // ...
   ```

9. Given this generic interface

```
interface IGenIF<T, V extends T> { // ...
```

show the declaration of a class called **MyClass** that implements **IGenIF**.

The solution is

```
class MyClass<T, V extends T> implements IGenIF<T, V> { // ...
```

10. Given a generic class called **Counter<T>**, show how to create an object of its raw type.

To obtain **Counter<T>**'s raw type, simply use its name without any type specification, as shown here:

```
Counter x = new Counter();
```

11. Do type parameters exist at run time?

No. All type parameters are erased during compilation, and appropriate casts are substituted. This process is called erasure.

12. Convert your solution to question 10 of the Self Test for Chapter 9 so that it is generic. In the process, create a stack interface called **IGenStack** that generically defines the operations **push( )** and **pop( )**.

```
// A generic stack.

interface IGenStack<T> {
  void push(T obj) throws StackFullException;
  T pop() throws StackEmptyException;
}

// An exception for stack-full errors.
class StackFullException extends Exception {
  int size;

  StackFullException(int s) { size = s; }

  public String toString() {
   return "\nStack is full. Maximum size is " +
          size;
  }
}

// An exception for stack-empty errors.
class StackEmptyException extends Exception {

  public String toString() {
   return "\nStack is empty.";
  }
}
```

```java
// A stack class for characters.
class GenStack<T> implements IGenStack<T> {
  private T[] stck; // this array holds the stack
  private int tos; // top of stack

  // Construct an empty stack given its size.
  GenStack(T[] stckArray) {
    stck = stckArray;
    tos = 0;
  }

  // Construct a stack from a stack.
  GenStack(T[] stckArray, GenStack<T> ob) {
    tos = ob.tos;
    stck = stckArray;

    try {
      if(stck.length < ob.stck.length)
        throw new StackFullException(stck.length);
    }
    catch(StackFullException exc) {
      System.out.println(exc);
    }

    // Copy elements.
    for(int i=0; i < tos; i++)
      stck[i] = ob.stck[i];
  }

  // Construct a stack with initial values.
  GenStack(T[] stckArray, T[] a) {
    stck = stckArray;

    for(int i = 0; i < a.length; i++) {
      try {
        push(a[i]);
      }
      catch(StackFullException exc) {
        System.out.println(exc);
      }
    }
  }

  // Push objects onto the stack.
  public void push(T obj) throws StackFullException {
    if(tos==stck.length)
      throw new StackFullException(stck.length);
```

```java
      stck[tos] = obj;
      tos++;
    }

    // Pop an object from the stack.
    public T pop() throws StackEmptyException {
      if(tos==0)
        throw new StackEmptyException();

      tos--;
      return stck[tos];
    }
  }

// Demonstrate the GenStack class.
class GenStackDemo {
  public static void main(String[] args) {
    // Construct 10-element empty Integer stack.
    Integer[] iStore = new Integer[10];
    GenStack<Integer> stk1 = new GenStack<Integer>(iStore);

    // Construct stack from array.
    String[] name = {"One", "Two", "Three"};
    String[] strStore = new String[3];
    GenStack<String> stk2 =
        new GenStack<String>(strStore, name);

    String str;
    int n;

    try {
      // Put some values into stk1.
      for(int i=0; i < 10; i++)
        stk1.push(i);
    } catch(StackFullException exc) {
      System.out.println(exc);
    }

    // Construct stack from another stack.
    String[] strStore2 = new String[3];
    GenStack<String> stk3 =
        new GenStack<String>(strStore2, stk2);

    try {
      // Show the stacks.
      System.out.print("Contents of stk1: ");
      for(int i=0; i < 10; i++) {
        n = stk1.pop();
        System.out.print(n + " ");
      }
```

```
      System.out.println("\n");

      System.out.print("Contents of stk2: ");
      for(int i=0; i < 3; i++) {
        str = stk2.pop();
        System.out.print(str + " ");
      }

      System.out.println("\n");

      System.out.print("Contents of stk3: ");
      for(int i=0; i < 3; i++) {
        str = stk3.pop();
        System.out.print(str + " ");
      }

    } catch(StackEmptyException exc) {
      System.out.println(exc);
    }

    System.out.println();
  }
}
```

13. What is < >?

   The diamond operator.

14. How can the following be simplified?

   ```
   MyClass<Double,String> obj = new MyClass<Double,String>(1.1,"Hi");
   ```

   It can be simplified by use of the diamond operator as shown here:

   ```
   MyClass<Double,String> obj = new MyClass<>(1.1,"Hi");
   ```

   Assuming a local variable declaration and beginning with JDK 10, it can also be simplified by use of local variable type inference, like this:

   ```
   var obj = new MyClass<Double, String>(1.1, "Hi");
   ```

Chapter 14: Lambda Expressions and Method References

1. What is the lambda operator?

   The lambda operator is –>.

2. What is a functional interface?

   A functional interface is an interface that contains one and only one abstract method.

3. How do functional interfaces and lambda expressions relate?

A lambda expression provides the implementation for the abstract method defined by the functional interface. The functional interface defines the target type.

4. What are the two general types of lambda expressions?

The two types of lambda expressions are expression lambdas and block lambdas. An expression lambda specifies a single expression, whose value is returned by the lambda. A block lambda contains a block of code. Its value is specified by a **return** statement.

5. Show a lambda expression that returns **true** if a number is between 10 and 20, inclusive.

```
(n) -> (n > 9 && n < 21)
```

6. Create a functional interface that can support the lambda expression you created in question 5. Call the interface **MyTest** and its abstract method **testing( )**.

```
interface MyTest {
  boolean testing(int n);
}
```

7. Create a block lambda that computes the factorial of an integer value. Demonstrate its use. Use **NumericFunc**, shown in this chapter, for the functional interface.

```
interface NumericFunc {
  int func(int n);
}

class FactorialLambdaDemo {
  public static void main(String[] args)
  {

    // This block lambda computes the factorial of an int value.
    NumericFunc factorial = (n) -> {
      int result = 1;

      for(int i=1; i <= n; i++)
        result = i * result;

      return result;
    };

    System.out.println("The factorial of 3 is " + factorial.func(3));
    System.out.println("The factorial of 5 is " + factorial.func(5));
    System.out.println("The factorial of 9 is " + factorial.func(9));
  }
}
```

8. Create a generic functional interface called **MyFunc<T>**. Call its abstract method **func( )**. Have **func( )** return a reference of type **T**. Have it take a parameter of type **T**. (Thus, **MyFunc** will be a generic version of **NumericFunc** shown in the chapter.) Demonstrate its use by rewriting your answer to 7 so it uses **MyFunc<T>** rather than **NumericFunc**.

```
interface MyFunc<T> {
  T func(T n);
}


class FactorialLambdaDemo {
  public static void main(String[] args)
  {

    // This block lambda computes the factorial of an int value.
    MyFunc<Integer> factorial = (n) ->  {
      int result = 1;

      for(int i=1; i <= n; i++)
        result = i * result;

      return result;
    };

    System.out.println("The factorial of 3 is " + factorial.func(3));
    System.out.println("The factorial of 5 is " + factorial.func(5));
    System.out.println("The factorial of 9 is " + factorial.func(9));
  }
}
```

9. Using the program shown in Try This 14-1, create a lambda expression that removes all spaces from a string and returns the result. Demonstrate this method by passing it to **changeStr( )**.

Here is the lambda expression that removes spaces. It is used to initialize the **remove** reference variable.

```
StringFunc remove = (str) ->  {
  String result = "";

  for(int i = 0; i < str.length(); i++)
    if(str.charAt(i) != ' ') result += str.charAt(i);

  return result;
};
```

Here is an example of its use:

```
outStr = changeStr(remove, inStr);
```

10. Can a lambda expression use a local variable? If so, what constraint must be met?

Yes, but the variable must be effectively **final**.

11. If a lambda expression throws a checked exception, the abstract method in the functional interface must have a **throws** clause that includes that exception. True or False?

    True

12. What is a method reference?

    A method reference is a way to refer to a method without executing it.

13. When evaluated, a method reference creates an instance of the _____ _____ supplied by its target context.

    functional interface

14. Given a class called **MyClass** that contains a **static** method called **myStaticMethod( )**, show how to specify a method reference to **myStaticMethod( )**.

    ```
    MyClass::myStaticMethod
    ```

15. Given a class called **MyClass** that contains an instance method called **myInstMethod( )** and assuming an object of **MyClass** called **mcObj**, show how to create a method reference to **myInstMethod( )** on **mcObj**.

    ```
    mcObj::myInstMethod
    ```

16. To the **MethodRefDemo2** program, add a new method to **MyIntNum** called **hasCommonFactor( )**. Have it return **true** if its **int** argument and the value stored in the invoking **MyIntNum** object have at least one factor in common. For example, 9 and 12 have a common factor, which is 3, but 9 and 16 have no common factor. Demonstrate **hasCommonFactor( )** via a method reference.

    Here is **MyIntNum** with the **hasCommonFactor( )** method added:

    ```
    class MyIntNum {
      private int v;

      MyIntNum(int x) { v = x; }

      int getNum() { return v; }

      // Return true if n is a factor of v.
      boolean isFactor(int n) {
        return (v % n) == 0;
      }

      boolean hasCommonFactor(int n) {
        for(int i=2; i < v/i; i++)
          if( ((v % i) == 0) && ((n % i) == 0) ) return true;

        return false;
      }
    }
    ```

Here is an example of its use through a method reference:

```
ip = myNum::hasCommonFactor;
result = ip.test(9);
if(result) System.out.println("Common factor found.");
```

17. How is a constructor reference specified?

A constructor reference is created by specifying the class name followed by **::** followed by **new**. For example, **MyClass::new**.

18. Java defines several predefined functional interfaces in what package?

java.util.function

Chapter 15: Modules

1. In general terms, modules give you a way to specify when one unit of code depends on another. True or False?

True

2. A module is declared using what keyword?

module

3. The keywords that support modules are context sensitive. Explain what this means.

A context-sensitive keyword is recognized as a keyword only in specific situations that relate to its use and not elsewhere. As it relates to the module keywords, they are recognized as keywords only within a module declaration.

4. What is **module-info.java** and why is it important?

A **module-info.java** file contains a module declaration.

5. To declare that one module depends on another module, what keyword do you use?

requires

6. To make the public members of a package accessible outside the module in which it is contained, it must be specified in an _____ statement.

exports

7. When compiling or running a module-based application, why is the module path important?

The module path specifies where the modules for the application will be found.

8. What does **requires transitive** do?

By using **requires transitive** you enable one module to pass along its dependence on another module so that any module that relies on the current module also relies on the one specified in the **requires transitive** statement. This is called implied dependence or implied readability.

9. Does an **exports** statement export another module, or does it export a package?

An **exports** statement exports a package.

10. In the first module example, if you remove

```
exports appfuncs.simplefuncs;
```

from the **appfuncs** module-info file and then attempt to compile the program, what error do you see?

The compiler will report that the **SimpleMathFuncs** package does not exist. Since this package is required by **MyModAppDemo**, it will not compile.

11. Module-based services are supported by what keywords?

provides, **uses**, and **with**

12. A service specifies the general form of a unit of program functionality using either an interface or abstract class. True or False?

True

13. A service provider _____ a service.

implements

14. To load a service, what class do you use?

ServiceLoader

15. Can a module dependency be made optional at run time? If so, how?

Yes, by using an **exports static** statement.

16. Briefly describe what **open** and **opens** do.

Modifying a module declaration with the keyword **open** enables access to its packages at run time, including by reflection, whether or not they have been exported. An **opens** statement enables run-time access to a package, including for the purposes of reflection.

Chapter 16: Switch Expressions, Records, and Other Recently Added Features

1. Rewrite the following sequence so that it uses a constant list:

```
case 3: prime = true;
          break;
case 5: prime = true;
          break;
case 7: prime = true;
          break;
```

Here is the answer:

```
case 3, 5, 7: prime = true;
          break;
```

2. When using an arrow **case**, does execution fall through to the next **case**?

No.

3. Given this **switch**, show the **yield** statement that returns the value 98.6:

```
double val = switch(x) {
  case "temp":  // produce the value 98.6
// ...
```

```
case "temp": yield 98.6;
```

4. Assuming the **switch** in Question 3, show how to use an arrow **case** to yield the value 98.6.

```
case "temp" -> 98.6;
```

5. Can you mix an arrow **case** and a colon **case** in the same **switch**?

No.

6. Can the target of an arrow **case** be a block?

Yes.

7. A **record** is commonly referred to as a/an _____ type.

aggregate

8. Given this record declaration, what are its components? What elements are implicitly created?

```
record MyRec(Double highTemp, Double lowTemp, String location) { }
```

The components are **highTemp**, **lowTemp**, and **location**. Private final fields with the same names are implicitly created. Getter methods called **highTemp( )**, **lowTemp( )**, and **location( )** are also implicitly created.

9. Does a **record** have a default constructor? If not, what type of constructor does a **record** automatically have?

No, a **record** does not have a default constructor. Instead, a **record** automatically defines a canonical constructor.

10. Given **MyRec** from Question 8, show the compact canonical constructor that removes leading and trailing spaces from the **location** string.

```
public MyRec {
  // Remove leading and trailing spaces from location
  location = location.trim();
}
```

11. If you were to override a **record** getter method, in what way would you need to be very careful?

A **record** is immutable. Thus, to preserve the immutable semantics of a **record**, your getter must not return a value other than that contained in the record.

12. In Try This 13-1 you created a generic queue class. Can this class be used to store **record** objects without any changes? If so, demonstrate its use to store the **Item** records used in the **Item record** examples.

```
class RecordQDemo {
  public static void main(String[] args) {
    // Create a queue for Item records.
    Item[] items = new Item[4];
    GenQueue<Item> q = new GenQueue<Item>(items);

    // Create some Item records.
    items[0] = new Item("Hammer", 257, 10.99);
    items[1] = new Item("Wrench", 18, 19.29);
    items[2] = new Item("Drill", 903, 22.25);
    items[3] = new Item("Saw", 27, 34.59);

    // Put records into the queue.
    try {
      for(int i=0; i < items.length; i++) {
        System.out.println("Adding " + items[i].name() + " to queue.");
        q.put(items[i]); // add record to q

      }
    }
    catch (QueueFullException exc) {
      System.out.println(exc);
    }
    System.out.println();

    // Retrieve records from the queue.
    try {
      Item r;

      for(int i=0; i < items.length; i++) {
        System.out.print("Getting next record from queue: ");
        r = q.get();
        System.out.println(r.name() + ", Item Number " + r.itemNum() +
                          ", " + " Price: " + r.price());
      }
    }
    catch (QueueEmptyException exc) {
      System.out.println(exc);
    }
  }
}
```

13. Rework the **Item** record so that the **price** component is generic, with an upper bound of **Number**.

```
record Item<T extends Number>(String name, int itemNum, T price) {}
```

14. In the **BlockArrowCaseDemo** program, the **switch** expression yields the shipping method, but the variable **extraCharge** is set separately inside each **case**. This program can be improved by having the **switch** yield a **record** that contains both the shipping method and the **extraCharge** value. In essence, the use of a **record** enables the **switch** to yield two or more values when it returns its result. Rework the **BlockArrowCaseDemo** program to demonstrate this approach.

Here is one way to code the improved **BlockArrowCaseDemo** program. It uses a **record** called **ShippingInfo** that holds both the shipping method and the **extraCharge** value.

```java
// Demonstrate a switch expression that yields a record.
class SwitchWithRecord {

  enum ShipMethod { STANDARD, TRUCK, AIR, OVERNIGHT }

  record ShippingInfo (ShipMethod how, boolean extraCharge) { }

  public static void main(String[] args) {

    int productID = 5099;

    // Here, the switch expression uses a record to efficiently
    // yield two values.
    ShippingInfo shipInfo = switch(productID) {
      case 1774, 8708, 6709 -> new ShippingInfo(ShipMethod.TRUCK, true);
      case 4657, 2195, 1887, 3621 ->
                          new ShippingInfo(ShipMethod.AIR, true);
      case 2907, 5099 -> new ShippingInfo(ShipMethod.OVERNIGHT, true);
      default-> new ShippingInfo(ShipMethod.STANDARD, false);
    };

    System.out.println("Shipping method for product number " +
                      productID + " is " + shipInfo.how());
    if(shipInfo.extraCharge())
              System.out.println("Extra charge required.");
  }
}
```

15. Show the general form of **instanceOf** when using pattern matching.

    objref instanceof *type pattern-var*

16. Given:

    ```java
    Object myOb = "A test string";
    ```

    fill in the blank in the following **if** statement that uses **instanceof** to determine whether **myOb** refers to a **String**.

    ```java
    if(myObj instanceof String str) System.out.println("Is a string: " + str);
    ```

17. A **sealed** class explicitly specifies the subclasses that can inherit it. True or false?

True.

18. Given the following:

```
public sealed class MyClass permits Alpha, Beta, Gamma { // ...
```

which of the following declarations are legal?

A. `public final class Alpha extends MyClass { // ...`

B. `public final class Beta { // ...`

C. `public class Gamma extends MyClass { // ...`

D. `public non-sealed SomeOtherClass extends MyClass { // ...`

Only A is legal. B does not extend **MyClass**. C needs to be modified by either **final**, **sealed**, or **non-sealed**. In D, **SomeOtherClass** is not permitted by **MyClass**.

19. Can an interface be sealed? If so, what effect does sealing an interface have?

Yes, an interface can be sealed. Only those interfaces that are listed in its **permits** clause can extend a sealed interface. Only those classes listed in its **permits** clause can implement the interface.

20. A preview feature is a new feature that is fully developed, but not yet formally part of Java. True or False?

True.

21. A preview feature is subject to change or may even be withdrawn. True or False?

True.

Chapter 17: Introducing Swing

1. In general, AWT components are heavyweight and Swing components are *lightweight*.

2. Can the look and feel of a Swing component be changed? If so, what feature enables this?

Yes. Swing's pluggable look and feel is the feature that enables this.

3. What is the most commonly used top-level container for an application?

JFrame

4. Top-level containers have several panes. To what pane are components added?

Content pane

5. Show how to construct a label that contains the message "Select an entry from the list".

`JLabel("Select an entry from the list")`

6. All interaction with GUI components must take place on what thread?

event-dispatching thread

7. What is the default action command associated with a **JButton**? How can the action command be changed?

The default action command string is the text shown inside the button. It can be changed by calling **setActionCommand( )**.

8. What event is generated when a push button is pressed?

ActionEvent

9. Show how to create a text field that has 32 columns.

```
JTextField(32)
```

10. Can a **JTextField** have its action command set? If so, how?

Yes, by calling **setActionCommand( )**.

11. What Swing component creates a check box? What event is generated when a check box is selected or deselected?

JCheckBox creates a check box. An **ItemEvent** is generated when a check box is selected or deselected.

12. **JList** displays a list of items from which the user can select. True or False?

True

13. What event is generated when the user selects or deselects an item in a **JList**?

ListSelectionEvent

14. What method sets the selection mode of a **JList**? What method obtains the index of the first selected item?

setSelectionMode( ) sets the selection mode. **getSelectedIndex( )** obtains the index of the first selected item.

15. Add a check box to the file comparer developed in Try This 17-1 that has the following text: Show position of mismatch. When this box is checked, have the program display the location of the first point in the files at which a mismatch occurs.

```
/*
    Try This 17-1

    A Swing-based file comparison utility.

    This version has a check box that causes the
    location of the first mismatch to be shown.

*/
```

```java
import java.awt.*;
import java.awt.event.*;
import javax.swing.*;
import java.io.*;

public class SwingFC implements ActionListener {

  JTextField jtfFirst;  // holds the first file name
  JTextField jtfSecond; // holds the second file name

  JButton jbtnComp; // button to compare the files

  JLabel jlabFirst, jlabSecond; // displays prompts
  JLabel jlabResult; // displays results and error messages

  JCheckBox jcbLoc; // check to display location of mismatch

  SwingFC() {

    // Create a new JFrame container.
    JFrame jfrm = new JFrame("Compare Files");

    // Specify FlowLayout for the layout manager
    jfrm.setLayout(new FlowLayout());

    // Give the frame an initial size.
    jfrm.setSize(200, 220);

    // Terminate the program when the user closes the application.
    jfrm.setDefaultCloseOperation(JFrame.EXIT_ON_CLOSE);

    // Create the text fields for the file names..
    jtfFirst = new JTextField(14);
    jtfSecond = new JTextField(14);

    // Set the action commands for the text fields.
    jtfFirst.setActionCommand("fileA");
    jtfSecond.setActionCommand("fileB");

    // Create the Compare button.
    JButton jbtnComp = new JButton("Compare");

    // Add action listener for the Compare button.
    jbtnComp.addActionListener(this);

    // Create the labels.
    jlabFirst = new JLabel("First file: ");
```

```java
    jlabSecond = new JLabel("Second file: ");
    jlabResult = new JLabel("");

    // Create check box.
    jcbLoc = new JCheckBox("Show position of mismatch");

    // Add the components to the content pane.
    jfrm.add(jlabFirst);
    jfrm.add(jtfFirst);
    jfrm.add(jlabSecond);
    jfrm.add(jtfSecond);
    jfrm.add(jcbLoc);
    jfrm.add(jbtnComp);
    jfrm.add(jlabResult);

    // Display the frame.
    jfrm.setVisible(true);
  }

  // Compare the files when the Compare button is pressed.
  public void actionPerformed(ActionEvent ae) {
    int i=0, j=0;
    int count = 0;

    // First, confirm that both file names have
    // been entered.
    if(jtfFirst.getText().equals("")) {
      jlabResult.setText("First file name missing.");
      return;
    }
    if(jtfSecond.getText().equals("")) {
      jlabResult.setText("Second file name missing.");
      return;
    }

    // Compare files. Use try-with-resources to manage the files.
    try (FileInputStream f1 = new FileInputStream(jtfFirst.getText());
         FileInputStream f2 = new FileInputStream(jtfSecond.getText()))
    {
      // Check the contents of each file.
      do {
        i = f1.read();
        j = f2.read();
        if(i != j) break;
        count++;
      } while(i != -1 && j != -1);
```

```
      if(i != j) {
        if(jcbLoc.isSelected())
          jlabResult.setText("Files differ at location " + count);
        else
          jlabResult.setText("Files are not the same.");
      }
      else
        jlabResult.setText("Files compare equal.");

    } catch(IOException exc) {
      jlabResult.setText("File Error");
    }
  }

  public static void main(String[] args) {
    // Create the frame on the event dispatching thread.
    SwingUtilities.invokeLater(new Runnable() {
      public void run() {
        new SwingFC();
      }
    });
  }
}
```

16. Change the **ListDemo** program so that it allows multiple items in the list to be selected.

```
// Demonstrate multiple selection in a JList.

import javax.swing.*;
import javax.swing.event.*;
import java.awt.*;
import java.awt.event.*;

public class ListDemo implements ListSelectionListener {

  JList<String> jlst;
  JLabel jlab;
  JScrollPane jscrlp;

  // Create an array of names.
  String[] names = { "Sherry", "Jon", "Rachel",
                     "Sasha", "Josselyn", "Randy",
                     "Tom", "Mary", "Ken",
                     "Andrew", "Matt", "Todd" };

  ListDemo() {
    // Create a new JFrame container.
    JFrame jfrm = new JFrame("JList Demo");
```

```
      // Specify a flow Layout.
      jfrm.setLayout(new FlowLayout());

      // Give the frame an initial size.
      jfrm.setSize(200, 160);

      // Terminate the program when the user closes the application.
      jfrm.setDefaultCloseOperation(JFrame.EXIT_ON_CLOSE);

      // Create a JList.
      jlst = new JList<String>(names);

      // By removing the following line, multiple selection (which
      // is the default behavior of a JList) will be used.
//      jlst.setSelectionMode(ListSelectionModel.SINGLE_SELECTION);

      // Add list to a scroll pane.
      jscrlp = new JScrollPane(jlst);

      // Set the preferred size of the scroll pane.
      jscrlp.setPreferredSize(new Dimension(120, 90));

      // Make a label that displays the selection.
      jlab = new JLabel("Please choose a name");

      // Add list selection handler.
      jlst.addListSelectionListener(this);

      // Add the list and label to the content pane.
      jfrm.add(jscrlp);
      jfrm.add(jlab);

      // Display the frame.
      jfrm.setVisible(true);
    }

    // Handle list selection events.
    public void valueChanged(ListSelectionEvent le) {
      // Get the indices of the changed item.
      int[] indices = jlst.getSelectedIndices();

      // Display the selections, if one or more items
      // were selected.
      if(indices.length != 0) {
        String who = "";
```

```
      // Construct a string of the names.
      for(int i : indices)
        who += names[i] + " ";

      jlab.setText("Current selections: " + who);
    }
    else // Otherwise, reprompt.
      jlab.setText("Please choose a name");
  }

  public static void main(String[] args) {
    // Create the frame on the event dispatching thread.
    SwingUtilities.invokeLater(new Runnable() {
      public void run() {
        new ListDemo();
      }
    });
  }
}
```

Appendix B

Using Java's Documentation Comments

As explained in Chapter 1, Java supports three types of comments. The first two are the **//** and the **/* */**. The third type is called a *documentation comment*. It begins with the character sequence **/****. It ends with ***/**. Documentation comments allow you to embed information about your program into the program itself. You can then use the **javadoc** utility program (supplied with the JDK) to extract the information and put it into an HTML file. Documentation comments make it convenient to document your programs. You have almost certainly seen documentation that uses such comments, because that is the way the Java API library was documented. Beginning with JDK 9, **javadoc** includes support for modules.

The javadoc Tags

The **javadoc** utility recognizes several tags, including those shown here:

Tag	Meaning
@author	Identifies the author.
{@code}	Displays information as-is, without processing HTML styles, in code font.
@deprecated	Specifies that a program element is deprecated.
{@docRoot}	Specifies the path to the root directory of the current documentation.
@exception	Identifies an exception thrown by a method or constructor.
@hidden	Prevents an element from appearing in the documentation.
{@index}	Specifies a term for indexing.
{@inheritDoc}	Inherits a comment from the immediate superclass.
{@link}	Inserts an in-line link to another topic.
{@linkplain}	Inserts an in-line link to another topic, but the link is displayed in a plain-text font.
{@literal}	Displays information as-is, without processing HTML styles.
@param	Documents a parameter.
@provides	Documents a service provided by a module.
@return	Documents a method's return value.
@see	Specifies a link to another topic.
@serial	Documents a default serializable field.
@serialData	Documents the data written by the **writeObject( )** or **writeExternal( )** methods.
@serialField	Documents an **ObjectStreamField** component.
@since	States the release when a specific change was introduced.
{@summary}	Documents a summary of an item.
{@systemProperty}	States that a name is a system property.
@throws	Same as **@exception**.
@uses	Documents a service needed by a module.
{@value}	Displays the value of a constant, which must be a **static** field.
@version	Specifies the version of a program element.

Document tags that begin with an "at" sign (@) are called *block* tags (also called *stand-alone* tags), and they must be used at the beginning of their own line. Tags that begin with a brace, such as {**@code**}, are called *inline* tags, and they can be used within a larger description. You may also use other, standard HTML tags in a documentation comment. However, some tags such as headings should not be used, because they disrupt the look of the HTML file produced by **javadoc**.

As it relates to documenting source code, you can use documentation comments to document classes, interfaces, fields, constructors, methods, packages, and modules. In all cases, the documentation comment must immediately precede the item being documented. Some tags, such as **@see**, **@since**, and **@deprecated**, can be used to document any element. Other tags apply to only the relevant elements. Several key tags are examined next.

NOTE

As one would expect, the capabilities of **javadoc** and the documentation comment tags have evolved over time, often in response to new Java features. You will want to refer to the **javadoc** documentation for information on the latest **javadoc** features.

@author

The **@author** tag documents the author of a program element. It has the following syntax:

@author *description*

Here, *description* will usually be the name of the author. You will need to specify the **-author** option when executing **javadoc** in order for the **@author** field to be included in the HTML documentation.

{@code}

The {**@code**} tag enables you to embed text, such as a snippet of code, into a comment. That text is then displayed as-is in code font, without any further processing such as HTML rendering. It has the following syntax:

{ @code *code-snippet*}

@deprecated

The **@deprecated** tag specifies that a program element is deprecated. It is recommended that you include **@see** or {**@link**} tags to inform the programmer about available alternatives. The syntax is the following:

@deprecated *description*

Here, *description* is the message that describes the deprecation. The **@deprecated** tag can be used in documentation for fields, methods, constructors, classes, modules, and interfaces.

{@docRoot}

{**@docRoot**} specifies the path to the root directory of the current documentation.

@exception

The **@exception** tag describes an exception to a method. Today, **@throws** is the preferred alternative, but **@exception** is still supported. It has the following syntax:

@exception *exception-name explanation*

Here, the fully qualified name of the exception is specified by *exception-name,* and *explanation* is a string that describes how the exception can occur. The **@exception** tag can be used only in documentation for a method or constructor.

@hidden

The **@hidden** tag prevents an element from appearing in the documentation.

{@index}

The {**@index**} tag specifies an item that will be indexed, and thus found when using the search feature. It has the following syntax:

{@index *term usage-str* }

Here, *term* is the item (which can be a quoted string) to be indexed. *usage-str* is optional. Thus, in the following **@throws** tag, {**@index**} causes the term "error" to be added to the index:

```
@throws IOException On input {@index error}.
```

Note that the word "error" is still displayed as part of the description. It's just that now it is also indexed. If you include the optional *usage-str*, then that description will be shown in the index and in the search box to indicate how the term is used. For example, {**@index error Serious execution failure**} will show "Serious execution failure" under "error" in the index and in the search box.

{@inheritDoc}

This tag inherits a comment from the immediate superclass.

{@link}

The {**@link**} tag provides an in-line link to additional information. It has the following syntax:

{@link *mod-name/pkg-name.class-name#member-name text*}

Here, *mod-name/pkg-name.class-name#member-name* specifies the name of a class or method to which a link is added, and *text* is the string that is displayed. The *text* field is optional. If not included, *member-name* is displayed as the link. Notice that the module name (if present) is separated from the package name with a /. For example,

{@link java.base/java.io.Writer#write}

defines a link to the **write( )** method of **Writer** in **java.io**, in the module **java.base**.

{@linkplain}

The {**@linkplain**} tag inserts an in-line link to another topic. The link is displayed in plain-text font. Otherwise, it is similar to {**@link**}.

{@literal}

The {**@literal**} tag enables you to embed text into a comment. That text is then displayed as-is, without any further processing such as HTML rendering. It has the following syntax:

{@literal *description*}

Here, *description* is the text that is embedded.

@param

The **@param** tag documents a parameter. It has the following syntax:

@param *parameter-name explanation*

Here, *parameter-name* specifies the name of a parameter. The meaning of that parameter is described by *explanation*. The **@param** tag can be used only in documentation for a method, a constructor, or a generic class or interface.

@provides

The **@provides** tag documents a service provided by a module. It has the following syntax:

@provides *type explanation*

Here, *type* specifies a service provider type and *explanation* describes the service provider.

@return

The **@return** tag describes the return value of a method. It has two forms. The first is the block tag show here.

@return *explanation*

Here, *explanation* describes the type and meaning of the value returned by a method. Thus, the tag can be used only in documentation for a method. JDK 16 added an inline tag version:

{@return explanation}

This form must be at the top of the method's documentation comment.

@see

The **@see** tag provides a reference to additional information. Two commonly used forms are shown here:

@see *anchor*

@see *mod-name/pkg-name.class-name#member-name text*

In the first form, *anchor* is a link to an absolute or relative URL. In the second form, *mod-name/pkg-name.class-name#member-name* specifies the name of the item, and *text* is the text displayed for that item. The text parameter is optional, and if not used, then the item specified by *mod-name/pkg-name.class-name#member-name* is displayed. The member name, too, is optional. Thus, you can specify a reference to a module, package, class, or interface in addition to a reference to a specific method or field. The name can be fully qualified or partially qualified. However, the dot that precedes the member name (if it exists) must be replaced by a hash character. There is a third form of **@see** that lets you simply specify a text-based description.

@since

The **@since** tag states that an element was introduced in a specific release. It has the following syntax:

@since *release*

Here, *release* is a string that designates the release or version in which this feature became available.

{@summary}

The **{@summary}** tag explicitly specifies a summary for an item. It must be the first tag in the documentation for the item. It has the following syntax:

@summary *explanation*

Here, *explanation* provides a summary of the tagged item, which can span multiple lines. Without the use of **{@summary}**, the first line in an item's documentation comment is used as the summary.

@throws

The **@throws** tag has the same meaning as the **@exception** tag, but is now the preferred form.

@uses

The **@uses** tag documents a service provider needed by a module. It has the following syntax:

@uses *type explanation*

Here, *type* specifies a service provider type and *explanation* describes the service.

{@value}

{@value} has two forms. The first displays the value of the constant that it precedes, which must be a **static** field. It has this form:

{@value}

The second form displays the value of a specified **static** field. It has this form:

{ @value *pkg.class#field*}

Here, *pkg.class#field* specifies the name of the **static** field.

@version

The **@version** tag specifics the version of a program element. It has the following syntax:

@version *info*

Here, *info* is a string that contains version information, typically a version number, such as 2.2. You will need to specify the **-version** option when executing **javadoc** in order for the **@version** field to be included in the HTML documentation.

The General Form of a Documentation Comment

After the beginning /**, the first line or lines become the main description of your class, interface, field, constructor, method, or module. After that, you can include one or more of the various @ tags. Each @ tag must start at the beginning of a new line or follow one or more asterisks (*) that are at the start of a line. Multiple tags of the same type should be grouped together. For example, if you have three **@see** tags, put them one after the other. In-line tags (those that begin with a brace) can be used within any description.

Here is an example of a documentation comment for a class:

```
/**
 * This class draws a bar chart.
 * @author Herbert Schildt
 * @version 3.2
 */
```

What javadoc Outputs

The **javadoc** program takes as input your Java program's source file and outputs several HTML files that contain the program's documentation. Information about each class will be in its own HTML file. **javadoc** will also output an index and a hierarchy tree. Other HTML files can be generated. Beginning with JDK 9, a search box feature is also included.

An Example That Uses Documentation Comments

Following is a sample program that uses documentation comments. Notice the way each comment immediately precedes the item that it describes. After being processed by **javadoc**, the documentation about the **SquareNum** class will be found in **SquareNum.html**.

```
import java.io.*;

/**
 * This class demonstrates documentation comments.
 * @author Herbert Schildt
```

```
 * @version 1.2
 */
public class SquareNum {
  /**
   * This method returns the square of num.
   * This is a multiline description. You can use
   * as many lines as you like.
   * @param num The value to be squared.
   * @return num squared.
   */
  public double square(double num) {
    return num * num;
  }

  /**
   * This method inputs a number from the user.
   * @return The value input as a double.
   * @throws IOException On input error.
   * @see IOException
   */
  public double getNumber() throws IOException {
    // create a BufferedReader using System.in
    InputStreamReader isr = new InputStreamReader(System.in);
    BufferedReader inData = new BufferedReader(isr);
    String str;

    str = inData.readLine();
    return (new Double(str)).doubleValue();
  }

  /**
   * This method demonstrates square().
   * @param args Unused.
   * @throws IOException On input error.
   * @see IOException
   */
  public static void main(String[] args)
    throws IOException
  {
    SquareNum ob = new SquareNum();
    double val;

    System.out.println("Enter value to be squared: ");
    val = ob.getNumber();
    val = ob.square(val);

    System.out.println("Squared value is " + val);
  }
}
```

Appendix C

Compile and Run Simple Single-File Programs in One Step

n Chapter 1, you were shown how to compile a Java program into bytecode using the **javac** compiler and then run the resulting **.class** file(s) using the Java launcher **java**. This is how Java programs have been compiled and run since Java's beginning, and it is the method that you will use when developing applications. However, beginning with JDK 11, it is possible to compile and run some types of simple Java programs directly from the source file without having to first invoke **javac**. To do this, pass the name of the source file, using the **.java** file extension, to **java**. This causes **java** to automatically invoke the compiler and execute the program.

For example, the following automatically compiles and runs the first example in this book:

```
java Example.java
```

In this case, the **Example** class is compiled and then run in a single step. There is no need to use **javac**. Be aware, however, that no **.class** file is created. Instead, the compilation is done behind the scenes. As a result, to rerun the program, you must execute the source file again. You can't execute its **.class** file, because one wasn't created.

One use of the source-file launch capability is to facilitate the use of Java programs in script files. It can also be useful for short one-time-use programs. In some cases, it makes it a little easier to run simple example programs when you are experimenting with Java. It is not, however, a general-purpose substitute for Java's normal compilation/execution process.

Although this new ability to launch a Java program directly from its source file is appealing, it comes with some restrictions. First, the entire program must be contained in a single source file. However, most real-world programs use multiple source files. Second, it will always execute the first class it finds in the file, and that class must contain a **main( )** method. If the first class in the file does not contain a **main( )** method, the launch will fail. This means that you must follow a strict organization for your code, even if you would prefer to organize it otherwise. Third, because no **.class** files are created, using **java** to run a single-file program does not result in a class file that can be reused, possibly by other programs. As a result of these restrictions, using **java** to run a single-file source program can be useful, but it constitutes what is, essentially, a special-case technique.

As it relates to this book, it is possible to use the single source-file launch feature to try many of the examples; just be sure that the class with the **main( )** method is first in your file. That said, it is not, however, applicable or appropriate in all cases. Furthermore, the discussions (and many of the examples) in the book assume that you are using the normal compilation process of invoking **javac** to compile a source file into bytecode and then using **java** to run that bytecode. This is the mechanism that is used for real-world development, and understanding this process is an important part of learning Java. It is imperative that you are thoroughly familiar with it. For these reasons, when trying the examples in this book, it is strongly recommended that in all cases you use the normal approach to compiling and running a Java program. Doing so ensures that you have a solid foundation in the way Java works. Of course, you might find it fun to experiment with the single source-file launch option!

NOTE

It is possible to execute a single-file program from a file that does not use the **.java** extension. To do so, you must specify the **--source** *APIVer* option, where *APIVer* specifies the JDK version number.

Appendix D

Introducing JShell

Beginning with JDK 9, Java has included a tool called JShell. It provides an interactive environment that enables you to quickly and easily experiment with Java code. JShell implements what is referred to as *read-evaluate-print loop* (REPL) execution. Using this mechanism, you are prompted to enter a fragment of code. This fragment is then read and evaluated. Next, JShell displays output related to the code, such as the output produced by a **println( )** statement, the result of an expression, or the current value of a variable. JShell then prompts for the next piece of code, and the process continues (i.e., loops). In the language of JShell, each code sequence you enter is called a *snippet*.

A key point to understand about JShell is that you do not need to enter a complete Java program to use it. Each snippet you enter is simply evaluated as you enter it. This is possible because JShell handles many of the details associated with a Java program for you automatically. This lets you concentrate on a specific feature without having to write a complete program, which makes JShell especially helpful when you are first learning Java.

As you might expect, JShell can also be useful to experienced programmers. Because JShell stores state information, it is possible to enter multiline code sequences and run them inside JShell. This makes JShell quite useful when you need to prototype a concept because it lets you interactively experiment with your code without having to develop and compile a complete program.

This appendix introduces JShell and explores several of its key features, with the primary focus being on those features most useful to beginning Java programmers.

JShell Basics

JShell is a command-line tool. Thus, it runs in a command-prompt window. To start a JShell session, execute **jshell** from the command line. After doing so, you will see the JShell prompt:

```
jshell>
```

When this prompt is displayed, you can enter a code snippet or a JShell command.

In its simplest form, JShell lets you enter an individual statement and immediately see the result. To begin, think back to the first example Java program in this book. It is shown again here.

```
class Example {
  // A Java program begins with a call to main().
  public static void main(String[] args) {
    System.out.println("Java drives the Web.");
  }
}
```

In this program, only the **println( )** statement actually performs an action, which is displaying its message on the screen. The rest of the code simply provides the required class and method declarations. In JShell, it is not necessary to explicitly specify the class or method in order to execute the **println( )** statement. JShell can execute it directly on its own. To see how, enter the following line at the JShell prompt:

```
System.out.println("Java drives the Web.");
```

Then, press ENTER. This output is displayed:

```
Java drives the Web.

jshell>
```

As you can see, the call to **println( )** is evaluated and its string argument is output. Then, the prompt is redisplayed.

Before moving on it is useful to explain why JShell can execute a single statement, such as the call to **println( )**, when the Java compiler, **javac**, requires a complete program. JShell is able to evaluate a single statement because JShell automatically provides the necessary program framework for you, behind the scenes. This consists of a *synthetic class* and a *synthetic method*. Thus, in this case, the **println( )** statement is embedded in a synthetic method that is part of a synthetic class. As a result, the preceding code is still part of a valid Java program even though you don't see all of the details. This approach provides a very fast and convenient way to experiment with Java code.

Next, let's look at how variables are supported. In JShell, you can declare a variable, assign the variable a value, and use it in any valid expressions. For example, enter the following line at the prompt:

```
int count;
```

After doing so you will see the following response:

```
count ==> 0
```

This indicates that **count** has been added to the synthetic class and initialized to zero. Furthermore, it has been added as a **static** variable of the synthetic class.

Next, give **count** the value 10 by entering this statement:

```
count = 10;
```

You will see this response:

```
count ==> 10
```

As you can see, **count**'s value is now 10. Because **count** is **static**, it can be used without reference to an object.

Now that **count** has been declared, it can be used in an expression. For example, enter this **println( )** statement:

```
System.out.println("Reciprocal of count: " + 1.0 / count);
```

JShell responds with

```
Reciprocal of count: 0.1
```

Here, the result of the expression **1.0 / count** is 0.1 because **count** was previously assigned the value 10.

In addition to demonstrating the use of a variable, the preceding example illustrates another important aspect of JShell: it maintains state information. In this case, **count** is assigned the value 10 in one statement, and then this value is used in the expression **1.0 / count** in the subsequent call to **println( )** in a second statement. Between these two statements, JShell stores **count**'s value. In general, JShell maintains the current state and effect of the code snippets that you enter. This lets you experiment with larger code fragments that span multiple lines.

Before moving on, let's try one more example. In this case, we will create a **for** loop that uses the **count** variable. Begin by entering this line at the prompt:

```
for(count = 0; count < 5; count++)
```

At this point, JShell responds with the following prompt:

```
...>
```

This indicates that additional code is required to finish the statement. In this case, the target of the **for** loop must be provided. Enter the following:

```
System.out.println(count);
```

After entering this line, the **for** statement is complete and both lines are executed. You will see the following output:

```
0
1
2
3
4
```

In addition to statements and variable declarations, JShell lets you declare classes and methods and use import statements. Examples are shown in the following sections. One other point: Any code that is valid for JShell will also be valid for compilation by **javac**, assuming the necessary framework is provided to create a complete program. Thus, if a code fragment can be executed by JShell, then that fragment represents valid Java code. In other words, JShell code *is* Java code.

List, Edit, and Rerun Code

JShell supports a large number of commands that let you control the operation of JShell. At this point, three are of particular interest because they let you list the code that you have entered, edit a line of code, and rerun a code snippet. As the subsequent examples become longer, you will find these commands to be very helpful.

In JShell, all commands start with a **/** followed by the command. Perhaps the most commonly used command is **/list**, which lists the code that you have entered. Assuming that you have followed along with the examples shown in the preceding section, you can list your code by entering **/list** at this time. Your JShell session will respond with a numbered list of the snippets you entered. Pay special attention to the entry that shows the **for** loop. Although it consists of two lines, it constitutes one statement. Thus, only one snippet number is used. In the language of JShell, the snippet numbers are referred to as *snippet IDs*. In addition to the basic form of **/list** just shown, other forms are supported, including those that let you list specific snippets by name or number. For example, you can list the **count** declaration by using **/list count**.

You can edit a snippet by using the **/edit** command. This command causes an edit window to open in which you can modify your code. Here are three forms of the **/edit** command that you will find helpful at this time. First, if you specify **/edit** by itself, the edit window contains all of the lines you have entered and lets you edit any part of it. Second, you can specify a snippet to edit by using **/edit** *n*, where *n* specifies the snippet's number. For example, to edit snippet 3, use **/edit 3**. Finally, you can specify a named element, such as a variable. For example, to change the value of **count**, use **/edit count**.

As you have seen, JShell executes code as you enter it. However, you can also rerun what you have entered. To rerun the last fragment that you entered, use **/!**. To rerun a specific snippet, specify its number using this form: **/n**, where *n* specifies the snippet to run. For example, to rerun the fourth snippet, enter **/4**. You can rerun a snippet by specifying its position relative to the current fragment by use of a negative offset. For example, to rerun a fragment that is three snippets before the current one, use **/-3**.

Before moving on, it is helpful to point out that several commands, including those just shown, allow you to specify a list of names or numbers. For example, to edit lines 2 and 4, you could use **/edit 2 4**. For recent versions of JShell, several commands allow you specify a range of snippets. These also include the **/list**, **/edit**, and **/n** commands just described. For example, to list snippets 4 through 6, you would use **/list 4-6**.

There is one other important command that you need to know about now: **/exit**. This terminates JShell.

Add a Method

You first learned about methods in Chapter 4. As you saw there, methods occur within classes. However, when using JShell, it is possible to experiment with a method without having to *explicitly* declare it within a class. As explained earlier, this is because JShell automatically wraps code fragments within a synthetic class. As a result, you can easily and quickly write a method without having to provide a class framework. You can also call the method without having to create an object. This feature of JShell is especially beneficial when learning the basics of methods in Java or when prototyping new code. To understand the process, we will work through an example.

To begin, start a new JShell session and enter the following method at the prompt:

```
double reciprocal(double val) {
  return 1.0/val;
}
```

This creates a method that returns the reciprocal of its argument. After you enter this, JShell responds with the following:

```
|   created method reciprocal(double)
```

This indicates the method has been added to JShell's synthetic class and is ready for use.

To call **reciprocal( )**, simply specify its name, without any object or class reference. For example, try this:

```
System.out.println(reciprocal(4.0));
```

JShell responds by displaying 0.25.

You might be wondering why you can call **reciprocal( )** without using the dot operator and an object reference. Here is the answer. When you create a stand-alone method in JShell, such as **reciprocal( )**, JShell automatically makes that method a **static** member of the synthetic class. As you know from Chapter 5, **static** methods are called relative to their class, not on a specific object. So, no object is required. This is similar to the way that stand-alone variables become **static** variables of the synthetic class, as described earlier.

Another important aspect of JShell is its support for a *forward reference* inside a method. This feature lets one method call another method, even if the second method has not yet been defined. This enables you to enter a method that depends on another method without having to worry about which one you enter first. Here is a simple example. Enter this line in JShell:

```
void myMeth() { myMeth2(); }
```

JShell responds with the following:

```
|  created method myMeth(), however, it cannot be invoked until myMeth2()
   is declared
```

As you can see, JShell knows that **myMeth2( )** has not yet been declared, but it still lets you define **myMeth( )**. As you would expect, if you try to call **myMeth( )** at this time, you will see an error message since **myMeth2( )** is not yet defined, but you are still able to enter the code for **myMeth( )**.

Next, define **myMeth2( )** like this:

```
void myMeth2() { System.out.println("JShell is powerful."); }
```

Now that **myMeth2( )** has been defined, you can call **myMeth( )**.

In addition to its use in a method, you can use a forward reference in a field initializer in a class.

Create a Class

Although JShell automatically supplies a synthetic class that wraps code snippets, you can also create your own class in JShell. Furthermore, you can instantiate objects of your class. This allows you to experiment with classes inside JShell's interactive environment. The following example illustrates the process.

Start a new JShell session and enter the following class, line by line:

```
class MyClass {
  double v;

  MyClass(double d) { v = d; }
```

```
  // Return the reciprocal of v.
  double reciprocal() { return 1.0 / v; }
}
```

When you finish entering the code, JShell will respond with

```
|  created class MyClass
```

Now that you have added **MyClass**, you can use it. For example, you can create a **MyClass** object with the following line:

```
MyClass ob = new MyClass(10.0);
```

JShell will respond by telling you that it added **ob** as a variable of type **MyClass**. Next, try the following line:

```
System.out.println(ob.reciprocal());
```

JShell responds by displaying the value 0.1.

As a point of interest, when you add a class to JShell, it becomes a **static** nested member of a synthetic class.

Use an Interface

Interfaces are supported by JShell in the same way as classes. Therefore, you can declare an interface and implement it by a class within JShell. Let's work through a simple example. Before beginning, start a new JShell session.

The interface that we will use declares a method called **isLegalVal( )** that is used to determine if a value is valid for some purpose. It returns **true** if the value is legal and **false** otherwise. Of course, what constitutes a legal value will be determined by each class that implements the interface. Begin by entering the following interface into JShell:

```
interface MyIF {
  boolean isLegalVal(double v);
}
```

JShell responds with

```
|  created interface MyIf
```

Next, enter the following class, which implements **MyIF**:

```
class MyClass implements MyIF {

  double start;
  double end;

  MyClass(double a, double b) { start = a; end = b; }
```

```
  // Determine if v is within the range start to end, inclusive.
  public boolean isLegalVal(double v) {
    if((v >= start) && (v <= end)) return true;
    return false;
  }

}
```

JShell responds with

```
|  created class MyClass
```

Notice that **MyClass** implements **isLegalVal( )** by determining if the value **v** is within the range (inclusive) of the values in the **MyClass** instance variables **start** and **end**.

Now that both **MyIF** and **MyClass** have been added, you can create a **MyClass** object and call **isLegalVal( )** on it, as shown here:

```
MyClass ob = new MyClass(0.0, 10.0);

System.out.println(ob.isLegalVal(5.0));
```

In this case, the value **true** is displayed because 5 is within the range 0 through 10.

Because **MyIF** has been added to JShell, you can also create a reference to an object of type **MyIF**. For example, the following is also valid code:

```
MyIF ob2 = new MyClass(1.0, 3.0);
boolean result = ob2.isLegalVal(1.1);
```

In this case, the value of **result** will be **true** and will be reported as such by JShell.

One other point: Enumerations and annotations are supported in JShell in the same way as classes and interfaces.

Evaluate Expressions and Use Built-in Variables

JShell includes the ability to directly evaluate an expression without it needing to be part of a full Java statement. This is especially useful when you are experimenting with code and don't need to execute the expression in a larger context. Here is a simple example. Using a new JShell session, enter the following at the prompt:

```
3.0 / 16.0
```

JShell responds with:

```
$1 ==> 0.1875
```

As you can see, the result of the expression is computed and displayed. However, note that this value is also assigned to a temporary variable called **$1**. In general, each time an expression is evaluated directly, its result is stored in a temporary variable of the proper type. Temporary variable names all begin with a **$** followed by a number, which is increased each time a new temporary variable is needed. You can use these temporary variables like any other variable. For example, the following displays the value of **$1**, which is 0.1875 in this case.

```
System.out.println($1);
```

Here is another example:

```
double v = $1 * 2;
```

Here, the value **$1** times 2 is assigned to **v**. Thus, **v** will contain 0.375.

You can change the value of a temporary variable. For example, this reverses the sign of **$1**:

```
$1 = -$1
```

JShell responds with

```
$1 ==> -0.1875
```

Expressions are not limited to numeric values. For example, here is one that concatenates a **String** with the value returned by **Math.abs($1)**.

```
"The absolute value of $1 is " + Math.abs($1)
```

This results in a temporary variable that contains the string

```
The absolute value of $1 is 0.1875
```

Importing Packages

As described in Chapter 8, an **import** statement is used to bring members of a package into view. Furthermore, any time you use a package other than **java.lang**, you must import it. The situation is much the same in JShell except that by default, JShell imports several commonly used packages automatically. These include **java.io** and **java.util**, among several others. Since these packages are already imported, no explicit **import** statement is required to use them.

For example, because **java.io** is automatically imported, the following statement can be entered:

```
FileInputStream fin = new FileInputStream("myfile.txt");
```

Recall that **FileInputStream** is packaged in **java.io**. Since **java.io** is automatically imported, it can be used without having to include an explicit **import** statement. Assuming that you actually have a file called **myfile.txt** in the current directory, JShell will respond by adding the variable **fin** and opening the file. You can then read and display the file by entering these statements:

```
int i;
do {
  i = fin.read();
  if(i != -1) System.out.print((char) i);
} while(i != -1);
```

This is the same basic code that was discussed in Chapter 10, but no explicit **import java.io** statement is required.

Keep in mind that JShell automatically imports only a handful of packages. If you want to use a package not automatically imported by JShell, then you must explicitly import it as you do with a normal Java program. One other point: You can see a list of the current imports by using the **/imports** command.

Exceptions

In the I/O example shown in the preceding section on imports, the code snippets also illustrate another very important aspect of JShell. Notice that there are no **try/catch** blocks that handle I/O exceptions. If you look back at the similar code in Chapter 10, the code that opens the file catches a **FileNotFoundException**, and the code that reads the file watches for an **IOException**. The reason that you don't need to catch these exceptions in the snippets shown earlier is because JShell automatically handles them for you. More generally, JShell will automatically handle checked exceptions in many cases.

Some More JShell Commands

In addition to the commands discussed earlier, JShell supports several others. One command that you will want to try immediately is **/help**. It displays a list of the commands. You can also use **/?** to obtain help. Several other commonly used commands are examined here.

You can reset JShell by using the **/reset** command. This is especially useful when you want to change to a new project. By use of **/reset** you avoid the need to exit and then restart JShell. Be aware, however, that **/reset** resets the entire JShell environment, so all state information is lost.

You can save a session by using **/save**. Its simplest form is shown here:

/save *filename*

Here, *filename* specifies the name of the file to save into. By default, **/save** saves your current source code, but it supports several options, of which two are of particular interest. By specifying **-all** you save all lines that you enter, including those that you entered incorrectly. You can use the **-history** option to save your session history (i.e., the list of the commands that you have entered).

You can load a saved session by using **/open**. Its form is shown next:

/open *filename*

Here, *filename* is the name of the file to load.

JShell provides several commands that let you list various elements of your work. They are shown here:

Command	Effect
/types	Shows classes, interfaces, and enums.
/imports	Shows import statements.
/methods	Shows methods.
/vars	Shows variables.

For example, if you entered the following lines:

```
int start - 0;
int end = 10;
int count = 5;
```

and then entered the **/vars** command, you would see

```
|   int start = 0;
|   int end = 10;
|   int count = 5;
```

Another often useful command is **/history**. It lets you view the history of the current session. The history contains a list of what you have typed at the command prompt.

Exploring JShell Further

The best way to get proficient with JShell is to work with it. Try entering several different Java constructs and watching the way that JShell responds. As you experiment with JShell you will find the usage patterns that work best for you. This will enable you to find effective ways to integrate JShell into your learning or development process. Also, keep in mind that JShell is not just for beginners. It also excels when prototyping code. Thus, as you advance in your study of Java, you will still find JShell helpful whenever you need to explore new areas. Simply put: JShell is an important tool that further enhances the overall Java development experience.

Appendix E

More Java Keywords

There are five Java keywords not discussed elsewhere in this book. They are:

- **transient**
- **volatile**
- **native**
- **strictfp**
- **assert**

These keywords are most often used in programs more advanced than those found in this book. However, an overview of each is presented so that you will know their purpose. In addition, another form of **this** is described.

The transient and volatile Modifiers

The **transient** and **volatile** keywords are type modifiers that handle somewhat specialized situations. When an instance variable is declared as **transient**, then its value need not persist when an object is stored. Thus, a **transient** field is one that does not affect the persisted state of an object.

The **volatile** modifier tells the compiler that a variable can be changed unexpectedly by other parts of your program. One of these situations involves multithreaded programs. In a multithreaded program, sometimes two or more threads will share the same variable. For efficiency considerations, each thread can keep its own, private copy of such a shared variable, possibly in a register of the CPU. The real (or *master*) copy of the variable is updated at various times, such as when a **synchronized** method is entered. While this approach works fine, there may be times when it is inappropriate. In some cases, all that really matters is that the master copy of a variable always reflects the current state, and that this current state is used by all threads. To ensure this, declare the variable as **volatile**.

strictfp

One of the more esoteric keywords is **strictfp**. When Java 2 was released several years ago, the floating-point computation model was relaxed slightly. Specifically, the new model did not require the truncation of certain intermediate values that occur during a computation. This prevented overflow or underflow in some cases. By modifying a class, method, or interface with **strictfp**, you could ensure that floating-point calculations (and thus all truncations) took place precisely as they did in earlier versions of Java. When a class was modified by **strictfp**, all of the methods in the class were also **strictfp** automatically. However, beginning with JDK 17, all floating-point computations are now strict and **strictfp** is obsolete and no longer required. Its use will now generate a warning message.

REMEMBER

Beginning with JDK 17, the keyword **strictfp** is obsolete and its use will now generate a warning message.

assert

The **assert** keyword is used during program development to create an *assertion,* which is a condition that is expected to be true during the execution of the program. For example, you might have a method that should always return a positive integer value. You might test this by asserting that the return value is greater than zero using an **assert** statement. At run time, if the condition actually is true, no other action takes place. However, if the condition is false, then an **AssertionError** is thrown. Assertions are often used during testing to verify that some expected condition is actually met. They are not usually used for released code.

The **assert** keyword has two forms. The first is shown here:

assert *condition*;

Here, *condition* is an expression that must evaluate to a Boolean result. If the result is true, then the assertion is true and no other action takes place. If the condition is false, then the assertion fails and a default **AssertionError** object is thrown. For example,

```
assert n > 0;
```

If **n** is less than or equal to zero, then an **AssertionError** is thrown. Otherwise, no action takes place.

The second form of **assert** is shown here:

assert *condition* : *expr*;

In this version, *expr* is a value that is passed to the **AssertionError** constructor. This value is converted to its string format and displayed if an assertion fails. Typically, you will specify a string for *expr,* but any non-**void** expression is allowed as long as it defines a reasonable string conversion.

To enable assertion checking at run time, you must specify the **-ea** option. For example, to enable assertions for **Sample**, execute it using this line:

```
java -ea Sample
```

Assertions are quite useful during development because they streamline the type of error checking that is common during testing. But be careful—you must not rely on an assertion to perform any action actually required by the program. The reason is that normally, released code will be run with assertions disabled and the expression in an assertion will not be evaluated.

Native Methods

Although rare, there may occasionally be times when you will want to call a subroutine that is written in a language other than Java. Typically, such a subroutine will exist as executable code for the CPU and environment in which you are working—that is, native code. For example, you may wish to call a native code subroutine in order to achieve faster execution time. Or you may want to use a specialized, third-party library, such as a statistical package. However, since Java programs are compiled to bytecode, which is then interpreted (or compiled on the fly) by the Java run-time system, it would seem impossible to call a native code subroutine from within your Java program. Fortunately, this conclusion is false. Java provides the **native** keyword, which is used to declare native code methods. Once declared, these methods can be called from inside your Java program just as you call any other Java method.

To declare a native method, precede the method with the **native** modifier, but do not define any body for the method. For example:

```
public native int meth() ;
```

Once you have declared a native method, you must provide the native method and follow a rather complex series of steps in order to link it with your Java code.

Another Form of this

There is another form of **this** that enables one constructor to invoke another constructor within the same class. You saw an example of this usage in Chapter 16 where it was required when creating a non-canonical record constructor. Its use is, however, not limited to that situation. It is often used to reduce code duplication. The general form of this use of **this** is shown here:

this(*arg-list*)

When **this( )** is executed, the overloaded constructor that matches the parameter list specified by *arg-list* is executed first. Then, if there are any statements inside the original constructor, they are executed. The call to **this( )** must be the first statement within the constructor. Here is a simple example:

```
class MyClass {
  int a;
  int b;

  // Initialize a and b individually.
  MyClass(int i, int j) {
    a = i;
    b = j;
  }

  // Use this() to initialize a and b to the same value.
  MyClass(int i) {
    this(i, i); // invokes MyClass(i, i)
  }
}
```

In **MyClass**, only the first constructor actually assigns a value to **a** and **b**. The second constructor simply invokes the first. Therefore, when this statement executes:

```
MyClass mc = new MyClass(8);
```

the call to **MyClass(8)** causes **this(8, 8)** to be executed, which translates into a call to **MyClass(8, 8)**.

As mentioned, invoking overloaded constructors through **this( )** can be useful because it can prevent the unnecessary duplication of code. However, you need to be careful. Constructors that call **this( )** may execute a bit slower than those that contain all of their initialization code in-line. This is because the call and return mechanism used when the second constructor is invoked adds overhead. Remember that object creation affects all users of your class. If your class will be used to create large numbers of objects, then you must carefully balance the benefits of smaller code against the increased time it takes to create an object. As you gain more experience with Java, you will find these types of decisions easier to make.

There are two restrictions you need to keep in mind when using **this( )**. First, you cannot use any instance variable of the constructor's class in a call to **this( )**. Second, you cannot use **super( )** and **this( )** in the same constructor because each must be the first statement in the constructor.

Index

K

L